CW00591211

Oxford Lecture Series in
Mathematics and its Applications 18

OXFORD LECTURE SERIES
IN MATHEMATICS AND ITS APPLICATIONS

1. J. C. Baez (ed.): *Knots and quantum gravity*
2. I. Fonseca and W. Gangbo: *Degree theory in analysis and applications*
3. P. L. Lions: *Mathematical topics in fluid mechanics, Vol. 1: Incompressible models*
4. J. E. Beasley (ed.): *Advances in linear and integer programming*
5. L. W. Beineke and R. J. Wilson (eds): *Graph connections: Relationships between graph theory and other areas of mathematics*
6. I. Anderson: *Combinatorial designs and tournaments*
7. G. David and S. W. Semmes: *Fractured fractals and broken dreams*
8. Oliver Pretzel: *Codes and algebraic curves*
9. M. Karpinski and W. Rytter: *Fast parallel algorithms for graph matching problems*
10. P. L. Lions: *Mathematical topics in fluid mechanics, Vol. 2: Compressible models*
11. W. T. Tutte: *Graph theory as I have known it*
12. Andrea Braides and Anneliese Defranceschi: *Homogenization of multiple integrals*
13. Thierry Cazenave and Alain Haraux: *An introduction to semilinear evolution equations*
14. J. Y. Chemin: *Perfect incompressible fluids*
15. Giuseppe Buttazzo, Mariano Giaquinta and Stefan Hildebrandt: *One-dimensional variational problems: an introduction*
16. Alexander I. Bobenko and Ruedi Seiler: *Discrete integrable geometry and physics*
17. Doina Cioranescu and Patrizia Donato: *An introduction to homogenization*
18. E. J. Janse van Rensburg: *The statistical mechanics of interacting walks, polygons, animals and vesicles*

The Statistical Mechanics of Interacting Walks, Polygons, Animals and Vesicles

E. J. JANSE van RENSBURG

Associate Professor of Mathematics
York University, Toronto

Great Clarendon Street, Oxford OX2 6DP

Oxford University Press is a department of the University of Oxford.
It furthers the University's objective of excellence in research, scholarship,
and education by publishing worldwide in

Oxford New York

Athens Auckland Bangkok Bogotá Buenos Aires Calcutta
Cape Town Chennai Dar es Salaam Delhi Florence Hong Kong Istanbul
Karachi Kuala Lumpur Madrid Melbourne Mexico City Mumbai
Nairobi Paris São Paulo Singapore Taipei Tokyo Toronto Warsaw

with associated companies in Berlin Ibadan

Oxford is a registered trade mark of Oxford University Press
in the UK and in certain other countries

Published in the United States
by Oxford University Press, Inc., New York

First published 2000

A catalogue record for this book is available from the British Library

Library of Congress Cataloging in Publication Data
(Data available)

ISBN 0 19 850561 2 (Hbk)

Typeset by Newgen Imaging Systems (P) Ltd., Chennai, India
Printed in Great Britain
on acid-free paper by
T.J. International Ltd., Padstow

Preface

The self-avoiding walk and its relatives (lattice polygons, trees and animals, surfaces and vesicles) have received considerable attention as models of large molecules (polymers) in solution. The interesting and important fact in all these models is the significant contribution of conformational degrees of freedom to the free energy, and this may have an important effect on the thermodynamical properties of the model. There have been significant developments in both the underlying theory and in the study of a variety of interacting models. In this monograph I examine in particular the extent to which the thermodynamic properties of an interacting model can be obtained by rigorous methods. This poses many interesting problems in combinatorics and analysis, which makes this an interesting and active area from a mathematical point of view. The thermodynamic properties of a given model may reflect the behaviour of a polymer, and so it is not surprising that many of the results presented here first appeared in the physics literature, and that many physicists still actively investigate some of the interacting models presented here.

The most important object in an interacting model is the limiting free energy, and in interesting models it is a non-analytic function from which the phase behaviour in the model may be derived. In addition to the limiting free energy, there is the (microcanonical) density function, which will be examined generally and also in some specific models. The phase diagram of an interacting model may include a tricritical point which is characterized by certain scaling regimes in its vicinity. The theory of tricritical points contains a variety of scaling assumptions and little mathematical rigour. However, there are also significant numerical and experimental results which strongly support its basic assumptions, and there seems little reason to deviate from the standards it set. I review it in Chapter 2, and present models and results in later chapters in the context of tricriticality. The basic notions surrounding the limiting free energy, and the density function, are presented in Chapter 3. Exact models of directed walks and polygons are examined in Chapter 4. These models are related to various models in enumerative combinatorics, and in the present context are only used to illustrate the concepts in tricritical scaling introduced in Chapters 2 and 3. Interacting models of walks and polygons, trees and animals, and surfaces and vesicles, are presented in Chapters 5, 6 and 7. The basic starting point in all cases is a pattern theorem, and its relation to the density function and limiting free energy.

Acknowledgements

I have benefitted much by interactions with Neal Madras, Carla Tesi, Enzo Orlandini, and especially Stu Whittington, and they have contributed significantly to the ideas expressed in this monograph. I am also indebted to Stu Whittington, Carla Tesi and

Enzo Orlandini for comments and suggestions made while preparing the manuscript. This book is dedicated to my wife Katherine, and also to Elisabeth, Ellen and Margaret, who unfailingly supported me throughout its preparation.

Bloemfontein E.J.J. van R.
July 1999

Contents

1
Introduction

1.1 Lattice models of polymers and vesicles

The general approach to lattice models of walks, polygons, trees and animals is rooted in statistical mechanics, and particularly in critical phenomena. It is from this vantage point that questions about these models are asked, and answers are also interpreted within this framework. In this chapter a very brief, and perhaps incomplete, overview is given. The motivation for studying these models comes from the chemistry and physics of macromolecules, and the models are attempts at representing, in a simple manner, the entropic contribution to the free energy made by the conformational degrees of freedom in these molecules. The mathematical description of almost all the models in this monograph is based on a basic theorem of subadditive functions, which I discuss in Section 1.2.1. The scaling properties and critical exponents are discused in Section 1.3, and a Flory argument for estimating the metric exponent of linear polymers, and of branched polymers, is given in Sections 1.2.3 and 1.3.3. For the most part I shall accept the Flory values of exponents as adequate; while they are known to be accurate in some cases, it is also the case that in some dimensions (three dimensions for walks) the Flory argument gives a value which is incorrect (but not far off the mark).

A linear polymer is a large molecule, consisting of a backbone of atoms or groups of atoms (called monomers) which are joined in a sequence by covalent bonds. There may also be side-groups or side-chains attached to the mononers; if these are also polymers, then the molecule is a branched polymer. Branched polymers may have a variety of different topologies[1] which give rise to classes of molecules referred to as "brushes", "combs", "stars" or "trees".

A defining feature of polymers is the large number of rotational degrees of freedom about the covalent bonds between the monomers. These conformational degrees of freedom make an important entropic contribution to the free energy of the polymer, and have many effects on the chemical and physical properties of the polymer [64, 123]. A lattice random walk has been used as a model of a linear polymer. This model will take into account the contributions of the conformational degrees of freedom to entropy, but it fails to explain the asymptotic properties of a linear polymer in a good solvent. In particular, it ignores the effects of excluded volume, which exercises control over the asymptotic properties of the polymer. The self-avoiding walk in a lattice is a better and

[1] The word *topology* has a broader meaning in the chemistry literature when it is applied to molecules. For example, a polymer which consists of three linear polymers of equal length joined into a three-armed star at one end has the same topology (in the mathematical sense) as a single linear polymer with one short side-chain, but chemists may consider these to be of different topologies.

much more successful model for linear polymers [64, 123]. It can be defined as a path graph embedded in the hypercubic lattice, with only one parameter (its length). Despite its simplicity, this model is very successful in predicting the asymptotic properties of linear polymers, and it has been studied extensively in the mathematical and physical literature; see for example the book by N. Madras and G. Slade [249].

The self-avoiding walk can be generalized to serve as a model for polymers of other topologies. For example, there are branched models of the self-avoiding walk which serve as lattice models of stars, brushes or combs [237]; and lattice trees (acyclic and connected subgraphs of the lattice) have been investigated as models for branched polymers [183, 219, 367]. A polygon is a model for a ring polymer. Ring polymers in three dimensions may be knotted (as is often observed in the study of DNA), and knotted polygons have been used to study the incidence of knots in ring polymers. The Frisch–Wasserman–Delbruck conjecture [65] states that the probability that a ring polymer is a knot approaches one as the length of the polymer increases to infinity. In the case of lattice polygons this is indeed the case [292, 336]. The statistical mechanics of topological effects in ring polymers was considered by S.F. Edwards [101, 102]. A self-avoiding walk, a polygon, and a lattice tree in the square lattice are illustrated in Fig. 1.1.

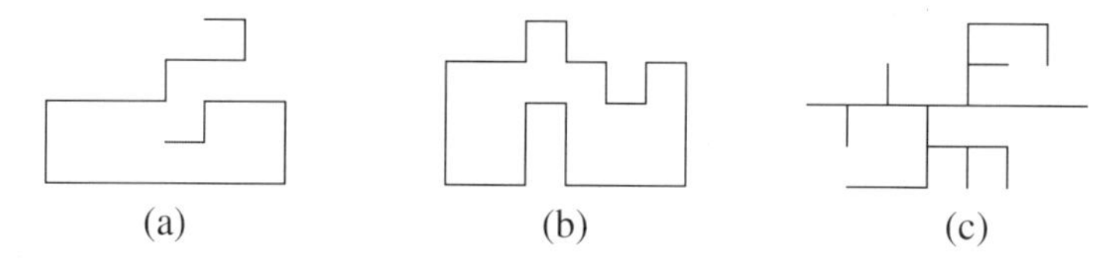

Fig. 1.1: (a) A walk in the square lattice. (b) A polygon in the square lattice. (c) A tree in the square lattice.

Membranes around cells or other organelles in living tissue are composed of a bilayer of phospholipids. A phospholipid consists of two parts: a hydrophilic head, and a hydrophobic (hydrocarbon) tail. The hydrophilic head is the polar part of the molecule, and dissolves in polar liquids, such as water. The hydrophobic tail consists of hydrocarbon chains. When mixed with water, the phospholipids form a lipid bilayer. The hydrophilic heads of the molecule are arranged on the surface of the layer, and interact with the water molecules, while the tails are isolated from the water by being hidden in a lipid layer between the layers of hydrophilic heads (Fig. 1.2). Artificial bilayer membranes have been made in the laboratory [310]. The membrane is in a fluid state, and is called a fluid membrane. A lipid bilayer which encloses a spherical volume is also

Fig. 1.2: A schematic drawing of the cross-section of a bilayer fluid membrane. The hydrophobic hydrocarbon chains are shielded from water by the layer of hydrophilic heads.

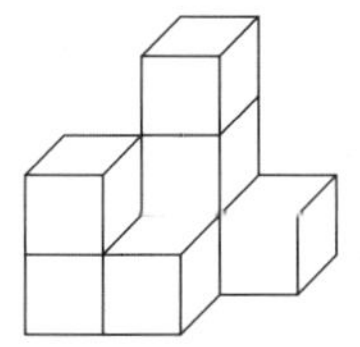

Fig. 1.3: A self-avoiding surface in the cubic lattice can be used as a model of a vesicle.

called a *vesicle*. There is a critical temperature below which the hydrocarbon chains in the membrane become ordered, in the sense that they are all in *trans*-conformations, while *cis*-conformations are rare. This affects the properties (such as stiffness or curvature) of the bilayer. Depending on the temperature, the chemical properties of the phospholipid, and the solvent, it seems that these vesicles will have a rich phase diagram. This turns out to be the case; see for example reference [310].

There are also membranes which are not in a fluid state, but which consist of cross-linked polymer chains. Such membranes are also called *tethered surfaces*, or polymerized membranes [209]. In this monograph I shall only be considering fluid membranes and surfaces, where local constraints, such as those introduced by cross-linked polymer chains, will be ignored. The statistical mechanics of surfaces and membranes is reviewed in reference [267].

Vesicles can assume many shapes. These include the ball shape of some vacuoles in cells, the disk-like shape of red blood cells, and the sickle shape of red blood cells in sickle cell aneamia. Red blood cells can also be induced to change shape into asymmetric and spiked shapes by treatment with certain drugs [230]. This phenomenon is called echinocytosis, and it is also connected to pH changes, depletion of ATP, aging, or anaesthetics. In the case of drugs, it is conjectured that these are preferentially absorbed into one of the monolayers in the bilipid. This adds a curvature contribution to the free energy, which, if strong enough, will change the shape of the vesicle. A review of biological vesicles from a statistical mechanics point of view was done by S. Leibler [231].

A lattice polygon with fixed area may be used as a model for a two-dimensional vesicle [116]. This model shows that there is an inflated, and a "flaccid" phase, for the vesicle (these phases are separated by a first-order transition), depending on an area energy. A good lattice model for a three-dimensional vesicle with a fluid membrane is a closed self-avoiding surface which is a (topological) sphere. This surface is constructed of squares (commonly called plaquettes) in the lattice $\mathcal{Z}^3$, glued at their edges into a surface (see Fig. 1.3) [98]. The introduction of energy terms in the model gives an interacting surface; such models will be studied in Chapter 7. The rich phase diagrams of polymers, branched polymers and vesicles suggest that models of these objects, such as walks, trees and surfaces, should also have interesting phase behaviour. From a physical and mathematical point of view this looks interesting; not least because an understanding of the phase behaviour in a model might illuminate the phase diagram of the object as measured in experimental work; see for example experiments done on the θ-point of polymers [337].

1.2 Walks and polygons

1.2.1 Growth constants

The most fundamental quantity in the statistical mechanics of models of walks, polygons, trees or vesicles is the number of different conformations that one of these objects may assume. In the hypercubic lattice $\mathcal{Z}^d$ a self-avoiding walk of length n is defined as a sequence of $(n+1)$ distinct *vertices* $\{v_i\}_{i=0}^n$ such that the pair of vertices $(v_{i-1}, v_i) \equiv v_{i-1}v_i$ are adjacent for each $i = 1, 2, \ldots, n$. The adjacent pair $v_{i-1}v_i$ is also the i-th edge in the walk. Two walks are considered equivalent if one is a translation of the other; it is convenient to call the resulting equivalence classes walks, which redefines the notion of walks presented above (I shall also say that walks are counted modulo translation, or up to a translation). The number of walks of n edges is denoted by c_n. The number c_n has been enumerated to $n = 51$ in the square lattice due to a heroic effort of A.R. Conway and A.J. Guttmann [57]. In three dimensions c_n is known to $n = 23$ [258].

Any walk has two end-vertices. If one of these vertices is chosen as a first vertex in the walk, then the walk is said to be *oriented*. The number of oriented walks is $2c_n$. If the vertices v_0 and v_n in an oriented walk are identified, then an oriented *closed walk* is obtained. The equivalence class consisting of all cyclic permutations of an oriented closed walk is called an *oriented polygon*. If in addition the orientation on the polygon is ignored, then a *polygon* is obtained. Define p_n to be the number of all polygons of length n, counted modulo a translation in the lattice. Notice that $p_n = 0$ in the hypercubic lattice if n is an odd number. The number of polygons is known to $n = 21$ in the cubic lattice [151], and polygons have been counted to $n = 82$ in the hexagonal lattice [104], and more recently to $n = 90$ (I. Jensen and A.J. Guttmann, private communication).

It is relatively easy to find bounds on c_n. If only those walks which step in positive directions are counted, then $2c_n \geq d^n$. On the other hand, if the i-th step has been given, then the $(i+1)$-th step of a walk cannot retrace its previous step. Thus, there are at most $(2d-1)$ choices for the $(i+1)$-th step. This shows that $2c_n \leq 2d(2d-1)^{n-1}$ (where the first step can be in one of $2d$ possible directions). These observations show that c_n grows exponentially with n.

Each oriented self-avoiding walk of length $n+m$ can be divided into two walks of length n and of length m respectively by cutting the walk at its $(n+1)$-th vertex (see Fig. 1.4). Since each walk gives a unique outcome, this shows that $2c_n$ satisfies a basic

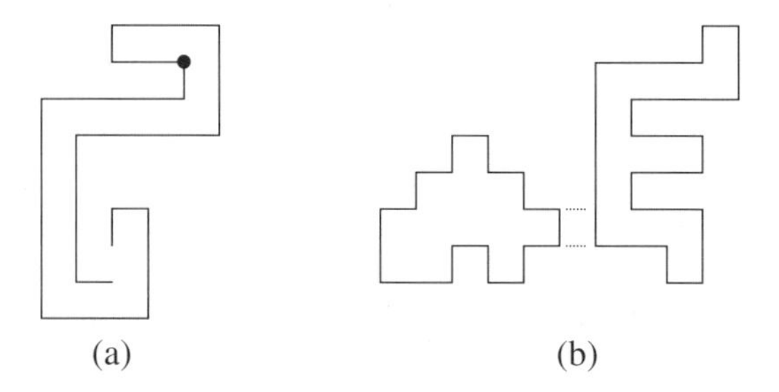

Fig. 1.4: (a) A walk can be divided into two walks by cutting it into two at the vertex denoted by $\bullet$. (b) The concatenation of two polygons.

submultiplicative[2] inequality: $(2c_n)(2c_m) \geq 2c_{n+m}$. Thus

$$c_n c_m \geq c_{n+m}/2 \tag{1.1}$$

An immediate consequence is the following theorem, due to J.M. Hammersley and K.W. Morton [164].

Theorem 1.1 *There exists a positive number μ_d such that*

$$\lim_{n\to\infty} \frac{1}{n} \log c_n = \inf_{n>0} \frac{1}{n} \log c_n = \log \mu_d.$$

Moreover, $c_n \geq \mu_d^n/2$ and $d \leq \mu_d \leq 2d-1$.

Proof The existence of the limit follows from eqn (1.1) and the theory of subadditive functions (see Lemma A.1 in Appendix A). The bounds on μ_d follows from the fact that $d^n/2 \leq c_n \leq d(2d-1)^{n-1}$, as explained above. □

The number μ_d is called a *growth constant* (while $\kappa_d = \log \mu_d$ is the *connective constant* [36, 158, 159, 163]). It is also known that $c_n \leq c_{n+1}$ [275], but ideally, one would like to strengthen Theorem 1.1 to $\lim_{n\to\infty}[c_{n+1}/c_n] = \mu_d$. There are some lattices for which this limit does not exist [159], but the situation for the cubic lattice remains unresolved. The limit

$$\lim_{n\to\infty} \frac{c_{n+2}}{c_n} = \mu_d^2, \tag{1.2}$$

is known to exist; this result is due to H. Kesten [210]; see also the paper by N. Madras [244]. The value of μ_d has been estimated using a variety of different techniques [117, 150, 355]. Exact enumeration of walks in the square lattice gives the result $\mu_2 = 2.638159 \pm 0.000002$ [57] (see also references [104, 151, 152] for estimates in other lattices). In high dimensions the expansion $\mu_d = 2d - 1 - (2d)^{-1} - 3(2d)^{-2} + O(d^{-3})$ becomes very accurate [115, 173, 174, 268, 316], see also references [175, 211]. The rate of convergence of $[\log c_n]/n$ to $\log \mu_d$ was studied by J.M. Hammersley [160, 166].

The *top vertex* and the *bottom vertex* of a polygon are found by a lexicographic ordering of its vertices with respect to their coordinates.[3] The *top and bottom edges* of a polygon are those edges with lexicographic most and least midpoints in a polygon. Naturally, the top edge is incident with the top vertex, and is perpendicular to $\hat{\imath}$, and similarly, the bottom edge is perpendicular to $\hat{\imath}$ and incident with the bottom vertex. Notice that both the top and bottom edges in a polygon are parallel to one of $(d-1)$ possible directions.

Translate a polygon of length n and a polygon of length m, and rotate the polygon of length m about the X-axis, until the midpoint of the top edge of the first polygon has

[2] In other words, $\log(2c_n)$ is a subadditive function. The self-avoiding walk will be the only model in this book which will be submultiplicative; all the other models will be supermultiplicative.

[3] The canonical orthonormal basis vectors are $\{\hat{\imath}, \hat{\jmath}, \hat{k}\}$ in $\mathcal{Z}^3$ and $\{\hat{\imath}, \hat{\jmath}\}$ in $\mathcal{Z}^2$. A lexicographic ordering will be first with repect to the direction $\hat{\imath}$, then $\hat{\jmath}$, and then $\hat{k}$, unless specifically stated otherwise. I shall also refer to the X-direction, Y-direction and Z-direction as being parallel to $\hat{\imath}$, $\hat{\jmath}$ and $\hat{k}$ respectively.

first coordinate exactly one less than the first coordinate of the midpoint of the bottom edge of the second polygon (but all other coordinates of these midpoints are the same).[4] *Concatenate* the two polygons by first removing the top edge of the first polygon and the bottom edge of the second polygon (see Fig. 1.4(b)). The operation is then completed by adding back two edges to join the two polygons into a single new polygon, consisting of all the vertices of the original pair. If there are p_n choices for the polygon of length n, then there are $p_m/(d-1)$ choices for the polygon of length m. The resulting polygon has length $n+m$, and it is not possible to construct every polygon of length $n+m$ in this way. Thus

$$p_n p_m \leq (d-1) p_{n+m}. \tag{1.3}$$

Taking the logarithm of eqn (1.3) proves that $\log[p_n/(d-1)]$ is a superadditive function, and $-\log[p_n/(d-1)]$ is a subadditive function. Then Lemma A.1 (if the limit is taken through even numbers) proves the existence of a growth constant for polygons [161].

Theorem 1.2 *There exists a positive number μ_p such that if $n \to \infty$ through even numbers, then*

$$\lim_{n\to\infty} \frac{1}{n} \log p_n = \sup_{n>0} \frac{1}{n} \log p_n = \log \mu_p.$$

Moreover, $\mu_p \leq \mu_d$ while $p_n \leq (d-1)\mu_p^n$.

Proof The existence of the limit and the definition $\log \mu_p = \sup_{n>0}(1/n) \log p_n$, while the fact that $p_n \leq (d-1)\mu_p^n$, follows from Lemma A.1. Since $p_n \leq c_{n-1}$ (delete the bottom edge in a polygon to see this), it follows that $\mu_p \leq \mu_d$. □

It is implicitly understood that the limit in Theorem 1.2 will be taken through even numbers, and I shall not dwell further on this in any other model where this issue arises. It is also known that

$$\mu_d = \mu_p; \tag{1.4}$$

this was proven by J.M. Hammersley [161], see also J.M. Hammersley and D.J.A. Welsh [166] (see Chapter 5).

1.2.2 Generating functions

The generating function of a sequence of integers will be an important tool for us. In the case of lattice polygons it is defined by

$$G_p(x) = \sum_{n=0}^{\infty} p_n x^n, \tag{1.5}$$

where n only takes even values. The variable x will be called an *activity*, and it is *conjugate* to n.[5] Theorem 1.2 implies that $p_n = \mu_d^{n+o(n)}$ (assuming that $\mu_d = \mu_p$),

[4] If $\hat{\imath}$ is added to the midpoint of the top edge of the first polygon, then the midpoint of the bottom edge of the second polygon is found.

[5] It is really $\log x$ which is conjugate to n. The parameter $\alpha = \log x$ is the *fugacity*, and from a statistical mechanics point of view, it may be interpreted as an "inverse temperature", or a "chemical potential". Although

and on substitution into (1.5) the radius of convergence of $G_p(x)$ is found to be μ_d^{-1}. If $x < \mu_d^{-1}$, then $G_p(x)$ is finite, and it is dominated by polygons of finite length. If $x > \mu_d^{-1}$, then $G_p(x)$ is infinite *because* of the contributions of polygons of arbrarily large n; this can be called a "phase" of infinite polygons. $G_p(x)$ is also viewed as a formal power series in x (where x is a complex number), and its singularities will be important in the description of the scaling properties of p_n. The above shows that the radius of convergence of $G_p(x)$ is $x_c = \mu_d^{-1}$, and that $G_p(x)$ has a singularity at $x = x_c$.

The subleading behaviour (to the exponential) of p_n is generally assumed to be described by a power-law correction. In particular, there is evidence that (where A_p is some constant)[6]

$$p_n \simeq A_p n^{\alpha-3} \mu_d^n, \tag{1.6}$$

for large n. The exponent $\alpha - 3$ is also called the "entropic exponent" of polygons, and it is generally believed that $p_n \sim A_p n^{\alpha-3} \mu_d^n$. Theorem 1.2 shows that $\alpha - 3 \leq 0$. It is in fact known that $\alpha - 3 \leq -1/2$ in two dimensions, $\alpha - 3 \leq -1$ in three dimensions and $\alpha - 3 < -1$ in more than three dimensions [245], see also [247]. This assumption is enough to show that

$$G_p(x) \simeq A_p (x_c - x)^{2-\alpha}, \quad \text{as } x \to x_c^-. \tag{1.7}$$

It is conjectured that $G_p(x) \sim A_p(x_c - x)^{2-\alpha}$. It is possible to use rooted polygons to show that $\alpha - 2 \leq 0$; this implies that $G_p(x_c)$ is finite. In fact, as $x \to x_c$, $G_p(x)$ is dominated by analytic terms, which was not explicitly included in eqn (1.7). On the other hand, eqn (1.7) does not imply that eqn (1.6) is true; this would require that the coefficients p_n are computable from the singularity in $G_p(x)$. This can only be done if some additional information is available, giving rise to a so-called Tauberian theorem (see for example [177]). In the case of walks it is generally assumed that

$$c_n \simeq A_w n^{\gamma-1} \mu_d^n, \tag{1.8}$$

which, similarly to eqn (1.7), gives a generating function for walks with conjectured asymptotic behaviour

$$G_w(x) \sim A_w (x_c - x)^{-\gamma}. \tag{1.9}$$

Theorem 1.1 shows that $\gamma - 1 \geq 0$; this demonstrates that $G_w(x)$ diverges as x approaches x_c, in contrast with the polygon generating function. Coulomb gas techniques give the "exact values" of the exponents: $\alpha = 1/2$ and $\gamma = 43/32$ in two

---

these notions may sometimes be useful, I shall not generally attempt to make connections between activities and real thermodynamic parameters. Instead, the parameters of any model will be the activities, without any intentional thermodynamic meaning.

[6] The use of "$\simeq$" here is taken to mean that there exists positive constants C_0 and C_1 such that

$$C_0 n^{\alpha-3} \mu_d^n \leq p_n \leq C_1 n^{\alpha-3} \mu_d^n$$

for all $n \geq N_0$, and where N_0 is a large constant. The use of "$\approx$" in $f(n) \approx g(n)$ will mean that $f(n)$ and $g(n)$ have the same asymptotic behaviour in some unspecified way. The symbol $\sim$ will be reserved for functions which are asymptotic to one another: $f(n) \sim g(n)$ if and only if $\lim_{n\to\infty} f(n)/g(n) = 1$.

dimensions [271], but these values are not rigorously known. Nevertheless, these numbers indicate that the generating functions of polygons and walks may have branch points at x_c. The behaviour of $G_w(x)$ and $G_p(x)$ mimics critical behaviour in thermodynamic systems. For example, if the activity x is thought of as a temperature-dependent parameter which controls the number of edges in the walk or polygon, then the regime $x < x_c$ may be interpreted as a "phase" of finite polygons or walks, while $x > x_c$ is a "phase" of infinite polygons or walks. These phases are separated by the radius of convergence of the generating function, which may be called a *critical point*. The exponents α and γ describe the scale-invariant behaviour of the generating function close to the critical point, and are often called *critical exponents* or *scaling exponents*.

1.2.3 The metric exponent

In the previous paragraphs the exponents γ and α were introduced in a description of the combinatorial nature and scaling of walks and polygons. In this section a different aspect of walks will be discussed, namely their dimensions. In particular, if a walk has n edges, what is its expected dimension, as measured by (say) the end-to-end distance between its end-points? In particular, if there are $c_n(r)$ walks with end-points a distance r apart, then $\langle r \rangle_n = \sum_r r\ c_n(r)/c_n$ is an interesting quantity. In the case of a random walk, it is known that $\langle r \rangle_n \sim C\sqrt{n}$, and since $c_n(r) = 0$ if $r > n$ one may expect that there are non-zero constants C_0 and C_1 such that $C_0\sqrt{n} \leq \langle r \rangle_n \leq C_1 n$ for walks (the lower bound is still an open question). In other words, an assumption that $\langle r \rangle_n \sim n^\nu$, where ν is the *metric exponent*, seems not unreasonable.

It turns out that a very simple argument, due to P.J. Flory [123], can give remarkably good estimates for ν in two, three and four dimensions. Moreover, this argument has been fruitfully applied in other situations, such as the scaling of a polymer in confined geometries [64]. The argument goes as follows. Let $t_n(r)$ be the number of random walks of length n and with end-points a distance r apart. Let $P_n(r)$ be the probability that a random walk, starting from the origin, has stepped a distance r away from the origin in n steps. Then $t_n(r) = P_n(r)t_n$, and the entropy in the random walk is $\log t_n(r)$. The distribution $P_n(r)$ can be approximated by the solution of the heat equation: if n is treated as a time variable, then

$$\frac{\partial P_n(r)}{\partial n} = D \triangle P_n(r), \tag{1.10}$$

with solution

$$P_n(r) = \frac{1}{\sqrt{4\pi Dn}} e^{-r^2/4Dn}. \tag{1.11}$$

Thus, the total entropy can be estimated as

$$S_n(r) = -A\frac{r^2}{4Dn} + C, \tag{1.12}$$

where A is a constant and where C is a term independent of r.

The second part of the Flory argument is a "mean field" estimate of the energy density. The density of monomers is proportional to n/r^d, so that each monomer is in a mean field of density n/r^d. Summing over the monomers indicates that the energy density is proportional to n^2/r^d:

$$E_n(r) = B\frac{n^2}{r^d}, \tag{1.13}$$

where B is a constant. The free energy of the walk is then

$$F_n(r) = E_n(r) - TS_n(r) = B\frac{n^2}{r^d} + AT\frac{r^2}{4Dn} - CT. \tag{1.14}$$

The minimum in $F_n(r)$ is attained when $\partial F_n(r)/\partial r = 0$, which gives

$$\langle r \rangle_n = A' n^{3/(d+2)}, \tag{1.15}$$

and so it seems that the metric exponent is

$$\nu = 3/(d+2). \tag{1.16}$$

This estimate for ν is remarkably accurate: in two and four dimensions it is generally accepted that $\nu = 3/4$ [272] and $\nu = 1/2$ respectively. In three dimensions numerical simulations [43, 233, 318] predict that $\nu = 0.5877 \pm 0.0006$ and the $N = 0$ limit of the N-vector model [149, 226, 227] shows that $\nu = 0.5880 \pm 0.0015$. These results are remarkably close to the predicted $\nu = 0.6$ from eqn (1.16). A more probabilistic approach to the Flory formula can be found in the book by N. Madras and G. Slade [249]. The Flory formula was also discussed by P.-G. de Gennes [64], and it is noted that it works despite flaws in both estimates (both contributions to the free energy are overestimated). While the Flory values are exact in some cases, they are non-rigorous estimates of the values of exponents. In this book the Flory values will be adequate for our purposes.

The exponent ν is also associated with any property of a self-avoiding walk with units of length. For example, one may either consider the mean square end-to-end distance $\langle R_e^2 \rangle_n \simeq n^{2\nu}$, or the mean square radius of gyration $\langle R_g^2 \rangle_n$, or the mean square average distance of a monomer from the end-points of the walk $\langle R_m^2 \rangle_n$. These measures of size are often used to define *amplitude ratios*; for example, one may define $A_n = \langle R_g^2 \rangle_n / \langle R_e^2 \rangle_n$, and $B_n = \langle R_m^2 \rangle_n / \langle R_e^2 \rangle_n$, and it is believed that $A_n \to A_\infty$ and $B_n \to B_\infty$ as $n \to \infty$, where A_∞ and B_∞ are finite numbers. A remarkably result is that these limiting amplitude ratios satisfies the relation $182B_\infty - 246A_\infty = 91/2$ [45, 48, 49] in two dimensions.

1.3 Scaling

1.3.1 The correlation length

In the last section a number of critical exponents were introduced in an attempt to describe the asymptotic behaviour of c_n, p_n and the mean end-to-end distance of a walk. In eqns (1.6) and (1.7) the proposed functional form for p_n is seen to be connected with the idea that there is a certain type of singularity in the generating function of polygons. Indeed, there seems to be a scale invariance in the singular part of $G_p(x)$. Define $\tau = (x_c - x)$. Then the singular part of the generating function $G_p(x)$ has the assumed behaviour $G_p(\tau) \simeq A_p\tau^{2-\alpha}$ and it seems that $G_p(c\tau) = c^{2-\alpha}G_p(\tau)$. In other words, if $G_p(\tau)$ is known at one value of τ, then its value can be computed anywhere else by using the "scaling property" above. This observation does not seem very useful in the one-parameter theory here, but it will play a key role when theories with more than one parameter are discussed in Chapter 2.

A suitable starting point for the discussion here is the number of walks of length n, $c_n(i, j)$, with their initial vertex at the site i, and their final vertex at site j. The mean end-to-end distance can be computed as follows

$$\langle r \rangle_n = \frac{1}{c_n} \sum_{j \in \mathcal{Z}^d} |j| c_n(0, j), \tag{1.17}$$

where $|j|$ is a norm (the Euclidean norm is sufficient). The generating function of $c_n(i, j)$

$$C_{ij}(x) = \sum_{n=0}^{\infty} c_n(i, j) x^n, \tag{1.18}$$

is also called the two-point function or a correlation function. It is known that the two-point function has Ornstein–Zernike behaviour:

$$C_{ij}(x) \sim A_x |i - j|^{-(d-1)/2} e^{-|i-j|/\xi(x)}. \tag{1.19}$$

This asymptotic behaviour is rigorous if i and j are separated along a coordinate axis [52], and is discussed at length in the book by N. Madras and G. Slade [249]. An important feature in eqn (1.19) is the exponential decay; the distance $|i - j|$ is measured in terms of a length $\xi(x)$. The mean end-to-end distance can also be estimated from eqn (1.19); note that this is in an ensemble where the lengths of the walks can change:

$$\begin{aligned} \langle r \rangle_x &= \frac{\sum_{j \in \mathcal{Z}^d} |j| C_{0j}(x)}{\sum_{j \in \mathcal{Z}^d} C_{0j}(x)} \\ &= \xi(x) \left[\frac{\sum_{j \in \mathcal{Z}^d} [\xi(x)/|j|]^{(d+1)/2} e^{-|j|/\xi(x)}}{\sum_{j \in \mathcal{Z}^d} [\xi(x)/|j|]^{(d-1)/2} e^{-|j|/\xi(x)}} \right]. \end{aligned} \tag{1.20}$$

On the other hand, in this ensemble the expected value of $\langle r \rangle_n$ can also be computed from eqn (1.17):

$$\langle r \rangle_x = \frac{\sum_{n \geq 0} \langle r \rangle_n c_n x^n}{\sum_{n \geq 0} c_n x^n} \approx [\log(\mu_d x)]^{-\nu}, \tag{1.21}$$

where the sums have been treated as Riemann sum approximations to integrals. If the sums in eqn (1.20) are also considered to be Riemann sums, then a comparison of eqn (1.20) and (1.21) indicates that a suitable scaling hypothesis for $\xi(x)$ is

$$\xi(x) \simeq [\log(\mu_d x)]^{-\nu} \simeq |x_c - x|^{-\nu}. \tag{1.22}$$

The function $\xi(x)$ is called the correlation length, and eqn (1.22) shows that it diverges as $x \to x_c^-$. The Ornstein–Zernike behaviour in eqn (1.19) indicates that $\xi(x)$ controls the rate of decay in the two-point function; it is in other words a length scale in the problem, and the connection to $\langle r \rangle_n$ via eqn (1.20) should not be surprising. The important notion which underlies all of the above is that there is only one length scale, the correlation length, which serves as a yard-stick for all other quantities which have units of length. The correlation length seems to vary smoothly with x away from (but close to) the critical point; hence the term "scaling".

The *zero-field susceptibility* is defined by

$$\chi(x) = \sum_{j \in \mathcal{Z}^d} C_{0j}(x) = \sum_{n \geq 0} c_n x^n, \tag{1.23}$$

and eqn (1.8) suggests the following scaling assumption:

$$\chi(x) \simeq |x_c - x|^{-\gamma}. \tag{1.24}$$

γ is the *susceptibility exponent*, and it describes the singularity in the susceptibility as the critical point is approached. The *specific heat* $\mathcal{C}(x)$ of this model is defined by the second derivative of $G_p(x)$, where the generating function is interpreted as a loop-gas representation of the free energy in the zero component $O(n)$ model. It is given by

$$\mathcal{C}(x) = \sum_{n \geq 0} n^2 p_n x^n, \tag{1.25}$$

and from eqn (1.6) it seems reasonable to assume that it will obey the scaling law

$$\mathcal{C}(x) \simeq |x_c - x|^{-\alpha}. \tag{1.26}$$

The entropic exponent α is also called the *specific heat exponent.*

1.3.2 Scaling relations

The Ornstein–Zernike behaviour in eqn (1.19) is valid only asymptotically. For shorter distances, the decay of the two-point function is assumed to follow a power-law. In particular, on distances which approximate the correlation length ($|i - j| \approx \xi(x)$), $C_{ij}(x) \simeq |i-j|^{2-d-\eta}$, where η is yet another critical exponent (it is called the *anomolous dimension*). For increasing $|i - j|$ this decay is expected to cross over into eqn (1.19). It is then not unreasonable to define a *universal scaling function* g such that

$$C_{ij}(x) \sim \frac{1}{|i-j|^{d-2+\eta}} g\left(\frac{|i-j|}{\xi(x)}\right). \tag{1.27}$$

The scaling function g is expected to be a constant function for small values of its argument, but at large argument it gives the exponential decay with the power-law correction in eqn (1.19). The susceptibility can be estimated from eqn (1.27), if it is assumed that the main contribution to it is when $|i - j| \approx \xi(x)$. This gives

$$\chi(x) \simeq \sum_{j \in \mathcal{Z}^d} \frac{1}{|i-j|^{d-2+\eta}} g\left(\frac{|i-j|}{\xi(x)}\right) \simeq \xi^{2-\eta} \simeq |x_c - x|^{-(2-\eta)\nu}. \tag{1.28}$$

Comparison with eqn (1.26) gives Fisher's relation

$$\gamma = (2 - \eta)\nu. \tag{1.29}$$

There is also a scaling relation involving ν and α and which includes the dimension d explicitly. Relations of this type are called hyperscaling relations, and they are much more controversial than scaling relations such as eqn (1.29). An important relation of this type is Josephson's hyperscaling relation, given by

$$2 - \alpha = d\nu. \tag{1.30}$$

Since nothing is certain about hyperscaling relations, it is also necessary to attempt a derivation of eqn (1.30) within the walks and polygons context. The following argument is due to N. Madras and G. Slade [249]. If A_{2n} is a polygon of length $2n$ containing the origin (and it is rooted), then it can be decomposed into two walks, each starting from the origin and terminating at j. Take the walks to be B_n and C_n, each of length n, and only intersecting each other at their end-points. Now make a couple of assumptions. (1) Assume that the end-points of B_n and C_n are at most a distance proportional to n^ν from the origin. Thus j may be in a volume proportional to $n^{d\nu}$. (2) Assume that the number of walks B_n and C_n is equal to $(c_n(0, j))^2$ multiplied by the probability that two walks, starting at the the same point, avoid each other (and square this to get the effect at both the origin and at j). The probability that two walks each of length n, starting at the origin, avoid each other is roughly $c_{2n}/c_n^2 \simeq n^{1-\gamma}$. (3) Assume that $c_n(0, j) \simeq c_n n^{-d\nu}$, since j can explore a volume of size $n^{d\nu}$. Since there are np_n polygons containing the origin, the above may be collected to obtain

$$2np_{2n} \simeq n^{d\nu}\left[n^{1-\gamma}\right]^2\left[n^{\gamma-1-d\nu}\mu_d^n\right]^2. \tag{1.31}$$

Substitute eqn (1.6) in eqn (1.31) to obtain eqn (1.30).

In a strict sense, there is no thermodynamic content in the exponents in this section. They simply describe the static and entropic properties of walks and polygons. For example, the exponent ν, which was connected to a correlation length above, simply describes the scaling of the mean end-to-end length with n and the exponents α and γ describe the entropic properties of polygons and walks (and are also sometimes called "entropic exponents"). Exponents for walks and polygons have been computed using a variety of techniques, including numerical simulations [18, 43, 151, 233], Coulomb gas techniques [271, 272, 273, 274], conformal field theory [13, 47, 78, 79, 95, 96], and renormalization group analysis of $O(N)$-vector models [149, 226, 227, 229]. The exponents are also stated to $O(\epsilon^5)$ in an ϵ-expansion in reference [228]. In two dimensions the transfer matrix is also very accurate, see J.M. Yeomans [374] for a discussion, as well as reference [72]. Exponents have also been computed in an ϵ-expansion of the Edwards model [100]. The results listed in Table 1.1 are the conformal values in two dimensions, and the field theoretic values in three dimensions. The mean field exponents are expected to be exact above the critical dimensions (this is four dimensions), and it is also known that some of these exponents exists in five and more dimensions [168, 171, 315]; this is obtained using the lace expansion (see also [249, 345]). Experimental values of the critical exponents, using neutron diffraction studies, were obtained by J. des Cloisseaux and G. Jannink [76].

1.3.3 Branched polymers

Branched polymers are usually modelled by models of lattice trees, which are acyclic connected subgraphs of the lattice. As for polygons and walks, trees are counted up to

Table 1.1: Linear and ring polymer exponents in dilute solution

Exponent	*Two dimensions*	*Three dimensions*	*Mean field*
α	$\frac{1}{2}$	0.237(2)	0
γ	$\frac{43}{32}$	1.1575(6)	1
ν	$\frac{3}{4}$	0.5877(6)	$\frac{1}{2}$
η	$\frac{5}{24}$	0.031(4)	0

translational equivalence; define t_n to be the number of lattice trees composed of n edges in the hypercubic lattice. There is evidence [27, 113, 137, 222, 290] that the number of (unrooted) trees behaves as

$$t_n \simeq A_t n^{-\theta} \tau_d^n. \tag{1.32}$$

The existence of the limit $\lim_{n\to\infty}[\log t_n]/n = \log \tau_d$ can be shown by a concatenation argument (see Chapter 6), but there is a rigorous proof of eqn (1.32) only in high dimensions [169, 172]. *Lattice animals* (connected subgraphs of the lattice) are also considered a model of branched polymers in dilute solution, with the same exponents as lattice trees (that is, they are in the same universality class). However, it is also known that lattice trees and lattice animals do not have the same growth constant (see reference [133] for a discussion of the growth constants). The mean field value of θ is $5/2$. The definition of a metric exponent for lattice trees is similar to that of walks and polygons. Lattice animals were studied in cell growth problems [215]. They were also identified as weighted percolation clusters and are also a natural model for branched polymers [242]. Animals are generally accepted to be in the same universality class as trees, with the same critical exponents.

It is a remarkable fact that a dimensional reduction makes a connection between d-dimensional lattice trees and an Ising model in an imaginary magnetic field in $d-2$ dimensions. In particular, there is a relation between the exponent σ associated with the Yang–Lee edge singularity in the Ising model in an imaginary external field, and the entropic exponent θ and the metric exponent ν of lattice animals [290]. These relations are

$$\nu = (\sigma + 1)/(d-2), \qquad \theta = \sigma + 2. \tag{1.33}$$

Since the Ising model can be solved exactly in zero and in one dimensions, "exact values" for θ can be obtained in two dimensions, and for θ and ν in three dimensions. Furthermore, since $\nu = (\theta - 1)/(d-2)$, estimating ν will also produce an estimate for θ. Monte Carlo simulations have been used to estimate ν in two and in four dimensions, to give the estimates for branched polymer exponents in Table 1.2 [194, 195, 291]. Simulations were also done by S. Redner [307], and W.A. Seitz and D.J. Klein [311]; see also references [44, 137] for more Monte Carlo approaches to trees and animals. Renormalization group approaches can be found in references [73, 75]. Series techniques have been used in reference [3].

Table 1.2: Branched polymer exponents in dilute solution

Exponent	*Two dimensions*	*Three dimensions*	*Four dimensions*	*Mean field*
ν	0.637(12)	$\frac{1}{2}$	0.420(17)	$\frac{1}{4}$
θ	1	$\frac{3}{2}$	1.840(34)	$\frac{5}{2}$

There is a Flory argument for branched polymers in dilute solution [183]. Let $t_n(\{0, x\})$ be the number of trees of size n containing the origin and the lattice site x. Assume that the probability of a tree to extend a distance $r = |x|$ is asymptotically Gaussian with covariance $\sigma^2 \sim n^{2\nu}$. In other words,

$$\frac{t_n(\{0, x\})}{t_n(\{0\})} \to e^{-r^2/a\sqrt{n}}, \tag{1.34}$$

where the fact that the mean field value of $\nu = 1/4$ is used. Thus, the entropy contribution to the free energy is $S \approx C_0 - r^2/a\sqrt{n}$. Secondly, the density of monomers in a lattice tree is proportional to n/r^d, and each monomer finds itself in a background field of monomers of density n/r^d. Thus, the energy is proportional to n^2/r^d. Thus, the free energy is

$$F = \frac{bn^2}{r^d} + \frac{r^2}{a\sqrt{n}} + \mathrm{C}, \tag{1.35}$$

and if the derivative of this with respect to r is taken, then the minimum in F is determined by $r^{d+2} \simeq An^{5/2}$. In other words, the Flory value of the critical exponent ν is

$$\nu = \frac{5}{2(d+2)}. \tag{1.36}$$

By utilizing eqns (1.33), a Flory value can also be assigned for the entropic exponent:

$$\theta = \frac{7d-6}{2(d+2)}. \tag{1.37}$$

It is also known that the scaling limit of lattice trees in high dimensions is a process known as super-Brownian excursions [71, 169, 172].

2 Tricriticality

2.1 Interacting models of polygons

The phase diagram of a model of interacting polygons (or walks) may include a tricritical point, which makes an understanding of tricriticality essential if the thermodynamic nature of interacting polygons is considered. In the case of the models described in this monograph, the theory of tricriticality will give a framework for the description of the (intuitive) phase behaviour expected from a typical interacting model. In this respect it is very useful, since it guides numerical and mathematical approaches to the models in general. It is important not to forget that tricriticality is not mathematically rigorous, but is based on a large number of assumptions which form a belief-system within which models are described. The basic theory of tricriticality was reviewed extensively by I.D. Lawrie and S. Sarlbach [225], see also R. Brak, A.L. Owczarek and T. Prellberg [34].

As a matter of convenience, I shall refer to models which satisfy a supermultiplicative inequality, such as in eqn (1.3), as "supermultiplicative models". These include the vast majority of interesting lattice models, including polygons, trees and animals, or surfaces and vesicles. The archetypical supermultiplicative model is a model of polygons, and I shall use the term "polygon" synonymously with any supermultiplicative model (except when a specific model of polygons is considered). The self-avoiding walk satisfies eqn (1.1), and is a "submultiplicative model". This model is in a different class, but the notions in this chapter apply to it as well.

The starting point is the generating functions $G_p(x)$ and $G_w(x)$ in eqns (1.7) and (1.9), and the "finite size scaling" assumptions in eqns (1.6) and (1.8). These assumptions were used to argue that $G_p(x) \simeq A_p(x_c - x)^{2-\alpha}$, and $G_w(x) \simeq A_w(x_c - x)^{-\gamma}$. In an interacting model each polygon or walk will have an energy associated with it, and the result is that the generating functions will depend on a new paramater which will be conjugate to the energy.

The basic quantity in any interacting model of polygons will be the number $p_n^{\#}(m)$, which is the number of a family of polygons of length (size) n and counted with respect to some property of size m.[1] The *energy* of polygons counted by $p_n^{\#}(m)$ is m. The *canonical*

[1] In other words, polygons are counted with respect to size n (the number of edges in the case of walks, polygons, trees or animals, or the number of plaquettes in models of vesicles and surfaces), and with respect to the number m of occurrences of a certain arrangement of edges or plaquettes (for example, the number of right angles between successive edges), or with respect to the size m of some other property of the polygon. Polygons counted by $p_n^{\#}(m)$ are also said to have *energy* m.

partition function of the model is

$$p_n^{\#}(z) = \sum_{m \geq 0} p_n^{\#}(m) z^m, \tag{2.1}$$

where z is an *activity* "conjugate" to the energy (if $z = e^{\alpha}$, then it is really the fugacity α which is conjugate to m, but I shall freely abuse this notion). The partition function is also called the "z-transform" of $p_n^{\#}(m)$ [242, 250]. The free energy density is defined by

$$F_n(z) = \frac{1}{n} \log p_n^{\#}(z). \tag{2.2}$$

The limiting free energy density (per edge, or per plaquette) must be shown to exist explicitly for a given z. If it exists (in this chapter it is assumed that it does), then it is defined by

$$\mathcal{F}_{\#}(z) = \lim_{n \to \infty} F_n(z) = \lim_{n \to \infty} \frac{1}{n} \log p_n^{\#}(z). \tag{2.3}$$

In many models it will be seen that $\mathcal{F}_{\#}(z)$ is asymptotic to a line $S_{\mathcal{F}} + M \log z$ as $z \to \infty$. The number $S_{\mathcal{F}}$ will be called the *limiting entropy* of the model; this definition is very natural if $\log z$ is interpreted as an inverse temperature, and M as an energy density. The function $\log p_n^{\#}(z)$ is a convex function of $\log z$; this follows immediately from the Cauchy–Schwartz inequality:

$$\begin{aligned} p_n^{\#}(z_1) p_n^{\#}(z_2) &= \sum_{m_1 \geq 0} p_n^{\#}(m_1) z_1^{m_1} \sum_{m_2 \geq 0} p_n^{\#}(m_2) z_2^{m_2} \\ &\geq \left(\sum_{m \geq 0} p_n^{\#}(m) \left[\sqrt{z_1 z_2}\, \right]^m \right)^2 \\ &= \left(p_n^{\#}(\sqrt{z_1 z_2}) \right)^2, \end{aligned} \tag{2.4}$$

so that $\log p_n^{\#}(z_1) + \log p_n^{\#}(z_2) \geq 2 \log p_n^{\#}(\sqrt{z_1 z_2})$. This result is quite general, and applies to all systems with a partition function such as in eqn (2.1). In addition, $p_n^{\#}(z)$ is a polynomial in z, and is analytic in $(0, \infty)$. Thus, $F_n(z)$ is continuous, analytic, and is a convex function of $\log z$ in $(0, \infty)$. If $\mathcal{F}_{\#}(z)$ exists, then it is the limit of a sequence of convex functions. By Lemmas B.1, B.2 and B.4 in Appendix B, $\mathcal{F}_{\#}(z)$ is a convex function of $\log z$. Moreover, if it is finite, then its right- and left-derivatives exist everywhere in $(0, \infty)$, and they are non-decreasing functions of z (Lemma B.3). Theorem B.6 also implies that $\mathcal{F}_{\#}(z)$ is differentiable almost everywhere, and wherever $d\mathcal{F}_{\#}(z)/dz$ exists it is given by $\lim_{n \to \infty}(dF_n(z)/dz)$, by Theorem B.7.

The *generating function* of a model with partition function $p_n^{\#}(z)$ (sometimes called the "grand partition function") is

$$G_{\#}(x, z) = \sum_{n \geq 0} p_n^{\#}(z) x^n = \sum_{n \geq 0} \sum_{m \geq 0} p_n^{\#}(m) z^m x^n. \tag{2.5}$$

The function $R_{\#}(x, z) = \log G_{\#}(x, z)$ is the *thermodynamic potential* of the model. The radius of convergence $x_c(z)$ of $G_{\#}(x, z)$ defines a *critical curve* in the *singularity*

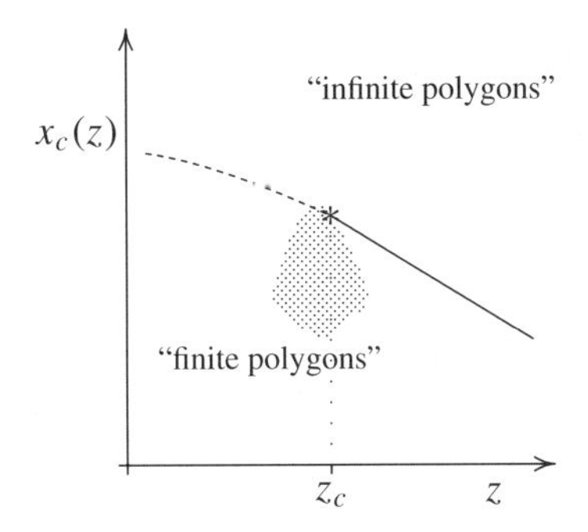

Fig. 2.1: The critical curve $x_c(z)$ in a typical model. The point $x_c(z_c)$, marked by a $*$ on the critical line, separates a line of branch points (the solid line) from a line of essential singularities (the broken line) in the generating function $G_\#(x, z)$. This point is called a tricritical point. The tricritical scaling region is the shaded area surrounding the tricritical point.

diagram (or *phase diagram*) of the model. Derivatives of $R_\#(x, z)$ are the thermodynamic quantities in the model. These are singular along the critical curve. If the limiting free energy exists, then it can be computed from the radius of convergence of $G_\#(x, z)$ by using eqn (2.3):

$$[x_c(z)]^{-1} = \lim_{n\to\infty} \left[p_n^\#(z)\right]^{1/n} = e^{\mathcal{F}_\#(z)}. \tag{2.6}$$

(If the limiting free energy does not exist, then the limit above is replaced by a lim sup.) Non-analyticities in $\mathcal{F}_\#(z)$ correspond to non-analyticities in $x_c(z)$. These are called *multicritical points*, and they separate phases on either sides of the *critical curve* $x_c(z)$. A typical critical curve with a single non-analyticity at z_c is illustrated in Fig. 2.1. The multicritical point divides the critical curve in two, each part corresponding to a different type of singularity in the generating function (at its radius of convergence). If the critical curve is a line of branch points, or poles, which meets a line of essential singularities at the multicritical point, then the multicritical point is called *tricritical*.[2]

The region below the critical curve is labelled "finite polygons", since the generating function is dominated by polygons of finite length in this regime. Similarly, the region above the critical curve is labelled "infinite polygons", since the generating function is dominated by polygons of infinite length (and is infinite as well). The phase of "infinite polygons" may consist not of polygons at all, but rather of a class of related but infinite objects in the lattice. This phase is by no means unphysical, but studying it is (for the most part) beyond the scope of the methods in this monograph. At best, the critical curve is accessible by the methods described here, and then only in a limited sense.

[2] This is a loose definition of a tricritical point. A more accurate definition in terms of scaling laws will be given in Section 2.5. However, a meeting point of a line of essential singularities and simpler singularities will always be referred to as a tricritical point. This follows the tradition in the literature, even if this is not strictly consistent with a more rigorous definition.

2.2 Classical tricriticality

The notion of tricriticality is surprisingly complex. It is classically defined as the endpoint of a line of triple points where three coexisting phases become critical simultaneously. In a phase diagram, the topology of the tricritical point is as follows. The line of triple points is the junction of three critical surfaces of first-order transitions which separate three phases. These phases coexist on the triple line. The critical surfaces each have a boundary which is a line of continuous transitions (called *critical isotherms*), and which meet at the tricritical point. The line of triple points is usually called the τ-line, while the critical isotherms of continuous transitions are called the λ-lines. This situation is illustrated in Fig. 2.2. In this diagram there are three fields, one of which is assumed to be the activity z, related to temperature in some way, and two "external fields", H_1 and H_2.

Tricritical points are found in the phase diagrams of the spin-1 Ising model and the antiferromagnetic Ising model. In almost all the models in this monograph, there will be only one external field. In that case, Fig. 2.2 collapses to a diagram such as that in Fig. 2.3, but the full geometry in Fig. 2.2 should be kept in mind; the introduction of an extra activity conjugate to some other feature in the model might make the phase diagram in Fig. 2.2 explicitly visible. The activities can be interpreted as external fields or a temperature variable, but I shall not present them as such. Instead, they will be considered as parameters of the model, and also as (complex) variables in the formal power series definitions of the generating functions.

In the classical description of tricriticality, it is assumed that the critical curve is differentiable in a neighbourhood of the tricritical point. In particular, the tangent to

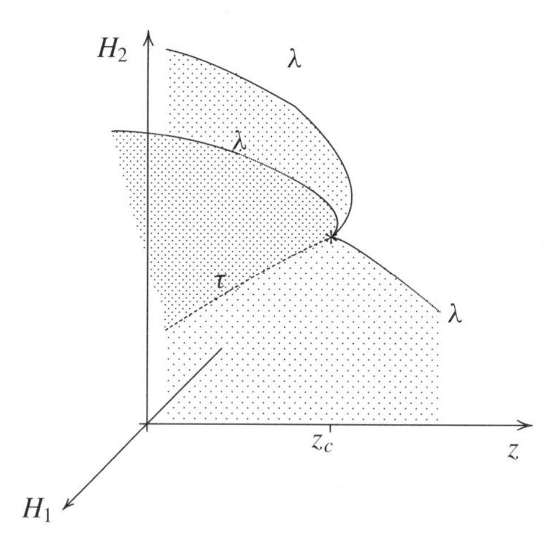

Fig. 2.2: The geometry of a tricritical phase diagram. The horizontal axis labelled by z may be imagined to be a temperature axis, while the other two axes are external fields. The τ-line of triple points ends at the tricritical point, where it branches into three λ-lines of continuous transitions. Each λ-line and the τ-line bounds a critical surface of first-order transitions. The critical surfaces are indicated by the shaded areas.

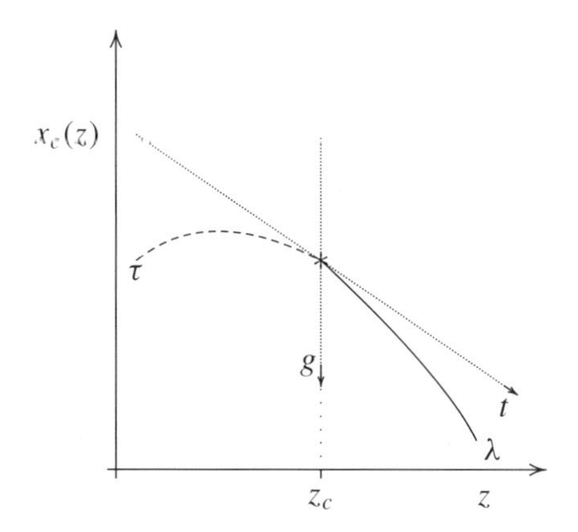

Fig. 2.3: The critical curve $x_c(z)$ and the scaling fields at the tricritical point.

$x_c(z)$ is assumed to exist in an open interval containing z_c.[3] (In some models this is not the case; it is found that the tangents of the τ- and λ-lines meet at an angle at the tricritical point. In this case much of the discussion below still applies, albeit in a slightly altered form.) The shape of the critical curve around the tricritical point is the starting point of a phenomological description of tricriticality. In Fig. 2.3 a generic tricritical point in a critical curve $x_c(z)$ is illustrated.

The critical curve can be expanded around the tricritical point. Suppose that a_λ is the gradient of the tangent line to $x_c(z)$ at the tricritical point (it exists by our assumption). Then

$$x_\tau(z) = x_\lambda(z_c) - a_\lambda(z - z_c) - b_\tau(z_c - z)^{\psi_\tau}, \tag{2.7}$$

$$x_\lambda(z) = x_\lambda(z_c) - a_\lambda(z - z_c) - b_\lambda(z - z_c)^{\psi_\lambda}. \tag{2.8}$$

Here, $x_\lambda(z)$ is the critical curve on the λ-side of the tricritical point, and $x_\tau(z)$ is the critical curve on the τ-side of the tricritical point. The exponents ψ_λ and ψ_τ are called *shift-exponents*, and it is assumed that they exist.[4] These exponents are true tricritical exponents, and are closely associated with the shape of the critical curve in the vicinity of the tricritical point. The basic idea is to assume the existence of a natural set of coordinates for expressing the shape of the critical curve near the tricritical point. Let these coordinates be (g, t), and choose the axis $g = 0$ (this is the t-axis) tangent to the λ-line at the tricritical point, and the axis $t = 0$ (this is the g-axis) transverse to the critical curve and through the tricritical point. Good definitions for t and g follows from eqns (2.7) and (2.8):

$$t = (z - z_c), \qquad g = -x + x_\lambda(z_c) - a_\lambda t. \tag{2.9}$$

[3] Convexity of the free energy, and thus of $-\log x_c(z)$, implies that the tangent vector exists almost everywhere.

[4] If the τ-line is straight, then the situation in eqn (2.7) is slightly different. One may argue that either $b_\tau = 0$ with ψ_τ undefined, or that $\psi_\tau = 1$ in which case b_τ can be absorbed into a_λ. This often occurs if the (straight) τ-line and the tangent of the λ-line meet at an angle at the tricritical point. Models in which the τ-line is straight are refered to as "asymmetric models".

The critical curve in the (g, t)-plane close to the tricritical point is then described by

$$g_\tau(t) = b_\tau(z_c - z)^{\psi_\tau} = b_\tau|t|^{\psi_\tau}, \quad \text{if } z \leq z_c, \tag{2.10}$$

$$g_\lambda(t) = b_\lambda(z - z_c)^{\psi_\lambda} = b_\lambda t^{\psi_\lambda}, \quad \text{if } z \geq z_c. \tag{2.11}$$

If $g = 0$ in eqn (2.9), then the tangent line to the critical curve at the tricritical point is described. The t-axis is thus the tangent to the critical curve at the tricritical point. On the other hand, if $t = 0$, then $g = -x + x_c(z_c)$, and the g-axis is a vertical axis in Fig. 2.3. If the critical curve is plotted in the (g, t)-plane, then Fig. 2.4 is obtained, and it is the *tricritical phase diagram* of the model. This transformation to the (g, t)-coordinates will make the phenomenological description of tricriticality simpler.

The generating function is singular along the critical curve in Fig. 2.4. Along the τ-line, the transition to the infinite phase for fixed z is a first-order transition; this may manifest itself as a jump discontinuity in the first derivative of $\log G_\#(x, z)$. More generally, we interpret this to mean that there is an essential singularity in the generating function along the τ-line. In fact, a stronger assumption will be made: assume that $G_\#(x, z)$ is finite on the τ-line, and that the restriction of $G_\#(x, z)$ to the τ-line is an analytic function of t. Along the λ-line, the transition to the infinite phase is assumed to be continuous. This implies a continuous first derivative of $\log G_\#(x, z)$, which is interpreted to imply a simpler singularity (such as a branch point), in $G_\#(x, z)$ along the λ-line. The singularity along the λ-line is characterized by assuming that, in analogy with eqn (1.7),

$$G_\#(x, z) \simeq A_+(t)(g - g_\lambda(t))^{2-\alpha_+}, \tag{2.12}$$

where α_+ is an exponent which describes the nature of the singularity. It is important to remember that *only* the singular part of $G_\#(x, z)$ is assumed to scale as in eqn (2.12). In particular, there is a "background" analytic contribution to the generating function

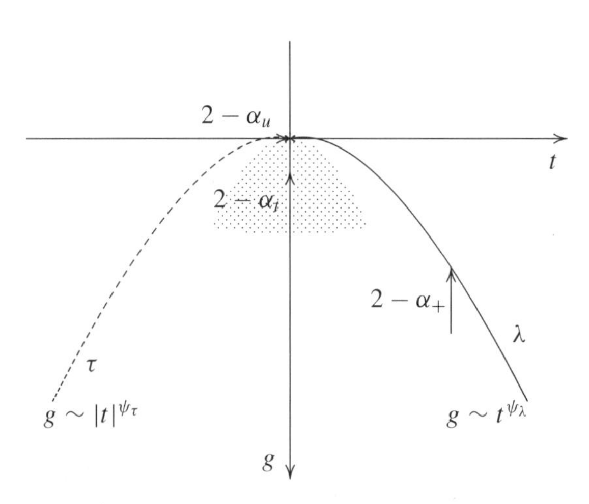

Fig. 2.4: The generic tricritical scaling diagram. The tricritical scaling region is shaded.

which is ignored by this assumption. I shall also assume that the model has the same character as the polygon generating function $G_p(x)$ (eqn (1.5)); that is, $2 - \alpha_+ \geq 0$, so that the singular part of $G_\#(x, z)$ approaches zero as g approaches the λ-line. The analytic contribution can be important; close to the critical point it is the dominant part of the generating function, and if the ratio with $G_\#(x, z)$ in the denominator is computed, then the singular term must be ignored, and the analytic term kept.

A hyperscaling relation which relates α_+ to a metric exponent ν_+ may also be expected. This is Josephson's relation (eqn (1.30)),

$$2 - \alpha_+ = d\nu_+, \tag{2.13}$$

and this is assumed in analogy with the arguments leading to eqn (1.30). The exponent $1/\nu_+$ can be interpreted as a fractal dimension of the polygon near the critical line. The nature of the singularity in $G_\#(x, z)$ along the λ-line is assumed to be a branch point, consistent with a continuous transition in the model. Note that $(g - g_\lambda(t)) = (x_\lambda(t) - x)$, and that the direction of approach to the critical curve is parallel to the g-axis in Fig. 2.4.

There are models in which the singular part of $G_\#(x, z)$ diverges along the λ-line. These models exhibit a character reminiscent of the generating function $G_w(x)$ of walks in eqn (1.9). The appropriate scaling assumption is then

$$G_\#(x, z) \simeq A_+(t)(g - g_\lambda(t))^{-\gamma_+}, \tag{2.14}$$

where γ_+ is non-negative. These models differ in one respect from those described by assumption (2.12); the singular part of the generating function dominates the analytic part close to the critical point, and so the analytic contribution to $G_\#(x, z)$ can be safely ignored.

The classical nature of tricriticality is further described by assuming that $G_\#(x, z)$ is finite along the τ-line, and that it is singular if the tricritical point is approached along the τ-line: the usual scaling assumption is

$$G_\#(x, z) \simeq S_-|t|^{2-\alpha_u}, \qquad \text{along the } \tau\text{-line } (t \to 0^-). \tag{2.15}$$

The exponent α_u is the third tricritical exponent (the other two are the shift-exponents). The singular behaviour of $G_\#(x, z)$ is yet again different if the tricritical point is approached along the g-axis. It is still assumed to be a branch point, but with a different exponent than the exponent in eqn (2.12),

$$G_\#(x, z) \simeq A_t g^{2-\alpha_t}, \qquad \text{along the } g\text{-axis } (g > 0). \tag{2.16}$$

where A_t is assumed to be a constant.[5]

---

[5] Again, it is assumed that both $2 - \alpha_u \geq 0$ and $2 - \alpha_t \geq 0$. In models where these are negative the divergence is again assumed to be similar to that of eqn (1.9). In particular, the assumptions will be

$$G_\#(x, z) \simeq S_-|t|^{-\gamma_u}, \qquad \text{along the } \tau\text{-line } (t \to 0^-);$$

$$G_\#(x, z) \simeq A_t g^{-\gamma_t}, \qquad \text{along the } g\text{-axis } (g > 0).$$

With eqn (2.14), these assumptions are the tricritical assumptions for a model with a generating function more akin to eqn (1.9).

The behaviour of $G_\#(x, z)$ in eqns (2.15) and (2.16) must be reconciled. This is done by assuming *cross-over* behaviour as one moves from the t-axis to the g-axis, controlled by a *cross-over exponent* ϕ. In particular, assume that there exists a universal scaling function $f(x)$ such that

$$G_\#(x, z) \simeq A_t\, g^{2-\alpha_t} f(g^{-\phi} t), \tag{2.17}$$

where $f(0) = 1$ (this recovers (2.16)).[6] If it is assumed that $f(-x) \approx x^u$ as $x \to \infty$, then a comparison with eqn (2.15) shows that $u = 2-\alpha_u$, and that $2-\alpha_t-\phi(2-\alpha_u) = 0$, which gives the relation[7]

$$\phi = \frac{2-\alpha_t}{2-\alpha_u}. \tag{2.18}$$

The assumption in eqn (2.17) is expected to be a good approximation to the generating function near the tricritical point, for example in a region which may qualitatively look like the shaded area in Fig. 2.4 (this is the "tricritical scaling region"). Derivatives of $G_\#(x, z)$ with respect to g shows that the n-th derivative with respect to g should scale as $G_\#^{(n)}(x, z) \simeq S_-|t|^{2-\alpha_u-n\Delta}$, where $\Delta = 1/\phi$ is the *gap exponent.*

The cross-over exponent ϕ is the fifth tricritical exponent. Along the λ-line, the singularity in $G_\#(x, z)$ is expressed by eqn (2.12), and this means that the argument of the tricritical scaling function in eqn (2.17) cannot grow unbounded along the λ-line, lest it spoils the required scaling behaviour. Thus $g^{-\phi}t \simeq$ constant along the λ-line, which shows that $g_\lambda(t) \approx b_\lambda t^{1/\phi}$. Comparing this to eqn (2.11) shows that

$$\psi_\lambda = 1/\phi. \tag{2.19}$$

This is an interesting conclusion. The shift-exponent is controlled by the shape or geometry of the λ-line close to the tricritical point. Hence, the crossover behaviour is controlled by the shift exponent, as is the ratio in eqn (2.18).

A similar analysis can be performed along the τ-line in some models, by replacing λ by τ in the above. In many applications it is concluded from (2.19) that $\psi_\lambda = \psi_\tau = 1/\phi$. These are the *symmetric* models. In this monograph some models will be encountered where the τ-line is straight (the exponent ψ_τ is not defined); these are the *asymmetric* models. In an asymmetric model, the λ- and τ-lines may meet at an angle in the tricritical point. In that case the t-axis is chosen as the tangent to the λ-line at the tricritical point, and the g-axis is perpendicular to the τ-line. It is also possible that the τ- and λ-lines meet at an angle, with different shift exponents. In that case the crossover behaviour is more complicated, as can be seen by repeating the analysis above.

There is also the possibility that two lines of branch points meet in a critical point (which is a non-analyticity in the radius of convergence of the generating function). In

---

[6] The scaling assumption in eqn (2.17) assumes that the shape of the λ-line close to the tricritical point is invariant under rescaling of the coordinates by $g \to Ag$ and $t \to A^\phi t$.

[7] In terms of the exponents γ_u and γ_t this relation reads as

$$\phi = \gamma_t/\gamma_u.$$

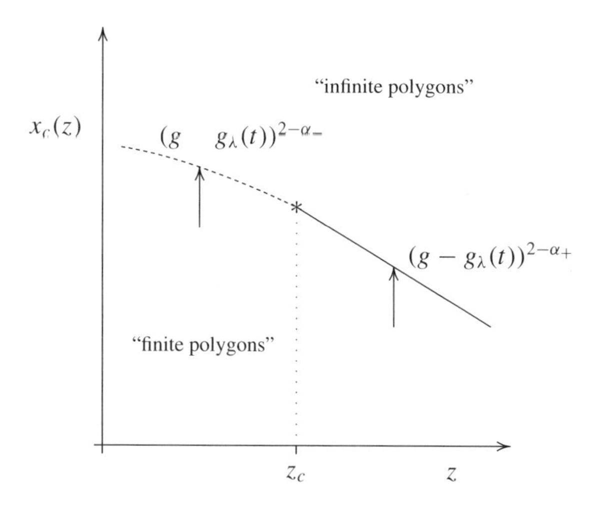

Fig. 2.5: Two λ-lines may meet at a critical point $(z_c, x_c(z_c))$ in the critical curve $x_c(z)$. Both these λ-lines are branch points in the generating function $G_\#(x, z)$, and they are described by the exponents α_+ and α_-, as in eqn (2.12). In particular, if $z < z_c$, then $G_\#(x, z) \simeq A_-(g - g_\lambda(t))^{2-\alpha_-}$ and if $z > z_c$, then $G_\#(x, z) \simeq A_+(g - g_\lambda(t))^{2-\alpha_+}$. If $G_\#(x, z)$ is divergent along the critical curve, then the exponents γ_- and γ_+ are instead defined similar to eqn (2.14).

these models the τ-line is not a line of essential singularities, but is a line of branch points, described by a scaling assumption similar to eqn (2.12) or (2.14); $G_\#(g, t) \simeq A_-(t)(g - g_\tau(t))^{2-\alpha_-}$, and similarly if the exponent γ_- is used instead. Observe that the exponent α_u may still exist if $G_\#(g, t)$ is finite along the τ-line. Otherwise, if the generating function is infinite on the τ-line, the existence of γ_u is not obvious. This situation is illustrated in Fig. 2.5.

2.3 Finite size scaling

In Section 2.2 I described scaling for the generating function $G_\#(x, z)$. This description does not give a full picture of tricriticality; there is an important alternative point of view called "finite size scaling" [9, 300]. This may be considered as scaling of the partition function (or the free energy, as opposed to the generating function), and valuable insights are made when it is compared with the results obtained in the previous section. The basic assumption is that the convergence of $F_n(z) \to \mathcal{F}_\#(z)$, as $n \to \infty$, is controlled by a rescaling of the z-axis about z_c. (In effect, this assumes that the curves $F_n(z)$ are all scaled images of the same curve.) Thus, $F_n(z)$, and thus $p_n^\#(z)$, is a function of a single variable $n^{\phi_c}(z - z_c)$, where ϕ_c is called the "finite size cross-over exponent", and it controls the rescaling of the z-axis. This assumption is reminiscent of the assumption in eqn (2.17), but now the parameters of the theory are assumed to be $t = (z - z_c)$ and n (or $1/n$). The finite size cross-over exponent extends the influence of the non-analyticity in $\mathcal{F}_\#(z)$ to the finite size approximations to it, and it is thus a tricritical exponent, belonging to

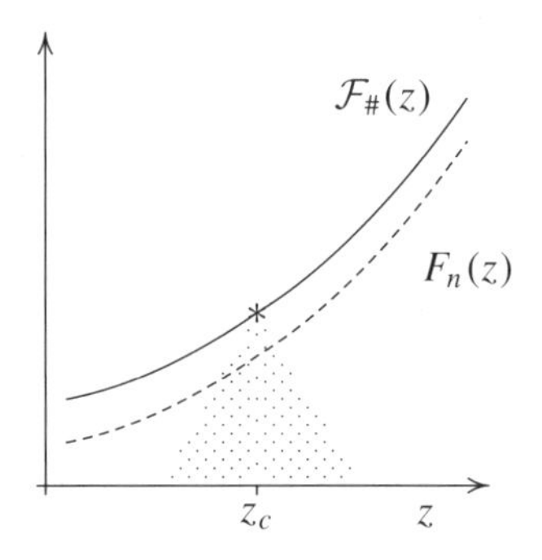

Fig. 2.6: The convergence of $F_n(z)$ to $\mathcal{F}_{\#}(z)$. The tricritical point is assumed to control this convergence by rescaling the z-axis around z_c with n^{ϕ_c}, where ϕ_c is the finite size cross-over exponent which controls the magnitude of the rescaling. The shaded area corresponds to the finite size cross-over scaling regime.

the class of tricritical exponents defined in Section 2.2. In Fig. 2.6 a typical limiting free energy is illustrated. So, assume that $p_n^{\#}(z)$ is a function of only the scaled variable $n^{\phi_c}t$: $p_n^{\#}(z) = \hat{f}(n^{\phi_c}t)$ (where $t = (z - z_c)$).

Since $p_n^{\#}(z)$ is generated by $G_{\#}(x, z)$ (which has different singularities along the λ-line, at the tricritical point, and along the τ-line), the asymptotic behaviour of $p_n^{\#}(z)$, as $n \to \infty$, should also be different for $z > z_c$, $z = z_c$ and $z < z_c$. From eqns (2.12) and (2.16) one can infer that[8] $p_n^{\#}(z) \simeq B_+ n^{\alpha_+ - 3}[x_c(z)]^{-n}$ along the λ-line ($z > z_c$),[9] and $p_n^{\#}(z) \simeq B_t n^{\alpha_t - 3}[x_c(z_c)]^{-n}$ at the tricritical point ($z = z_c$). [10] The rescaling of the t-axis describes the cross-over between these asymptotic regimes. On the other hand, I have argued above that $p_n^{\#}(z)$ is a function of only the scaled variable $n^{\phi_c}t$, and a natural assumption of a functional form for the partition function will be

$$p_n^{\#}(z) \simeq B_\lambda h_\lambda(n^{\phi_c}t) n^{\alpha_t - 3} \mu_+^{-[n^{\phi_c}t]^{1/\phi_c}}, \tag{2.20}$$

where $h_\lambda(x)$ is a function to be determined. The asymptotic form of $h_\lambda(x)$ can be guessed by examining the behaviour of $p_n^{\#}(z)$ along the λ-line and comparing it with the behaviour at z_c: in particular, $h_\lambda(x) \approx$ constant if x is small, and $h_\lambda(x) \approx x^{(\alpha_+ - \alpha_t)/\phi_c}$ as $x \to \infty$.

Along the τ-line similar arguments to the above can be made to argue that

$$p_n^{\#}(z) \sim B_\tau h_\tau(n^{\phi_c}|t|) n^{\alpha_t - 3} \mu_-^{-[n^{\phi_c}|t|]^{1/\phi_c}} \tag{2.21}$$

where $h_\tau(x) \sim |x|^{(\alpha_- - \alpha_t)/\phi_c}$ as $x \to -\infty$. In the case of asymmetric models, $\mu_- = 1$, but it may take other values in symmetric models. However, this is not a complete

[8] Under these assumptions, and eqns (2.20), (2.22) and (2.23), it is only necessary to replace $2 - \alpha_*$ by $-\gamma_*$ to obtain the finite size scaling assumptions for models which have a divergent generating function like $G_w(x)$ (eqn (1.9)), where $*$ is any of $\{+, -, t\}$.

[9] The choice of the exponent $(\alpha_+ - 3)$ makes it compatible with eqn (2.12). In particular,

$$G_{\#}(x, z) = \sum_{n \geq 0} p_n^{\#}(z) x^n \approx \sum_{n \geq 0} n^{\alpha_+ - 3}[x/x_x(z)]^n \approx \int n^{\alpha_+ - 3}[x/x_x(z)]^n dn \approx (x_c(z) - x)^{2 - \alpha_+}.$$

[10] These assumptions should also be compared with eqn (1.6), which has the same general appearance.

description; in models where the polygon collapses to a compact shape, an entropic contribution to the free energy is expected from the surface of the collapsed object. In particular, the surface area of the collapsed object is expected to grow as $n^{(d-1)/d}$ and a factor of the form $\mu_s^{n^\sigma}$, with $\sigma = (d-1)/d$, is expected to account for this contribution [33]. The number μ_s is an effective free energy per unit area, while the critical exponent σ is defined along the τ-line. Generally, the exponent σ may not be equal to $1 - 1/d$, if the collapsed polygon is not compact, and has a "surface" with a fractal dimension. To take the above together, absorb the factors in μ_+ and μ_- in the scaling functions h_λ and h_τ, and define the suitable finite size scaling assumption (compare this to eqn (1.6))

$$p_n^{\#}(z) \simeq \hat{h}(n^{\phi_c} t) n^{\alpha_t - 3}, \tag{2.22}$$

where

$$\hat{h}(x) \simeq \begin{cases} B_\lambda x^{(\alpha_+ - \alpha_t)/\phi_c} \mu_+^{x^{1/\phi_c}}, & \text{if } x \to \infty, \\ \text{constant}, & \text{if } x = 0, \\ B_\tau |x|^{(\alpha_- - \alpha_t)/\phi_c} \mu_s^{|x|^{\sigma/\phi_c}} \mu_-^{|x|^{1/\phi_c}}, & \text{if } x \to -\infty. \end{cases} \tag{2.23}$$

The scaling behaviour of the limiting free energy can be obtained from eqn (2.23). By taking logarithms and dividing by n, then for large n,

$$F_n(z) \simeq \frac{1}{n} \mu_d(n^{\phi_c} t), \tag{2.24}$$

with

$$\mu_d(x) \approx \begin{cases} x^{1/\phi_c}, & \text{as } x \to \infty, \\ |x|^{\sigma/\phi_c}, & \text{if } x \to -\infty \text{ and } \mu_- = 1, \\ |x|^{1/\phi_c}, & \text{if } x \to -\infty \text{ and } \mu_- > 1. \end{cases} \tag{2.25}$$

Asymmetric models are usually found to have $\mu_- = 1$ if $z \leq z_c$, in which case the surface term dominates the behaviour as $n \to \infty$ (if $t < 0$) [290]. In symmetric models it is generally expected that bulk contributions will dominate the surface terms, and the surface correction will only be seen as a second-order effect.[11]

---

[11] This situation is analogous to the scaling of the perimeter length in percolation. Define the partition function of percolation clusters of size n at edge probability p (and let $q = 1 - p$): $p_n^P(p) = \sum_{c,k} a_n(c,k) p^n q^{s+k}$, where $a_n(c,k)$ is the number of lattice animals of size n with c cycles and k nearest-neighbour contacts (and s contacts between occupied and empty sites). The total perimeter of a cluster is $s + k$, and direct computation shows that

$$\langle s + k \rangle = \frac{q}{p} n + q \frac{d}{dq} \log p_n^P(p).$$

The usual finite size scaling assumption $p_n^P(p) = \hat{f}((p_c - p) n^{\sigma_P})$, where σ_P is a cross-over exponent, then gives

$$\langle s + k \rangle \approx \frac{q}{p} n + q n^{\sigma_P} \hat{f}'(0) \qquad \text{as } p \to p_c^-.$$

The term $(q/p)n$ is a bulk contribution to this expectation, and it dominates the slower growing contribution $q n^{\sigma_P} \hat{f}'(0)$ with increasing n.

In the $n \to \infty$ case, if $t > 0$, it is known that $F_n(z)$ approaches a limit, and from eqns (2.24) and (2.25), it appears that

$$\mathcal{F}_\#(z) \simeq C_t t^{1/\phi_c}, \qquad \text{as } t \to 0^+, \tag{2.26}$$

where analytic contributions and lesser non-analycicities are neglected. The signal that a phase transition occurs in the model was pointed out to be a non-analyticity in $x_c(z)$ in Section 2.1, and this non-analyticity is also present in $\mathcal{F}_\#(z)$. If it is visible as a singularity in the first derivative of $\mathcal{F}_\#(z)$, then the transition is classified as *first order*, and if it is in a higher derivative, then the transition is classified as *continuous*. It is generally assumed that the non-analyticity in the free energy is described by

$$\mathcal{F}_\#(z) \simeq C_t t^{2-\alpha}, \qquad \text{as } t \to 0^+, \tag{2.27}$$

where α is called the *specific heat exponent*,[12] since it describes the behaviour of the specific heat (which is the second derivative of $\mathcal{F}_\#(z)$ to z) close to the critical point (compare this to eqn (1.26)). If the transition is first order, then $\alpha = 1$, or α is not defined. All other values of $\alpha < 1$ are consistent with a continuous transition. Values of α larger than one seem to be ruled out by convexity of the free energy. Comparing eqn (2.26) with (2.27), it is seen that the value of α is determined by the value of the finite size cross-over exponent.

$$2 - \alpha = 1/\phi_c. \tag{2.28}$$

On the other hand, $x_c(z) = e^{-\mathcal{F}_\#(z)}$. Consequently, $x_c(z) \sim 1 - \mathcal{F}_\#(z) + \cdots \sim t^{2-\alpha}$, since this singular term is the slowest to approach zero, and by eqns (2.8) and (2.19)

$$2 - \alpha = \psi_\lambda = 1/\phi = 1/\phi_c. \tag{2.29}$$

Hence $\phi = \phi_c$, and this is the connection between scaling of the generating and partition functions. Equation (2.29) is a *hyperscaling relation*. In this case, this relation makes an extraordinary claim: it connects a thermodynamic exponent α to the shift exponent ψ_λ, or to the shape of the critical curve close to the tricritical point. In other words, it claims a direct relation between the combinatorial properties of the model and its thermodynamic behaviour. The dimension d is usually explicitly present in a hyperscaling relation. In this case it will seem to be present as a fractal dimension.

2.4 Homogeneity of the generating function

The points of view taken in Sections 2.2 and 2.3 do not completely describe tricriticality. Tricriticality has yet to be examined from a metric point of view, where the correlation length is used as a basic quantity, similar to the approach taken in Section 1.3.

At the basis of the analysis in the preceding sections lies the idea of *scale invariance*. In a system undergoing a continuous phase transition, it is assumed that there is only one

[12] Note that this α is not the same as α_t, α_u or α_+. The specific heat is the second derivative of $\mathcal{F}_\#(z)$ with respect to z, and the singularity in it is generally assumed to be of the form $|z - z_c|^{-\alpha}$.

length scale determined by the correlation length. Thus, the system is invariant under a spatial dilation followed by a rescaling of the unit length to compensate for the dilation.[13] In other words, a change in the correlation length (due to changes in the fields t or g), may be considered equivalent to a dilation of space. Fix the scaling field t and consider the correlation length as a function of g, as the λ-line is approached. Equation (1.21) then suggests that, along the λ-line,

$$\xi(g) \simeq (g - g_\lambda(t))^{-\nu_+}. \tag{2.30}$$

ν_+ is a critical exponent which describes the divergence of the correlation function as criticality is approached. As was seen in eqn (2.13), there is also a hyperscaling relation which relates ν_+ and α_+, namely $2 - \alpha_+ = d\nu_+$. I shall now argue that similar relations are valid at the tricritical point.

If the tricritical point is approached along the g-axis, then the divergence in $\xi(g)$ is described by the assumption that

$$\xi(g) \simeq g^{-\nu_t}, \tag{2.31}$$

where ν_t is the *tricritical metric exponent*. A rescaling of space by a factor s is achieved by rescaling all vectors $\mathbf{r}$ by $\mathbf{r} \to \mathbf{r}/s$. The assumption of invariance in the system means that the scaling fields t and g must be rescaled to compensate for the change in length scale. To do this, two scaling exponents y_g and y_t are introduced to describe the scaling of t and g with s:

$$g \to s^{y_g} g \qquad t \to s^{y_t} t. \tag{2.32}$$

In other words, changes in g are compensated for by changes in s through $s \simeq g^{-1/y_g}$. In the vicinity of the tricritical point, $\mathbf{r}$ is measured in terms of $\xi(g)$, and a change in the correlation length is compensated for by a corresponding change in s. In particular, since $\xi(g) \simeq g^{-\nu_t}$, and since it scales with g exactly as s does, the exponents may be identified:

$$y_g = 1/\nu_t. \tag{2.33}$$

Homogeneity is a mathematical assumption about the behaviour of the generating function $G_\#(g, t)$ close to criticality. In particular, it is assumed that the singular part of the generating function can be rescaled by

$$G_\#(g, t) \simeq s^{-d} G_\#(s^{y_g} g, s^{y_t} t) \tag{2.34}$$

---

[13] Of course, I also assume invariance under translations and rotations. This assumption of scale invariance can be generalized to conformal invariance, which leads to powerful techniques for describing two-dimensional models in statistical mechanics.

under a dilation of space, in d dimensions. Assume that $s > 1$ and that y_g and y_t are positive exponents.[14] If s is eliminated from eqn (2.34) by using the relation $s \simeq g^{-1/y_g}$, then

$$G_{\#}(g, t) \simeq g^{d\nu_t} G_{\#}(c, g^{-y_t/y_g} t), \tag{2.35}$$

where c is a constant and where t is kept fixed. If it is assumed that eqn (2.35) describes the singular part of the generating function, and that it can be compared to eqn (2.17), then

$$2 - \alpha_t = d\nu_t, \qquad \phi = y_t/y_g. \tag{2.36}$$

It is similarly possible to show that

$$2 - \alpha_u = d/y_t. \tag{2.37}$$

It is important not to confuse these relations with the hyperscaling law in eqn (2.29). In fact, replacing ϕ in eqn (2.29) by the result in eqn (2.36) produces $2 - \alpha = 1/(y_t \nu_t)$, and $1/\nu_t$ may be interpreted as a "fractal dimension" of the polygon at the tricritical point. This interpretation justifies the fact that eqn (2.29) is referred to as a hyperscaling relation. In addition, the gap exponent is the ratio $\Delta = y_g/y_t$, and from eqn (2.35) it appears that derivatives of $G_{\#}(g, t)$ to g will obey power-law relations with exponents separated by Δ, as claimed when eqn (2.17) was considered.[15]

If homogeneity of the generating function is accepted as an hypothesis, then the scaling form in eqn (2.17) is the result. At a more fundamental level, homogeneity accepts that one can classify the singularities in the generating function using a set of exponents, and that there is a connection between the thermodynamic properties of the model and the nature of the singularities.

2.5 Uniform asymptotics and the finite size scaling function

2.5.1 Asymmetric tricriticality and ϵ-asymptotics

I have considered various aspects of tricritical scaling in Sections 2.1–2.4. In this section the assumptions will be considered in a more formal way, with the goal that the asymptotics of the generating function be examined and extended to a uniform asymptotic description. I shall follow the arguments as presented by R. Brak and A.L. Owczarek

[14] Since $s > 1$, and the exponents y_g and y_t are positive, the scaling of the arguments with powers of s drives the generating function away from the tricritical point, and so it becomes larger (since $2 - \alpha_t \geq 0$, the singular part of $G_{\#}(g, t)$ vanishes at the tricritical point). The exponents y_g and y_t are defined such that a factor of s^{-d} will restore $G_{\#}(s^{y_g} g, s^{y_t} t)$ to its previous value. In some models it may be the case that the singular part of $G_{\#}(g, t)$ diverges as the tricritical point is approached. In this case the exponent $\gamma_t \geq 0$ is used, see footnote 5, and the appropriate assumption is that $G_{\#}(g, t) \simeq s^d G_{\#}(s^{y_g} g, s^{y_t} t)$.

[15] In the case that $G_{\#}(g, t)$ is divergent at the tricritical point, then the hyperscaling relations

$$\gamma_t = d\nu_t, \qquad \gamma_u = d/y_t$$

are found instead. The hyperscaling relation $2 - \alpha = 1/\phi$ with $\phi = \nu_t y_t$ is still valid, as is the definition of the gap exponent Δ.

[33], who developed the theory in this section. Only a model with a straight τ-line and a critical line which is differentiable in an interval which contains the tricritical point will be considered. Models in which the τ-line is not straight are also covered by the results here, as long as the critical line is differentiable in an interval which contains the tricritical point. In particular, let $G_{\#}(x,z) = \sum_{n\geq 0} p_n^{\#}(z)x^n$ be the generating function of a model with a tricritical scaling phase diagram, and let the radius of convergence of $G_{\#}(x,z)$ be $x_c(z)$. Let the tricritical point be at $z = z_c$ and suppose that the τ-line is given by $x_c(z)$ with $z < z_c$. Define the function

$$\omega(z) = \begin{cases} x_c(z), & \text{if } z < z_c; \\ x_c(z_c), & \text{otherwise.} \end{cases} \tag{2.38}$$

Define the modified generating function

$$G_A(y,z) = \sum_{n\geq 0} p_n^{\#}(z)[\omega_c(z)y]^n. \tag{2.39}$$

Then the radius of convergence of $G_A(y,z)$ is

$$y_c(z) = \begin{cases} 1, & \text{if } z < z_c; \\ x_c(z)/x_c(z_c), & \text{otherwise.} \end{cases} \tag{2.40}$$

In other words, $G_A(y,z)$ has an asymmetric phase diagram, with a straight τ-line. A schematic drawing of the phase diagram is in Fig. 2.7. Since the τ-line is parallel to the z-axis, and since it is assumed that it is differentiable at $z = z_c$, it is possible to use eqns (2.7)–(2.9) to choose the (g,t)-coordinates with $g = 1 - y$ and $t = z - z_c$.

The discussions in the previous sections were centred around imprecise scaling assumptions. For example, the tricritical scaling assumption in eqn (2.17) does not claim

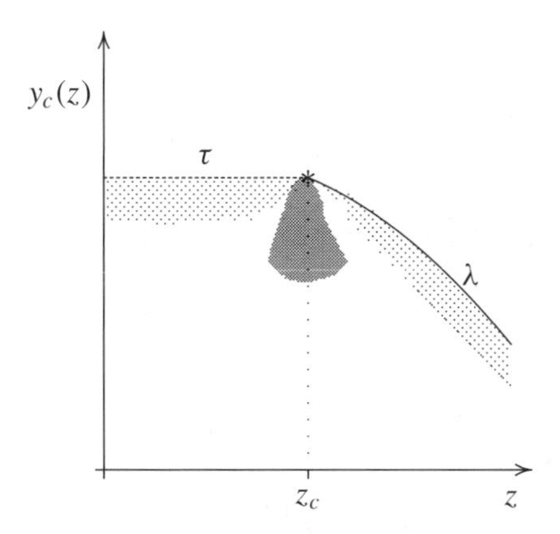

Fig. 2.7: The tricritical scaling diagram of an asymmetric model. The tricritical scaling region is the shaded area around the tricritical point. Scaling regions close to the λ- and τ-lines are also shaded. The cross-over occurs around the tricritical point.

any interval over which the assumption is good. Thus, it is important that the "scaling regimes" in Fig. 2.7 where scaling assumptions are good to a "tolerance" of say ϵ, are carefully defined. Such areas will be called ϵ-asymptotic regions.

Definition 2.1 (ϵ-Asymptotic regions) Suppose that $u(x) \sim p(x)$ as $x \to x_0^-$. Then the ϵ-asymptotic region of $u(x)$ with respect to $p(x)$, $\Delta_p^u(\epsilon)$, is defined as follows. Let $x_1 = \min\{x \big| |u(y)/p(y) - 1| < \epsilon,\ \forall y \in (x, x_0)\}$. Then $\Delta_p^u(\epsilon) = (x_1, x_0)$ and the *width* of the ϵ-asymptotic interval is $\delta_p^u(\epsilon) = x_0 - x_1$. □

Thus, $\Delta_p^u(\epsilon)$ is that interval close to x_0 where $p(x)$ is a good approximation to $u(x)$. In Fig. 2.7 the ϵ-asymptotic regions for scaling on approach of the τ-line, the λ-line (eqn (2.12)), and the tricitical scaling area (eqn (2.17)), are indicated. These regions may, or may not, overlap. If these scaling areas do not overlap, then there are cusps (or wedges) around the tricritical point which are not covered by any of the scaling regions. This can be a problem if they contain singularities of the generating function not accounted for by our scaling hypotheses. It is therefore important that the scaling regions around the tricritical point in Fig. 2.7 overlap, and another assumption (called "asymptotic completeness") is needed for the description of tricriticality.

2.5.2 *Extended tricriticality*

In this section two issues are considered. In the first instance a complete definition (and thus a summary) of tricriticality will be given. In the second instance, it will be shown that the definition can be generalized to "extended tricriticality", which will eventually allow asymptotic completeness to be considered. An interval of the activity $z \in (z_0, z_1)$ will be considered, with the tricritical value $z_c \in (z_0, z_1)$, and with z_0 and z_1 far enough away from z_c so that scaling in the vicinity of the τ- and λ-lines is visible. In addition, only the asymptotics of the non-analytic part of the generating function will be considered in what follows; in effect, consider the analytic terms in the generating function to have been subtracted from it. In addition, the use of the symbol "$\sim$" will be relaxed here, in the sense that $f \sim g$ will mean that the singular parts of f and g are asymptotic. With this, consider the following definition of tricriticality.

Definition 2.2 (Tricriticality) Let $G_A(y, z)$ be the generating function defined in eqn (2.39), and let its radius of convergence be $y_c(z)$ in eqn (2.40). Let $0 < z_0 < z_c < z_1$, and define the intervals $I_- = (z_0, z_c)$, $I_+ = (z_c, z_1)$ and $I = (z_0, z_1)$. Moreover, define the λ- and τ-lines as before, and the coordinates g and t as in eqn (2.9). Let the λ-line be given by $g = g_\lambda(t)$ and the τ-line by $g = g_\tau(t)$ in these coordinates. Tricritical scaling is said to occur at the point $(1, z_c)$ if

(1) $G_A(y, z)$ has a radius of convergence $y_c(z) > 0$, with $y_c(z) = 1$ if $z \le z_c$ and $y_c(z) < 1$ and analytic if $z \in I_+$.
(2) There exists positive numbers D and ϕ such that

$$g_\lambda(t) \sim D t^{1/\phi}, \qquad \text{as } t \to 0^+. \tag{2.41}$$

This defines the cross-over exponent as a shift-exponent which describes the λ-line as it approaches the critical point (see eqn (2.11)).

(3) There exists a number $2-\alpha_+$ and a function $A_+(t)$, analytic in I_+, such that

$$G_A(y,z) \sim A_+(t)(g-g_\lambda(t))^{2-\alpha_+} \qquad \text{as } g \to g_\lambda(t)^+. \tag{2.42}$$

Compare this to eqn (2.12).

(4) There exist numbers A_t and $2-\alpha_t$ such that if $z=z_c$ then

$$G_A(y,z_c) \sim A_t g^{2-\alpha_t}, \qquad \text{as } g \to 0^+. \tag{2.43}$$

Compare this to eqn (2.16).

(5) Assume that $G_A(y,z)$ is finite on the τ-line; in fact, assume the existence of an analytic function $A_-(t)$ in I_- such that

$$G_A(y,z) \sim A_-(t) \qquad \text{on the } \tau\text{-line}. \tag{2.44}$$

Moreover, there exist numbers S_- and $2-\alpha_u$ (compare this to eqn (2.15)) such that

$$A_-(t) \sim S_-|t|^{2-\alpha_u}, \tag{2.45}$$

and where $2-\alpha_u=(2-\alpha_t)/\phi$.

(6) There exists a number S_+ such that

$$A_+(t) \sim S_+ t^{[(\alpha_+-2)/\phi]+(2-\alpha_u)}, \qquad \text{as } t \to 0^+. \tag{2.46}$$

(7) There exists a tricritical scaling function $f(x)$ such that

$$G_A(y,z) \sim A_t g^{2-\alpha_t} f(A_s g^{-\phi} t). \tag{2.47}$$

This tricritical scaling assumption is the same as eqn (2.17). The scaling function $f(x)$ must be analytic in an interval $(-\infty, x_0)$, where $x_0 = A_s D^{-\phi}$. Moreover, one can check the following for consistency against items (1)–(6). In the first place,

$$f(0)=1. \tag{2.48}$$

Secondly, the following asymptotic formulas must hold for $f(x)$:

$$f(x) \sim G_-(-x)^{2-\alpha_u}, \qquad \text{as } x \to -\infty \tag{2.49}$$

see the arguments following eqn (2.17), and

$$f(x) \sim G_+(x_0-x)^{2-\alpha_+}, \qquad \text{as } x \to x_0^-. \tag{2.50}$$

The constants G_+ and G_- should be given by $G_+ = S_+ D^{\alpha_t-\alpha_+}(x_0\phi)^{(\alpha_+-2)} A_t^{-1}$, and $G_- = S_- D^{\alpha_t-2} x_0^{\alpha_u-2} A_t^{-1}$, to be compatible with the scaling assumptions in eqns (2.42) and (2.45).

(8) Assume that neither (2.42) nor (2.45) need be uniform, and observe that eqn (2.47) cannot be uniform either. □

Our first goal is to develop a scaling assumption which gives uniform asymptotics for $G_A(y, z)$ in the interval I defined in Definition 2.2. The definition above is still inadequate; in particular, eqn (2.47) must be adapted to account for the scaling in eqn (2.45) and especially in eqn (2.42). This can be achieved by slightly changing the assumption in eqn (2.47) to "extended tricritical scaling". The extended tricritical scaling assumption will give the asymptotic behaviour of the generating function as the critical line is approached for any fixed $z \in I$.

Theorem 2.3 (Extended tricritical scaling) *Let $G_A(y, z)$ exhibit tricritical scaling as in Definition 2.2. Suppose that $z \in I$ is fixed, but arbitrary. Then there exists real-valued functions $d(t)$ and $h(t)$ such that*

$$G_A(y, z) \sim A_t d(t) g^{2-\alpha_t} f(A_s g^{-\phi} h(t)),$$

where

$$\begin{aligned} d(t) &\sim 1 \qquad t \to 0, \\ h(t) &\sim t \qquad t \to 0. \end{aligned}$$

Moreover, $h(t)$ is monotonic, and has an inverse in I.

Proof Define

$$h(t) = \begin{cases} D^{-\phi}[g_\lambda(t)]^\phi, & \text{if } z \in I_+; \\ t, & \text{if } z \in I_- \cup \{z_c\}. \end{cases}$$

$g_\lambda(t)$ is analytic in I_+, thus $h(t)$ is analytic in I_- and in I_+. By eqn (2.41) $g_\lambda(t) \sim Dt^{1/\phi}$; this shows that $h(t) \sim t$ as $t \to 0$. If $g \to [g_\lambda(t)]^+$, then the argument of f in eqn (2.47) approaches $x_0 = A_s D^{-\phi}$, and by eqn (2.50) the correct behaviour is found along the λ-line. Since $g_\lambda(t)$ is monotonic in I_+, so is $h(t)$ and therefore $h(t)$ is invertible in I. The function $d(t)$ can also be explicitly defined:

$$d(t) = \begin{cases} A_+(t)(x_0\phi)^{(\alpha_+ - 2)} A_t^{-1} G_+^{-1} [g_\lambda(t)]^{\alpha_t - \alpha_+}, & \text{if } z \in I_+; \\ 1, & \text{if } z = z_c; \\ A_-(t) A_s^{\alpha_u - 2} |t|^{\alpha_u - 2} A_t^{-1} G_-^{-1}, & \text{if } z \in I_-. \end{cases}$$

Now use eqns (2.41), (2.45) and (2.46) to show that the assumed asymptotic form has the required behaviour along the τ-line, and in its approach to the tricritical point. □

In Theorem 2.3 the approach to the tricritical point was with $t = 0$, as $g \to 0^+$, and the scaling is then given by eqn (2.44). It will also be useful to be able to approach the tricritical point along other trajectories, in particular, consider the family of curves indexed by $q \in (-\infty, x_0]$:

$$A_s g^{-\phi} h(t) = q, \qquad \text{as } g \to 0^+. \tag{2.51}$$

If q approaches $x_0 = A_s D^{-\phi}$, then the approach to the tricritical point is along the λ-line. If q approaches $-\infty$, then the approach will be along the τ-line. Since $h(t)$ is invertible in I, solve for t in eqn (2.51) and consider the curve $(g, h^{-1}(qg^\phi/A_s))$ parametrized by g and indexed by q. If g is sufficiently small, then $h^{-1}(qg^\phi/A_s) \in I$, and as $g \to 0^+$, the tricritical point is approached. By Theorem 2.3, and eqn (2.47):

$$\begin{aligned} G_A(g, h^{-1}(qg^\phi/A_s)) &\sim A_t d(h^{-1}(qg^\phi/A_s))g^{2-\alpha_t} f(A_s g^{-\phi} h(h^{-1}(qg^\phi/A_s))) \\ &\sim A_t g^{2-\alpha_t} f(q), \end{aligned} \tag{2.52}$$

as $g \to 0^+$. Thus, for any $q \in (-\infty, x_0)$ the correct scaling behaviour is found on approaching the tricritical point along any of these curves.

Lemma 2.4 *Let q be a fixed number in $(-\infty, x_0]$, and consider the curve $A_s g^{-\phi} h(t) = q$, as $g \to 0^+$. Then*

$$G_A(g, h^{-1}(qg^\phi/A_s)) \sim A_t d(h^{-1}(qg^\phi/A_s))g^{2-\alpha_t} f(q),$$

as $g \to 0^+$. □

2.5.3 Uniform asymptotics for the generating function

It is generally accepted that the asymptotic behaviour of the generating function around the tricritical point can be completely described by a set of critical exponents, and a scaling function. This may not be the case if the ϵ-asymptotic regions in Fig. 2.7 do not overlap. In particular, if there are cusps or wedges around the tricritical point which are not covered by the ϵ-asymptotic regions of either the tricritical point, or of the λ- and τ-lines, then the belief above may be false (for example, there may be additional singularities, and critical behaviour, in the wedges). To avoid these possibilities, assume that the description of tricritical scaling is *asymptotically complete*; by this I mean that the ϵ-asymptotic regions overlap around the tricritical point. To make this more precise, consider the level curve defined by eqn (2.51), and suppose that $q \in (0, x_0)$. Fix a point $z \in I_+$, and consider the separation between the corresponding g-coordinates for points on the curve in eqn (2.51), and the critical curve. This separation is given by

$$\left[\frac{A_s h(t)}{q}\right]^{1/\phi} - \left[\frac{A_s h(t)}{x_0}\right]^{1/\phi}. \tag{2.53}$$

Asymptotic completeness requires that the width of the ϵ-asymptotic region of the λ-line at this value of z be wider than this separation, for some fixed value of q, and for all $z \in I_+$. In other words, the wedge formed by the λ-line and the level curve must be completely contained in the ϵ-asymptotic region of the λ-line, for some finite and fixed value of q, and for all $z \in I_+$. The same requirement is made on the ϵ-asymptotic region of the τ-line.

Definition 2.5 (Asymptotic completeness) Let $G_A(g, t)$ be a generating function which exhibits tricritical scaling as set out in Definition 2.2 and which exhibits extended tricritical scaling as set out in Theorem 2.3 and Lemma 2.4. Let the ϵ-asymptotic regions of the τ-line be $\Delta^-(\epsilon)$ and of the λ-line be $\Delta^+(\epsilon)$ with corresponding widths $\delta^-(\epsilon, t)$

and $\delta^+(\epsilon, t)$. Then $G_A(g, t)$ satisfies a condition of asymptotic completeness if there exists a $q_0 \in (0, x_0)$ and a $q_1 \in (-\infty, 0)$, both dependent on ϵ, such that for all $z \in I_+$

$$\delta^+(\epsilon, t) \geq \left[\frac{A_s h(t)}{q_0}\right]^{1/\phi} - \left[\frac{A_s h(t)}{x_0}\right]^{1/\phi},$$

and for all $z \in I_-$

$$\delta^-(\epsilon, t) \geq \left[\frac{A_s h(t)}{q_1}\right]^{1/\phi}. \qquad \square$$

It is now possible to prove that if $G_A(g, t)$ satisfies the hypothesis in Definition 2.2, and if it is asymptotically complete, then there exists a uniform tricritical scaling for $G_A(g, t)$. It will be helpful to define the function

$$H(g, t) = A_t d(t) g^{2-\alpha_t} f(A_s g^{-\phi} h(t)), \tag{2.54}$$

and this will be used in the next theorem.

Theorem 2.6 *Let $G_A(g, t)$ be a generating function which exhibits tricritical scaling as in Definition 2.2 and is also asymptotically complete in the sense of Definition 2.5. Then $G_A(g, t)$ has extended tricritical scaling uniformly for all $z \in I$. In particular*

$$G_A(g, t) \sim A_t d(t) g^{2-\alpha_t} f(A_s g^{-\phi} h(t)),$$

as g approaches the critical line for all $z \in I$ uniformly, and with $d(t)$ and $h(t)$ as defined in Theorem 2.3.

Proof Define the ϵ-asymptotic regions of $H(g, t)$ by $\Delta_H^+(\epsilon)$, $\Delta_H^-(\epsilon)$ on I_+ and I_- respectively. By asymptotic completeness, $\Delta_H^+(\epsilon)$ contains the region between the curve $A_s g^{2-\alpha_t} h(t) = p_0$, the critical curve $g = g_\lambda(t)$ and the curve $g - g_\lambda(t) =$ constant, for some non-zero constant, where p_0 is close enough to x_0; say for p_0 in the interval (Q_0, x_0). Since $f(x) \sim G_+(x_0 - x)^{2-\alpha_+} = f_+(x)$ (eqn (2.50)), there is an asymptotic region with $|f(x) - f_+(x)| \leq \epsilon' |f_+(x)| = \epsilon' f_+(x)$. Thus

$$\begin{aligned}
&H(g, t) \\
&\quad = A_t d(t) g^{2-\alpha_t} [f(A_s g^{-\phi} h(t)) - f_+(A_s g^{-\phi} h(t))] + A_t d(t) g^{2-\alpha_t} f_+(A_s g^{-\phi} h(t)), \\
&\quad < A_t d(t) g^{2-\alpha_t} \epsilon' f_+(A_s g^{-\phi} h(t)) + A_t d(t) g^{2-\alpha_t} f_+(A_s g^{-\phi} h(t)) \\
&\quad = (1 + \epsilon') A_+ (g - g_\lambda(t))^{2-\alpha_+} + (1 + \epsilon') O((g - g_\lambda(t))^{3-\alpha_+}),
\end{aligned}$$

where the fact that $A_t d(t) g^{2-\alpha_t} f_+(A_s g^{-\phi} h(t)) = A_+(g - g_\lambda(t))^{2-\alpha_+} + O((g - g_\lambda(t))^{3-\alpha_+})$ as $g \to [g_\lambda(t)]^+$ was used.[16] Thus, by rearranging terms,

$$\frac{H(g,t)}{A_+(g - g_\lambda(t))^{2-\alpha_+}} - 1 < \epsilon' + (1+\epsilon')O(g - g_\lambda(t)).$$

Now choose $\epsilon' = (\epsilon - O(g - g_\lambda(t)))/(1 + O(g - g_\lambda(t)))$ and g sufficiently close to $g_\lambda(t)$ such that $\epsilon' > 0$. Then $H(g,t) \sim A_+(g - g_\lambda(t))^{2-\alpha_+}$ as g approaches the λ-line. This is valid if p_0 is chosen in the interval (Q_0, x_0) and such that $f(x)$ and $f_+(x)$ are ϵ'-asymptotic. In particular, let the ϵ-asymptotic region of $f(x)$ and $f_+(x)$ have width $\delta_f^{f+}(\epsilon)$. Then the level curve determined by p_0 must be in the region determined by the interval $(g_\lambda(t) + \delta_f^{f+}(\epsilon), g_\lambda(t))$, for all $z \in I_+$. But this means that p_0 must be chosen in the interval $(A_s(g_\lambda(t) + \delta_f^{f+}(\epsilon))^{-\phi} h(t), x_0)$. The intersection of this interval with (Q_0, x_0) gives possible values for p_0. Let q_2 be one such value.

Next, it must be made certain that the region between the level curve defined by q_2, and $g_\lambda(t)$, is contained in some ϵ-asymptotic region of $G_A(g,t)$ and $H(g,t)$. By Lemma 2.4 there is a q_0, so that if q_3 is the maximum of q_2 and q_0, then the region between the level curve determined by q_3 and $g_\lambda(t)$ is in the (say) 3ϵ-asymptotic region of $G_A(g,t)$ and $H(g,t)$ (this is $\Delta_G^H(3\epsilon)$), and $H(g,t) \sim A_+(g - g_\lambda(t))^{2-\alpha_+}$ as g approaches the λ-line. Thus it follows that $G_A(g,t) \sim H(g,t) \sim A_+(g - g_\lambda(t))^{2-\alpha_+}$ as g approaches the λ-line within the region determined by the level curve indexed by q_3 and $g_\lambda(t)$, for all $z \in I_+$.

The next step is to show that the asymptotics in the previous paragraph are uniform. By Lemma 2.4 for each $q \in (0, x_0)$, there exists a width $\delta_G^H(\epsilon)$ such that $|G_A(g,t)/H(g,t) - 1| < \epsilon$ if $g - g_\lambda(t) < \delta_G^H(\epsilon)$, and where $\delta_G^H(\epsilon)$ may be dependent on $z \in I_+$. $\delta_G^H(\epsilon) > 0$ for all $q \in (0, x_0)$, by asymptotic completeness, and $\delta_G^H(\epsilon)$ is strictly positive for all $q \in (0, q_3]$. This ϵ-asymptotic region is different from that region defined by the ϵ-asymptotic region of $f(x)$ and $f_+(x)$; but they are both contained in $\Delta_G^H(3\epsilon)$. Thus, they have a common neighbourhood. So choose some $\delta'(\epsilon)$ independent of z so that the region defined by $|g - g_\lambda(t)| < \delta'(\epsilon)$ is contained in $\delta_G^H(3\epsilon)$ for all $z \in I_+$. In other words, there is a $\delta'(\epsilon) > 0$ independent of z, so that $G_A(g,t) \sim H(g,t)$ uniformly in I_+.

The proof follows similar arguments for $z \in I_-$ and the τ-line. □

---

[16] This is seen by using eqn (2.50). Notice first that

$$A_t d(t) g^{2-\alpha_t} f_+(A_s g^{-\phi} h(t)) = A_+(g/g_\lambda(t))^{2-\alpha_t} (g_\lambda(t))^{2-\alpha_+} \phi^{\alpha_+ - 2} (1 - [g_\lambda(t)/g]^\phi)^{2-\alpha_+}.$$

Expand the exponents:

$$1 - [g_\lambda(t)/g]^\phi \approx \phi \log(g/g_\lambda(t))$$
$$(g/g_\lambda(t))^{2-\alpha_t} \approx 1 + (2-\alpha_t)\log(g/g_\lambda(t))$$

Substitute these in the previous expressions to obtain the following:

$$A_+(g - g_\lambda(t))^{2-\alpha_+}[1 + (2-\alpha_t)\log(g/g_\lambda(t))] = A_+(g - g_\lambda(t))^{2-\alpha_+} + O((g - g_\lambda(t))^{3-\alpha_+}).$$

Theorem 2.6 is due to R. Brak and A.L. Owzcarek [33], and its significance is that with the assumption of asymptotic completeness, and the properties of tricritical scaling as set out in Definition 2.2, it is possible to find extending functions which will give a uniform asymptotic approximation to the generating function.

2.5.4 The finite size scaling function

I have described the tricritical scaling of the generating function in the previous sections, without reference to finite size scaling. In particular, the connection between the finite size cross-over exponent in eqn (2.22), and the cross-over exponent in eqn (2.29) cannot be taken too seriously based on the imprecise arguments made there. In this section I aim to use the results on scaling of the generating function to see what may be said about finite size scaling. It appears that the connection will be through the generating function

$$G_A(g,t) = \sum_{n \geq 0} p_n^{\#}(z) x^n, \tag{2.55}$$

which suggests that

$$p_n^{\#}(z) = \oint_C G_A(g,t) \frac{dy}{y^{n+1}}, \tag{2.56}$$

where C is a contour which circles the origin and which contains no singularities in $G_A(g,t)$.

The main problem in this approach is that not enough is known about $G_A(g,t)$ to make it possible to execute this integral. Suppose that $z \in I$, and suppose now that $G_A(g,t)$ satisfies the conditions in Definition 2.2, and the uniform asymptotics in theorem 2.6 ($G_A(g,t) \sim H(g,t)$). Then one might want to execute the integral in eqn (2.56) by replacing $G_A(g,t)$ by $H(g,t)$. This can be done, if another assumption about the singularity structures of $G_A(g,t)$ and $H(g,t)$ is made. This assumption is a Darboux condition, which will allow the use of Darboux's theorem [278] to find a representation of $p_n^{\#}(z)$ in terms of $H(g,t)$. Since much more is known about $H(g,t)$ (than $G_A(g,t)$), it becomes possible to do the integral in eqn (2.56) (with $G_A(g,t)$ replaced by $H(g,t)$) and to find a finite size scaling function. Darboux's theorem states the following.

Theorem 2.7 (Darboux's theorem, see [278]) *Let $\psi(t)$ be an analytic function with Laurent expansion $\psi(t) = \sum_{n=-\infty}^{\infty} a_n t^n$ in the annulus $0 < |t| < r$. Let $\chi(t)$ be a function which is analytic in $0 < |t| < r$ and with Laurent expansion $\chi(t) = \sum_{n=-\infty}^{\infty} b_n t^n$ in $0 < |t| < r$. Suppose that the difference of the m-th derivatives of $\psi(t)$ and $\chi(t)$ has a finite number of singularities at $t = t_j$ such that*

$$\psi^{(m)}(t) - \chi^{(m)}(t) = O([t - t_j]^{\sigma_j - 1}),$$

for some positive constants σ_j, as $t \to t_j$. Then

$$a_n = b_n + o(r^{-n} n^{-m}), \qquad n \to \infty. \qquad \square$$

Since both $G_A(g,t)$ and $H(g,t)$ are analytic in the annulus $0 < |y| < y_c(z)$, it is only necessary to assume that the difference of derivatives has a finite number of singularities (at y_i), such that

$$G_A^{(m)}(g,t) - H^{(m)}(g,t) = O([y - y_i]^{\sigma_j - 1}), \tag{2.57}$$

where the σ_j are positive constants. With this assumption, Darboux's theorem states that

$$p_n^{\#}(z) = \frac{1}{2\pi i}\oint_C H(g,t)\frac{dy}{y^{n+1}} + o(y_c(z)^{-n}n^{-m}), \tag{2.58}$$

where the contour circles the origin, and excludes all singularities of $H(g,t)$. If $H(g,t)$ is substituted from eqn (2.54), then this becomes

$$p_n^{\#}(z) = \frac{A_t d(t)}{2\pi i}\oint_C (1-y)^{2-\alpha_t} f(A_s(1-y)^{-\phi}h(t))\frac{dy}{y^{n+1}} + o(y_c(z)^{-n}n^{-m}), \tag{2.59}$$

where the fact that $g = 1 - y$ was used. Define $1 - y = q/n$, and substitute. Note that the contour will now circle the point $q = n$; denote this by C_n:

$$p_n^{\#}(z) = -\frac{A_t d(t) n^{\alpha_t - 3}}{2\pi i}\oint_{C_n} q^{2-\alpha_t}\left(1 - \frac{q}{n}\right)^{1-n} f(A_s n^{\phi} h(t)/q^{\phi})dq + o(y_c(z)^{-n}n^{-m}). \tag{2.60}$$

Put $n^{\phi}h(t) = \zeta$. Then $z = z_c + h^{-1}(\zeta n^{\phi})$; n should be increased in eqn (2.60), while ζ should remain fixed. This implies that z approaches z_c as $z_c + h^{-1}(\zeta n^{\phi})$. Hence $y_c(z)^{-n} = [y_c(z_c + h^{-1}(\zeta n^{\phi}))]^{-n}$ in eqn (2.60), and as $n \to \infty$ this approaches a constant. Thus, the correction term can be ignored asymptotically, and only the evaluation of the integral in eqn (2.60) should be of concern.

Equations (2.49) and (2.50) show that the scaling function $f(x)$ has radius of convergence $x_0 = A_s D^{-\phi}$; therefore, it has a Taylor expansion about $x = 0$; in particular, with $\zeta = n^{\phi}h(t)$, then $A_s n^{\phi}h(t)/q^{\phi} = A_s\zeta q^{-\phi}$ and

$$f(A_s\zeta q^{-\phi}) = \sum_{m=0}^{\infty} f^{(m)}(0)\frac{[A_s\zeta q^{-\phi}]^m}{m!} \tag{2.61}$$

if $\zeta < q^{\phi}/A_s$. The contour in eqn (2.60) can be chosen such that the series is uniformly convergent in a compact domain of C_n. Parametrize C_n by $n + Re^{i\theta}$, and choose r_0 such that $0 < R < r_0 < n$. Then

$$\left| f^{(m)}(0)\frac{[A_s\zeta q^{-\phi}]^m}{m!}\right| \le \left| f^{(m)}(0)\frac{[A_s\zeta (n - r_0)^{-\phi}]^m}{m!}\right|,$$

and the series

$$\sum_{m=0}^{\infty}\left| f^{(m)}(0)\frac{[A_s\zeta (n - r_0)^{-\phi}]^m}{m!}\right|$$

is convergent if $|\zeta| < |x_0(n-r)^\phi / A_s|$. Pick ζ appropriately, substitute (2.61) in (2.60), interchange the series and the integral, and evaluate the integral (it is a β-function). The result is

$$p_n^\#(\zeta) = -A_t d(t) \sum_{m=0}^{\infty} f^{(m)}(0)[A_s\zeta]^m n^{-m\phi} \frac{\Gamma(m\phi + \alpha_t - 4 + n)}{\Gamma(m\phi + \alpha_t - 2)\Gamma(n-1)}. \tag{2.62}$$

From eqn (2.22) the finite size scaling function can be computed by

$$\hat{h}(\zeta) = \lim_{n\to\infty} n^{3-\alpha_t} p_n^\#(\zeta). \tag{2.63}$$

Since

$$\lim_{n\to\infty} n^{3-\alpha_t - m\phi} \frac{\Gamma(m\phi + \alpha_t - 4 + n)}{\Gamma(n-1)} = 1, \tag{2.64}$$

the series in eqn (2.62) is absolutely convergent if $|\zeta| < |x_0(n-r)^\phi / A_s|$, in which case the sum and the limit can be interchanged in eqn (2.63). Evaluating the limit then shows that

$$\hat{h}(\zeta) = -A_t d(t) \sum_{m=0}^{\infty} f^{(m)}(0) \frac{(A_s\zeta)^m}{\Gamma(m\phi + \alpha_t - 2)}. \tag{2.65}$$

Notice that $\hat{h}(\zeta)$ is an entire function: it is a convergent power series, which also converges absolutely. Since $x_0 = \limsup_{m\to\infty} |f^{(m)}(0)|^{-1/m}$, the radius of convergence of eqn (2.65) is determined by

$$\limsup_{m\to\infty} \left| f^{(m)}(0) \frac{(A_s\zeta)^m}{\Gamma(m\phi + \alpha_t - 2)} \right|^{-1/m} = 0 \tag{2.66}$$

provided that $\phi > 0$ (this is guaranteed by the second point in Definition 2.2). In other words, the introduction of a finite size scaling cross-over exponent equal to the cross-over exponent in eqn (2.29) is justified, as is the finite size scaling assumption in eqn (2.20).

3
Density functions and free energies

3.1 The density function

In this chapter I shall revisit and extend the ideas first introduced in Section 2.1. The main concern will be the limiting free energy in an interacting model of polygons, and its relation to the *density function* of the model, which will be defined as the exponential of the Legendre transform of the free energy $\mathcal{F}_{\#}(z)$. The results in eqns (2.1)–(2.6) will be reformulated in terms of the density function, and its properties will have consequences for thermodynamic properties of the model. Legendre transforms[1] have been used effectively in a self-avoiding walk problem [165], see also the book by R.S. Ellis [103] and reference [250]. The density function formulates the thermodynamic problem in Section 2.1 into a purely combinatorial question. The reformulated problem contains no activities, but it is still possible to give answers to some thermodynamic questions regarding the original model.

The same terminology as in Chapter 2 will be used in this chapter. A supermultiplicative model of polygons will be considered, where the term "polygon" is a stand-in for polygons, trees, animals or vesicles, and other supermultiplicative models. The results in this chapter are also valid for a submultiplicative model of walks with suitable assumptions; I shall not consider a generic submultiplicative model, since it will duplicate (with adjustments) much of what will be said about supermultiplicative models. In addition, a process of "unfolding" will make it possible to work with supermultiplicative models of walks in some cases, as I shall show in Chapter 5.

3.1.1 Density functions

I shall be primarily interested in the number $p_n^{\#}(m)$ of polygons of length n and which has energy m. For example, polygons may be counted with respect to the number of visits to a hyperplane, or animals with respect to their cyclomatic index, or vesicles with respect to surface area and volume. The number $p_n^{\#}(\lfloor \epsilon n \rfloor)$ will be most important in this chapter; this is the number of polygons with energy $\lfloor \epsilon n \rfloor$ (or which contains a given property $\lfloor \epsilon n \rfloor$ times); the *energy density* is said to be ϵ (this is not strictly true, the density is $\lfloor \epsilon n \rfloor / n$, and it only becomes ϵ in the limit as $n \to \infty$). The growth constant of $p_n^{\#}(\lfloor \epsilon n \rfloor)$ is the density function of a model, denoted by $\mathcal{P}_{\#}(\epsilon)$.[2] The first important issue is the existence of the density function in a particular model. It will soon become

[1] Legendre transforms are also sometimes called *max-transforms* [165].

[2] It is appropriate to think of $\log \mathcal{P}_{\#}(\epsilon)$ as the microcanonical density of the model.

apparent that this is closely related to the supermultiplicative properties of $p_n^\#(\lfloor \epsilon n \rfloor)$, and so a natural starting point is a few assumptions about $p_n^\#(\lfloor \epsilon n \rfloor)$.

Assumptions 3.1 Let $p_n^\#(m)$ be the number of polygons of length n, and which has energy m. Assume that $p_n^\#(m)$ satisfies the following properties:

(1) There exists a constant $K > 0$ such that $0 \leq p_n^\#(m) \leq K^n$ for each value of n and of m.

(2) There exist a finite constant $C > 0$, and numbers A_n and B_n such that $0 \leq A_n \leq B_n \leq Cn$ and $p_n^\#(m) > 0$ if $A_n \leq m \leq B_n$, and $p_n^\#(m) = 0$ otherwise.

(3) $p_n^\#(m)$ satisfies a supermultiplicative inequality of the type

$$p_{n_1}^\#(m_1) p_{n_2}^\#(m_2) \leq p_{n_1+n_2}^\#(m_1 + m_2). \tag{3.1}$$

□

Assumption 3.1(2) may seem unnecessarily strong, but by replacing $p_n^\#(m)$ by $p_n^\#(m) + 1$ for all $A_n \leq m \leq B_n$, this may be relaxed to $p_n^\#(m) \geq 0$ if $A_n \leq m \leq B_n$ with $p_n^\#(A_n) > 0$ and $p_n^\#(B_n) > 0$, and $p_n^\#(m) = 0$ otherwise.

The maximal and minimal densities in the model can be computed from A_n and B_n in Assumption 3.1(2). These are

$$\epsilon_M = \limsup_{n \to \infty} [B_n/n], \qquad \epsilon_m = \liminf_{n \to \infty} [A_n/n]. \tag{3.2}$$

Since C in Assumption 3.1(2) is finite, these densities are also finite. Since $p_n^\# = \sum_{m \geq 0} p_n^\#(m)$ is the total number of polygons, it is also possible to show that there exists a growth constant.

Theorem 3.2 $\lim_{n\to\infty}[(\log p_n^\#)/n] = \log \mu_\#$ *exists, where* $\mu_\#$ *is the growth constant of* $p_n^\#$.

Proof Sum eqn (3.1) over both m_1 and m_2; this gives

$$p_{n_1}^\# p_{n_2}^\# \leq (1 + B_{n_1+n_2} - A_{n_1+n_2}) p_{n_1+n_2}^\#.$$

Thus, by Assumption 3.1(1) and Theorem A.3, there is a growth constant. □

The growth constant of $p_n^\#(\lfloor \epsilon n \rfloor)$ will be examined next. Supermultiplicativity is the most important feature in the class of models under consideration; eqn (3.1) provides a starting point and a general background for a sensible description of the statistical mechanics of these models. The first important fact is that it is enough to show that there exists a density function (and a limiting free energy).

Lemma 3.3 *There exists a function* z_{n_1,n_2}*, dependent on* ϵ*, and with* $|z_{n_1,n_2}| \leq 1$*, such that*

$$p_{n_1}^\#(\lfloor \epsilon n_1 \rfloor) p_{n_2}^\#(\lfloor \epsilon n_2 \rfloor + z_{n_1,n_2}) \leq p_{n_1+n_2}^\#(\lfloor \epsilon(n_1+n_2) \rfloor).$$

Proof The fact that either $\lfloor a \rfloor + \lfloor b \rfloor = \lfloor a+b \rfloor$ or $\lfloor a \rfloor + \lfloor b \rfloor = \lfloor a+b \rfloor - 1$, means that for some given values of ϵ, n_1, and n_2, $\lfloor \epsilon n_1 \rfloor + \lfloor \epsilon n_2 \rfloor = \lfloor \epsilon(n_1+n_2) \rfloor - 1$. In this case the supermultiplicative inequality in eqn (3.1) gives $p^\#_{n_1}(\lfloor\epsilon n_1\rfloor)p^\#_{n_2}(\lfloor\epsilon n_2\rfloor + 1) \leq p^\#_{n_1+n_2}(\lfloor\epsilon(n_1+n_2)\rfloor)$. On the other hand, if $\lfloor\epsilon n_1\rfloor + \lfloor\epsilon n_2\rfloor = \lfloor\epsilon(n_1+n_2)\rfloor$, then $p^\#_{n_1}(\lfloor\epsilon n_1\rfloor)p^\#_{n_2}(\lfloor\epsilon n_2\rfloor) \leq p^\#_{n_1+n_2}(\lfloor\epsilon(n_1+n_2)\rfloor$. Define $z_{n_1,n_2} = 1$ if $\lfloor\epsilon n_1\rfloor + \lfloor\epsilon n_2\rfloor = \lfloor\epsilon(n_1+n_2)\rfloor - 1$, and zero otherwise. Then these inequalities can be combined into $p^\#_{n_1}(\lfloor\epsilon n_1\rfloor)p^\#_{n_2}(\lfloor\epsilon n_2\rfloor + z_{n_1,n_2}) \leq p^\#_{n_1+n_2}(\lfloor\epsilon(n_1+n_2)\rfloor)$ for any n_1 and n_2. □

Lemma 3.3 can now be used to show that the density function exists.

Theorem 3.4 *If $\epsilon \in (\epsilon_m, \epsilon_M)$ then the density function $\mathcal{P}_\#(\epsilon)$ is defined by the limit*

$$\log \mathcal{P}_\#(\epsilon) = \lim_{n\to\infty} \frac{1}{n} \log p^\#_n(\lfloor\epsilon n\rfloor).$$

Moreover, there exists a number η_n in $\{0,1\}$ such that for each value of n, $p^\#_n(\lfloor\epsilon n\rfloor + \eta_n) \leq [\mathcal{P}_\#(\epsilon)]^n$.

Proof The proof is a generalization of the proof of Lemma A.2 in Appendix A. Choose $n_1+n_2 = n$, $n_1 = n-k$ and $n_2 = k$ in the supermultiplicative inequality in Lemma 3.3. Then

$$p^\#_{n-k}(\lfloor\epsilon(n-k)\rfloor)p^\#_k(\lfloor\epsilon k\rfloor + z_{n-k,k}) \leq p^\#_n(\lfloor\epsilon n\rfloor). \qquad (\dagger)$$

Let m be a fixed (large) integer, and let $n = Nm + r$, with r chosen such that $N_0 \leq r < N_0 + m$, for some fixed large N_0, and where n is assumed to be much bigger than N_0. If $k = r$ is chosen, then eqn (†) becomes

$$p^\#_{mN}(\lfloor\epsilon mN\rfloor)p^\#_r(\lfloor\epsilon r\rfloor + z_{mN,r}) \leq p^\#_{mN+r}(\lfloor\epsilon(mN+r)\rfloor).$$

Increase N_0 if necessary, until $\lfloor\epsilon r\rfloor \geq A_r + z_{mN,r}$. Since $\epsilon > \epsilon_m$, and $|z_{mN,r}| \leq 1$, this is always possible. Apply eqn (†) to $p^\#_{mN}(\lfloor\epsilon mN\rfloor)$, with $k = m$, and repeat the application with $n = mN, m(N-1), \ldots$. The result is

$$\begin{aligned}
p^\#_n(\lfloor\epsilon n\rfloor) &= p^\#_{mN+r}(\lfloor\epsilon(mN+r)\rfloor) \\
&\geq p^\#_{mN}(\lfloor\epsilon(mN)\rfloor)p^\#_r(\lfloor\epsilon r\rfloor + z_{mN,r}) \\
&\geq \cdots \\
&\geq \left[\prod_{j=1}^{N} p^\#_m(\lfloor\epsilon m\rfloor + z_{m(N-j),m})\right] p^\#_r(\lfloor\epsilon r\rfloor + z_{mN,r}).
\end{aligned}$$

Choose η_m to be that value of j in the set $\{0,1\}$ which minimizes $p^\#_m(\lfloor\epsilon m\rfloor + j)$. Then $|\eta_m| \leq 1$ and the $z_{m(N-j),m}$ in the above may be replaced by η_m; this gives

$$p^\#_{mN+r}(\lfloor\epsilon(mN+r)\rfloor) \geq \left[p^\#_m(\lfloor\epsilon m\rfloor + \eta_m)\right]^N p^\#_r(\lfloor\epsilon r\rfloor + z_{mN,r}).$$

Take logarithms of this, and divide by $n = mN + r$. Fix the value of m and take the lim inf as $n \to \infty$ of the left-hand side. Then $N \to \infty$, while $N_0 \leq r < N_0 + m$.

This gives

$$\liminf_{n\to\infty} \frac{1}{n} \log p_n^\#(\lfloor \epsilon n \rfloor) \geq \frac{1}{m} \log p_m^\#(\lfloor \epsilon m \rfloor + \eta_m). \tag{‡}$$

Now take the lim sup as $m \to \infty$ of the right-hand side. This gives

$$\liminf_{n\to\infty} \frac{1}{n} \log p_n^\#(\lfloor \epsilon n \rfloor) \geq \limsup_{m\to\infty} \frac{1}{m} \log p_m^\#(\lfloor \epsilon m \rfloor + \eta_m). \tag{¶}$$

Next, choose $n_1 + n_2 = m$, $n_1 = m - k$, $n_2 = k$, with $m_1 + m_2 = \lfloor \epsilon m \rfloor + \eta_m$ and $m_1 = \lfloor \epsilon(m-k) \rfloor$ in eqn (3.1). Then

$$p_{m-k}^\#(\lfloor \epsilon(m-k) \rfloor) p_k^\#(\lfloor \epsilon m \rfloor - \lfloor \epsilon(m-k) \rfloor + \eta_m) \leq p_m^\#(\lfloor \epsilon m \rfloor + \eta_m).$$

Define $\delta_{m,k}$ by $\lfloor \epsilon k \rfloor + \delta_{m,k} = \lfloor \epsilon m \rfloor - \lfloor \epsilon(m-k) \rfloor$. Then $|\delta_{m,k}| \leq 1$, and

$$p_{m-k}^\#(\lfloor \epsilon(m-k) \rfloor) p_k^\#(\lfloor \epsilon k \rfloor + \delta_{m,k} + \eta_m) \leq p_m^\#(\lfloor \epsilon m \rfloor + \eta_m).$$

Choose k large enough that $\lfloor \epsilon k \rfloor \geq A_k + 2 \geq A_k + \delta_{m,k} + \eta_m$. Since $\epsilon > \epsilon_m$ this is always possible. Fix k, take logarithms and divide the above by m, and take the lim sup on both sides. This gives

$$\limsup_{m\to\infty} \frac{1}{m} \log p_m^\#(\lfloor \epsilon m \rfloor + \eta_m) \geq \limsup_{m\to\infty} \frac{1}{m} \log p_m^\#(\lfloor \epsilon m \rfloor). \tag{ℵ}$$

Comparison with eqn (¶) establishes the existence of the limit. The fact that there exists an $\eta_n \in \{0, 1\}$ such that $p_n^\#(\lfloor \epsilon n \rfloor + \eta_n) \leq [\mathcal{P}_\#(\epsilon)]^n$ for each value of n follows from eqn (‡). □

The density function is also continuous. This is seen by showing that $\log \mathcal{P}_\#(\epsilon)$ is concave in (ϵ_m, ϵ_M).

Theorem 3.5 *$\log \mathcal{P}_\#(\epsilon)$ is a concave function of $\epsilon \in (\epsilon_m, \epsilon_M)$. Therefore, $\mathcal{P}_\#(\epsilon)$ is continuous in (ϵ_m, ϵ_M), has right- and left-derivatives everywhere in (ϵ_m, ϵ_M), and is differentiable almost everywhere in (ϵ_m, ϵ_M).*

Proof Let $\epsilon_m < \epsilon < \delta < \epsilon_M$, and let η_n be that value of j in the set $\{0, 1\}$ which minimizes $p_n^\#(\lfloor \delta n \rfloor + j)$. In eqn (3.1), let $n_1 = n_2 = n$, $m_1 = \lfloor \epsilon n \rfloor$ and $m_1 + m_2 = \lfloor (\epsilon + \delta) n \rfloor$. Then $m_2 = \lfloor (\epsilon + \delta) n \rfloor - \lfloor \epsilon n \rfloor$, and by arguing as in Lemma 3.3, there is a number z_n, dependent on δ and ϵ, such that $m_2 = \lfloor \delta n \rfloor + z_n$ and $|z_n| \leq 1$. Thus $p_n^\#(\lfloor \epsilon n \rfloor) p_n^\#(\lfloor \delta n \rfloor + z_n) \leq p_{2n}^\#(\lfloor (\epsilon + \delta) n \rfloor)$. Replace z_n by η_n, divide by n, and take the lim sup on both sides as $n \to \infty$:

$$\log \mathcal{P}_\#(\epsilon) + \limsup_{n\to\infty} \frac{1}{n} \log p_n^\#(\lfloor \delta n \rfloor + \eta_n) \leq 2 \log \mathcal{P}_\#((\epsilon + \delta)/2).$$

Finally, eqn (ℵ) in the proof of Theorem 3.4 gives

$$\log \mathcal{P}_\#(\epsilon) + \log \mathcal{P}_\#(\delta) \leq 2 \log \mathcal{P}_\#((\epsilon + \delta)/2),$$

so that $\log \mathcal{P}_\#(\epsilon)$ is concave in (ϵ_m, ϵ_M). □

In many applications the number $p_n^\#(\delta_n)$ will be encountered, where δ_n is a sequence such that $\lim_{n\to\infty}[\delta_n/n] = \delta$ with $\epsilon_m < \delta < \epsilon_M$. Thus, the growth constant of $p_n^\#(\delta_n)$ should be investigated, which intuitively should be equal to $\mathcal{P}_\#(\delta)$. I show that this is indeed the case in Theorem 3.6.

Theorem 3.6 *Let δ_n be a sequence of integers such that $A_n \leq \delta_n \leq B_n$ for all $n \geq N_0$ (where N_0 is a fixed integer). Suppose that $\lim_{n\to\infty}[\delta_n/n] = \delta$ with $\epsilon_m < \delta < \epsilon_M$. Then*

$$\lim_{n\to\infty} \frac{1}{n} \log p_n^\#(\delta_n) = \log \mathcal{P}_\#(\delta).$$

Proof Since $\delta < \epsilon_M$ there is a $\beta > \delta$ such that $\beta < \epsilon_M$. Moreover, since $\delta > \epsilon_m$ there is an N_1 and a fixed value of k so that $B_k \geq \delta_n - \delta_{n-k} \geq A_k$ for all $n \geq N_1$. Increase N_1, if necessary, until $\lfloor \beta n \rfloor - \delta_n \geq 0$ for all $n \geq N_1$. Then $\lfloor \beta n \rfloor - \delta_{n-k} \geq A_k$. Now choose $n_1 = n - k$ and $n_2 = k$, with $m_1 = \delta_{n-k}$ and $m_1 + m_2 = \lfloor \beta n \rfloor$ in eqn (3.1). Then

$$p_{n-k}^\#(\delta_{n-k}) p_k^\#(\lfloor \beta n \rfloor - \delta_{n-k}) \leq p_n^\#(\lfloor \beta n \rfloor).$$

Take the logarithm of this, divide by n, and take the lim sup as $n \to \infty$ with k fixed. This gives

$$\limsup_{n\to\infty} \frac{1}{n} \log p_n^\#(\delta_n) \leq \log \mathcal{P}_\#(\beta),$$

for any $\delta \leq \beta < \epsilon_M$. But $\mathcal{P}_\#(\beta)$ is a continuous function, thus take $\beta \to \delta^+$ to find that

$$\limsup_{n\to\infty} \frac{1}{n} \log p_n^\#(\delta_n) \leq \log \mathcal{P}_\#(\delta^+). \qquad (\dagger)$$

Next, since $\delta > \epsilon_m$, choose $\beta' < \delta$ and $\beta' > \epsilon_m$. Choose a large N_2, and increase k and N_2 if necessary, so that $\delta_n - \delta_{n-k} \geq A_k$ for all $n \geq N_2$. Increase N_2 if necessary, until $\delta_{n-k} - \lfloor \beta'(n-k) \rfloor \geq 0$ for all $n \geq N_2$. Then $\delta_n - \lfloor \beta'(n-k) \rfloor \geq A_k$ for all $n \geq N_2$. Choose $n_1 = n - k$ and $n_1 + n_2 = n$ in eqn (3.1). In addition, put $m_1 = \lfloor \beta'(n-k) \rfloor$ and $m_1 + m_2 = \delta_n$. This gives

$$p_{n-k}^\#(\lfloor \beta'(n-k) \rfloor) p_k^\#(\delta_n - \lfloor \beta'(n-k) \rfloor) \leq p_n^\#(\delta_n).$$

Take logarithms, divide by n and take the lim inf as $n \to \infty$. This gives

$$\log \mathcal{P}_\#(\beta') \leq \liminf_{n\to\infty} \frac{1}{n} \log p_n^\#(\delta_n),$$

for any $\epsilon_m < \beta' \leq \delta$. Let $\beta' \to \delta^-$ (where the continuity of $\mathcal{P}_\#(\beta')$ is used); then

$$\log \mathcal{P}_\#(\delta^-) \leq \liminf_{n\to\infty} \frac{1}{n} \log p_n^\#(\delta_n).$$

But $\mathcal{P}_\#(\delta^-) = \mathcal{P}_\#(\delta^+)$ everywhere, since $\mathcal{P}_\#(\epsilon)$ is continuous. Comparison with eqn (†) shows that the limit exists and equals $\log \mathcal{P}_\#(\delta)$. □

Remark 3.7 In some models it can be shown that the limits

$$\lim_{n\to\infty} \frac{1}{n} \log p_n^{\#}(A_n) = \log \mathcal{P}_{\#}(\epsilon_m)$$

and

$$\lim_{n\to\infty} \frac{1}{n} \log p_n^{\#}(B_n) = \log \mathcal{P}_{\#}(\epsilon_M)$$

exist as well. The concavity of $\mathcal{P}_{\#}(\epsilon)$ gives that

$$\log \mathcal{P}_{\#}(\epsilon_m^+) \geq \limsup_{n\to\infty} \frac{1}{n} \log p_n^{\#}(A_n), \tag{3.3}$$

a result which is a direct consequence of the arguments in Theorem 3.4 and using Theorem 3.5. Similarly, it follows that

$$\log \mathcal{P}_{\#}(\epsilon_M^-) \geq \limsup_{n\to\infty} \frac{1}{n} \log p_n^{\#}(B_n). \tag{3.4}$$

It is possible that there is a discontinuity in the density function at $\epsilon = \epsilon_m$ or at $\epsilon = \epsilon_M$ (or both). Such discontinuities correspond to phase transitions at zero or infinite temperature in the phase diagram, and are not very interesting from that perspective. Thus, I shall often redefine the density function by putting $\mathcal{P}_{\#}(\epsilon_m) = \mathcal{P}_{\#}(\epsilon_m^+)$ and $\mathcal{P}_{\#}(\epsilon_M) = \mathcal{P}_{\#}(\epsilon_M^-)$. Then, $\mathcal{P}_{\#}(\epsilon)$ is log-concave and continuous in $[\epsilon_m, \epsilon_M]$. □

Let σ_n be the minimum value of m which maximizes $p_n^{\#}(m)$. Then

$$p_n^{\#}(\sigma_n) \leq p_n^{\#} \leq \sum_{m=A_n}^{B_n} p_n^{\#}(m) \leq (1 + B_n - A_n) p_n^{\#}(\sigma_n), \tag{3.5}$$

and therefore, by Theorem 3.2,

$$\lim_{n\to\infty} \frac{1}{n} \log p_n^{\#}(\sigma_n) = \mu_{\#}.$$

In particular, if ς_n is the maximum value of m which maximizes $p_n^{\#}(m)$, and

$$\epsilon_0 = \liminf_{n\to\infty} [\sigma_n/n], \qquad \epsilon_1 = \limsup_{n\to\infty} [\varsigma_n/n]; \tag{3.6}$$

then $\mathcal{P}_{\#}(\epsilon) = \mu_{\#}$ for all values of $\epsilon \in [\epsilon_0, \epsilon_1]$. The general appearance of $\mathcal{P}_{\#}(\epsilon)$ is illustrated in Fig. 3.1.

3.1.2 Integrated density functions

The *integrated density functions* are close relatives of the density function $\mathcal{P}_{\#}(\epsilon)$. The primary reason for introducing integrated density functions is that they will show that Assumptions 3.1 can be relaxed somewhat. In particular, Assumptions 3.1(2) and 3.1(3)

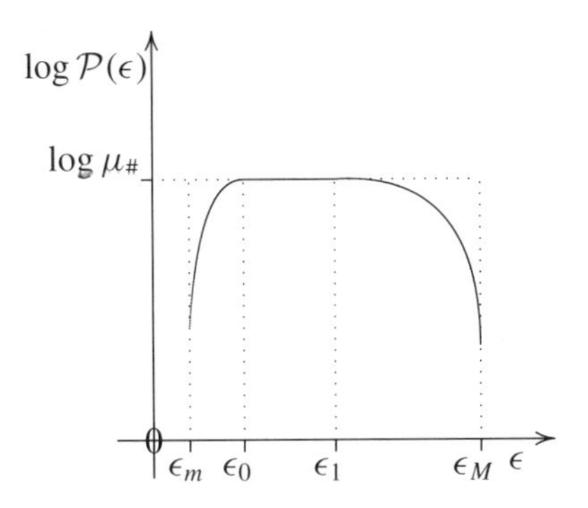

Fig. 3.1: The general appearance of the density function $\mathcal{P}(\epsilon)$.

can be made more general, while it will still be possible to show that a density function exists. The definition of $\mathcal{P}_\#(\epsilon)$ as in Theorem 3.4 will only be valid if a further condition (regularity) is assumed; this is explicitly shown in Remark 3.9 below. In the assumptions below I relax those made in Assumptions 3.1 considerably; in this section this will be the starting point. Consider the number of polygons counted by $p_n^\#(m)$ again, but now subject to the following set of assumptions.

Assumptions 3.8 Suppose that $p_n^\#(m)$ is the number of polygons of length n and energy m. Assume that $p_n^\#(m)$ satisfies the following.

(1) There exists a constant $K > 0$ such that $0 \leq p_n^\#(m) \leq K^n$ for each value of n and of m.

(2) There exist a finite constant $C > 0$, and numbers A_n and B_n such that $p_n^\#(A_n) > 0$ and $p_n^\#(B_n) > 0$, and $p_n^\#(m) \geq 0$, when $0 \leq A_n < m < B_n \leq Cn$. Moreover, suppose that $p_n^\#(m) = 0$ if $m < A_n$ or $m > B_n$.

(3) $p_n^\#(m)$ satisfies a generalized supermultiplicative inequality of the type

$$p_{n_1}^\#(m_1) p_{n_2}^\#(m_2) \leq \sum_{i=-q}^{q} p_{n_1+n_2}^\#(m_1 + m_2 + i), \tag{3.7}$$

where q is a constant.

Assumption 3.8(3) can also be relaxed to Assumption 3.8(3′); this will be used in Section 3.3.

(3′) $p_n^\#(m)$ satisfies a generalized supermultiplicative inequality of the type

$$p_{n_1}^\#(m_1) p_{n_2}^\#(m_2) \leq \sum_{i=-q}^{q} p_{n_1+n_2}^\#(m_1 + m_2 + i), \tag{3.8}$$

where q is a function of $n_1 + n_2$ such that $q(n) = o(n)$. □

In this section, however, the basic starting point will be the assumptions in Assumptions (3.8), with eqn (3.7). Define ϵ_m and ϵ_M as in eqn (3.2).

Notice that the existence of a growth constant $\mu_\#$ follows immediately from Assumptions 3.8, since summing eqn (3.7) over m_1 and m_2 gives

$$p^\#_{n_1} p^\#_{n_2} \leq (2q+1)(1 + B_{n_1+n_2} - A_{n_1+n_2}) p^\#_{n_1+n_2}. \tag{3.9}$$

The existence then follows from Assumption 3.8(1) and Theorem A.3. This defines the growth constant

$$\lim_{n\to\infty} \frac{1}{n} \log p^\#_n = \log \mu_\#. \tag{3.10}$$

Remark 3.9 (Regularity) A supermultiplicative model satisfying Assumptions 3.8 will be called *regular* if $p^\#_n(m)$ satisfies the following condition. There exists a constant $k \geq 0$ such that

$$p^\#_n(m) \leq p^\#_{n+k}(m), \qquad p^\#_n(m) \leq p^\#_{n+k}(m+1), \tag{3.11}$$

where both $p^\#_{n+k}(m) > 0$ and $p^\#_{n+k}(m+1) > 0$. With an assumption of regularity, Assumption 3.8(3) leads to

$$p^\#_{n_1}(m_1) p^\#_{n_2}(m_2) \leq (2q+1) p^\#_{n_1+n_2+2qk}(m_1+m_2+q). \tag{3.12}$$

Thus, $[p^\#_{n-2qk}(m-q)]/(2q+1)$ satisfies Assumptions 3.1(3). In other words, by Theorems 3.4, 3.5 and 3.6, there exists a density function defined by $\log \mathcal{P}_\#(\epsilon) = \lim_{n\to\infty} [\log p^\#_n(\lfloor \epsilon n \rfloor)]/n$, and with all the properties shown in Section 3.1.1. Thus, Assumptions 3.8 with regularity may be viewed as being equivalent to Assumptions 3.1.

□

If the model is not regular, then proceed as follows. Define ϵ_m and ϵ_M as in eqn (3.2), and let

$$p^\#_n(\leq m) = \sum_{i \leq m} p^\#_n(i), \qquad p^\#_n(\geq m) = \sum_{i \geq m} p^\#_n(i), \tag{3.13}$$

Then $p^\#_n(\leq m) + p^\#_n(\geq m) = p^\#_n + p^\#_n(m)$. Assumptions 3.8 can also be used to show that $p^\#_n(\leq m)$ and $p^\#_n(\geq m)$ satisfy generalized supermultiplicative inequalities.

Lemma 3.10 *Suppose that $p^\#_n(m)$ satisfies Assumptions 3.8(1), (2) and (3). Then there exists a function $f_C(n_1, n_2) = (2q+1)C(n_1+n_2+1)$ such that $p^\#_n(\leq m)$ and $p^\#_n(\geq m)$ satisfy the following generalized supermultiplicative inequalities:*

$$p^\#_{n_1}(\leq(m_1-q)) p^\#_{n_2}(\leq(m_2-q)) \leq f_{C_0}(n_1,n_2)\, p^\#_{n_1+n_2}(\leq(m_1+m_2-q)),$$
$$p^\#_{n_1}(\geq(m_1+q)) p^\#_{n_2}(\geq(m_2+q)) \leq f_{C_1}(n_1,n_2)\, p^\#_{n_1+n_2}(\geq(m_1+m_2+q)).$$

Notice that $f_C(n_1, n_2) = O(n_1 + n_2)$.

Proof Notice that $p_n^\#(\leq m) \leq p_n^\#(\leq(m+1))$. In addition, note that $p_n^\#(m) = 0$ if $m > Cn$. These facts show that if $p_{n_1}^\#(i_1)p_{n_2}^\#(i_2) \leq \sum_{i=-q}^{q} p_{n_1+n_2}^\#(i_1+i_2+i)$ is summed for all $i_1 \leq m_1$ and $i_2 \leq m_2$, then

$$p_{n_1}^\#(\leq m_1)p_{n_2}^\#(\leq m_2) \leq (2q+1)C(n_1+n_2+1)p_{n_1+n_2}^\#(\leq m_1+m_2+q).$$

Now replace m_1 by $m_1 - q$ and m_2 by $m_2 - q$ to find the first inequality. The second inequality is proved using similar arguments. □

Lemma 3.10 and Assumption 3.8(1) are enough to prove that the integrated density functions exists as limits, given Assumptions 3.8(1), (2) and (3).

Theorem 3.11 *Let* $\epsilon \in (\epsilon_m, \epsilon_M)$. *Then the integrated density functions are defined by*

$$\log \mathcal{P}_\#(\leq\epsilon) = \lim_{n\to\infty} \frac{1}{n} \log p_n^\#(\leq\lfloor\epsilon n\rfloor - q),$$

$$\log \mathcal{P}_\#(\geq\epsilon) = \lim_{n\to\infty} \frac{1}{n} \log p_n^\#(\geq\lceil\epsilon n\rceil + q).$$

Proof Choose $m_1 = \lfloor\epsilon n_1\rfloor$ and $m_2 = \lfloor\epsilon n_2\rfloor$ in the first inequality in Lemma 3.10. Since $\lfloor a\rfloor+\lfloor b\rfloor = \lfloor a+b\rfloor-1$ or $\lfloor a\rfloor+\lfloor b\rfloor = \lfloor a+b\rfloor$, the fact that $p_n^\#(\leq m) \leq p_n^\#(\leq m+1)$ gives $p_{n_1+n_2}^\#(\leq\lfloor\epsilon n_1\rfloor + \lfloor\epsilon n_2\rfloor - q) \leq 2p_{n_1+n_2}^\#(\leq\lfloor\epsilon(n_1+n_2)\rfloor - q)$. In other words,

$$p_{n_1}^\#(\leq\lfloor\epsilon n_1\rfloor - q)p_{n_2}^\#(\leq\lfloor\epsilon n_2\rfloor - q) \leq 2f(n_1,n_2)p_{n_1+n_2}^\#(\leq\lfloor\epsilon(n_1+n_2)\rfloor - q).$$

Thus $p_n^\#(\leq\lfloor\epsilon n\rfloor - q)$ satisfies a generalized supermultiplicative inequality, and by Theorem A.3 the limit exists as claimed. A similar argument shows that the function $\mathcal{P}_\#(\geq\epsilon)$ is well defined. □

It will be important to make a connection between the integrated density functions and the density function defined in the previous section. Assumptions 3.8 are much more relaxed compared with those in the previous section (Assumptions 3.1), and the existence of a density function is not known (yet). However, I shall show that the existence of integrated density functions implies the existence of a density function. The next two theorems are important steps to realize this goal.

Theorem 3.12 *The functions* $\log \mathcal{P}_\#(\leq\epsilon)$ *and* $\log \mathcal{P}_\#(\geq\epsilon)$, *defined in Theorem 3.11, are concave.*

Proof Observe that either $\lfloor a\rfloor + \lfloor b\rfloor = \lfloor a+b\rfloor - 1$ or that $\lfloor a\rfloor + \lfloor b\rfloor = \lfloor a+b\rfloor$. Use this result in Lemma 3.10 with $n_1 = n_2 = n$, $m_1 = \lfloor\epsilon n\rfloor$ and $m_2 = \lfloor\delta n\rfloor$ (argue as in the proof of Theorem 3.11) to find

$$p_n^\#(\leq\lfloor\epsilon n\rfloor - q)p_n^\#(\leq\lfloor\delta n\rfloor - q) \leq 2f(n,n)p_{2n}^\#(\leq\lfloor(\epsilon+\delta)n\rfloor - q).$$

Take logarithms, divide by n and let $n \to \infty$. By Theorem 3.11 this gives

$$\log \mathcal{P}_\#(\leq\epsilon) + \log \mathcal{P}_\#(\leq\delta) \leq 2\log \mathcal{P}_\#(\leq(\epsilon+\delta)/2).$$

The proof that $\log \mathcal{P}_\#(\geq\epsilon)$ is concave follows similar lines. □

I next show that the integrated density functions are robustly defined. The sequence $[\log p_n^{\#}(\leq\delta_n)]/n$ will converge to $\log \mathcal{P}_{\#}(\leq\delta)$ as $n \to \infty$ if $[\delta_n/n] \to \delta$ (and similarly for $[\log p_n^{\#}(\geq\delta_n)]/n$).

Theorem 3.13 *Let δ_n be a sequence of integers such that $A_n \leq \delta_n \leq B_n$, for all $n \geq N_0$ (where N_0 is a fixed integer). Suppose that $\lim_{n\to\infty}[\delta_n/n] = \delta$ exists, where $\epsilon_m < \delta < \epsilon_M$. Then*

$$\lim_{n\to\infty} \frac{1}{n} \log p_n^{\#}(\leq\delta_n) = \log \mathcal{P}_{\#}(\leq\delta),$$

$$\lim_{n\to\infty} \frac{1}{n} \log p_n^{\#}(\geq\delta_n) = \log \mathcal{P}_{\#}(\geq\delta).$$

Proof Since $\delta < \epsilon_M$ there is a β such that $\delta < \beta < \epsilon_M$. Choose n large enough that $\lfloor\beta n\rfloor - q \geq \delta_n$ for all n larger than (say) N_1. Then $p_n^{\#}(\leq\lfloor\beta n\rfloor - q) \geq p_n^{\#}(\leq\delta_n)$, which gives $\log \mathcal{P}_{\#}(\leq\beta) \geq \limsup_{n\to\infty}[\log p_n^{\#}(\leq\delta_n)]/n$. Now take $\beta \to \delta^+$. Then

$$\log \mathcal{P}_{\#}(\leq\delta^+) \geq \limsup_{n\to\infty} \frac{1}{n} \log p_n^{\#}(\leq\delta_n). \tag{†}$$

Next, since $\delta > \epsilon_m$ there is a β' such that $\epsilon_m < \beta' < \delta$. Increase n until $\lfloor\beta' n\rfloor - q \leq \delta_n$ for all n larger than (say) N_2. Then $p_n^{\#}(\leq\lfloor\beta' n\rfloor - q) \leq p_n^{\#}(\leq\delta_n)$, so that $\log \mathcal{P}_{\#}(\leq\beta) \leq \liminf_{n\to\infty}[\log p_n^{\#}(\leq\delta_n)]/n$. Take $\beta' \to \delta^-$:

$$\log \mathcal{P}_{\#}(\leq\delta^-) \leq \liminf_{n\to\infty} \frac{1}{n} \log p_n^{\#}(\leq\delta_n).$$

Compare this to eqn (†) and use the concavity, and thus continuity of $\log \mathcal{P}_{\#}(\leq\epsilon)$, from Theorem 3.12 to establish the first claim. The second claim is similarly proven. □

As a result, define the integrated density functions as follows, and note that the numbers ϵ_0 and ϵ_1 in eqn (3.6) can be defined by $\epsilon_0 = \inf\{\epsilon \,|\, \mathcal{P}_{\#}(\leq\epsilon) = \mu_{\#}\}$ and $\epsilon_1 = \sup\{\epsilon \,|\, \mathcal{P}_{\#}(\geq\epsilon) = \mu_{\#}\}$.

Corollary 3.14 *Given Assumptions 3.8(1), (2) and (3), the integrated density functions can be defined by*

$$\log \mathcal{P}_{\#}(\leq \epsilon) = \lim_{n\to\infty} \frac{1}{n} \log p_n^{\#}(\leq \lfloor\epsilon n\rfloor),$$

$$\log \mathcal{P}_{\#}(\geq \epsilon) = \lim_{n\to\infty} \frac{1}{n} \log p_n^{\#}(\geq \lfloor\epsilon n\rfloor).$$

□

Note that $p_n^{\#} = p_n^{\#}(\leq\lfloor\epsilon n\rfloor) + p_n^{\#}(\geq\lfloor\epsilon n + 1\rfloor)$, so that the following corollary follows from Theorem 3.13.

Corollary 3.15 *Either $\mathcal{P}_{\#}(\leq\epsilon)$, or $\mathcal{P}_{\#}(\geq\epsilon)$, or both, is equal to $\mu_{\#}$. That is,*

$$\max\{\mathcal{P}_{\#}(\leq\epsilon), \mathcal{P}_{\#}(\geq\epsilon)\} = \mu_{\#}.$$

□

In other words, there are two values of ϵ in $[\epsilon_m, \epsilon_M]$, defined by

$$\epsilon_0 = \inf\{\epsilon \,|\, \mathcal{P}_\#(\leq\epsilon) = \mu_\#\}, \qquad \epsilon_1 = \sup\{\epsilon \,|\, \mathcal{P}_\#(\geq\epsilon) = \mu_\#\}. \tag{3.14}$$

By concavity, $\mathcal{P}_\#(\leq\epsilon)$ is strictly increasing in (ϵ_m, ϵ_0) and equal to $\mu_\#$ in (ϵ_0, ϵ_M), while $\mathcal{P}_\#(\geq \epsilon)$ is equal to $\mu_\#$ in (ϵ_m, ϵ_1) while it is strictly decreasing in (ϵ_1, ϵ_M). Given Assumptions 3.8, it can now be shown that a density function exists, without regularity as defined in Remark 3.9. It is only necessary to slightly change the definition given in Theorem 3.4.

Theorem 3.16 *Let $\epsilon \in (\epsilon_m, \epsilon_M)$. Then there exists a sequence of integers δ_n, such that a density function may be defined by*

$$\log \mathcal{P}_\#(\epsilon) = \lim_{n\to\infty} \frac{1}{n} \log p_n^\#(\delta_n) = \min\{\log \mathcal{P}_\#(\leq\epsilon), \log \mathcal{P}_\#(\geq\epsilon)\}.$$

Moreover, if ϵ_0 and ϵ_1 are defined as in eqns (3.6) or (3.14), then, if $\epsilon \leq \epsilon_0$, δ_n may be selected to be that value of m in $\{A_n, A_n + 1, \ldots, \lfloor \epsilon n \rfloor\}$ which maximizes $p_n^\#(m)$. Alternatively, if $\epsilon \geq \epsilon_1$, then δ_n is that value of m in $\{\lfloor \epsilon n \rfloor, \lfloor \epsilon n \rfloor + 1, \ldots, B_n\}$ which maximizes $p_n^\#(m)$. Finally, if $\epsilon_0 < \epsilon < \epsilon_1$, then δ_n can be selected as any number in $\{\lfloor \epsilon_0 n \rfloor, \lfloor \epsilon_0 n \rfloor + 1, \ldots, \lfloor \epsilon_1 n \rfloor\}$. Note also that the limit $\lim_{n\to\infty}[\delta_n/n] = \epsilon$ exists, so that $\delta_n = \lfloor \epsilon n \rfloor + \sigma_n$, where $\sigma_n = o(n)$.

Proof Suppose first that $\epsilon \in (\epsilon_m, \epsilon_0]$. Let δ_n be that value of m, $A_n \leq \delta_n \leq \lfloor \epsilon n \rfloor$, which maximizes $p_n^\#(m)$. Then

$$\begin{aligned} p_n^\#(\leq\lfloor \epsilon n \rfloor) &\leq (\lfloor \epsilon n \rfloor - A_n + 1) p_n^\#(\delta_n) \\ &\leq (\lfloor \epsilon n \rfloor - A_n + 1) p_n^\#(\leq\delta_n) \\ &\leq (\lfloor \epsilon n \rfloor - A_n + 1) p_n^\#(\leq\lfloor \epsilon n \rfloor). \end{aligned}$$

This shows that

$$\lim_{n\to\infty} \frac{1}{n} \log p_n^\#(\delta_n) = \log \mathcal{P}_\#(\leq\epsilon)$$

exists if $\epsilon \in (\epsilon_m, \epsilon_0]$. Next, I show that $\lim_{n\to\infty}[\delta_n/n] = \epsilon$. Notice that $\delta_n \leq \lfloor \epsilon n \rfloor$, and suppose that $\liminf_{n\to\infty}[\delta_n/n] = \kappa < \epsilon$. Suppose that this lim inf is approached through the sequence of integers $\{n_i\}$. But then

$$(\lfloor \epsilon n_i \rfloor - A_{n_i} + 1) p_{n_i}^\#(\delta_{n_i}) \geq p_{n_i}^\#(\leq\lfloor \epsilon n_i \rfloor),$$

for each i. By Theorem 3.13, this shows that

$$\log \mathcal{P}_\#(\leq\kappa) = \lim_{i\to\infty} \frac{1}{n_i} \log p_{n_i}^\#(\delta_{n_i}) \geq \log \mathcal{P}_\#(\leq\epsilon).$$

But $\mathcal{P}_\#(\leq\epsilon)$ is strictly increasing in $(\epsilon_m, \epsilon_0]$ and so $\kappa \geq \epsilon$. This is a contradiction, and so $\liminf_{n\to\infty}[\delta_n/n] = \epsilon$, and thus $\lim_{n\to\infty}[\delta_n/n] = \epsilon$ exists.

If $\epsilon \in [\epsilon_1, \epsilon_M)$, then a similar argument to the above works. Lastly, if $\epsilon \in (\epsilon_0, \epsilon_1)$, then eqn (3.7) shows that $\log \mathcal{P}_\#(\epsilon) = \log \mu_\#$ (first show that $\log \mathcal{P}_\#(\epsilon)$ is concave (use eqn (3.7) and Lemma B.2)), and then show that $\log \mathcal{P}_\#(\epsilon_0) = \log \mathcal{P}_\#(\epsilon_1) = \log \mu_\#$. Thus, the choice $\delta_n = \lfloor \epsilon n \rfloor$ may be made. □

In other words, a density function can still be defined given Assumptions 3.8, albeit in a slightly different form (as compared with Assumptions 3.1 and Theorem 3.4). A contrived example is the following. Suppose that $0 \le m \le n-1$ and let $p_n^\#(m) = e^n$ if m is odd, and $p_n^\#(m) = 1$ if m is even. Then surely $p_{n_1}(m_1)p_{n_2}(m_2) \le p_{n_1+n_2}(m_1+m_2) + p_{n_1+n_2}(m_1+m_2+1)$ so that $p_n^\#(m)$ satisfies Assumptions (3.8). However, $\lim_{n\to\infty}[\log p_n^\#(\lfloor \epsilon n \rfloor)]/n$ does not exist (take $n \to \infty$ through a subsequence such that $\lfloor \epsilon n \rfloor$ is odd, and then again even, to see this). However, there is a sequence $\{\sigma_n\}$, with $|\sigma_n| \le 1$, so that $\lim_{n\to\infty}[\log p_n^\#(\lfloor \epsilon n \rfloor + \sigma_n)]/n$ exists.

3.2 Density functions and free energies

The partition function $p_n^\#(z)$ has been defined in eqn (2.1). The activity z is a parameter of the model, and what is of interest is the thermodynamic behaviour of the model (in the limit $n \to \infty$) as z is changed. In this section the relation between the density function and the free energy (see eqn (2.2)) is explored. This is an important idea, since it suggests that the thermodynamic properties of the model, as defined by the limiting free energy, are in fact directly related to the combinatorial properties of the model, as expressed through properties of the density function. First consider the existence of a limiting free energy, given the existence of integrated density functions.

Theorem 3.17 *Suppose that $p_n^\#(m)$ satisfies Assumptions 3.8(1), (2) and (3). Then the integrated density functions $\mathcal{P}_\#(\le\epsilon)$ and $\mathcal{P}_\#(\ge\epsilon)$ exist as defined in Theorem 3.13, and are concave. Define the density function $\mathcal{P}_\#(\epsilon) = \min\{\mathcal{P}_\#(\le\epsilon), \mathcal{P}_\#(\ge\epsilon)\}$. If $p_n^\#(z)$ is defined as in eqn (2.1), then the limits*

$$\mathcal{F}_\#(z) = \lim_{n\to\infty} \frac{1}{n} \log p_n^\#(z),$$

$$\log \mathcal{P}_\#(\epsilon) = \lim_{n\to\infty} \frac{1}{n} \log p_n^\#(\delta_n),$$

exist for a sequence of numbers $\delta_n = \lfloor \epsilon n \rfloor + \sigma_n$, where $\sigma_n = o(n)$. Moreover,

$$\mathcal{F}_\#(z) = \sup_{\epsilon_m \le \epsilon \le \epsilon_M} \{\log \mathcal{P}_\#(\epsilon) + \epsilon \log z\}.$$

Proof Since $p_n^\#(m)$ satisfies Assumptions 3.8(1), (2) and (3), it follows from Theorem 3.16 that for any $\epsilon \in (\epsilon_m, \epsilon_M)$ there is a sequence of integers δ_n, such that $\lim_{n\to\infty}[\delta_n/n] = \epsilon$ and

$$\log \mathcal{P}_\#(\epsilon) = \lim_{n\to\infty} \frac{1}{n} \log p_n^\#(\delta_n).$$

Next, for any such ϵ, it follows that $p_n^\#(z) \ge p_n^\#(\delta_n) z^{\delta_n}$, and therefore

$$\liminf_{n\to\infty} \frac{1}{n} \log p_n^\#(z) \ge \log \mathcal{P}_\#(\epsilon) + \epsilon \log z. \qquad (\dagger)$$

Consider $\rho_n^{\#}(m) = p_n^{\#}(m) z^m$. Multiply eqn (3.7) by $z^{m_1+m_2}$; this shows that

$$\rho_{n_1}^{\#}(m_1)\rho_{n_2}^{\#}(m_2) \leq \left[\sum_{i=-q}^{q}(z+z^{-1})^i\right]\sum_{i=-q}^{q}\rho_{n_1+n_2}^{\#}(m_1+m_2+i),$$

and so $\rho_n^{\#}(m)/\sum_{i=-q}^{q}(z+z^{-1})^i$ satisfies Assumptions 3.8(1), (2) and (3) as well. Choose κ_n to be that least value of m (dependent on z) which maximizes $\rho_n^{\#}(m)$. By Theorem 3.16, $\lim_{n\to\infty}[\kappa_n/n] = \kappa$ exists, and so by Theorem 3.13,

$$\begin{aligned}\limsup_{n\to\infty}\frac{1}{n}\log\rho_n^{\#}(\kappa_n) &\leq \min\left\{\lim_{n\to\infty}\frac{1}{n}\log p_n^{\#}(\leq\kappa_n),\ \lim_{n\to\infty}\frac{1}{n}\log p_n^{\#}(\geq\kappa_n)\right\} + \kappa\log z\\ &= \log\mathcal{P}_{\#}(\kappa) + \kappa\log z.\end{aligned}$$

Thus, since $\rho_n^{\#}(\kappa_n) \leq p_n^{\#}(z) \leq (B_n - A_n + 1)\rho_n^{\#}(\kappa_n)$,

$$\limsup_{n\to\infty}\frac{1}{n}\log\rho_n^{\#}(\kappa_n) \leq \log\mathcal{P}_{\#}(\kappa) + \kappa\log z.$$

Comparing this to eqn (†) above completes the proof. □

I shall next show that if the limiting free energy exists, then there is a density function. There is also an opportunity for further relaxation of Assumptions 3.8 here. In particular, replace Assumption 3.8(3) (eqn (3.7)) by Assumption 3.8(3′) (eqn (3.8)). Then there is still a limiting free energy in the model.

Theorem 3.18 *Let $p_n^{\#}(m)$ be a model of polygons which satisfies assumptions 3.8(1), 3.8(2) and 3.8(3′). Then*

$$\mathcal{F}_{\#}(z) = \lim_{n\to\infty}\frac{1}{n}\log p_n^{\#}(z)$$

exists for every $z \in (0,\infty)$, and is a convex function of $\log z$. *Lastly, notice that $\mathcal{F}_{\#}(z) \geq$* $\max\{\epsilon_m\log z, \epsilon_M\log z\}$.

Proof Multiply eqn (3.8) by $z^{m_1+m_2}$ and sum over m_1 and m_2. This gives

$$p_{n_1}^{\#}(z)p_{n_2}^{\#}(z) \leq (2q+1)(1+B_n-A_n)p_{n_1+n_2}^{\#}(z).$$

Thus, $p_n^{\#}(z)$ is a generalized supermultiplicative function. In addition, by Assumption 3.8(1), $p_n^{\#} \leq K_0^n$ for some constant $K_0 > 0$, and thus $p_n^{\#}(z) \leq K_0^n$ if $z \leq 1$ while $p_n^{\#}(z) \leq (zK_0)^n$ if $z > 1$. Since $B_n - A_n \leq Cn$ for some contant C by Assumption 3.8(2), Theorem A.3 can be used to show the existence of the limit. To show that $\mathcal{F}_{\#}(z)$ is a convex function of $\log z$, consider eqn (2.3). That $\mathcal{F}_{\#}(z) \geq \max\{\epsilon_m\log z, \epsilon_M\log z\}$ follows from the fact that $p_n^{\#}(z) \geq \max\{p_n^{\#}(A_n)z^{A_n}, p_n^{\#}(B_n)z^{B_n}\}$. □

Moreover, the Legendre transform of the limiting free energy is the density function, and this shows that a density function may also be defined under the more general assumptions in Assumptions 3.8(3′). In fact, given the existence of a limiting free energy, and Assumptions 3.8(1) and 3.8(2), it may be shown that there is a density function.

Theorem 3.19 *Suppose that $p_n^{\#}(m)$ satisfies Assumptions 3.8(1) and 3.8(2), and that the free energy $\mathcal{F}_{\#}(z)$ exists and is finite in $(0, \infty)$ and is a convex function of $\log z$, such that $\mathcal{F}_{\#}(z) \geq \max\{\epsilon_m \log z, \epsilon_M \log z\}$. Then the density function $\mathcal{P}_{\#}(\epsilon)$ exists in (ϵ_m, ϵ_M) and is defined by*

$$\log \mathcal{P}_{\#}(\epsilon) = \inf_{0<z<\infty} \{\mathcal{F}_{\#}(z) - \epsilon \log z\}.$$

Moreover, there is a sequence of integers $\{\sigma_n\}_{n=0}^{\infty}$ such that $\sigma_n = o(n)$ and the limit

$$\log \mathcal{P}_{\#}(\epsilon) = \lim_{n\to\infty} \frac{1}{n} \log[p_n^{\#}(\lfloor \epsilon n \rfloor + \sigma_n)]$$

exists (and is finite and concave in (ϵ_m, ϵ_M)). Lastly, note also that $\delta_n = \lfloor \epsilon n \rfloor + \sigma_n$ is that least value of m which maximizes $p_n^{\#}(m) z^m$.

Proof Let δ_n be that least value of m, dependent on z, such that $p_n^{\#}(m) z^m$ is a maximum. Then $A_n \leq \delta_n \leq B_n$, and

$$p_n^{\#}(\delta_n) z^{\delta_n} \leq p_n^{\#}(z) \leq (1 + B_n - A_n) p_n^{\#}(\delta_n) z^{\delta_n}. \tag{$\dagger$}$$

In other words, by taking logarithms, dividing by n and letting $n \to \infty$,

$$\mathcal{F}_{\#}(z) = \lim_{n\to\infty} \frac{1}{n} \log p_n^{\#}(\delta_n) z^{\delta_n}.$$

Notice that $\epsilon_m \leq \liminf_{n\to\infty}[\delta_n/n] \leq \limsup_{n\to\infty}[\delta_n/n] \leq \epsilon_M$. Choose an $\epsilon \in (\epsilon_m, \epsilon_M)$ and multiply eqn ($\dagger$) by $z^{-\lfloor \epsilon n \rfloor}$. Take logarithms, divide by n, and let $n \to \infty$. This gives

$$\lim_{n\to\infty} \frac{1}{n} \log[p_n^{\#}(\delta_n) z^{\delta_n - \lfloor \epsilon n \rfloor}] = \mathcal{F}_{\#}(z) - \epsilon \log z. \tag{$\ddagger$}$$

But $\mathcal{F}_{\#}(z) \geq \max\{\epsilon_m \log z, \epsilon_M \log z\}$; thus, if $\epsilon \in (\epsilon_m, \epsilon_M)$, then the infimum over z of $\mathcal{F}_{\#}(z) - \epsilon \log z$ is found at a value of say $z = z_1$ (since $\mathcal{F}_{\#}(z)$ is a convex function of $\log z$). But then the left-hand side of the equation above is finite if the infimum over z is taken, This is only possible if $\lim_{n\to\infty}[\delta_n/n] = \epsilon$, and if this limit exists. This implies that $\delta_n = \lfloor \epsilon n \rfloor + \sigma_n$ where $\sigma = o(n)$. Lastly, take the infimum over z in eqn ($\ddagger$) (and use the fact that $\delta_n = \lfloor \epsilon n \rfloor + \sigma_n$) to complete the proof. □

It only need be shown that the density function exists to conclude that a convex free energy exists. In fact, much can be learned about the free energy by considering the differentiability and other properties of the density function. Notice that if $\log \mathcal{P}_{\#}(\epsilon)$ is a strictly concave function, then the supremum in $\mathcal{F}_{\#}(z) = \sup_{\epsilon m \leq \epsilon \leq \epsilon M} \{\log \mathcal{P}_{\#}(\epsilon) + \epsilon \log z\}$ is found either as $\epsilon \to \epsilon_m$, or $\epsilon \to \epsilon_M$, or at a unique $\epsilon_* \in (\epsilon_m, \epsilon_M)$. Thus, the supremeum is found at a unique volue of ϵ. This is not necessarily the case if $\log \mathcal{P}_{\#}(\epsilon)$ is not strictly concave. Notice that as $z \to \infty$, the supremum above exists at larger and larger values of ϵ. In fact,

$$\mathcal{F}_{\#}(z) \sim \log \mathcal{P}_{\#}(\epsilon_M^-) + \epsilon_M^- \log z \tag{3.15}$$

as $z \to \infty$. This shows that the limiting entropy of the model is the left-limit of $\log \mathcal{P}_{\#}(\epsilon)$ as $\epsilon \to \epsilon_M$.

3.3 Properties of the density function

Features in the density function have implications for the free energy (and the thermodynamic properties) of a model. In particular, I shall be interested in the various manifestations of non-analyticities in the density function, including jump discontinuities at the end-points of the interval $[\epsilon_m, \epsilon_M]$ and discontinuous first derivatives. Finite left- and right-derivatives at the end-points of $[\epsilon_m, \epsilon_M]$ will also be significant. As before, I shall use the notation $\mathcal{P}_\#(\epsilon^-)$ and $\mathcal{P}_\#(\epsilon^+)$ to indicate a left- and a right-limit at ϵ; the left- and right-derivatives of the density function at ϵ_0 will be denoted by

$$\left[\frac{d^-}{d\epsilon}\mathcal{P}_\#(\epsilon)\right]\Bigg|_{\epsilon=\epsilon_0} \quad \text{and} \quad \left[\frac{d^+}{d\epsilon}\mathcal{P}_\#(\epsilon)\right]\Bigg|_{\epsilon=\epsilon_0}.$$

Figure 3.2 is plot of a density function which has some of the properties above; it is discontinuous at ϵ_m, with a constant first derivative in $[\epsilon_0, \epsilon_1]$. It also has a singularity in a higher derivative at ϵ_0 and a jump discontinuity in its first derivative at ϵ_1, while its right-derivative is finite at ϵ_m.

Since $\mathcal{F}_\#(z)$ is differentiable almost everywhere, and is the limit of a sequence of continuously differentiable functions $[\log p_n^\#(z)]/n$ for $z \in (0, \infty)$, it follows that

$$z\frac{d}{dz}\log\mathcal{F}_\#(z) = \frac{1}{\mathcal{F}_\#(z)}\lim_{n\to\infty}\left[\frac{1}{n}\log\sum_{m=An}^{Bn} m p_n^\#(m) z^m\right], \tag{3.16}$$

almost everywhere (that is, wherever the function $\mathcal{F}_\#(z)$ is differentiable; see Theorem B.7). Suppose that $\mathcal{P}_\#(\epsilon)$ is strictly concave; then the supremum in Theorem 3.18 exists at either the end-points of the interval $[\epsilon_m, \epsilon_M]$, or at a unique value of $\epsilon \in (\epsilon_m, \epsilon_M)$ (say ϵ_*). In that case eqn (3.16) gives

$$z\frac{d}{dz}\log\mathcal{F}_\#(z) = \epsilon_*, \tag{3.17}$$

Thus, the first derivative of the free energy can be interpreted as the energy density of the energy (per edge). The mean density is a function of z, and by examining the

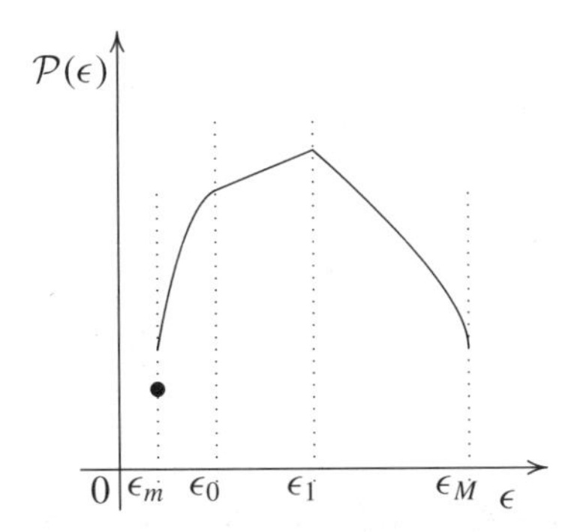

Fig. 3.2: A sensity function with a jump discontinuity at ϵ_m, a constant first derivative in $[\epsilon_0, \epsilon_1]$, a non- analyticity as ϵ_0 and a discontinuous first derivative at ϵ_1. In addition, the left-derivative is finite at ϵ_m, but infinite at ϵ_M.

density function, the energy density may be used as an order parameter to discover phase transitions.

3.3.1 Jump discontinuities at the end-points of $[\epsilon_m, \epsilon_M]$

A jump discontinuity at ϵ_m occurs when $\mathcal{P}_\#(\epsilon_m) < \mathcal{P}_\#(\epsilon_m^+)$. Thus $\mathcal{F}_\#(0) = \log \mathcal{P}_\#(\epsilon_m) < \log \mathcal{P}_\#(\epsilon_m^+) \leq \mathcal{F}_\#(z)$ if $z > 0$. Thus, there is a jump discontinuity in $\mathcal{F}_\#(z)$ at $z = 0$. This situation is quite pathological; in fact, a different model is found at $z = 0$ by requiring that just the term which contains none of the properties conjugate to z appears in the partition function, and if there are no such polygons, then a pathological situation is encountered at zero temperature. On the other hand, it may be argued that such jump discontinuities corresponds to first-order phase transitions at zero or infinite values of z (see Section 3.3.4), but I shall often ignore this possibility. It is therefore convenient if the discontinuity at ϵ_m (and at ϵ_M) is "repaired" by redefining the density function at these two points by requiring that

$$\mathcal{P}_\#(\epsilon_m) = \mathcal{P}_\#(\epsilon_m^+), \qquad \mathcal{P}_\#(\epsilon_M) = \mathcal{P}_\#(\epsilon_M^-). \tag{3.18}$$

In this case the density function is defined and continuous on the compact interval $[\epsilon_m, \epsilon_M]$. This does not rule out phase transitions at $z = 0$ or $z = \infty$, see for example Section 5.3.1.

3.3.2 Finite right- and left-derivatives at ϵ_m and ϵ_M

In this case there is a non-analyticity in the free energy at a critical value of the activity which is not zero or infinite. In other words, there is a phase transition in the model at finite temperature. It is also important to note that asymmetric models are obtained in these cases: the free energy is a linear function of $\log z$ until the critical point is reached, beyond which it increases faster than linear.

Lemma 3.20 *Suppose that*

$$-\infty < \left[\frac{d^+}{d\epsilon}\mathcal{P}_\#(\epsilon)\right]\bigg|_{\epsilon=\epsilon_m} < \infty.$$

Then there exists a finite z_c, such that for all $z < z_c$, $\mathcal{F}_\#(z) = \log \mathcal{P}_\#(\epsilon_m) + \epsilon_m \log z$, and $\mathcal{F}_\#(z) > \log \mathcal{P}_\#(\epsilon_m) + \epsilon_m \log z$ if $z > z_c$. In other words, there is a non-analyticity in $\mathcal{F}_\#(z)$ at z_c. In addition,

$$\log z_c = -\left[\frac{d^+}{d\epsilon}\log \mathcal{P}_\#(\epsilon)\right]\bigg|_{\epsilon=\epsilon_m^+}.$$

Similarly, if

$$-\infty < \left[\frac{d^-}{d\epsilon}\mathcal{P}_\#(\epsilon)\right]\bigg|_{\epsilon=\epsilon_M} < \infty,$$

then there exists a finite z_d, such that for all $z > z_d$, $\mathcal{F}_\#(z) = \log \mathcal{P}_\#(\epsilon_M) + \epsilon_M \log z$, and if $z < z_d$, then $\mathcal{F}_\#(z) > \log \mathcal{P}_\#(\epsilon_M) + \epsilon_M \log z$. In other words, there is a non-analyticity *in $\mathcal{F}_\#(z)$ at z_d. In addition*

$$\log z_d = -\left[\frac{d^-}{d\epsilon}\log \mathcal{P}_\#(\epsilon)\right]\bigg|_{\epsilon=\epsilon_M^-}.$$

Proof Define $Q(\epsilon) = \log \mathcal{P}_\#(\epsilon) + \epsilon \log z$. Then $Q(\epsilon)$ is concave, and

$$\left[\frac{d^+}{d\epsilon}Q(\epsilon)\right]\bigg|_{\epsilon=\epsilon_m^+} = \left[\frac{d^+}{d\epsilon}\log \mathcal{P}_\#(\epsilon)\right]\bigg|_{\epsilon=\epsilon_m^+} + \log z.$$

Define

$$\log z_o = -\left[\frac{d^+}{d\epsilon}\log \mathcal{P}_\#(\epsilon)\right]\bigg|_{\epsilon=\epsilon_m^+}.$$

Then $Q(\epsilon)$ is a monotonic non-increasing function of ϵ if $z < z_o$, and its (global) maximum is found at $\epsilon = \epsilon_m$. By Theorem 3.18, $\mathcal{F}_\#(z) = \log \mathcal{P}_\#(\epsilon_m) + \epsilon_m \log z$ if $z < z_o$. On the other hand, z can be increased until $\epsilon_M \log z + \log \mathcal{P}_\#(\epsilon_M) > \epsilon_m \log z + \log \mathcal{P}_\#(\epsilon_m)$. In that case note that $Q(\epsilon)$ is concave, and thus $\mathcal{F}_\#(z) \geq \log \mathcal{P}_\#(\epsilon_M) + \epsilon_M \log z > \log \mathcal{P}_\#(\epsilon_m) + \epsilon_m \log z$. Now choose

$$z_c = \sup\{z | \mathcal{F}_\#(z) = \log \mathcal{P}_\#(\epsilon_m) + \epsilon_m \log z\}.$$

The second part of the lemma follows from similar arguments. □

Observe that $Q(\epsilon)$ in the above proof has a global maximum at $\epsilon = \epsilon_m$ if $z < z_c$, which implies that the partition function is dominated by conformations where the density of the energy is ϵ_m. Once $z > z_c$, then $Q(\epsilon)$ has a global maximum at values of $\epsilon > \epsilon_m$. Define ϵ_* as in eqn (3.17). If ϵ_* is plotted as a function of z, then the result might be a graph such as Fig. 3.3. Notice that $\epsilon_* = \epsilon_m$ until $z = z_c$, where it may move away from ϵ_m either continuously (a continuous transition) or jump away discontinuously (a first-order transition), and this occurs if $\log \mathcal{P}_\#(\epsilon)$ is linear in an interval $[\epsilon_m, \epsilon']$, for some ϵ'. A similar situation is found if the left-derivative at ϵ_M is finite. The shape of the curve should be given by $|z - z_c|^{1-\alpha}$ (eqn (2.27)).

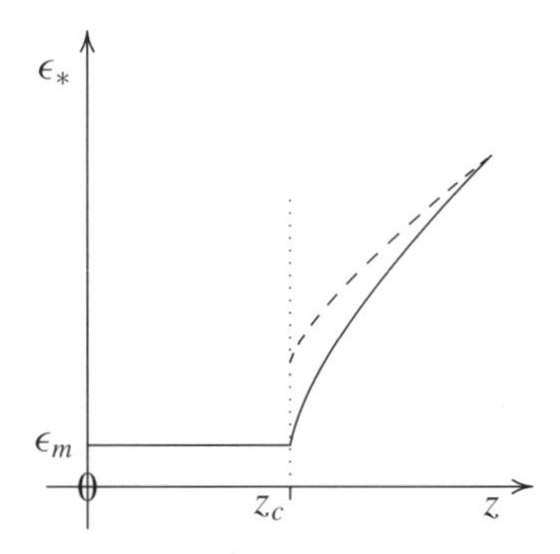

Fig. 3.3: The energy density as a function of z. In this figure it is assumed that the density function has a finite right-derivative at ϵ_m. The solid curve shows the situation if there is a continuous transition, while the dashed curve illustrates the situation if there is a first-order transition.

3.3.3 A jump discontinuity in $d\mathcal{P}_{\#}(\epsilon)/d\epsilon$

Suppose that the jump discontinuity occurs at ϵ'. An examination of the Legendre transform shows that in this case, the value of $\mathcal{F}_{\#}(z)$ will be determined by ϵ' for z in some interval $[z_1, z_2]$. In fact, it will follow that $\mathcal{F}_{\#}(z) = \log[\mathcal{P}_{\#}](\epsilon') + \epsilon' \log z$ for $z \in [z_1, z_2]$. Outside this interval this will not be true. Consequently, there are two non-analyticities in $\mathcal{F}_{\#}(z)$, one each at z_1 and z_2:

Lemma 3.21 *Suppose that there exists $\epsilon' \in (\epsilon_m, \epsilon_M)$ such that*

$$\left[\frac{d^-}{d\epsilon}\mathcal{P}_{\#}(\epsilon)\right]\Bigg|_{\epsilon=\epsilon'} > \left[\frac{d^+}{d\epsilon}\mathcal{P}_{\#}(\epsilon)\right]\Bigg|_{\epsilon=\epsilon'}.$$

Then there exists $z_1 < z_2$, both finite, such that $\mathcal{F}_{\#}(z) = \log \mathcal{P}_{\#}(\epsilon') + \epsilon' \log z$ for all $z_1 < z < z_2$, and moreover, $\mathcal{F}_{\#}(z)$ is not analytic at z_1 and z_2.

Proof Define $Q(\epsilon) = \log \mathcal{P}_{\#}(\epsilon) + \epsilon \log z$. $Q(\epsilon)$ is concave in $[\epsilon_m, \epsilon_M]$. Then,

$$\left[\frac{d^-}{d\epsilon}Q(\epsilon)\right]\Bigg|_{\epsilon=\epsilon'} = \left[\frac{d^-}{d\epsilon}\log \mathcal{P}_{\#}(\epsilon)\right]\Bigg|_{\epsilon=\epsilon'} + \log z.$$

Define

$$D^-\mathcal{P}_{\#}(\epsilon') = \left[\frac{d^-}{d\epsilon}\mathcal{P}_{\#}(\epsilon)\right]\Bigg|_{\epsilon=\epsilon'}, \qquad D^+\mathcal{P}_{\#}(\epsilon') = \left[\frac{d^+}{d\epsilon}\mathcal{P}_{\#}(\epsilon)\right]\Bigg|_{\epsilon=\epsilon'},$$

and then define

$$\log z_c(\lambda) = -\left[\lambda D^-\mathcal{P}_{\#}(\epsilon') + (1-\lambda)D^+\mathcal{P}_{\#}(\epsilon')\right]/\mathcal{P}_{\#}(\epsilon'),$$

where λ is a parameter between 0 and 1. Direct calculation shows that if $z = z_c(\lambda)$, then

$$D^-Q(\epsilon') = (1-\lambda)(D^-\mathcal{P}_{\#}(\epsilon') - D^+\mathcal{P}_{\#}(\epsilon'))/\mathcal{P}_{\#}(\epsilon') > 0,$$

and

$$D^+Q(\epsilon') = \lambda(D^+\mathcal{P}_{\#}(\epsilon') - D^-\mathcal{P}_{\#}(\epsilon'))/\mathcal{P}_{\#}(\epsilon') < 0.$$

Thus, there is a maximum in $Q(\epsilon)$ at $\epsilon = \epsilon'$, and thus for $z \in [z_c(0), z_c(1)]$, $\mathcal{F}_{\#}(z) = \log \mathcal{P}_{\#}(\epsilon') + \epsilon' \log z$. By decreasing z, note that eventually $\log \mathcal{P}_{\#}(\epsilon_m^+) + \epsilon_m \log z < \log \mathcal{P}_{\#}(\epsilon') + \epsilon' \log z$, and by increasing z, note that eventually $\log \mathcal{P}_{\#}(\epsilon_M^-) + \epsilon_M \log z > \log \mathcal{P}_{\#}(\epsilon') + \epsilon' \log z$. In other words, there exists z_1 and z_2 such that $\mathcal{F}_{\#}(z) = \log \mathcal{P}_{\#}(\epsilon') + \epsilon' \log z$ for $z \in (z_1, z_2)$, and $\mathcal{F}_{\#}(z)$ is not analytic at z_1 and z_2. □

Related to this case is the possibility that there is a point where the second derivative of $\mathcal{P}_{\#}(\epsilon)$ is singular or non-analytic (for example, the curvature of $\mathcal{P}_{\#}(\epsilon)$ could be infinite or undefined at a single point). In that case a second-order transition is found (which is classified as continuous). The same notions apply if a higher derivative of $\mathcal{P}_{\#}(\epsilon)$ is non-analytic.

3.3.4 $\log \mathcal{P}_{\#}(\epsilon) = \kappa\epsilon + \delta$ *in* $[\epsilon', \epsilon'']$

Since $\log \mathcal{P}_{\#}(\epsilon)$ is concave, $\log \mathcal{P}_{\#}(\epsilon) \leq \kappa\epsilon + \delta$. Thus, the supremum in $\mathcal{F}_{\#}(z) = \sup_{\epsilon_m \leq \epsilon_M}\{\log \mathcal{P}_{\#}(\epsilon) + \epsilon \log z\}$ is achieved at $\epsilon \leq \epsilon'$ if $\log z < -\kappa$, and at $\epsilon \geq \epsilon''$ if $\log z > -\kappa$. This implies a jump discontinuity in the first derivative of $\mathcal{F}_{\#}(z)$ at $\log z = -\kappa$, and thus a first-order transition.

Theorem 3.22 *Suppose that* $\log \mathcal{P}_{\#}(\epsilon) = \kappa\epsilon + \delta$ *for all* $\epsilon \in [\epsilon', \epsilon'']$. *Then there is a jump discontinuity in the first derivative of* $\mathcal{F}_{\#}(z)$ *at* $z_c = e^{-\kappa}$. □

Notice that a jump discontinuity at ϵ_m or ϵ_M in the density function may be seen as a limiting case of the above (see Section 3.3.1). Thus, such jump discontinuities may be interpreted as first-order phase transitions at zero or infinite z.

3.4 Examples

3.4.1 Chromatic polynomial of path graphs

The chromatic polynomial of a graph G is the number of ways of colouring the vertices of G with k colours (such that adjacent vertices have different colours). The chromatic polynomial of a complete graph on n vertices is $k!/(k-n)!$; if $k = n$ then this gives $n!$. In reality this overcounts the number of colourings by a factor of $n!$ (it is easily checked that a complete graph on n vertices can be coloured in only one way by n colours, if it is assumed that the graph is not labelled). Normalize the chromatic polynomial by division by $n!$. In the case of a path graph on n vertices, the normalised chromatic polynomial is $\chi_n(k) = k(k-1)^{n-1}/n!$. The generating function can be explicitly computed:

$$\begin{aligned} G_{\#}(x, z) &= \sum_{n>0}\sum_{k>1} \frac{k(k-1)^{n-1}}{n!} z^k x^n, = \sum_{k>1} \frac{k}{k-1} z^k e^{(k-1)x}, \\ &= 2z\left[\frac{z^2 e^{2x}}{4}\, {}_2F_1(2, 2; 3; ze^x) - \log(1 - ze^x)\right]. \end{aligned} \tag{3.19}$$

The hypergeometric function is convergent on the unit circle, including its boundary, except when $ze^x = 1$. This relation describes the critical curve of this model. In particular, it is now an easy matter to compute the limiting free energy: $\mathcal{F}_{\#}(z) = -\log(-\log z)$, for $z \in (0, 1)$. The free energy is not a convex function in this case, and this model is atypical. Curiously, the function $\mathcal{F}_{\#}(z) - \epsilon \log z$ has a local minimum for $\epsilon \in (1/e, 1)$, but a global minimum at $z = 0$ (where it is $-\infty$). If the density of colours is computed at the local minimum, then the density function $\mathcal{P}_{\#}(\epsilon) = 1 + \log \epsilon$ is found. The thermodynamics of this local minimum will be "metastable"; in other words, quantities such as the energy will be stable against small perturbations around this minimum, but a large perturbation will shift the system to the global minimum at $z = 0$.

3.4.2 Combinations

Let $p_n^{\#}(m) = \binom{n}{m}$ be the number of combinations of m objects in n objects. Then notice that the normalized chromatic polynomial of complete graphs on m vertices are also given by $p_n^{\#}(m)$. The identity

$$\sum_{p=0}^{m} \binom{n_1}{p} \binom{n_2}{m-p} = \binom{n_1+n_2}{m}, \tag{3.20}$$

implies that

$$\sum_{m_1=0}^{m} p_{n_1}^{\#}(m-m_1) p_{n_2}^{\#}(m_1) \leq p_{n_1+n_2}^{\#}(m), \tag{3.21}$$

so that $p_n^{\#}(m)$ is a supermultiplicative function as defined in Assumptions 3.1. Observe that $p_n^{\#}(\lfloor \epsilon n \rfloor) = 0$ if $\epsilon < 0$ or $\epsilon > 1$ (and n large enough). Thus, $\epsilon \in [0, 1]$ (or $\epsilon_m = 0$ and $\epsilon_M = 1$). The density function is given by

$$\log \mathcal{P}_{\#}(\epsilon) = \lim_{n\to\infty} \frac{1}{n} \log \binom{n}{\lfloor \epsilon n \rfloor} = -\log\left(\epsilon^{\epsilon}(1-\epsilon)^{1-\epsilon}\right). \tag{3.22}$$

The function $1/(\epsilon^{\epsilon}(1-\epsilon)^{1-\epsilon})$ will be seen again; some of its properties are described in the footnote below.[3] By Theorem 3.18,

$$\mathcal{F}_{\#}(\beta) = \sup_{\epsilon>0} \left[\epsilon \log z - \log(\epsilon^{\epsilon}(1-\epsilon)^{(1-\epsilon)})\right]. \tag{3.23}$$

Differentiation shows that the sup is attained if $(1-\epsilon)/\epsilon = 1/z$, from which it can be calculated that

$$\mathcal{F}_{\#}(\beta) = \log(1+z). \tag{3.24}$$

On the other hand, the generating function could instead be calcaluted:

$$G_{\#}(x, z) = \sum_{n=0}^{\infty} \sum_{m=0}^{n} \binom{n}{m} z^m x^n = \frac{1}{1 - x(1+z)}, \tag{3.25}$$

if $x(1+z) < 1$. The radius of convergence is $x_c(z) = 1/(1+z)$, from which the free energy in eqn (3.24) can again be obtained. The locus of points $(x_c(z), z)$ in the xz-plane is a line of simple poles in the generating function, and is the critical curve for this model. Below it the generating function is dominated by finite terms, while above it, the generating function is infinite, due to contributions of terms with arbitrarily large n. Comparison to eqn (2.14) shows that the assignment $\gamma_+ = 1$ may be made, and this interprets the entire critical curve as a λ-line. Fluctuations in m can be interpreted as a

[3] The function $E(\epsilon) = 1/(\epsilon^{\epsilon}(1-\epsilon)^{1-\epsilon})$ has infinite left- and right-derivatives at the end-points of $[0, 1]$. Moreover, $E(0) = E(1) = 1$, and the maximum of the function is achieved at $\epsilon = 1/2$ (where $E(1/2) = 2$). More generally, if $E(\epsilon, \delta) = \epsilon^{\epsilon} X^{\delta}/(\delta^{\delta}(\epsilon-\delta)^{\epsilon-\delta})$, then it has a maximum at $\delta = \epsilon X/(1+X)$ where its value is $(1+X)^{\epsilon}$.

specific heat $C_m(z)$ (in the limit as $n \to \infty$); this is the second derivative of $\mathcal{F}_\#(z)$ to $\log z$. Direct differentiation with respect to z gives

$$C_m(z) = \langle m^2 \rangle_z - \langle m \rangle_z^2 = \frac{z}{(1+z)^2}. \tag{3.26}$$

Here $\langle \cdot \rangle_z$ is the expected value at a given value of the fugacity z. A plot of $C_m(z)$ against $\log z$ shows that the fluctuations are largest when $z = 1$, and they decrease to 0 as $z \to 0$ or $z \to \infty$.

Consider a random walk in the hypercubic lattice of length n with exactly k corners. Let $r_n(k)$ be the number of such walks. Then

$$r_n(k) = \binom{n-1}{k} 2^{n-1-k}(2d-2)^k = 2^{n-1}\binom{n-1}{k}(d-1)^k. \tag{3.27}$$

By (3.22) and (3.23) note that

$$\begin{aligned} \log \mathcal{P}_\#(\epsilon) &= \log 2 + \frac{(d-1)^\epsilon}{\epsilon^\epsilon (1-\epsilon)^{1-\epsilon}}, \\ \mathcal{F}_\#(z) &= \log 2 + \log(1 + z(d-1)). \end{aligned} \tag{3.28}$$

Thus, combinations are also related to random walks.

3.4.3 *Directed or staircase walks in the square lattice*

Consider walks in the square lattice, starting at the origin and allowed only to step east or north. Let $p_n^\#(m)$ be the number of these walks of length n, where the first m edges is either in the north or east direction (call these the *choice edges*), but the remaining $n - m$ edges all point in the east direction. Then $p_n^\#(m) = 2^m$ if $m \leq n$, and the density function can be computed:

$$\log \mathcal{P}_\#(\epsilon) = \lim_{n\to\infty} \frac{1}{n} \log 2^{\lfloor \epsilon n \rfloor} = \epsilon \log 2. \tag{3.29}$$

By Theorem 3.18, $\mathcal{F}_\#(z) = \sup_{0<\epsilon<1} \{\epsilon \log z + \epsilon \log 2\}$ is the free energy. The supremum exists at $\epsilon = 0$ if $z \leq 1/2$, and $\epsilon = 1$ if $z > 1/2$, so that

$$\mathcal{F}_\#(z) = \begin{cases} 0, & \text{if } z \leq 1/2; \\ \log(2z), & \text{if } z > 1/2. \end{cases} \tag{3.30}$$

The radius of convergence of the generating function $G_\#(x, z)$ can be computed directly to be

$$x_c(z) = \begin{cases} 1, & \text{if } z \leq 1/2; \\ 1/2z, & \text{if } z > 1/2. \end{cases} \tag{3.31}$$

One can compute the partition function directly if $z < 1/2$:

$$p_n^{\#}(z) = \frac{1 - (2z)^{n+1}}{1 - 2z}, \tag{3.32}$$

and the generating function is

$$G_{\#}(x, z) = \frac{1}{(1 - x)(1 - 2zx)}, \tag{3.33}$$

provided that both $x < 1$ and $xz < 1/2$. In this example, two lines of simple poles meet in the xz-plane at the point $(1/2, 1)$. Since the derivative of $x_c(z)$ has a jump discontinuity at $z_c = 1/2$, there is a first-order transition in this model. If the derivatives of the free energy is taken, then the expected density of choice edges is found: if $z < 1/2$, then that density is 0, and asymptotically, the walk extends a mean distance n in the east direction. If $z > 1/2$, then the density is 1, and asymptotically, every edge is a choice edge. A walk of length n is expected to extend a mean distance $n/2$ in the east direction, since on average, half the edges will be in the east direction. The derivative does not exist at $z = 1/2$, and other techniques must be used to determine the density of choice edges: if the expected value of m for finite n is calculated directly from the partition function, then $\langle m \rangle = n/2$ if $z = 1/2$. The density of choice edges is $1/2$, and a walk of length n extends a mean distance $3n/4$ in the east direction. This gives an intermediate regime, distinct from the first two, at the critical point. The arguments in Chapter 2 can be applied to this model only in a limited sense. At the critical point $z_c = 1/2$ it appears that $G(x, 1/2) = 1/(1 - x)^2$ so that the assignment $\gamma_t = 2$ may be made. The general appearance of the singularity diagram is in Fig. 2.5, and the g-axis is given by the line $z = 1/2$, and the t-axis by $x = 2(1 - z)$. Thus, the (x, z) coordinates transform into (g, t) by $g = x - 2(1 - z)$ and $t = z - 1/2$. The assignments $\gamma_- = 1$ and $\gamma_+ = 1$ can be made, and the shift exponent ψ_λ can be computed by noting that $g_\lambda(t) = 1/2z - 2(1 - z) = (2z - 1)/z \simeq 2t/z$. Thus $\psi_\lambda = 1$ so that the cross-over exponent is $\phi = 1$. The exponent γ_u is not defined in this model, since $G(x, z)$ is infinite along the entire critical curve.

3.4.4 *Partitions*

In this example I count a class of polygons, called Ferrers diagrams, of area n, by their height m. Define the polygons as follows. Let a partition of n be $n = n_1 + n_2 + \cdots + n_m$, where $n_1 \geq n_2 \geq \cdots \geq n_m$. Starting from the left at the horizontal axis in the plane, place n_1 unit squares in a horizontal row along the x-axis. Continue by placing n_i unit squares from the left at the y-axis on top of the $(i - 1)$-th row of unit squares; for $i = 2, 3, \ldots, m$. The resulting placements of squares is called a Young diagram, and its perimeter will be called a partition polygon. $p_n^{\#}(m)$ is the number of partition polygons of area n and height m. One such polygon is illustrated in Fig. 3.4. It is known that (see for example [330])

$$\sum_{n=0}^{\infty} p_n^{\#}(m) x^n = \prod_{i=1}^{m} \frac{x}{1 - x^i}. \tag{3.34}$$

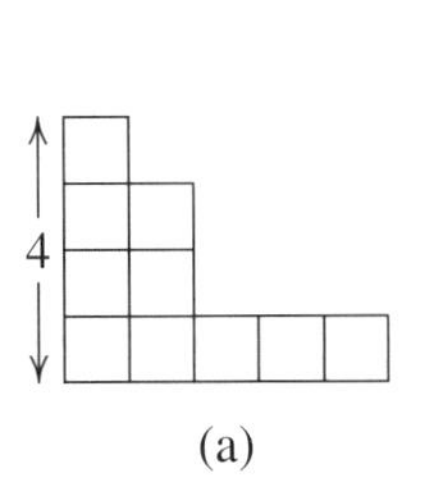

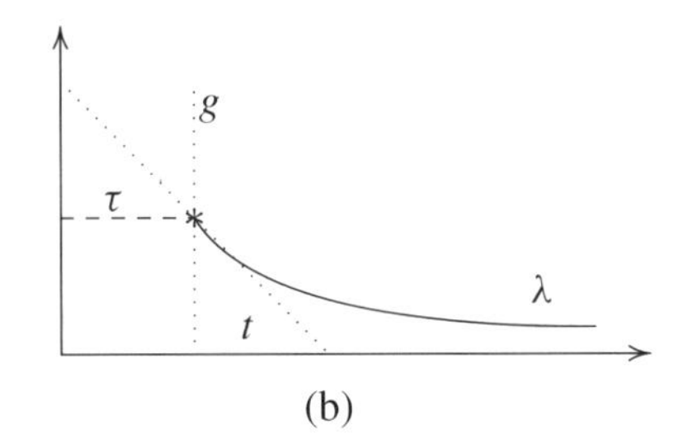

(a) (b)

Fig. 3.4: A Ferrers or partition polygon of area $n = 10$ and height 4 on the left (a). The phase diagram for this model is illustrated on the right (b). The t-axis is tangent to the λ-curve at the tricritical point $*$, and the g-axis is chosen to be at right angles to the τ-curve at the tricritical point. The generating function has a line of essential singularities along the dashed τ-line, but poles along the λ-curve.

Thus, the generating function is given by the infinite series

$$G_{\#}(x,z) = \sum_{m=1}^{\infty} \prod_{i=1}^{m} \frac{xz}{1-x^i}. \tag{3.35}$$

On the other hand, it can also be shown that[4]

$$G_{\#}(x,z) = \prod_{i=1}^{\infty} \frac{1}{(1-zx^i)} - 1, \tag{3.36}$$

so that a non-trivial partition identity is obtained.[5] There is a line of essential singularities in the xz-plane in $G_{\#}(x,z)$ along the line $x = 1$ (this is seen in eqn (3.35)). If $x < 1$, then the ratio test reveals a curve defined by $x = 1/z$ of simple poles (see eqn (3.36)) in $G_{\#}(x,y)$ (which implies that $\gamma_+ = 1$ in eqn (2.14)). Thus the critical curve is described by

$$x_c(z) = \begin{cases} 1, & \text{if } z < 1 \text{ (essential singularities in } G_{\#}(x,z)); \\ 1/z, & \text{if } z > 1 \text{ (simple poles in } G_{\#}(x,z)). \end{cases} \tag{3.37}$$

By eqn (3.36) $G_{\#}(x,z)$ is finite if $z = 1$ and $x < 1$, and it diverges as $x \to 1^-$.

[4] Sum over the columns in Fig. 3.4.

[5] The asymptotic behaviour of the number of partitions can be found by putting $z = 1$ in eqn (3.36) and then evaluating eqn (2.56). Note that an essential singularity appears in the integrand, which presents tremendous difficulties in the evaluation of the integral. These problems were overcome by G.H. Hardy and V. Ramanujan [179], who derived the celebrated asymptotic formula

$$p_n^{\#} \sim \frac{1}{4\sqrt{3}n} e^{\pi\sqrt{2n/3}}$$

for the number of partitions of the integer n.

The radius of convergence is related to the free energy as in eqn (2.6). Thus,

$$\mathcal{F}_{\#}(z) = \begin{cases} 0, & \text{if } \log z < 0; \\ \log z, & \text{if } \log z > 0. \end{cases} \tag{3.38}$$

The multicritical point at $z = 1$, $x_c(1) = 1$ is a tricritical point in the sense described in Chapter 2. This model is asymmetric, and moreover, the τ- and λ-lines meet at an angle. In particular, in the (g, t)-plane the critical curve has the following behaviour close to the tricritical point: $g = 1 - x_c(z) = 1 - 1/z = (z - 1)/z = t/z$. This $g \simeq t$ if $t \to 0^+$ and therefore $\psi_\lambda = 1/\phi = 1$. The exponent γ_t cannot be assigned a value, since there is an essential singularity at the tricritical point.

The density function of the height m can also be computed, using Theorem 3.19: $\log \mathcal{P}_{\#}(\epsilon) = \inf_{z>0}\{\mathcal{F}_{\#}(z) - \epsilon \log z\}$. If $\epsilon = 0$, then the infimum exists for any $z \leq 1$, and $\mathcal{P}_{\#}(0) = 1$. For $0 < \epsilon < 1$ the infimum exists at $z = 1$, and $\mathcal{P}_{\#}(\epsilon) = 1$. If $\epsilon = 1$, then the infimum exists for any $z > 1$, and $\mathcal{P}_{\#}(\epsilon) = 1$. Thus $\mathcal{P}_{\#}(\epsilon)$ is a constant function in $[0, 1]$. These results are reflected in Theorem 3.22. There is a first-order transition in this model; the derivative $zd\mathcal{F}_{\#}(z)/dz$ is the density of the height of the partition polygon (as a fraction of its area n), in the thermodynamic limit. If $z < 1$, then the expected value of the density is 0, and if $z > 1$, then the expected value is 1. In other words, if $z < 1$, then the average height of a partition polygon grows as $o(n)$, while if $z > 1$, it grows as $O(n)$.

3.4.5 Queens on a chessboard

The number of ways of putting m queens on a $n \times n$ chessboard is $\binom{n^2}{m}$. Let $p_n^{\#}(m) = \binom{n^2}{m}$; then the generating function is

$$G_{\#}(x, z) = \sum_{n=0}^{\infty} [(1 + z)x]^{n^2}. \tag{3.39}$$

The Ramanujan Θ-function [358] $\phi(y)$ is defined by

$$\phi(y) = \sum_{n=-\infty}^{\infty} y^{n^2}. \tag{3.40}$$

Thus,

$$G_{\#}(x, z) = \frac{1}{2}\left(1 + \phi((1 + z)x)\right). \tag{3.41}$$

The function $\phi(y)$ is related to the standard Θ-function $\theta_3(y, q)$, and in particular

$$\phi(y) = \theta_3(0, y) = \prod_{n=1}^{\infty} \left[(1 + y^{2n-1})^2(1 - y^{2n})\right]. \tag{3.42}$$

Thus,

$$G_{\#}(x,z) = \frac{1}{2}\left(1 + \prod_{n=1}^{\infty}\left[(1+(x+xz)^{2n-1})^2(1-(x+xz)^{2n})\right]\right). \tag{3.43}$$

This exhibits an essential singularity if $x = 1/(1+z)$, and the critical curve and free energy is given by

$$x_c(z) = \frac{1}{1+z}, \qquad \mathcal{F}_{\#}(z) = \log(1+z). \tag{3.44}$$

There is no transition in this model. Notice that the free energy is the same as the free energy obtained for combinations (eqn (3.23)). In this model the entire critical curve may be interpreted as a τ-line.

3.4.6 Adsorbing walks

Let $p_n^{\#}(k)$ be the number of self-avoiding walks with its first vertex at the origin, its first k edges in the hyperplane $z = 0$, and its remaining $n - k$ steps in the half-space $z > 0$. This is a model of a polymer which interacts with a surface, and which may adsorb on to the surface from one end-point. In this model walks in the half-space $z > 0$, with their first vertices in the hyperplane $z = 0$, are counted. Let f_n be the number of self-avoiding walks with the first vertex at the origin, and with n steps, and let f_n^+ be the number of self-avoiding walks in the half-space $z > 0$ with first vertex at the origin. Any walk counted by f_n can be translated until it is in the space $z > 0$, and with at least one vertex within one step from the hyperplane $z = 0$. By cutting the walk at this vertex and by adding an edge to each subwalk to join them to a vertex in the $z = 0$ hyperplane, the following inequalities are obtained:

$$f_n^+ \leq f_n \leq \sum_k f_{k+1}^+ f_{n-k+1}^+. \tag{3.45}$$

There is a value of k which maximizes the right-hand side; let the least value of such a k be k_*. Then

$$f_n^+ \leq f_n \leq (n+1) f_{k_*+1}^+ f_{n-k_*+1}^+. \tag{3.46}$$

Define $K_* = \liminf_{n\to\infty}(k_*/n)$. By eqn (3.45),

$$\limsup_{n\to\infty} \frac{1}{n}\log f_n^+ \leq \log \mu_d,$$

and by taking the lim inf instead,

$$\begin{aligned} \log \mu_d &\leq K_* \liminf_{m\to\infty} \frac{1}{m}\log f_{m+1}^+ + (1-K_*)\liminf_{m\to\infty}\frac{1}{m}\log f_{m+1}^+ \\ &= \liminf_{m\to\infty}\frac{1}{m}\log f_m^+. \end{aligned} \tag{3.47}$$

In other words, the number of walks with an end-point in $Z = 0$, and otherwise confined to the half-space $Z > 0$, has growth constant given by

$$\lim_{n\to\infty} \frac{1}{n} \log f_n^+ = \log \mu_d. \tag{3.48}$$

In our model, the walks counted by $p_n^\#(\lfloor \epsilon n \rfloor)$ consist of walks of length $\lfloor \epsilon n \rfloor$ in the $(d-1)$-dimensional hyperplane $Z = 0$, followed by a walk of length $n - \lfloor \epsilon n \rfloor$ in the d-dimensional half-space $Z > 0$. Thus,

$$\mathcal{P}_\#(\epsilon) = \lim_{n\to\infty} [p_n^\#(\lfloor \epsilon n \rfloor)]^{1/n} = \mu_{d-1}^{\epsilon} \mu_d^{1-\epsilon}. \tag{3.49}$$

Hence,

$$\log \mathcal{P}_\#(\epsilon) = \epsilon \log \mu_{d-1} + (1-\epsilon) \log \mu_d \tag{3.50}$$

so that this model is described by the situation in Section 3.3.4. The critical point is given by (see Theorem 3.22)

$$z_c = \mu_d/\mu_{d-1}. \tag{3.51}$$

If $z < z_c$, then the free energy can be computed from the density function by putting $\epsilon = 0$ (the density of edges in the hyperplane $Z = 0$ is zero), and if $z > z_c$ then the density of edges at $z = 0$ is $\epsilon = 1$, and this value of ϵ will give the free energy. Direct computation gives

$$\mathcal{F}_\#(z) = \begin{cases} \log \mu_d, & \text{if } z < z_c; \\ \log \mu_{d-1} + \log z, & \text{if } z > z_c. \end{cases} \tag{3.52}$$

The density ϵ of visits undergoes a discontinuous change at the critical point, which shows that this model has a first-order phase transition at which the walk adsorbs on the hyperplane. The singularity in the generating function is expected to be determined by self-avoiding walk exponents (that is, the exponents of a linear polymer in a dilute solution (Table 1.1)). If $z < z_c$ (the desorbed phase) then the exponents of a three-dimensional self-avoiding walk are expected (in particular, $\gamma_+ \approx 1.1575(6)$), and if $z > z_c$ (the adsorbed phase) then the exponents of a two-dimensional self-avoiding walk are expected ($\gamma_- = 43/32$). These exponents describe the singularity in $G_\#(x, z)$ along the critical curve, and also show that a τ-line is absent in the phase diagram. Instead, two λ-lines meet at the critical point. The cross-over exponent can also be computed by arguing as follows. The critical curve is given by $x_c(z) = 1/\mu_d$ of $z < z_c$ and $x_c(z) = 1/(z\mu_{d-1})$ if $z > z_c$. Thus, the critical curve for $z > z_c$ is described by $g = 1/\mu_d - 1/(z\mu_{d-1}) \approx t\mu_{d-1}/\mu_d$, where $t = z - \mu_d/\mu_{d-1}$. This gives a shift exponent equal to one, and so $\phi = 1$.

4

Exact models

4.1 Introduction

In this chapter a variety of directed models of walks, polygons and animals will be studied. These models include the partition polygons encountered in Chapter 3, as well as directed and partially directed walks. The fact that the models in this chapter are called "exact models" implies that the generating function can, in may cases, be explicitly written down, often as an infinite product or a q-analogue of a special function, and in this way an "exact solution" of the model is found. Careful analysis is then required to extract the free energies and other critical properties from the exact solutions of these models; this can be a challenging procedure. It is not possible to give more than an introduction to this active and vast area in a single chapter; instead, I shall focus on a few specific models which are directed versions of models considered elsewhere in this monograph, and whose solution will illuminate the general comments made in Chapters 2 and 3, as well as being related to models considered in later chapters.

The model of partition polygons examined in Section 3.4.4 is a good starting point. The expressions in eqns (3.35) and (3.36) can be written more compactly by using q-deformed special functions. The q-analogue of the factorial is defined by the following finite product:

$$(t;q)_n = \prod_{i=0}^{n-1}(1-tq^i). \tag{4.1}$$

This is called a *q-product*. The generating function of partition polygons (eqn (3.35)) can be expressed in terms of q-analogues of the factorial as

$$G(x,z) = \sum_{m=1}^{\infty} \frac{(xz)^m}{(x;x)_m}, \tag{4.2}$$

where x is the area activity while z is an activity conjugate to the height of the polygon. In that example, the partition polygons were counted by area, and had energy equal to the height of the first column (or equivalently, if each vertical edge has activity $\sqrt{z}$, then the expressions will be the same). Equation (3.36) gives a different expression for the generating function of partition polygons in terms of a q-product:

$$G(x,z) = \frac{1}{(zx;x)_\infty} - 1. \tag{4.3}$$

This expression $1/(zx;x)_\infty$ is also called the *q-exponential*; notice that it is equal to $\sum_{m=0}^{\infty}(xz)^m/(x;x)_m$ by eqn (4.2), and since $(x;x)_m$ is the q-factorial, the expression

"q-exponential" is natural (see for example reference [329]). The singularity structure of $G(x, z)$ can be readily seen in eqn (4.2). In particular, if $z > 1$, then the factor $1/(1 - zx)$ is a simple pole at $x_c(z) = 1/z$. For $z \leq 1$ there is an essential singularity in $G(x, z)$ at $x_c(z) = 1$. At the point $(x, z) = (1, 1)$ a curve of simple poles meets a curve of essential singularities, which is thus a tricritical point. The curve of simple poles is the λ-line, and from eqn (2.14) it may be concluded that $\gamma_+ = 1$.

The density function of this model is $\mathcal{P}(\epsilon) = 1$ for $\epsilon \in [0, 1]$. Since $\log \mathcal{P}(\epsilon) = 0$, it follows from Theorem 3.22 that there is a first-order transition at $z_c = 1$; the energy density (density of vertical edges) has a jump discontinuity at the critical point from 0 to 1. If $z < 1$, then the density of vertical edges is 0, and for $z > 1$ it is 1. This is seen by taking the derivative of the free energy with respect to $\log z$; it is

$$z \frac{d}{dz} \mathcal{F}(z) = \begin{cases} 0, & \text{if } z < 1; \\ 1, & \text{if } z > 1. \end{cases}$$

In particular, along the locus of essential singularities in the critical curve the density of vertical edges is 0, and along the poles the density is 1 (for $z > 1$). The intuitive result is that if $z < 1$, then the partition polygons are "fat", with area which "grows" faster than $O(m)$, if the height is m. On the other hand, if $z > 1$, then the simple pole $1/(1 - zx)$ dominates the generating function. It can be checked that this is the result of a long "skinny" first column in the partition polygon; the other columns stays short (the polygon will have area $O(m)$ if the height is m). It is not clear what the behaviour is at the point $(x, z) = (1, 1)$, but it is conceivable that both these "phases" make a comparable contribution to the generating function, and coexist there.

The tricritical cross-over exponent of this model can be found by examining the shape of the critical λ-line close to the tricritical point. In particular, since this curve is given by $x = 1/z$, it appears that $1 - x = 1 - 1/z = (z - 1)/z$ and by defining the g- and t-axes by $g = 1 - x$ and $t = z - 1$, this gives $g \sim t$. Equation (2.11) then implies that the shift exponent is $\psi_\lambda = 1$ and, by eqn (2.19), $\phi = 1$ and thus $\alpha = 1$. The asymptotic behaviour of the generating function around the tricritical point can be found from Lemma C.2. This gives

$$G(x, 1) \simeq \sqrt{\frac{-\log x}{2\pi}}\, e^{-\pi^2/6 \log x} \approx \sqrt{\frac{1 - x}{2\pi}}\, e^{\pi^2/6(1-x)}, \tag{4.4}$$

where the approximation $\log x = -1 + x$ as $x \to 1^-$ was used. This result indicates the existence of an essential singularity at the tricritical point.[1]

[1] This asymptotic expression for the generating function is (in a strict sense) not of the assumed scaling forms in eqn (2.17) or in Definition 2.2. This shows that assumptions such as in eqn (2.17) may be inadequate in some models; the "constant" A_t is not a constant at all, but a function of g. It is unclear to what extent the exponents, scaling relations and general framework of tricriticality is preserved in these models. For example, if $x = 1$ in eqn (4.3), then $G(1, z) = \infty$, and there seems to be no clear way of introducing a scaling assumption such as in eqn (2.15) or in footnote 5 in Chapter 2, or of assigning a value to the exponents γ_u and γ_t in relation to the cross-over exponent, which is derived here purely from its relation to the shift-exponent in eqn (2.19).

By inverting $x_c(z)$, the function $z_c(x)$ is obtained, and $-\log z_c(x)$ is a free energy associated with the height of the partition, where x is an area activity. In particular

$$-\log z_c(x) = \begin{cases} \log x, & \text{if } x < 1; \\ \infty, & \text{if } x > 1. \end{cases} \tag{4.5}$$

Derivatives of $\log z_c(x)$ with respect to $\log x$ gives the density of area; if $x < 1$ then it is 1, otherwise it is divergent. Thus, if $x < 1$, then the area of the partition polygon grows proportional to its height. It is important to note that this free energy (though infinite) is a convex function of x, and does not violate the general convexity principle[2] as set out in eqn (2.4). It is infinite for $x > 1$ because there is no exponential bound on the number of partition polygons of a given height.

In Section 4.2 I slightly alter this model by replacing the activity z with an activity (z again) which is conjugate to the total perimeter of the partition polygon. As opposed to the area–height generating function (eqn (4.2)), the "area–perimeter" generating function of this model will be calculated, and the critical behaviour will be analysed with more care. The slight change in this model will introduce a curious (and atypical) new feature in the phase diagram. Apart from the tricritical point, there will be a special point in the line of essential singularities, where the generating function $G(1, z)$ will become divergent. However, the most interesting point will still be the tricritical point.

4.2 Partition polygons (Ferrers diagrams)

Partition polygons were discussed in Section 3.4.4, and in this section a slightly changed version of that model will be considered. In particular, the polygons will be weighted with respect to area and perimeter (rather than height). Our motive is to gain some insight into the thermodynamic properties of the model. This is a limited objective, but it will soon become apparent that mathematical rigour will have to be abandoned if this is to be achieved. Partitions have been the subject of investigation by many mathematicians, including S. Ramanujan and G.H. Hardy [176, 179], and G. Pólya [293]. In the context as models in statistical mechanics, they have been investigated by T. Prellberg and A.L. Owczarek [297].

4.2.1 The area–perimeter generating function of partition polygons

The generating function can be found by cutting a partition polygon as in Fig. 4.1. First, imagine that it is made by stacking slabs on top of one another as in Fig. 4.1(a); the first slab has length i_1, the second, i_2, and so on, up the the h-th slab of length i_h. These slabs are stacked such that their left sides have the same horizontal coordinates (or are left-adjusted); and if $i_h \leq i_{h-1} \leq \cdots \leq i_1$, then there is exactly one way of stacking them. The polygons are counted by area and perimeter (the energy of any polygon is its

[2] The continuity of the free energy is a consequence of both its convexity and finiteness. If the free energy is infinite for some range of the parameter z, then its continuity is not guaranteed. In other words, $-\log z_c(x)$ in eqn (4.5) is a perfectly legitimate free energy. In the next section a model of partition polygons with a finite jump discontinuity, followed by an infinite jump discontinuity in the free energy, will be encountered.

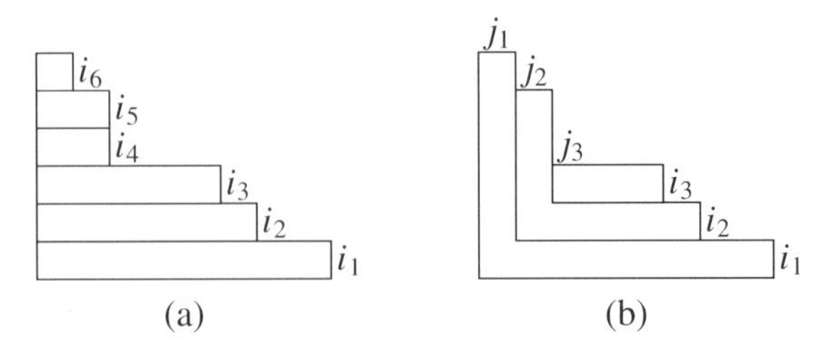

Fig. 4.1: The generating function of partition polygons with area and perimeter can be computed by either cutting the partition into slabs (a) or into L-shaped strips (b).

perimeter). Note that $i_1 > 0$ in Fig. 4.1 (otherwise there is no partition at all). The area of the partition polygon is $\sum_{m\geq 1} i_m$ and its perimeter is $2n + 2i_1$. As before, let x be the area activity, and let z be the perimeter activity. Then

$$\begin{aligned} G_f(x,z) &= \sum_{n=1}^{\infty}\sum_{i_n=1}^{\infty}\sum_{i_{n-1}=i_n}^{\infty}\cdots\sum_{i_2=i_3}^{\infty}\sum_{i_1=i_2}^{\infty} z^{2n+2i_1}x^{i_1+i_2+\cdots+i_n} \\ &= \sum_{n=1}^{\infty}\frac{z^{2n+2}x^n}{(1-z^2x)(1-z^2x^2)\cdots(1-z^2x^n)} = \sum_{n=1}^{\infty}\frac{z^{2n+2}x^n}{(z^2x;x)_n}, \end{aligned} \tag{4.6}$$

where the geometric series are evaluated starting at the sum over i_1 (on the right), and where the final expression is written using the q-notation of eqn (4.1). The relation of $G_f(x,z)$ to the q-exponential is manifest (see eqn (4.2)).

Alternatively, imagine a partition to be made from L-shaped strips sandwiched as in Fig. 4.1(b). If there are h such strips, and the generating function is computed from this figure instead, using the same techniques as in eqn (4.6), then

$$\begin{aligned} G_f(x,z) &= \sum_{h=1}^{\infty}\sum_{\substack{i_n\geq 0\\ j_n\geq 0}}\sum_{\substack{i_{n-1}\geq i_n+1\\ j_{n-1}\geq j_n+1}}\cdots\sum_{\substack{i_1\geq i_2+1\\ j_1\geq j_2+1}} z^{(2(i_1+j_1)+4)}x^{\left(\sum_k i_k+\sum_k j_k+n\right)} \\ &= \sum_{n=1}^{\infty}\frac{x^{n^2}z^{4n}}{(z^2x;x)_n^2}, \end{aligned} \tag{4.7}$$

and the result is a non-trivial partition identity (there are many others [329]).

Equations (4.6) and (4.7) are area–perimeter generating functions for partition polygons, and it is not difficult to extract the radius of convergence $x_c(z)$ from eqn (4.7): there are curves of poles defined by the factors $(1 - z^2x^n) = 0$ in $G_f(x,z)$; these curves accumulate on the line $x = 1$ if $0 \leq z \leq 1$, and this is a line of essential singularities in $G_f(x,z)$. Along this line it is possible to show, using eqn (4.6), that $G_f(z,1)$ is finite if $z^2 < 1/2$, and infinite if $1/2 \leq z^2 \leq 1$, by using D'Alembert's ratio test. If $z > 1$, then the radius of convergence is given by the factor $(1 - z^2 x)^2$ in the denominator of eqn (4.7). Therefore,

$$x_c(z) = \begin{cases} 1, & \text{if } 0 \leq z \leq 1; \\ 1/z^2, & \text{if } z > 1. \end{cases} \tag{4.8}$$

The phase diagram is illustrated in Fig. 4.2. The point $(x, z) = (1, 1)$ is the meeting point of a curve of poles with a line of essential singularities, and is by our definition a tricritical point.

The density function is given by

$$\mathcal{P}_f(\epsilon) = 1, \quad \forall \epsilon \in (0, 2), \tag{4.9}$$

and by Theorem 3.22 there is a first-order phase transition in this model. The limiting free energy is

$$\mathcal{F}_f(z) = -\log x_c(z) = \begin{cases} 0, & \text{if } z \leq 1; \\ 2 \log z, & \text{if } z > 1. \end{cases} \tag{4.10}$$

The phase transition is from an expanded partition polygon of large surface area to a deflated partition polygon with an L-shape. Derivatives of the free energy to $\log z$ gives

$$z \frac{d}{dz} \mathcal{F}_f(z) = \begin{cases} 0, & \text{if } z < 1; \\ 2, & \text{if } z > 1. \end{cases}$$

In other words, the density of perimeter edges is zero if $z < 1$, and 2 if $z > 1$ (thus, each unit square in the partition polygon has two edges in the perimeter in this phase; this is only possible if the appearance of the polygon is L-shaped). The factor $1/(1 - z^2 x)$

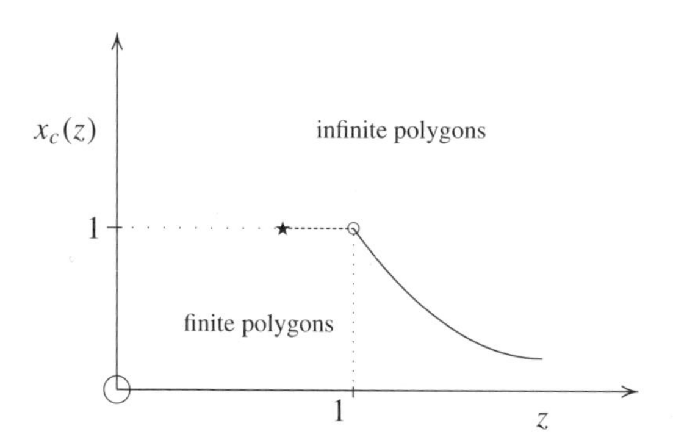

Fig. 4.2: The phase diagram of partition polygons counted by area and perimeter. There is a tricritical point at o, where a curve a simple poles meet a line of essential singularities. The point marked by $\star$ is a special point; the generating function is finite on the line $x = 1$ to the left of this point, but infinite at $\star$ and to the right.

in the generating function corresponds to a leg of the L-shaped polygon, and these factors diverge along the λ-line in the phase diagram. Inverting $x_c(z)$ gives the radius of convergence $z_c(x)$ of the generating function as a function of x. It can be seen that

$$z_c(x) = \begin{cases} 1/\sqrt{x}, & \text{if } x < 1; \\ 1/\sqrt{2}, & \text{if } x = 1; \\ 0, & \text{if } x > 1. \end{cases} \tag{4.11}$$

This indicates the presence of three possible regimes. If $x < 1$ then the deflated regime is encountered, with L-shaped polygons giving the dominant contribution to $G_f(x, z)$. If $x > 1$, then inflated partitions gives the dominant contribution. A curious situation occurs when $x = 1$. Neither deflated nor inflated vesicles dominate the generating function; the generating function diverges at a larger value of z than suggested by inflated partition polygons, and at a smaller value of z than suggested by deflated partition polygons; it may be supposed that the generating function is dominated by a third class of polygons intermediate between inflated and deflated polygons.

The "perimeter free energy" can be computed from eqn (4.11); it is given by

$$\mathcal{F}_f(x) = \begin{cases} (\log x)/2, & \text{if } x < 1; \\ (\log 2)/2, & \text{if } x = 1; \\ \infty, & \text{if } x > 1. \end{cases} \tag{4.12}$$

In other words, there is a finite jump discontinuity at $x = 1$, followed by an infinite jump discontinuity as the polygons inflate.

4.2.2 *Asymptotic analysis of the partition generating function*

The tricritical nature of the area–perimeter generating function of partition polygons can be examined by an asymptotic analysis. Thus, it may be expected that at least some of the assumptions in Chapter 2 may be applicable here, and that it may even be possible to assign values to some critical exponents. One important feature of the tricritical scaling assumption in eqns (2.15), (2.16) and (2.17) is the scaling function, and the fact that the behaviour of the singular part in the density function is controlled by the scaling exponent α_t (or γ_t, whichever is appropriate). In this model it will appear that this scaling assumption is not satisfied in a strict sense (see footnote 1, Chapter 4). The cross-over exponent will be found, but only through its relation to the shift exponent in eqn (2.19).

The area–perimeter generating function of partition polygons will be be examined asymptotically (as the critical curve is approached). There are three cases to consider: $z < 1$, $z = 1$ and $z > 1$. If $z > 1$, then $x_c(z) = 1/z^2 < 1$ and if $x < x_c(z)$, then by eqn (4.7),

$$\frac{xz^4}{(1-z^2x)^2} \le G_f(x,z) \le \frac{xz^4}{(1-z^2x)^2}\left(1 + \frac{1}{(z^2x^2;x)_\infty^2}\sum_{n=1}^{\infty} x^{n^2+2n} z^{4n}\right), \tag{4.13}$$

where it was noted that $(z^2x;x)_{n+1} = (1-z^2x)(z^2x^2;x)_n$, and $(z^2x^2,x)_n \geq (z^2x^2,x)_\infty$ if $x \leq 1$. Note that $x \leq x_c(z) < 1$ and that $z^2x^i < 1$ for all $i \geq 2$ (even if $x = x_c(z)$), thus, for all $0 < x \leq x_c(z)$, $(z^2x^2;x)_\infty > 0$ by Lemma C.5. Hence,

$$G_f(x,z) \simeq \frac{xz^4}{(1-z^2x)^2}, \tag{4.14}$$

as $x \to x_c^-(z)$ for $z > 1$ along the λ-line. This result gives the scaling of the generating function along the λ-line in the phase diagram. Comparing eqn (4.14) to eqn (2.14) indicates that $\gamma_+ = 2$.

The calculations are considerably more difficult if $z \leq 1$. Consider first the situation that $z = 1$, in which case the tricritical point is approached as $x \to x_c^-(z) = 1$. By eqn (4.6),

$$G_f(x,1) = \sum_{n=1}^{\infty} \frac{x^n}{(x;x)_n} = \sum_{n=1}^{\infty} \frac{x^n}{1-x^n} \frac{(x^n;x)_\infty}{(x;x)_\infty}. \tag{4.15}$$

Use Lemma C.2 and Corollary C.4 to approximate the q-factorials:

$$(x;x)_\infty \simeq \sqrt{\frac{2\pi}{-\log x}}\, e^{(\pi^2/6)/\log x}; \tag{4.16}$$

$$(x^n;x)_\infty \simeq \sqrt{1-x^n}\, e^{(\mathcal{L}i_2(x^n)-\pi^2/6)/\log x}, \tag{4.17}$$

where $\mathcal{L}i_2(x)$ is Euler's dilog function.[3] Substition of these approximations into $G_f(x,1)$ gives

$$G_f(x,1) \simeq \sqrt{\frac{-\log x}{2\pi}}\, e^{-(\pi^2/3)/\log x} \sum_{n=1}^{\infty} \frac{x^n}{\sqrt{1-x^n}}\, e^{\mathcal{L}i_2(x^n)/\log x}. \tag{4.18}$$

The infinite series can be approximated by an integral. If the substitution $s = x^n$ is also carried through, then the above becomes

$$G_f(x,1) \simeq \frac{1}{\sqrt{-2\pi \log x}}\, e^{-(\pi^2/3)/\log x} \int_0^1 \frac{1}{\sqrt{1-s}}\, e^{\mathcal{L}i_2(s)/\log x}\, ds \tag{4.19}$$

after some simplification. The integral cannot be evaluated in closed form; instead, a saddle-point method will have to suffice. By inserting a factor of s/s in the integrand, the above becomes

$$\int_0^1 \frac{1}{\sqrt{1-s}}\, e^{\mathcal{L}i_2(s)/\log x}\, ds = \int_0^1 \frac{1}{s\sqrt{1-s}}\, e^{(\mathcal{L}i_2(s)+\log s\cdot\log x)/\log x}\, ds. \tag{4.20}$$

[3] The dilogarithm is defined by the infinite series $\mathcal{L}i_2(x) = \sum_{m=1}^{\infty}[x^m/m^2]$. Notice that

$$\frac{d}{dx}\mathcal{L}i_2(x) = -[\log(1-x)]/x.$$

Generally, a saddle-point approximation relies on the following approximation formula:

$$\int_0^1 f(s)e^{\alpha g(s)}\,ds \approx \left[\sqrt{\frac{-2\pi}{\alpha g''(s_0)}}\right] e^{\alpha g(s_0)} f(s_0), \tag{4.21}$$

where s_0 is that value of s in the interval $(0, 1)$ such that $g'(s_0) = 0$ (this is the saddle point). To apply eqn (4.21) in our case, put $g(s) = \mathcal{L}i_2(s) + \log s \cdot \log x$, so that $g'(s) = (\log x - \log(1-s))/s = 0$ if $s = s_0 = 1 - x$. Thus $g''(s_0) = 1/(1-x)x$. Notice that $s_0 \in (0, 1)$, and that the approximation is valid. From eqn (4.21) it follows that

$$\int_0^1 \frac{1}{\sqrt{1-s}}\, e^{\mathcal{L}i_2(s)/\log x}\,ds \approx \sqrt{2\pi}\, e^{(-\mathcal{L}i_2(x)+\pi^2/6)/\log x}, \tag{4.22}$$

with the result that

$$G_f(x, 1) \approx \frac{1}{\sqrt{-\log x}}\, e^{-(\mathcal{L}i_2(x)+\pi^2/6)/\log x}. \tag{4.23}$$

Observe that $-\log x \approx (1-x)$ if $x \to 1^-$; making this approximation will simplify (4.23) somewhat, and I shall use this liberally.

Lastly, consider the case that $z < 1$. The starting point is again eqn (4.6), and it is approximated by using Lemma C.3 and Corollary C.4; in particular, note that

$$(t; q)_{n+1} \approx \sqrt{\frac{1-t}{1-tq^n}}\, e^{(\mathcal{L}i_2(t)-\mathcal{L}i_2(tq^n))/\log q}. \tag{4.24}$$

Then, by using this approximation in eqn (4.6);

$$G_f(x, z) \approx \sum_{n=1}^{\infty} \frac{x^n z^{2n+2}}{1-z^2x^n} \sqrt{\frac{1-z^2x^n}{1-z^2x}}\, e^{(\mathcal{L}i_2(z^2x^n)-\mathcal{L}i_2(z^2x))/\log x}, \tag{4.25}$$

which can be simplified, in the same manner as above, to

$$G_f(x, z) \approx \frac{z^2}{\sqrt{1-z^2x}}\, e^{-\mathcal{L}i_2(z^2x)/\log x}\, I(x, z), \tag{4.26}$$

where

$$I(x, z) \approx \int_0^1 \frac{1}{\sqrt{1-z^2s}}\, e^{(\mathcal{L}i_2(z^2s)+2\log s\cdot\log x)/\log x}\,ds. \tag{4.27}$$

The saddle-point method can be used to estimate the integral, using eqn (4.21) with $g(s) = \mathcal{L}i_2(z^2s) + 2\log s \cdot \log z$. Then $g'(s) = 0$ if $s = s_0 = (1-z^2)/z^2$, which is in the interval $(0, 1)$ only if $(1-z^2)/z^2 < 1$ or $1/2 \le z^2 < 1$. In other words, this approximation will only be proper along the τ-line where $G_f(x, z)$ diverges, between the special point ($\star$) and the tricritical point ($\circ$) in Fig. 4.2. In addition, $g''(s_0) = z^2/(1-z^2)$. Using this estimate, and simplifying the resulting equations, while using the identity

$$\mathcal{L}i_2(t) + \mathcal{L}i_2(1-t) = \pi^2/6 - \log t \log(1-t), \tag{4.28}$$

eventually gives

$$G_f(x,z) \approx \sqrt{\frac{-2\pi(1-z^2)}{\log x(1-z^2x)}}\, e^{-(\mathcal{L}i_2(z^2x)+\mathcal{L}i_2(z^2)+4(\log z)^2-\pi^2/6)/\log x}. \tag{4.29}$$

Of course, this approximation is only valid close to the τ-line with $1/\sqrt{2} < z < 1$ (in other words, between the special and the tricritical points in Fig. 4.2). For values of $z < 1/\sqrt{2}$ the generating function is finite at $x_c(z)$, and this approximation becomes invalid (the saddle-point estimate of the integral in eqn (4.27) fails).

The tricritical scaling axes can be identified explicitly in this model. The t-axis is taken tangent to the λ-line at the tricritical point, as explained in Chapter 2. In asymmetric models the g-axis is taken perpendicular to the τ-line at the tricritical point. In the gt-plane the origin is located at the tricritical point, and (x, z)-coordinates transform into (g, t)-coordinates linearly as follows:

$$t = \sqrt{3}(z-1), \qquad g = 3 - x - 2z. \tag{4.30}$$

These formulae were computed by explicitly finding the equations of the tangent line to the λ-line and the normal line to the τ-line. The λ-line has equation $z^2x = 1$ if $z > 1$, and by substituting for z and x from (4.30), $g_\lambda(t)$ can be solved for, which is the equation of the λ-line in (g, t)-coordinates:

$$g_\lambda(t) = -\frac{1}{3}\frac{t^2(9+2t\sqrt{3})}{3+2t\sqrt{3}+t^2} \approx -t^2 \quad \text{as } t \to 0. \tag{4.31}$$

Thus, by eqn (2.11), the shift-exponent ψ_λ has value 2, and the cross-over exponent is $\phi = 1/2$ (see eqn (2.19)) while the gap-exponent has value $\Delta = 2$. The specific heat exponent has value $\alpha = 0$ in this model, as can be seen from eqn (2.28). The equation of the τ-line (if $t < 0$) is

$$g_\tau(t) = 1 - 2t/\sqrt{3}. \tag{4.32}$$

Notice that $(1 - z^2x) = (g - g_\lambda(t))$ and, by using the approximation $-\log x \approx (1-x)$ if $1-x$ is small, the asymptotic expressions for $G_f(x,z)$ in eqns (4.14), (4.23) and (4.29) is

$$G_f(g,t) \simeq \begin{cases} (g - g_\lambda(t))^{-2}, & \text{if } t > 0; \\ \dfrac{1}{\sqrt{g}} e^{(\mathcal{L}i_2(1-g)+\pi^2/6)/g}, & \text{if } t = 0; \\ \dfrac{1}{\sqrt{g-g_\tau(t)}}\sqrt{\dfrac{2\pi(1-z^2)}{1-z^2x}}\, e^{P(g,t)/(g-g_\tau(t))}, & \text{if } t < 0, \end{cases} \tag{4.33}$$

where

$$P(g,t) = \mathcal{L}i_2(z^2x) + \mathcal{L}i_2(z^2) + 4(\log z)^2 - \pi^2/6, \tag{4.34}$$

and where some factors in x and z were left undisturbed to avoid unnecessarily complicated outcomes. This gives a complete account of the behaviour around the tricritical point. The value of the cross-over exponent is given by $\phi = 1/2$, but the scaling assumption in eqns (2.15) and (2.16) appear not to apply to this model, since $G(x, z)$ is infinite along the τ-line on approaching the tricritical point. Thus, it seems that a value cannot be assigned to γ_u or γ_t. The specific heat exponent α is given by $\alpha = 0$. The behaviour of $G_f(g, t)$ at the tricritical exponent is an essential singularity with a power-law factor and some scepticism should be exercised in assigning values to the exponents which were explicitly defined with an assumption of tricritical scaling.

4.3 Stack polygons

Stack polygons are close relatives of partition polygons. A typical stack polygon is illustrated in Fig. 4.3. In this section I shall find the generating function of stack polygons, and examine it asymptotically. In addition, the generating function of spiral walks, a model closely related to stack polygons, will be computed. Stacks were counted by M.E. Wright [371], see also reference [297].

4.3.1 The area–perimeter generating function of stack polygons

A stack polygon can be constructed by laying slabs on top of one another in a heap (the slabs are not left- or right-adjusted, and no overhangs are allowed). If the stack is cut along the dotted line in Fig. 4.3, then the result is two partition polygons, where the left polygon has greater height than the right. This observation indicates that the generating function of stack polygons is related to that of partition polyons. From Fig. 4.1(a) and eqn (4.6) the generating function of a partition polygon of height h is

$$G_f(h; x, z) = \frac{z^{2h+2}x^h}{(z^2x; x)_h}. \tag{4.35}$$

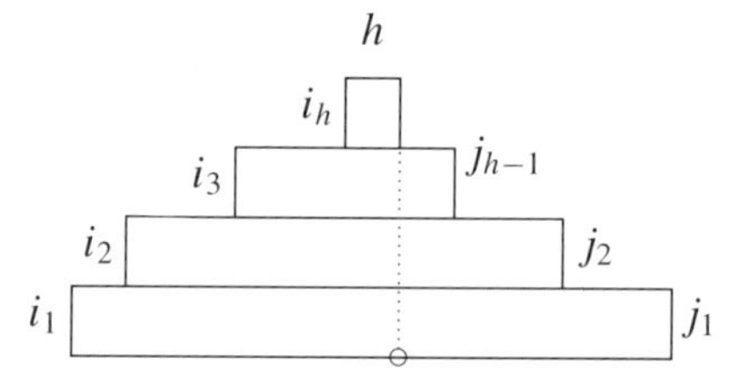

Fig. 4.3: A stack polygon of height area $\sum_k i_k + \sum_l j_l$ and perimeter $2h + 2i_1 + 2j_1$. If the polygon is cut along the dotted line, then the result is two partitions. On ther other hand, the generating function for stack polygons can be found by considering it to be constructed of horizontal layers (shown by the dashed lines) without overhangs.

The stack in Fig. 4.3 is either a partition of height h, or it is a partition of height h with a partition of height $m < h$ attached to it; this removes $2m$ boundary edges. Thus, the generating function for stack polygons of height h is

$$\begin{aligned} G_g(h; x, z) =& G_f(h; x, z) + \sum_{m=1}^{h-1} G_f(h; x, z) G_f(m; x, z) z^{-2m} \\ =& \frac{z^{2h+2} x^h}{(z^2 x; x)_h} + \sum_{m=1}^{h-1} \frac{z^{2m+2} x^m}{(z^2 x; x)_m} \frac{z^{2h+2} x^h}{(z^2 x; x)_h} z^{-2m}. \end{aligned} \tag{4.36}$$

It seems difficult to simplify for $G_g(h; x, z)$ in (4.36), but all this can be circumvented by using an approach similar to the direct approach used for partition polygons: sum over the i's and j's in Fig. 4.3. In that case

$$G_g(h; x, z) = \sum_{i_h=1}^{\infty} \sum_{\substack{i_{h-1}=i_h \\ j_{h-1}=0}}^{\infty} \sum_{\substack{i_{h-2}=i_{h-1} \\ j_{h-2}=j_{h-1}}}^{\infty} \cdots \sum_{\substack{i_1=i_2 \\ j_1=j_2}}^{\infty} z^{2h+2i_1+2j_1} x^{\sum_k i_k + \sum_l j_l}. \tag{4.37}$$

Evaluating the geometric series in the above from the right and summing over h gives the generating function for stacks:

$$G_g(x, z) = \sum_{h=1}^{\infty} \frac{z^{2h+2} x^h (1 - z^2 x^h)}{(z^2 x; x)_h^2}. \tag{4.38}$$

The singularities in $G_g(x, z)$ can be examined using the techniques of the previous section. The radius of convergence of $G_g(x, z)$ is best determined by D'Alembert's ratio test. For $z > 1$ there is a curve of double poles along the curve $x = 1/z^2$, and for $z \leq 1$ there a line of essential singularities along the line $x = 1$. Note that $G_g(1, z)$ is *finite* if $z < (\sqrt{5} - 1)/2$, and infinite for larger values of z. Thus, there is again a special point, and the phase diagram is similar to the phase diagram for partition polygons (Fig. 4.2), but with the special point moved to a different location. The point $(x, z) = (1, 1)$ is also a tricritical point. Since the critical λ-line has the same shape as the partition polygon model in the previous section, the value of the cross-over exponent is unchanged at $\phi = 1/2$. The free energy can be determined from the radius of convergence of $G_g(x, z)$;

$$\mathcal{F}_g(z) = \begin{cases} 0, & \text{if } z \leq 1; \\ 2 \log z, & \text{if } z > 1. \end{cases} \tag{4.39}$$

The density function is $\log \mathcal{P}_g(\epsilon) = \inf_{z>0}\{\mathcal{F}_g(z) - \epsilon \log z\} = 0$ if $0 < \epsilon < 2$. Thus

$$\mathcal{P}_g(\epsilon) = 1, \quad \text{if } \epsilon \in (0, 2). \tag{4.40}$$

The perimeter free energy is $-\log z_c(x)$, and by inverting $x_c(z)$

$$-\log z_c(x) = \begin{cases} (\log x)/2, & \text{if } x < 1; \\ \log[2/\sqrt{5} - 1], & \text{if } x = 1; \\ \infty, & \text{if } x > 1. \end{cases} \tag{4.41}$$

Thus the density of area per perimeter edge is $1/2$ if $x < 1$, and is infinite if $x > 1$.

4.3.2 *Asymptotic analysis of the stack generating function*

The phase diagram of a stack polygon includes a tricritical point, and as in the case of partition polygons, an asymptotic analysis of its generating function will show to what extent this model satisfies the assumptions of classical tricriticality. It will again appear that assumption (2.17) is not satisfied, but a value may be assigned to the cross-over exponent by examining the shift exponent.

If $z > 1$, then, as in the case of partition polygons, the inequalities $(z^2x, x)_1 \geq (z^2x, x)_n \geq (z^2x, x)_\infty$ imply that

$$\frac{1}{(z^2x, x)_1^2} \sum_{h=1}^{\infty} z^{2h+2} x^h (1 - z^2 x^h) \leq G_g(x, z) \leq \frac{1}{(z^2x, x)_\infty^2} \sum_{h=1}^{\infty} z^{2h+2} x^h, \tag{4.42}$$

and therefore

$$\frac{z^2}{(1 - z^2x)^3} \left(1 - \frac{z^2(1 - z^2x)}{(1 - z^2x^2)}\right) \leq G_g(x, z) \leq \frac{z^2}{(1 - z^2x)^3 (z^2x^2; x)_\infty^2}. \tag{4.43}$$

Thus, the divergence in $G_g(x, z)$ along the λ-line in this model is

$$G_g(x, z) \simeq \frac{z^2}{(1 - z^2x)^3}, \qquad \text{as } x \to (1/z^2)^- \text{ and } z > 1. \tag{4.44}$$

By eqn (2.14), $\gamma_+ = 3$. This result should be expected; the limiting shape of the stack polygon should be a ⊥-form, with three long and thin sequences of squares diverging as the limit as taken. Each such sequence gives a factor of $1/(1 - z^2x)$.

If $z = 1$, then a similar line of arguments as for partition polygons gives

$$G_g(x, 1) = \sum_{n=1}^{\infty} \frac{x^n(1 - x^n)}{(x; x)_n^2} = \sum_{n=1}^{\infty} \frac{x^n}{1 - x^n} \frac{(x^n; x)_\infty^2}{(x; x)_\infty^2}, \tag{4.45}$$

as $x \to 1^-$. Use the approximations in eqns (4.16) and (4.17). On substitution in the above, one obtains:

$$G_g(x, 1) \approx \left[\frac{-\log x}{2\pi}\right] e^{-2\pi^2/3 \log x} \sum_{n=1}^{\infty} x^n e^{2\mathcal{L}i_2(x^n)/\log x}. \tag{4.46}$$

The series will again be approximated by an integral, whose value will be estimated by a saddle-point approximation. Substituting $s = x^n$ gives the following:

$$\begin{aligned} \int_1^\infty x^n e^{2\mathcal{L}i_2(x^n)/\log x}\, dn &= -\frac{1}{\log x} \int_0^1 e^{2\mathcal{L}i_2(s)/\log x}\, ds \\ &= -\frac{1}{\log x} \int_0^1 s^{-2} e^{2(\mathcal{L}i_2(s) + \log s \cdot \log x)/\log x}\, ds. \end{aligned} \tag{4.47}$$

The saddle-point approximation is similar to that leading to eqn (4.23). Let $g(s) = \mathcal{L}i_2(s) + \log s \cdot \log x$, so that $g'(s) = (\log x - \log(1 - s))/s = 0$ if

$s = s_0 = 1 - x \in (0, 1)$. In addition, $g''(s_0) = 1/x(1-x)$, and thence, by (4.21),

$$\int_0^1 s^{-2} e^{2(\mathcal{L}i_2(s)+\log s \cdot \log x)/\log x} \approx \frac{\sqrt{\pi x}}{(1-x)} e^{2(\pi^2/6 - \mathcal{L}i_2(x))/\log x}. \tag{4.48}$$

On substition into (4.46), the approximation becomes

$$G_g(x, 1) \approx \frac{\sqrt{x/\pi}}{2(1-x)} e^{-2(\mathcal{L}i_2(x)+\pi^2/6)/\log x}. \tag{4.49}$$

The case of $z < 1$ remains. Approximating the sum by an integral and substituting $s = x^n$ as before, gives

$$\begin{aligned} G_g(x, z) \approx & \frac{z^2}{1 - z^2 x} e^{-2\mathcal{L}i_2(z^2 x)/\log x} \int_1^\infty (1 - z^2 x^n)^2 x^n z^{2n} e^{2\mathcal{L}i_2(z^2 x^n)/\log x} \, dn \\ \approx & \frac{-z^2}{(1 - z^2 x)\log x} e^{-2\mathcal{L}i_2(z^2 x)/\log x} I(x, z) \end{aligned} \tag{4.50}$$

where

$$I(x, z) = \int_0^1 (1 - z^2 s) e^{2(\mathcal{L}i_2(z^2 s)+\log z \cdot \log s)/\log x} \, ds \tag{4.51}$$

and where the substitution $s = x^n$ was made. A saddle-point approximation of the integral, with $g(s) = \mathcal{L}i_2(z^2 s) + \log z \cdot \log s$ indicates that $g'(s) = (\log z - \log(1 - z^2 s))/s = 0$ if $s_0 = s = (1-z)/z^2$; with s_0 in $(0, 1)$ if $(\sqrt{5} - 1)/2 < z < 1$. Also, $g(s_0) = \mathcal{L}i_2(1-z) + \log z . \log(1-z) - 2(\log z)^2$ and $g''(s_0) = z^3/(1-z)$. This approximation is valid only on the τ-line between the special point and the tricritical point (exactly as in the case of partition polygons). Using the approximation formula, and simplifying the result, finally gives

$$G_g(x, z) \approx \frac{z^3 \pi}{(1 - z^2 x)} \sqrt{\frac{z-1}{1-x}} e^{2(\mathcal{L}i_2(z)+\mathcal{L}i_2(z^2 x)+2(\log z)^2 - \pi^2/6)/\log x}. \tag{4.52}$$

This approximation also shows a line of essential singularities along the line $x = 1$ and $(\sqrt{5} - 1)/2 < z < 1$. The critical curve has the same shape as the critical curve for partition polygons (see Fig. 4.2). If the t-axis is tangent to the λ-line at the tricritical point, and the g-axis is chosen perpendicular to the τ-line at the tricritical point, then a point (x, z) will have (t, g)-coordinates given by eqn (4.30). The equations for the λ-line and the τ-line are

$$\begin{aligned} g_\lambda(t) = & -\frac{1}{3} \frac{t^2(9 + 2t\sqrt{3})}{3 + 2t\sqrt{3} + t^2} \approx -t^2 \quad \text{as } t \to 0, \\ g_\tau(t) = & 1 - 2t/\sqrt{3}. \end{aligned} \tag{4.53}$$

Noting that $(1-x) = (g - g_\tau(t))$ and $(1 - z^2 x) = (g - g_\lambda(t))$, the singularities in $G_g(x, z)$ become:

$$G_g(x,z) \approx \begin{cases} (g - g_\lambda(t))^{-3}, & \text{if } z > 1; \\ \dfrac{\sqrt{1-g}}{2g\sqrt{\pi}} e^{2(\mathcal{L}i_2(1-g)+\pi^2/6)/g}, & \text{if } z = 1; \\ \dfrac{\sqrt{\pi z^3(1-z)}}{(1-z^2x)\sqrt{g-g_\tau(t)}} e^{P(g,t)/(g-g_\tau(t))}, & \text{if } z < 1. \end{cases} \tag{4.54}$$

where

$$P(g,t) = 2(\mathcal{L}i_2(z) + \mathcal{L}i_2(z^2x) + 2(\log z)^2 - \pi^2/6). \tag{4.55}$$

Of course, all the z's and x's can be eliminated by using eqn (4.30), although that will give much more complicated expressions. Equation (4.53) gives a value for the shift exponents, and so the cross-over exponent can be obtained by using eqn (2.19), $\phi = 1/2$ (and so $\alpha = 0$). As for partition polygons, the generating function is infinite on the τ-line on approach to the tricritical point, and it appears that the scaling assumption in eqn (2.13) does not make much sense here.

4.3.3 *Spiral walks*

Let A be a self-avoiding walk from the origin in the square lattice. Then A is a spiral walk if every step along A is either in the same direction as the previous step, or follows

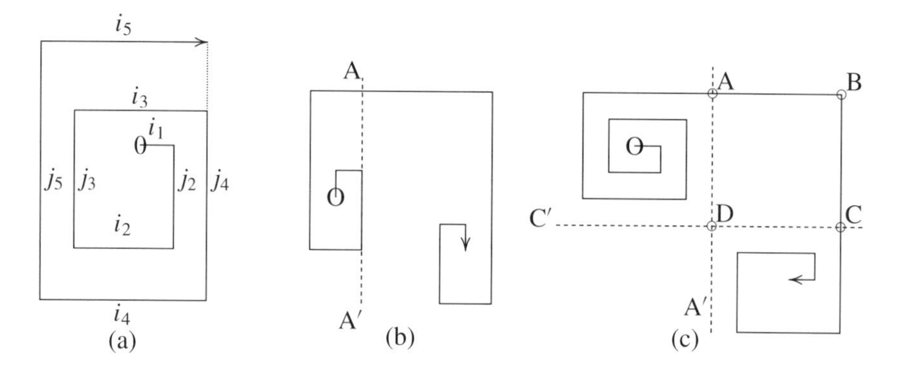

Fig. 4.4: (a) An out-spiral starts at the origin, and has the property that $i_1 < \cdots < i_5 < \cdots$ and $j_2 < j_3 < \cdots < j_5 < \cdots$. If $i_{m-1} = i_m$ (or $j_{m-1} = j_m$), when i_m or (j_m) is the last segment in the spiral, then this is a stopped spiral. (b) A spiral walk may be trapped if it is an out-spiral followed by an in-spiral. Such a trapped spiral always has a cutting line AA', which reduces it to a stopped spiral, followed by an in-spiral. (c) Some stopped spirals has two cutting line AA' and CC'. By reflecting the walk ABC to ADC, it appears that such a spiral walk can be decomposed into two stopped out-spirals, plus two extra edges.

a right turn after the previous step. Examples of some spiral walks are given in Fig. 4.4. An *out-spiral* is shown in Fig. 4.4(a); it has the property that the length of a given side is always strictly larger than the length of that side which precedes it by two in the spiral. If the arrow on an out-spiral is reversed, then an *in-spiral* is obtained. A spiral with side-lengths $i_1, j_2, i_2, j_3, \ldots, i_m, j_m$ is a *stopped spiral* if $j_m = j_{m-1}$ (or if $i_m = i_{m-1}$ if the last side in the spiral is i_m). A stopped spiral is illustrated in Fig. 4.4(a). There is a very natural map from out-spirals to stack polygons. Let the lengths of sides in an out-spiral be $i_1, j_2, i_2, j_3, \ldots, i_m, j_m$ with $i_1 < i_2 < \cdots < i_m$ and $j_2 < j_3 < \cdots < j_m$. Then build a stack polygon in the integer plane as follows. Let $\{-i_m, j_m\}$ be the coordinates of the end-points of the bottom slab in Fig. 4.4, $\{-i_{m-1}, j_{m-1}\}$ be the coordinates of the end-points of the second slab, and so on until the top slab has coordinates $\{-i_1, 0\}$. The area of the stack polygon is $\sum_{k>0} i_k + \sum_{k>1} j_k$, and this is also the length of the out-spiral. Not every stack polygon can be obtained in this way, since $i_1 < i_2 < \cdots < i_m$ and $j_2 < j_3 < \cdots < j_m$, and not every out-spiral corresponds to a stack polygon (this is because the length of the last side in an out-spiral is not restricted). However, it seems that it should be expected that this difference is not important as far as the thermodynamic properties of the model is concerned. Notice also that if the number of right turns in this out-spiral is $2m - 2$, and that the height of the stack polygon obtained is m.

Many spiral walks consist of an out-spiral followed by a line-segment of some length, which then becomes an in-spiral, as illustrated in Fig. 4.4(b). Such a spiral walk is a *trapped spiral*. A trapped spiral always has a *cutting line* which separates the trapped spiral such that the out-spiral is in one half-space and the in-spiral is in the other half-space. Moreover, the cutting line is chosen such that if the trapped spiral is cut into two spirals by it, then it gives a stopped out-spiral plus one edge on the one side, and an in-spiral on the other side. Some trapped spirals have two cutting lines, as illustrated in Fig. 4.4(c). These spirals can be related to stopped spirals by reflecting the walk ABC to ADC. This gives a map to two stopped spirals and two extra edges.

To find the generating function of spiral walks, focus first on a stopped out-spiral. If this spiral has n sides then its generating function can be directly computed using an approach similar to the one used for stack polygons. Indeed, the generating function for stopped out-spirals with exactly m sides is (where $m > 2$)

$$G_\perp(m;x) = \sum_{i_1=1}^{\infty}\sum_{j_1=1}^{\infty}\sum_{i_2=i_1+1}^{\infty}\sum_{j_2=j_1+1}^{\infty}\cdots$$

$$\cdots \sum_{i_{\lfloor m/2\rfloor}=i_{\lfloor m/2\rfloor-1}+1}^{\infty} \; \sum_{j_{\lceil m/2\rceil}=j_{\lceil m/2\rceil-1}+1}^{\infty} x^{i_{\lfloor m/2\rfloor}+\sum_k i_k+\sum_l j_l}. \tag{4.56}$$

If $m = 1$, then $G_\perp(1;x) = x/(1-x)$, and similarly, $G_\perp(2;x) = x^2/(1-x)^2$. Equation (4.56) can be evaluated to obtain

$$G_\perp(m;x) = \left[\frac{1-x}{x}\right]\prod_{i=1}^{\lceil m/2\rceil}\left[\frac{x^i}{1-x^i}\right]\prod_{j=1}^{\lfloor m/2\rfloor}\left[\frac{x^j}{1-x^j}\right]. \tag{4.57}$$

If an activity z is introduced conjugate to the number of sides in the stopped out-spiral, then the generating function of stopped out-spirals is

$$G_{\perp}(x,z) = 1 + \sum_{m=3}^{\infty} \frac{(1-x)z^m x^{\lceil m/2\rceil(\lfloor m/2\rfloor+1)}}{x(x;x)_{\lceil m/2\rceil}(x;x)_{\lfloor m/2\rfloor}} + \frac{z^2(1-2x)}{(1-x)^2}, \tag{4.58}$$

where the sum starts at $m = 3$ since every stopped out-spiral has at least three sides.

The generating function of an arbitrary out-spiral can similarly be computed. Using an expression similar to eqn (4.56) and simplifying it gives

$$G_T(x,z) = 1 + \left(\frac{x}{1-x}\right) \sum_{m=0}^{\infty} \frac{z^{m+1} x^{\lceil m/2\rceil(\lfloor m/2\rfloor+1)}}{(x;x)_{\lfloor m/2\rfloor}(x;x)_{\lceil m/2\rceil}}, \tag{4.59}$$

and this is related to a q-deformed Bessel function, see eqn (4.61) below. This shows a relationship with the generating function for partition polygons in eqn (4.7) as well. The generating function for out-spirals can be used to write down a generating function for spiral walks. This is done by noting that each spiral walk is either an out-spiral, or an in-spiral, or is a trapped spiral. The generating function of a trapped spiral is obtained by cutting it along a cutting line, see Fig. 4.4(b). This shows that $G_{\perp}(x,z)(x/z(1-x))G_{\perp}(x,z)$ generates trapped spirals, where the factor $x/(1-x)$ generates a line of length at least one connecting the spirals. Moreover, the division by z corrects for the number of sides in the trapped spiral. However, this expression double counts all trapped spirals with two cutting lines. This makes a correction necessary. The construction in Fig. 4.4(c) shows that any trapped spiral with two cutting lines can be mapped to two stopped spirals, and a connecting L-shaped walk of length at least two. Thus, these are generated by $G_{\perp}(x,z)[x^2/z^2(1-x)^2]G_{\perp}(x,z)$. So the full expression is

$$G_{\sqcup}(x,z) = 2G_T(x,z) + [G_{\perp}(x,z)]^2 \left[\frac{x(z(1-x)-x)}{z^2(1-x)^2}\right]. \tag{4.60}$$

There is an essential singularity in $G_{\perp}(x,z)$ as $x \to 1^-$, and the phase diagram is similar to that of partition and stack polygons.

Spiral walks in the square lattice were introduced by Z.A. Melzak [262], see also reference [299]. Their generating functions were studied to leading order by A.J. Guttmann and N.C. Wormald [157], see also references [20, 154]. The complete asymptotic expansion is due to G.S. Joyce [206]. Related work in other lattices can be found in references [206, 233, 235, 236, 239, 243, 359].

4.4 Staircase polygons

This model of polygons is also sometimes called *parallellogram polygons* [26, 69], and they have been studied at least since 1956 [340]. The area–perimeter generating function of staircase polygons can be found using a number of different techniques, and can be expressed in terms of Gaussian polynomials, or in terms of q-deformations of special functions. In particular, the q-deformed Bessel function defined by

$$J(p,q,t) = \sum_{m=0}^{\infty} \frac{(-t)^m q^{\binom{m}{2}}}{(p;q)_m (q;q)_m} \tag{4.61}$$

will appear in some cases. Five variants of techniques to compute the generating function of staircase polygons will be explored, and the perimeter generating function will be derived in two cases, while the area–perimeter generating function will be derived in three cases. These techniques have wider applicability, and they are used extensively in the calculation of generating functions of models of convex and partially convex polygons in enumerative combinatorics. The singularity structure of the generating function will be harder to discern in this model, compared with partition and stack polygons discussed in the last two sections. Thus, a less ambitious approach will be taken in the calculation of critical exponents and examination of scaling functions.

4.4.1 Perimeter generating function by counting paths

Staircase polygons of perimeter length $2n$ consist of two paths (these are directed walks, or staircase walks, see Section 3.4.3), each of length n, with a common first and last vertex. The internal vertices (other than the first and last) avoid each other, and the staircase walks step only in positive directions. A staircase polygon is illustrated in Fig. 4.5(a). To count these polygons, first assume that the first vertex on each of the staircase walks has coordinate $(0,0)$ and the last vertex has coordinate $(r, n-r)$. Then the number of staircase walks from $(0,0)$ to $(r, n-r)$ of length n is $\binom{n}{r}$ (since this is the number of ways of choosing r vertical edges from a total of n edges). The total number of pairs of staircase walks from $(0,0)$ to $(r, n-r)$ has generating function

$$g(z) = \sum_{n=0}^{\infty} \sum_{r=0}^{n} \binom{n}{r}^2 z^{2n} = \sum_{n=0}^{\infty} \binom{2n}{n} z^{2n} = \frac{1}{\sqrt{1-4z^2}}. \tag{4.62}$$

The combinatorial factor $\binom{2n}{n}$ shows a close relation to Catalan numbers defined by $C_n = \binom{2n}{n}/(n+1)$ (the integral of $g(\sqrt{z})$ should give the generating function of Catalan

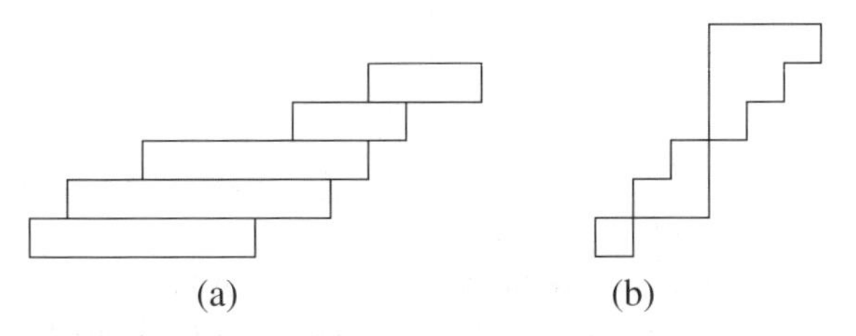

Fig. 4.5: (a) A staircase polygon may be imagined to be a stack of slabs, which are either right adjusted, or overhang on the right. (b) Two staircase walks with the same initial and terminal vertex will generate a sequence of staircase polygons (or double edges) joined in a chain at their top and bottom vertices.

numbers). Each conformation of the pairs of staircase walks is a chain with links which are staircase polygons or double edges. Also notice that the staircase walks can be interchanged, so that $g(z)$ double counts all the conformations which contributes to it, except for the single vertex conformation with $n = 0$ or the two double edge conformations. Let $h(z)$ be the generating function of a staircase polygon, or of a single double bond, double counted as in $g(z)$. Then

$$g(z) = 1 + h(z) + h(z)^2 + \cdots = \frac{1}{1 - h(z)}, \tag{4.63}$$

where the extra "1" counts the $n = 0$ contribution to $g(z)$. The perimeter generating function of staircase polygons, $G_\rho(z)$, is related to $h(z)$ by $G_\rho(z) = (h(z) - 2z^2)/2$, where z^2 is the weight of a double bond contribution to $h(z)$. Solving for $G_\rho(z)$ from $g(z)$ gives

$$G_\rho(z) = \left(1 - 2z^2 - \sqrt{1 - 4z^2}\right)\Big/2. \tag{4.64}$$

There is a branch point in $G_\rho(z)$ at $z_c = 1/2$. Notice that $G_\rho(z)$ can be written as

$$G_\rho(z) = \frac{2z^4}{1 - 2z^2 + \sqrt{1 - 4z^2}} \tag{4.65}$$

The factor in square brackets approaches zero as $z \to z_c^-$, and the singularity in $G_\rho(z)$ is a cusp. Comparison of eqn (4.64) to eqn (2.15) shows that the scaling assumption $G_\rho(z) \sim S_-(z_c - z)^{2-\alpha_u}$ is adequate for a description of a cusp singularity such as in eqn (4.64) (if S_- is assumed to be a constant and the analytic contribution to $G_\rho(z)$ is ignored). It will later be apparent that this critical point is the tricritical point of the area–perimeter generating function of staircase polygons, and that the approach is along the τ-line. Thus, the assignment $2 - \alpha_u = 1/2$ can be made, and here the exponent α_u is used (instead of γ_u) because this generating function appears to be similar to $G_p(x)$ in eqn (1.7).

4.4.2 *Perimeter generating function by algebraic languages*

An *algebraic language* is defined as follows. Let X be any finite non-empty set (it will be called the *alphabet*). Define X^* to be the free monoid generated by concatenation (the product in X^*) of the elements ("letters") in X. Elements of X^* are strings of letters in X, each called a "word", and the empty word e is the identity. A *language* is a subset of X^* (in other words, a set of words from all the possible words in X^*).

Let N and X be finite disjoint sets. A *production* on (N, X) is an ordered pair of elements $\alpha \in N$ and $\beta \in (N \cup X)^*$, written as $(\alpha \to \beta)$. For example, if $D \in N$ and $x \in X$, then $(D \to xDx)$ is a production on (N, X). An *algebraic grammar* can be defined by using productions on (N, X). In particular, it is a 4-tuple $\mathcal{G} = (N, X, P, s)$ where N and X are finite disjoint sets, P is a set of productions on (N, X) and s is an element in N. Starting at s, define the set of words in $(N \cup X)^*$ recursively by applying the productions in P. That is, if $(\alpha \to \beta)$ is a production, and $u\alpha v$ is in $(N \cup X)^*$, then the production indicates $u\alpha v \to u\beta v$, and so $u\beta v$ is generated by $\mathcal{G}$. The subset of words in X^* generated by $\mathcal{G}$ is called the *algebraic language* generated by $\mathcal{G}$, and it is indicated by $L(\mathcal{G})$.[4] If $w \in L(\mathcal{G})$ is a word, then $|w|_x$ is the number of occurrences of the letter x in w.

Example 4.1 (Dyck language) Let $N = \{D\}$, $X = \{O, C\}$, $P = \{(D \to ODCD), (D \to e)\}$ and $s = D$. Then

$$L(\mathcal{G}) = \{e, OC, OOCC, OCOC, \dots\}, \tag{4.66}$$

is the language generated on two letters. The Dyck language has two properties which are not hard to verify: (1) if u is a prefix (left factor) of a word w, then $|u|_O \geq |u|_C$, and (2) for any word w, $|w|_O = |w|_C$. □

Example 4.2 (Motzkin languages) Let $N = \{M\}$, $X = \{O, C, U\}$, $P = \{(M \to OMCM), (M \to UM), (M \to e)\}$. If $s = M$, the language of Motzkin words is found. This example has the same properties as the Dyck language above. If the set X is changed to $X = \{O, C, U, D\}$ and the production $(M \to DM)$ is added to P, then the two-coloured Motzkin words are found. □

Generating functions of algebraic languages are defined by enumerating all words of length n, and sending all letters to t: Let w_n be the number of words of length n. Then the generating function is $\sum_{n\geq 0} w_n t^n$. It is generally easier to use relations amongst words in order to derive the generating function; for example, every Dyck word w is either e, or is of the form $OwCw$. This can be made more precise by considering $L(\mathcal{G})$ to be a non-commutative algebra over integers with concatenation as the product. Then each word w is recursively defined by

$$w = e + OwCw. \tag{4.67}$$

Sending O and C to t and solving for w gives the generating function

$$G_D(t) = \left(1 - \sqrt{1 - 4t^2}\right) \Big/ 2t^2. \tag{4.68}$$

This factor $\sqrt{1 - 4t^2}$ already suggests a connection with staircase polygons and Catalan numbers. In a similar fashion, the generating functions of Motzkin languages are derived from

$$\begin{aligned} &w = e + Uw + OwCw, && \text{one coloured} \\ &w = e + Uw + Dw + OwCw, && \text{two coloured} \end{aligned} \tag{4.69}$$

respectively, which gives

$$G_1(t) = \left((1 - t) - \sqrt{1 - 2t - 3t^2}\right) \Big/ 2t^2, \tag{4.70}$$

$$G_2(t) = \left((1 - 2t) - \sqrt{1 - 4t}\right) \Big/ 2t^2, \tag{4.71}$$

---

[4] If $\mathcal{G} = (N, X, P, s)$ is an algebraic grammar, then N is called the *non-terminal set*, X the *terminal set*, P the *production rules* and s the *start element*.

as the respective generating functions. Notice that

$$G_D(t) = 1 + t^2 G_2(t^2), \tag{4.72}$$

so that a relation between two-coloured Motzkin words and Dyck words is expected. The morphism h (which factors over the product), defined by $h(O) = OO$, $h(C) = CC$, $h(U) = OC$ and $h(D) = CO$, defines a bijection $f(w) = Oh(w)C$ between two-coloured Motzkin words and Dyck words, as can be checked.

To find the generating function of staircase polygons, a bijection between conformations of the polygon and words in a particular language is established. This maps the generating function of the algebraic language on to the polygonal model, whose generating function can then be written down. In the case of simple polygonal models (for example, in the case of partitions, stacks and staircase polygons) this is not a very difficult process. This method is illustrated here for staircase polygons. In Fig. 4.6 a staircase polygon which is intersected by lines ℓ_i which have slope 135° with the positive horizontal, and which pass through points of integer coordinates, is shown. The staircase polygon is convex with respect to the lines ℓ_i, and is intersected at most twice by each. Generally, the lines cut any staircase polygon into a set of trapeziums and two triangles (which may be viewed as two degenerate trapeziums). These trapeziums are stretched and glued together to give the staircase polygon. Staircase polygons are mapped to an algebraic language by assigning a letter to each of these trapeziums as illustrated in Fig. 4.7. The first letter in any word obtained from a staircase polygon via the identification in Fig. 4.7 is an O, and the last letter in any such word is a C. These letters are superfluous, since they always correspond to the first and last triangles closing the stack of trapeziums into a staircase polygon. Therefore, a staircase polygon of length $2k + 2$ will be coded by a word of length $k - 1$.

The staircase polygon may be considered as two (non-intersecting paths) from its bottom vertex to its top vertex. Since the top staircase walk cannot intersect the bottom

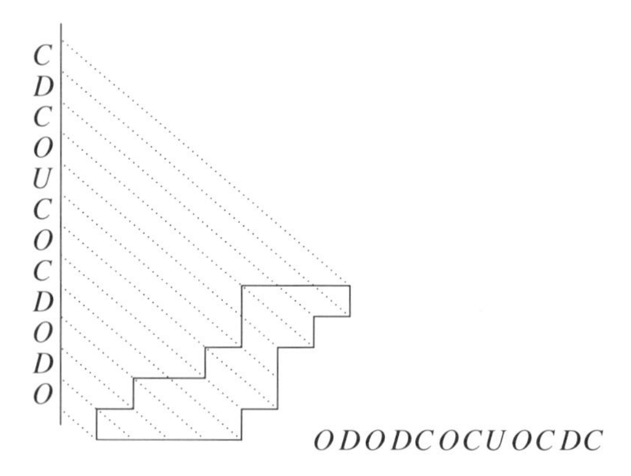

Fig. 4.6: A staircase polygon can be mapped to an algebraic language by cutting it into trapeziums along the dotted lines. There are four possible trapeziums, coded by O, C, D and U. The resulting sequence of letters is a two-coloured Motzkin word.

Fig. 4.7: The dissection of a staircase polygon into trapeziums in Fig. 4.5 results in a correspondence between staircase polygons and words in a two-coloured Motzkin algebraic language. The four letters $\{U, D, O, C\}$ in the algebraic language are associated with the trapeziums as indicated in Fig. 4.6.

staircase walk in a staircase polygon, except at the first and last vertices, the trapezium corresponding to an O (which moves the paths further apart), must occur more often than C in the prefix of any string of letters which corresponds to a staircase polygon. In other words, in any prefix u of the word for a given staircase polygon, $|u|_O > |u|_C$. In addition, the two paths always end at the same point, so that in any word w it is the case that $|w|_O = |w|_C$. In other words, these words are a subset of two-coloured Motzkin words (a subset because there is a strictly less than sign in the condition on the prefices). If the first O and last C are removed from the word, then the staircase polygon is still uniquely coded by a two-coloured Motzkin word. On the other hand; if w is a two-coloured Motzkin word of length $k-1$, then any prefix u of OwC satisfies $|u|_O > |u|_C$, and corresponds to a staircase polygon. Hence, there is a bijection between staircase polygons of length $2k+2$ and two-coloured Motzkin words of length $k-1$. Obviously, the generating function of words OwC, with w a two-coloured Motzkin word, is $t^2 G_2(t)$, and since each letter corresponds to two edges, the identification $t = z^2$ should give the perimeter generating function of staircase polygons:

$$G_\rho(z) = z^4 G_2(z^2) = \left(1 - 2z^2 - \sqrt{1-4z^2}\right)\Big/2 = z^2(G_D(z) - 1). \tag{4.73}$$

Attempts at using algebraic languages to find the area–perimeter generating function of staircase polygons has so far not been successful.

There is a natural correspondence between staircase polygons and random walks in an infinite sublattice of the square lattice. Suppose that O is a step in the positive Y-direction, and C a step in the negative Y-direction, while U is a step in the positive X-direction and D a step in the negative X-direction, all in the square lattice. Then any Motzkin word in these four letters (with the relations between the letters as above) is a random walk in the square lattice from the origin, and constrained to step only on vertices with non-negative Y-coordinates, and to terminate somewhere on the X-axis. The generating function of such walks is given by eqn (4.71). Similarly, the words in a Dyck algebraic language are words in the letters U and D where U occurs more often or as frequently as D in any prefix of a Dyck word. If U is a step on the integer line in the positive direction, and D is a step in the negative direction, then Dyck words correspond to random walks from zero (and terminating there) on the non-negative integers. The generating function of these random walks is given by eqn (4.68). This correspondence between Dyck words and random walks becomes even more interesting when it is noted that if the position of a random walk on the non-negative integers is plotted against time, then a staircase walk is obtained which is constrained to stay above the time axis; these walks are called Dyck paths (or Dyck walks), and they will be encountered again in Section 4.5.

The generating function of random walks on the integers from the origin is $1/(1-2z)$, if the steps are generated by z. Dyck walks can be turned into two-coloured Motzkin walks by first assuming that the Dyck walks are vertically oriented along the Y-axis. If horizontal steps are generated by $1/(1-2z)$, following every vertical step in a Dyck walk, then one should generate random walks on or above the X-axis starting from the origin, and terminating somewhere on the X-axis. In other words, is should be possible to obtain the generating function for staircase polygons from the Dyck generating function, as already seen in eqn (4.73). If $z \to z/(1-2z)$ in the Dyck generating function, and if an extra factor of $1/(1-2z)$ is supplied to generate horizontal steps at the first vertex, then one should expect that

$$G_2(t) = \frac{1}{1-2t} G_D \left(\frac{t}{1-2t} \right). \tag{4.74}$$

Direct computation shows that this relation is correct, and an alternative equality involving $G_D(t)$ and $G_2(t)$ is obtained.

The most important observation from the above is that models of staircase polygons are equivalent to random walks in subsets of the square lattice [110]. In some problems this can provide a very advantageous point of view. This will for example be of significant importance when the adsorption of staircase polygons is discussed. However, a simple connection between a model of (interacting) random walks in the square lattice, algebraic languages, and staircase polygons with area–perimeter activities, is not known. More on this can be found in the work of M.-P. Delest and G. Viennot [69].

4.4.3 *Area–perimeter generating function by a Temperley method*

The Temperley method [340] is an approach which classifies staircase polygons into a number of cases, and then uses this classification to derive a recursion relation. Let $H_n(x, z)$ be the generating function of a staircase polygon with base of length n, perimeter activity z and area activity x. The key to this method is the observation that a staircase polygon of base of unit length consists either of a single unit square (this gives a term xz^4), or of an arbitrary staircase polygon with a single square added to its base (see Fig. 4.8).

The addition of the single square gives a factor of xz^2 (two more edges, and one more unit of area) multiplied on to the generating function of the staircase polygon. From this

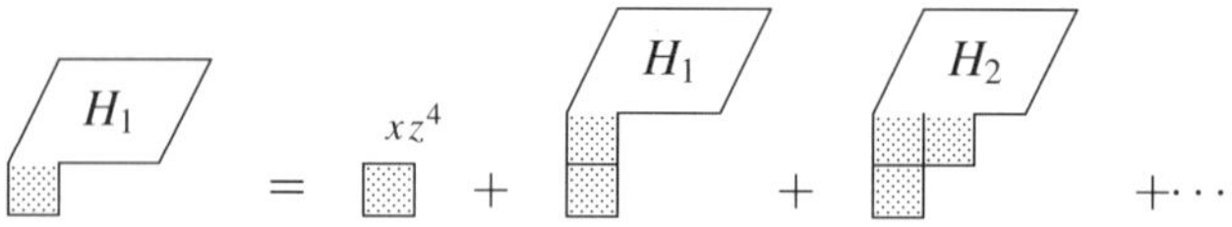

Fig. 4.8: A staircase polygon of unit base length is either a single square, or has its second row of length n for $n \geq 1$. This gives a recursion relation fo H_1.

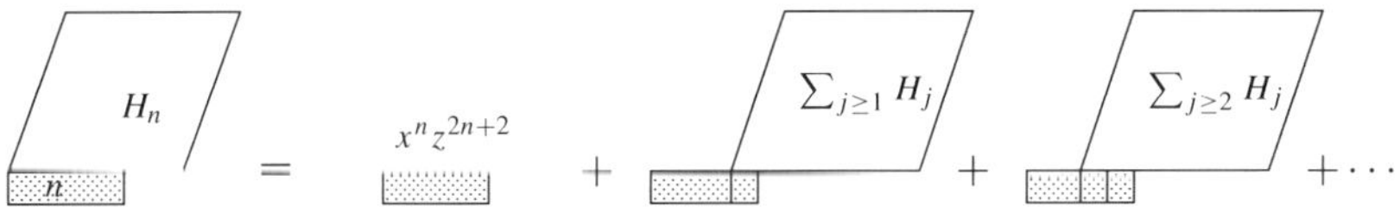

Fig. 4.9: A staircase polygon with base of length n can be decomposed in the classes illustrated here. This gives a recursion relation for H_n.

observation an equation can be written down for $H_1(x, z)$:

$$H_1 = xz^4 + xz^2 \sum_{j=1}^{\infty} H_j, \tag{4.75}$$

where the arguments of the H_j are suppressed. This situation becomes more difficult for $H_n(x, z)$: note that a staircase polygon of base n is in one of the classes illustrated in Fig. 4.9. The expansion in Fig. 4.9 indicates that a relation for $H_n(x, z)$ can be written down as follows:

$$\begin{aligned} H_n = x^n z^2 \Bigg(& z^{2n} + z^{2n-2} \sum_{j \geq 1} H_j + z^{2n-4} \sum_{j \geq 2} H_j + \cdots \\ & + z^2 \sum_{j \geq n-1} H_j + \sum_{j \geq n} H_j \Bigg), \end{aligned} \tag{4.76}$$

Inspection of this equation shows that it satisfies the recursion relation

$$H_{n+2} - x(1 + z^2(1 - x^{n+1}))H_{n+1} + x^2 z^2 H_n = 0. \tag{4.77}$$

This equation is difficult to solve, since it is a difference equation with non-constant coefficients. The general approach is to guess a very general form for the solution which includes a number of unknown functions, and to fix these unknown functions by using both the recursion and its initial conditions. The following assumption (see reference [304]) is general enough to solve eqn (4.77).

$$H_n(x, z) = A(x, y)\lambda^n \sum_{m=0}^{\infty} r_m(x) x^{mn}. \tag{4.78}$$

The normalizing function $A(x, y)$ will cancel on substitution in eqn (4.77), but it will have to be determined by initial or boundary conditions. The assumption contains several arbitrary functions, including λ and r_m, which (it is hoped) will be fixed by substitution into the recursion relation in eqn (4.77). Simplification then gives the following:

$$\begin{aligned} & \left[\lambda^2 x(1 + z^2)\lambda + x^2 z^2\right] \\ & + \sum_{m=1}^{\infty} \left[\lambda z^2 x^{1+m} r_{m-1}(x) + (\lambda^2 x^{2m} - \lambda(1 + z^2)x^{m+1} + x^2 z^2) r_m(x)\right] x^{mn} = 0. \end{aligned} \tag{4.79}$$

Consequently, λ and $r_m(x)$ are constrained by

$$\lambda^2 x(1+z^2)\lambda + x^2 z^2 = 0, \tag{4.80}$$

$$\lambda z^2 x^{1+m} r_{m-1}(x) + \left(\lambda^2 x^{2m} - \lambda(1+z^2)x^{m+1} + x^2 z^2\right) r_m(x) = 0. \tag{4.81}$$

Equation (4.80) has two solutions: $\lambda = x$ or $\lambda = xz^2$, while $r_m(x)$ can be found from the first-order recursion in eqn (4.81). Simplification gives

$$r_m(x) = \frac{(-\lambda)^m x^{m(m-1)/2}}{(\lambda; x)_m (\lambda z^{-2}; x)_m}. \tag{4.82}$$

This gives two possible solutions; one for each choice of λ. The correct solution is found by considering the behaviour of $H_n(x, z)$ at $x = 1$ and as $z \to 0$: then $H_n(1, z) \sim z^{2n+2}$, while $\lambda = 1$ or $\lambda = z^2$. Thus, the choice $\lambda = xz^2$ is appropriate; and

$$H_n(x, z) = A(x, z)(xz^2)^n \sum_{m=0}^{\infty} \frac{(-z^2)^m x^{m(m+1)/2} x^{mn}}{(xz^2; x)_m (x; x)_m}, \tag{4.83}$$

which in terms of a q-deformed Bessel function is given by (see eqn (4.61))

$$H_n(x, z) = A(x, z)(xz^2)^n J(xz^2, x, x^n z^2), \tag{4.84}$$

where $A(x, y)$ is an arbitrary function which cancelled on substitution of $H_n(x, z)$ in eqn (4.77). This function can be computed by noting that (at least formally) $H_0(x, z) = z^2$, and by substitution of $H_1(x, z)$ and $H_2(x, z)$ in eqn (4.77). Simplification of the resulting expression gives

$$\begin{aligned} [A(x, z)]^{-1} &= z^{-2} \sum_{m=0}^{\infty} \frac{(-z^2)^m x^{m(m+3)/2}}{(xz^2; x)_m (x; x)_m} (1 + z^2(1-x) - x^m z^2), \\ &= (z^{-2} + 1 + x) J(xz^2, x, x^2 z^2) - J(xz^2, x, x^3 z^2). \end{aligned} \tag{4.85}$$

Finally, the area–perimeter generating function of staircase polygons can be written down:

$$G_\rho(x, z) = \frac{z^2 \sum_{n=1}^{\infty} (xz^2)^n \sum_{m=0}^{\infty} \dfrac{(-z^2)^m x^{m(m+1)/2} x^{mn}}{(xz^2; x)_m (x; x)_m}}{\sum_{m=0}^{\infty} \dfrac{(-z^2)^m x^{m(m+3)/2}}{(xz^2; x)_m (x; x)_m} (1 + z^2(1-x) - x^m z^2)}. \tag{4.86}$$

There is an essential singularity in eqn (4.86) as $x \to 1^-$ with $z \leq 1$. I shall later show that the tricritical point is at $z_c = 1/2$ and $x_c = 1$. On the other hand, the curve $z^2 x = 1$ is a curve of poles in both $[A(x, z)]^{-1}$ and $\sum_{n \geq 1} H_n(x, z)$, but the possibility that these might cancel cannot be ruled out. A numerical investigation into eqn (4.86) can be found in reference [30].

4.4.4 Pólya's method for staircase polygons

The most natural method for obtaining the area–perimeter generating function for staircase polygons is due to G. Pólya [293]. The generating function for stack polygons (eqn (4.38)) can be decomposed into two partition polygons (eqn (4.36)); this observation hints at Polyà's method. *Gaussian binomial coefficients*, or the q-analogues of binomial coefficients, are defined in terms of q-notation by using the q-analogues of the factorial:

$$\begin{bmatrix} n \\ r \end{bmatrix} = \frac{(q;q)_n}{(q;q)_r (q;q)_{n-r}}. \tag{4.87}$$

These coefficients are in fact polynomials of degree $r(n-r)$ with positive integer coefficients. In other words, there are non-negative numbers $N_{nr\alpha}$ such that

$$\begin{bmatrix} n \\ r \end{bmatrix} = \sum_{\alpha=0}^{r(n-r)} N_{nr\alpha} q^{\alpha}. \tag{4.88}$$

In order to prove this fact, the following identities for Gaussian binomial coefficients are known [329], and they are not hard to verify:

$$\begin{bmatrix} n \\ r \end{bmatrix} = \begin{bmatrix} n-1 \\ r-1 \end{bmatrix} + q^r \begin{bmatrix} n-1 \\ r \end{bmatrix} \tag{4.89}$$

$$\begin{bmatrix} n \\ r \end{bmatrix} = \begin{bmatrix} n-1 \\ r \end{bmatrix} + q^{n-r} \begin{bmatrix} n-1 \\ r-1 \end{bmatrix}. \tag{4.90}$$

Lemma 4.3 *The Gaussian binomial coefficient $\left[{n \atop r} \right]$ is a polynomial of degree $r(n-r)$ with non-negative integer coefficients. In other words, there are integers $N_{nr\alpha} \geq 0$ such that*

$$\begin{bmatrix} n \\ r \end{bmatrix} = \sum_{\alpha=0}^{r(n-r)} N_{nr\alpha} q^{\alpha}.$$

Proof Notice that $\left[{n \atop 0} \right] = 1$ is a polynomial of degree 0. Suppose that $\left[{m \atop s} \right]$ is a polynomial with non-negative integer coefficients of degree $s(m-s)$ for all $m \leq n$ and $s \leq r$. By eqn (4.89)

$$\begin{aligned} \begin{bmatrix} n+1 \\ r \end{bmatrix} &= \begin{bmatrix} n \\ r-1 \end{bmatrix} + q^r \begin{bmatrix} n \\ r \end{bmatrix} \\ &= \sum_{\alpha=0}^{(r-1)(n-r+1)} N_{n(r-1)\alpha} q^{\alpha} + \sum_{\alpha=0}^{r(n-r)} N_{nr\alpha} q^{\alpha+r}. \end{aligned}$$

Inspection of the above shows that the highest power of q is $r(n-r+1)$ (as required), and that the coefficients $N_{(n+1)r\alpha}$ are sums over non-negative integers. In addition, eqn (4.89) indicates that

$$\begin{bmatrix} n \\ r+1 \end{bmatrix} = \begin{bmatrix} n-1 \\ r \end{bmatrix} + q^{r+1} \begin{bmatrix} n-1 \\ r+1 \end{bmatrix}.$$

Thus, if $\left[{n-1 \atop r+1} \right]$ is a polynomial with non-negative integer coefficients and of degree $(r+1)(n-r-2)$, then $\left[{n \atop r+1} \right]$ is a polynomial with non-negative integer coefficients and of degree $(r+1)(n-r-1)$. But notice that $\left[{r+1 \atop r+1} \right] = 1$. □

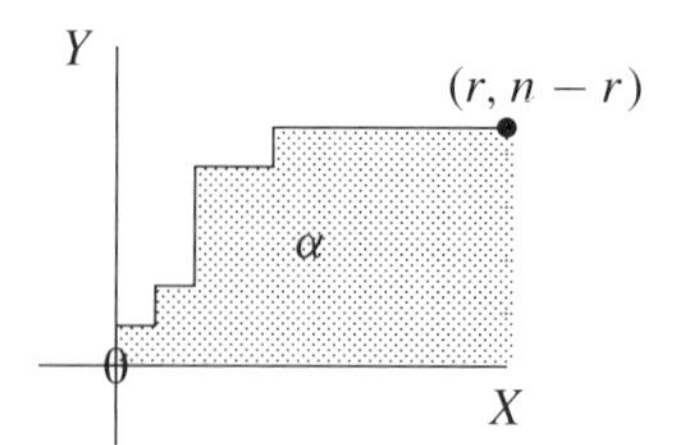

Fig. 4.10: A staircase walk from the origin to the point $(r, n - r)$ over as area α. The generating function of these walks is $\left[{n \atop r} \right]$.

The proof of Lemma 4.3 suggest something more profound about Gaussian binomial coefficients, which was observed by G. Polyà [293].

Lemma 4.4 *The number of square lattice walks of length n from* $(0, 0)$ *to* $(r, n - r)$, *and which enclose an area of size* α *between the* X*-axis, the line* $X = r$ *and the walk, (Fig. 4.10) is* $N_{nr\alpha}$.

Proof This is true for any α if $n = 1$ and $0 \le r \le 1$. Suppose that the lemma is true for any $m \le n$. Consider now $m = n + 1$. By eqn (4.90),

$$\sum_{\alpha=0}^{r(n+1-r)} N_{(n+1)r\alpha} q^{\alpha} = \sum_{\alpha=0}^{r(n-r)} N_{nr\alpha} q^{\alpha} + \sum_{\alpha=0}^{(r-1)(n-r+1)} N_{n(r-1)\alpha} q^{\alpha+n+1-r}.$$

If coefficients of q^{α} are compared, then

$$N_{(n+1)r\alpha} = N_{nr\alpha} + N_{n(r-1)(\alpha+r-n-1)}.$$

The meaning of this equation is illustrated in Fig. 4.11. By the induction hypothesis, the term $N_{nr\alpha}$ counts the number of walks of length n from $(0, 0)$ to $(r, n - r)$ enclosing area α. This can be turned into a subclass of walks counted by $N_{(n+1)r\alpha}$ if an extra (vertical) edge is appended from $(r, n - r)$ to $(r, n + 1 - r)$, without changing the area enclosed. In a similar fashion, the term $N_{n(r-1)(\alpha+r-n-1)}$ counts the number of walks from $(0, 0)$ to $(r - 1, n - r + 1)$, enclosing area $\alpha + r - n - 1$. If a horizontal edge is appended

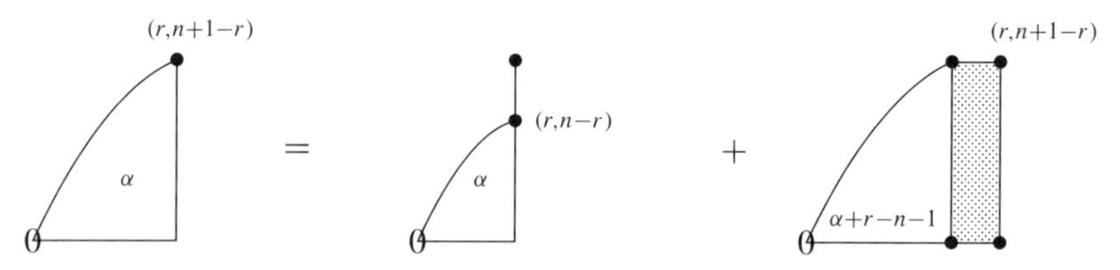

Fig. 4.11: Any staircase walk of length $(n+1)$ from the origin to $(r, n+1-1)$ enclosing area α either has its last step in a vertical direction, or a horizontal direction. In the first case the last edge can be deleted to fine $N_{nr\alpha}$, and in the second case one can truncate the last column to get $N_{n(r-1)(\alpha+r-n-1)}$.

from $(r-1, n-r+1)$ to $(r, n-r+1)$, then it encloses area α. Thus $N_{(n+1)r\alpha}$ counts the number of walks from $(0,0)$ to $(r, n-r+1)$ enclosing area α. □

Since the number of staircase walks with r horizontal edges and $n-r$ vertical edges is $\binom{n}{r}$, a corollary of Lemma 4.4 is

$$\binom{n}{r} = \sum_{\alpha=0}^{r(n-r)} N_{nr\alpha}. \tag{4.91}$$

Figure 4.11 suggests a close connection between Gaussian binomial coefficients and partition polygons. In particular, let the number of partitions of the integer α into r numbers, the largest of which is k, be $\rho_{k,r}(\alpha)$. Then $\rho_{k,r}(\alpha) = N_{(k+r)r\alpha}$. The generating function of $\rho_{k,r}(\alpha)$ for fixed k and r is

$$\sum_{\alpha=0}^{kr} \rho_{k,r}(\alpha) x^\alpha = \begin{bmatrix} k+r \\ r \end{bmatrix}_x, \tag{4.92}$$

where the subscript "x" on the Gaussian binomial coefficient indicates that x is the variable here (rather than q). This result can also be found in [329], where a different proof is given. If $x=1$, then eqn (4.92) is the number of partitions of perimeter $2(k+r)$. The generating function of all partitions is therefore

$$G_f(x,z) = \sum_{k=1}^{\infty}\sum_{r=1}^{\infty} \begin{bmatrix} k+r \\ r \end{bmatrix}_x z^{2(k+r)} = \sum_{k=1}^{\infty}\sum_{r=1}^{k-1} \begin{bmatrix} k \\ r \end{bmatrix}_x z^{2k}. \tag{4.93}$$

An expression for eqn (4.93) is given in eqns (4.6) or (4.7). This shows the connection between Gaussian binomial coefficients and partitions.

In order to apply the above to find a generating function for staircase polygons, consider first two staircase walks from $(0,0)$ to $(r, n-r)$. In general these will intersect in several places, generating a chain of double edges and staircase polygons which are attached at their lexicographic least and most vertices (Fig. 4.12). Let $H(x,z)$ be the

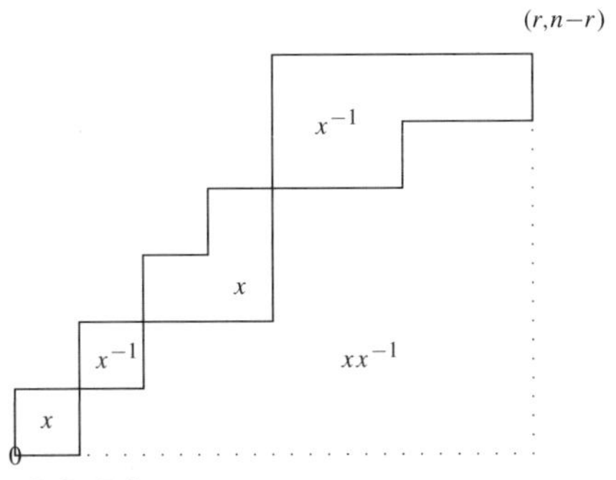

Fig. 4.12: The product $\begin{bmatrix} n \\ r \end{bmatrix}_x \begin{bmatrix} n \\ r \end{bmatrix}_{x^{-1}}$ generates a chain of staircase polygons with area fugacity which alternates between x and x^{-1} along the chain.

area–perimeter generating function of this string of double edges and staircase polygons (of infinite length). Each path and its enclosing area can be generated by a Gaussian binomial coefficient, but only the area *between* the two staircase walks should be included. This is achieved by generating one staircase walk with area activity x, and the second with area activity x^{-1}; in the area underneath both walks the activities cancel, and no area is generated. For staircase walks from $(0, 0)$ to $(r, n-r)$, the relevant terms in $H(x, z)$ are generated by

$$\begin{bmatrix} n \\ r \end{bmatrix}_x \begin{bmatrix} n \\ r \end{bmatrix}_{x^{-1}} = \begin{bmatrix} n \\ r \end{bmatrix}_x^2 x^{-r(n-r)}. \tag{4.94}$$

Summation over r and n should give $H(x, z)$; but care should be taken. Each staircase polygon in the string composing $H(x, z)$ has area once generated by x, or once by x^{-1}. Define

$$P_n(x) = \sum_{r=0}^{n} \begin{bmatrix} n \\ r \end{bmatrix}_x^2 x^{-r(n-r)}, \tag{4.95}$$

and observe the uncanny resemblance of this expression to the generating function of a pair of staircase walks in eqn (4.62), and thus to the generating function of Catalan numbers. Then

$$P_n(x) = P_n(x^{-1}), \tag{4.96}$$

and

$$H(x, z) = \sum_{n=1}^{\infty} (P_n(x) + P_n(x^{-1}))z^{2n}/2 = \sum_{n=1}^{\infty} P_n(x) z^{2n}. \tag{4.97}$$

Observe that if $G_\rho(x, z)$ is the area–perimeter generating function of a single staircase polygon, then

$$\begin{aligned} H(x, z) &= (2z^2 + G_\rho(x, z)) + (2z^2 + G_\rho(x, z))^2 + \cdots \\ &= \frac{2z^2 + G_\rho(x, z)}{1 - (2z^2 + G_\rho(x, z))}, \end{aligned} \tag{4.98}$$

from which an expression for $G_\rho(x, z)$ can be written down:

$$G_\rho(x, z) = 1 - 2z^2 - \frac{1}{1 + H(x, z)}. \tag{4.99}$$

This result agrees with that of G. Polyà, with the exception of the term $2z^2$, which is subtracted here to remove the two degenerate staircase polygons consisting of only two edges and no area.

4.4.5 Area–perimeter generating function from a functional equation

This method is due to T. Prellberg and R. Brak [295] and is a variant of the Temperley method. Its essential step is the classification of polygons into a finite number of cases, from which a functional recursion is derived. Before this is done for staircase polygons, consider instead a single column of squares in Fig. 4.13. If the generating function is indicated by a column of area generated by x, width generated by z, and height generated by y, then the generating function $G_c(x, y, z)$ contains terms which are either a single square (generated by xyz), or of a column with a single square appended to its top.

Thus,

$$G_c(x, y, z) = xyz + yG_c(x, y, zx), \tag{4.100}$$

and notice that z is replaced by zx to indicate that each edge in the base of the column (there is only one) generates one square at the top. This gives a functional equation which can be solved by assuming an infinite series solution. The answer is

$$G_c(x, y, z) = \frac{zxy}{1 - xy}. \tag{4.101}$$

The last observation is that the perimeter must be introduced. Since each z and y corresponds to two edges, they should be squared. Vertical and horizontal edges are treated as the same, so put $y = z$ as well. Thus, the generating function is

$$G_c(x, z) = \frac{xz^4}{1 - z^2x}. \tag{4.102}$$

In order to derive the generating function for staircase polygons, the steps above will be followed, but there are several technical difficulties. First observe that a given staircase polygon can be "fattened up" in the vertical direction by adding a layer of squares (as above, assume that the "vertical (height) activity" is y, the "horizontal activity" is z and the area activity is x). If $G_\rho(x, y, z)$ is the generating function for staircase polygons, then the fattened staircase polygons have generating function $yG_\rho(x, y, zx)$ (again, the number of new squares added is given by the horizontal dimension, and the extra factor of y takes care of the fact that the height has increased by one). In Fig. 4.14 it is indicated that all staircase polygons are counted by one of four possible cases. Either it is a single

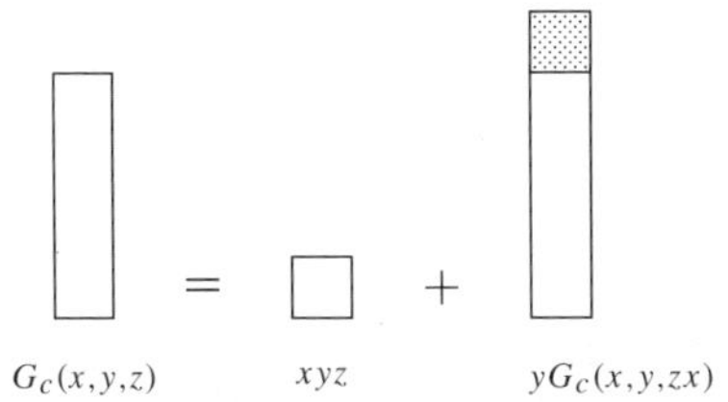

$G_c(x,y,z)$ = xyz + $yG_c(x,y,zx)$

Fig. 4.13: A column can be inflated by adding a new square on its top. Now observe that every column is either the single square, or an inflated column.

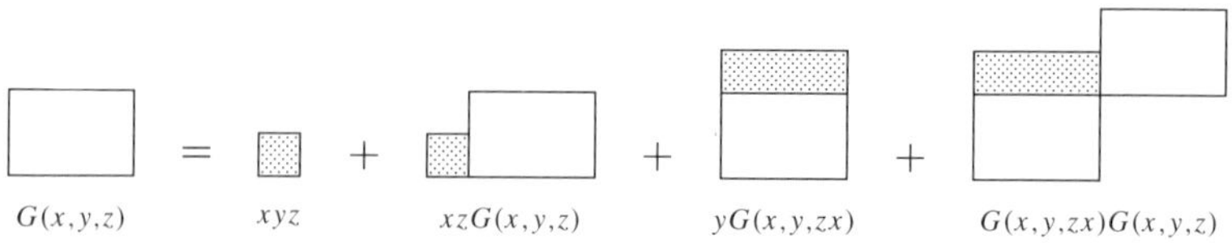

$G(x,y,z)$ = xyz + $xzG(x,y,z)$ + $yG(x,y,zx)$ + $G(x,y,zx)G(x,y,z)$

Fig. 4.14: The decomposition of staircase polygons into inflated staircase polygons: a staircase polygon is either a single square, or its first column is a single square, or it is inflated, or it can be factored into an inflated staircase polygon and a staircase polygon.

square, or its left-most column has height one. If it is not one of these, then its left-most column has height at least two, which means that either it is a "fattened" staircase polygon, or it is composed of a "fattened" staircase polygon concatenated to an arbitrary staircase polygon (that is, there is a first narrow "waist" in the polygon).

Notice that the first two terms are those with a first column of height one, and the last two terms are those with first column of height greater than one (and therefore can be obtained by inflating staircase polygons in height by one square). From Fig. 4.14 the following relation can be written down:

$$\begin{aligned} G_\rho(x, y, z) &= xyz + xzG_\rho(x, y, z) + yG\rho(x, y, zx) + G_\rho(x, y, zx)G_\rho(x, y, z) \\ &= (G_\rho(x, y, zx) + xz)(G_\rho(x, y, z) + y). \end{aligned} \tag{4.103}$$

Since only the z-argument of $G_\rho(x, y, z)$ is important here, the x-argument and y-arguments will be suppressed in $G_\rho(x, y, z)$, and just $G_\rho(z)$ will be written. It is difficult to solve (4.103). T. Prellberg and R. Brak (1995) observe that a linear functional equation of the form

$$\sum_{k=0}^{N} \alpha_k H(x^k z) + zH(xz) = 0 \qquad \text{with} \sum_{k=0}^{N} \alpha_k = 0, \tag{4.104}$$

where the α_k are constants independent of z, has a solution (regular at $z = 0$) given by

$$H(z) = \sum_{n=0}^{\infty} \frac{(-z)^n x^{n(n-1)/2}}{\prod_{m=1}^{n} \Lambda(x^m)} \qquad \text{where } \Lambda(t) = \sum_{k=0}^{N} \alpha_k t^k. \tag{4.105}$$

In order to use eqn (4.105) to solve for $G_\rho(z)$, first assume that

$$G_\rho(z) = \alpha \frac{F(xz)}{F(z)} - y. \tag{4.106}$$

Substitution of (4.106) in (4.103), followed by simplification, gives

$$\alpha^2 F(x^2 z) + \alpha(zx - 1 - y)F(xz) + yF(z) = 0. \tag{4.107}$$

If it is assumed that $F(z)$ is given by an infinite series such as in eqn (4.105), then $F(z) = 1 + \cdots$, and the $n = 0$ term in eqn (4.107) becomes

$$\alpha^2 + \alpha(xz - 1 - y) + y = 0, \tag{4.108}$$

and $\alpha = y$ is the only choice which cancels the terms in y (the terms in z will be cancelled by higher-order terms). Thus, with $\alpha = y$ in eqn (4.107), it appears that

$$\sum_{k=0}^{2} \alpha_k F(x^k z) + zF(xz) = 0, \tag{4.109}$$

and $F(z)$ satisfies eqn (4.104) if $\alpha_0 = 1/x$, $\alpha_1 = -(1+y)/x$ and $\alpha_2 = y/x$. Note that these α_i sum to zero, and that $\Lambda(t) = (1-t)(1-yt)/x$. Consequently, $F(z)$ is a q-deformed Bessel function given by

$$F(z) = \sum_{n=0}^{\infty} \frac{(-xz)^n x^{n(n-1)/2}}{(x;x)_n (yx;x)_n} = J(xy, x, xz). \tag{4.110}$$

Lastly, since each y and each z counts two edges, replace them by their squares, and put $y = z$ (since the horizontal and vertical perimeter edges are treated the same). This gives the area–perimeter generating function:

$$G_\rho(x,z) = z^2 \left[\frac{\sum_{n=0}^{\infty} \frac{(-z^2)^n x^{n(n+3)/2}}{(x;x)_n (z^2x;x)_n}}{\sum_{n=0}^{\infty} \frac{(-z^2)^n x^{n(n+1)/2}}{(x;x)_n (z^2x;x)_n}} - 1 \right] = z^2 \left[\frac{J(xz^2, x, x^2z^2)}{J(xz^2, x, xz^2)} - 1 \right]. \tag{4.111}$$

A great advantage of the functional recursion is that it can be used to obtain the tricritical exponents. The perimeter generating function (this is obtained when $x = 1$) in eqn (4.64) implies that $2 - \alpha_u = 1/2$ (see eqns (2.15) and (4.65)). To find the remaining exponents, the approach in reference [295] can be followed. In analogy with the results for partition polygons and stack polygons (eqns (4.23) and (4.49)), assume that $G_\rho(x,y,z)$ has an asymptotic expansion in $-\log x$ (or, equivalently, in $(1-x)$). Then

$$G_\rho(x,y,z) = \sum_{n=0}^{\infty} S_n(y,z)(-\log x)^n. \tag{4.112}$$

Inserting this expansion into (4.103) gives

$$S_0(y,z) = \left(1 - y - z - \sqrt{1 - 2y - 2z + y^2 - 2yz + z^2}\right)\Big/2, \tag{4.113}$$

and

$$S_n(y,z) = \frac{(y + S_0(y,z)) \sum_{m=1}^{n} S_{n-m}^{(m)}(y,z) x^m/m!}{1 - y - z - 2S_0(y,z)} + \frac{\sum_{k=1}^{n-1} S_k(y,z) \sum_{m=0}^{n-k} S_{n-m-k}^{(m)}(y,z) x^m/m!}{1 - y - z - 2S_0(y,z)}. \tag{4.114}$$

An examination of $S_1(y,z)$ shows that $S_1(z,z) \simeq 1/(1-4z)$, and since $S_0(z,z) \simeq \sqrt{1-4z}$ the value of the gap exponent is obtained: $\Delta = 3/2$. Therefore, the cross-over

Table 4.1: Exponents of area–perimeter staircase polygons

ϕ	α	α_t	α_u	ν_t	y_t	γ_+
$\frac{2}{3}$	$\frac{1}{2}$	$\frac{5}{3}$	$\frac{3}{2}$	$\frac{1}{6}$	4	$\frac{1}{2}$

exponent has value $2/3$, and from eqn (2.18) one notes that $2-\alpha_t = 1/3$. The remaining exponents can be found by using eqns (2.29), (2.36) and (2.37). These gives $\nu_t = 1/6$ and $y_g = 6$, while $y_t = 4$ and $\alpha = 1/2$.

The τ-line is given by $x = 1$ so that $G_\rho(x, z) = S_0(z^2, z^2) = (1-2z^2-\sqrt{1-4z^2})/2$, which was computed previously. Thus, $G_\rho(x, z)$ is finite along the τ-line, and moreover, $[-xdG_\rho(x, z)/dx]|_{x=1}$ is also finite along this line. In other words, the τ-line is a line of essential singularities, and the transition along it can be interpreted as a condensation of "droplets" along a line of first order transitions [5, 184]. This characterizes the tricritical nature of the critical point, with the exponents given in Table 4.1. Along the λ-line the staircase polygon will be deflated, and be in the class of staircase walks. From this it may be guessed that $\gamma_+ = 1/2$, from eqn (4.62).

4.4.6 Other models of convex polygons

A convex polygon is a polygon in the square lattice which intersects any vertical or horizontal line which cut it exactly twice. An example of a convex polygon is given in Fig. 4.5(a). All staircase polygons are convex. Convex polygons of various kinds have been studied extensively in the literature [14, 215, 216, 217, 218, 306]. For more recent work, see for example references [24, 23, 235].

The number of convex polygons of length $2n + 8$ is known to be given by

$$p_{2n+8} = (2n + 11)4^n - 4(2n + 1)\binom{2n}{n}. \tag{4.115}$$

The combinatorial term indicates again a close relation to Catalan numbers, just as was observed in the case of staircase polygons and walks. The result in eqn (4.115) is due to M.-P. Deleste and V. Viennot [69], and follows from an algebraic languages approach. A second proof of this fact can be found in reference [214]. The perimeter generating

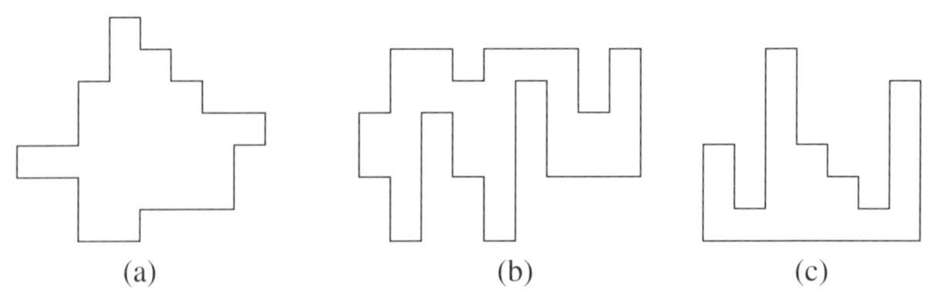

Fig. 4.15: (a) A convex polygon. (b) A column-convex polygon. (c) A histogram polygon.

function of convex polygons can be derived from eqn (4.115) by noting that $p_4 = 1$ and $p_6 = 2$ (see references [24, 152, 153, 235]). The result is

$$G(z) = \frac{z^4(1 - 6z^2 + 11z^4 - 4z^6)}{\sqrt{1-4z^2}^{\,4}} - \frac{4z^8}{\sqrt{1-4z^2}^{\,3}}. \tag{4.116}$$

The dominant singularity is given in the first term, and a comparison to eqn (2.15) shows that $\gamma_u = 2$ in this model. The second term is also singular, but is dominated by the first term in the scaling regime; it is referred to as a *confluent correction*.

Column-convex polygons are the most general class of polygons studied in the context of generating functions. These polygons intersect any vertical line in the square lattice exactly twice, and an example is given in Fig. 4.15(b). A more restricted model of column-convex polygons is the model of "bar-graph" or "histogram" polygons (an example of such a polygon is given in Fig. 4.15(c)). The area–perimeter generating function of histogram polygons can be found from a functional equation (see Section 4.4.5) [295]. This is done by noting that histogram polygons are in one of the cases illustrated in Fig. 4.16. The generating function for column-convex polygons was derived by M.-P. Delest [68], and the scaling behaviour of partially convex polygon models were examined in reference [35]. The area and number-of-columns generating function was determined by V. Domocos [83], and the generating functions of higher dimensional generalizations were considered in [25].

From Fig. 4.16 it is seen that the generating function of histogram polygons satisfies the following functional recursion equation (where x generates area, y is conjugate to one-half the number of vertical edges and z is conjugate to one-half the number of horizontal edges (that is, z is conjugate to the width of the histogram polygon):

$$G(x, y, z) = y\,G(x, y, xz) + [1 + G(x, y, xz)]xz[y + G(x, y, z)] \tag{4.117}$$

This functional recursion can be developed into an infinite fraction by

$$G(x, y, z) = \frac{y}{xz}\left[\frac{1}{1 - xz(1 + G(x, y, xz))} - (1 + xz)\right]. \tag{4.118}$$

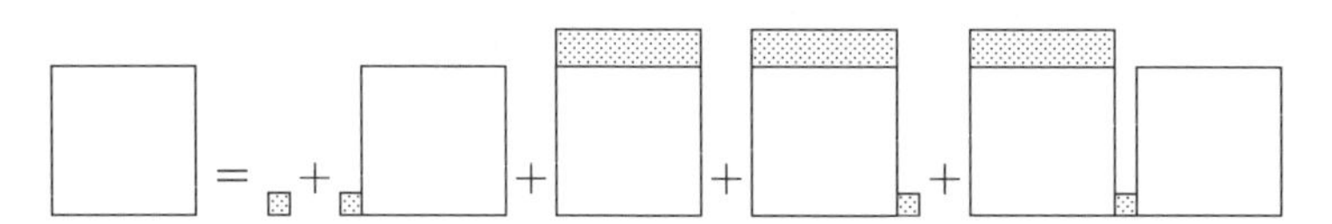

Fig. 4.16: A functional relation for the generating function of histogram polygons can be found by classifying all these polygons in one of the cases above. This graphical equation may be interpreted as follows. Every histogram polygons is either a single square or starts with a single square, otherwise it has thickness at least two everywhere, and it may be followed by nothing, or a single square, or a single square and then followed by another histogram polygon. This analysis will give a functional relation involving the area–perimeter generating function of histogram polygons.

The analysis of this generating function was done by T. Prellberg and R. Brak [295], and the critical exponents are the same as for staircase polygons. Models of column-convex polygons were also studied by using a functional equation for the generating functions, and in those cases the same set of exponents as for staircase polygons was obtained. A remarkable effort by A.J. Guttmann and R. Brak [30] produced an explicit formula for the area–perimeter generating function of column-convex polygons, in terms of q-deformed Bessel functions.

4.5 The adsorption of staircase walks on the main diagonal

Consider a directed or staircase walk (constrained to step only in the positive directions) in the square lattice which has its first vertex fixed at the origin, and is required to step only on vertices on or above the main diagonal. If an activity conjugate to the number of visits to the main diagonal is included, then this may be viewed as a directed model of a two-dimensional linear polymer adsorbing on an impenetrable wall [192]. The adsorption problem was considered in Section 3.4.6, and in this section the generating function of adsorbing directed walks will be considered from a variety of different points of view. There are numerous studies of the adsorption of directed and partially directed walks in the literature. Some examples can be found in references [29, 60, 124, 134, 263, 270, 302, 304, 338, 349, 362].

4.5.1 Staircase walks above the main diagonal

The generating function of (simple) random walks in the d-dimensional hypercubic lattice is given by

$$G(x) = \frac{1}{1-2dx}. \tag{4.119}$$

A staircase walk is a random walk in the square lattice, constrained to give steps only in the two positive directions. The generating function of these walks is

$$G(x) = \frac{1}{1-2x}, \tag{4.120}$$

and the similarity to the generating functions of simple random walks in eqn (4.119) is not accidental. The easiest connection is made by rotating a staircase walk (starting from the origin) through 45° as in Fig. 4.17.

The number of staircase walks from the origin to the vertex with coordinates $(r, n-r)$ is $\binom{n}{r}$. Since a model of staircase walks which interacts with the diagonal will be considered, walks of length $2n$ which end at the point (n, n) will be studied. The number of these walks is $\binom{2n}{n}$, and they are generated by eqn (4.62). An *excursion* is a staircase walk which starts at the origin and terminates on the main diagonal, but with all internal vertices above the main diagonal (see Fig. 4.17). Excursions can be mapped into an algebraic language by representing each vertical edge by a U and each horizontal edge by an R. Let w be a word in the letters $\{U, R\}$ which corresponds to an excursion. Removing the first and last edges in the excursion corresponds to removing

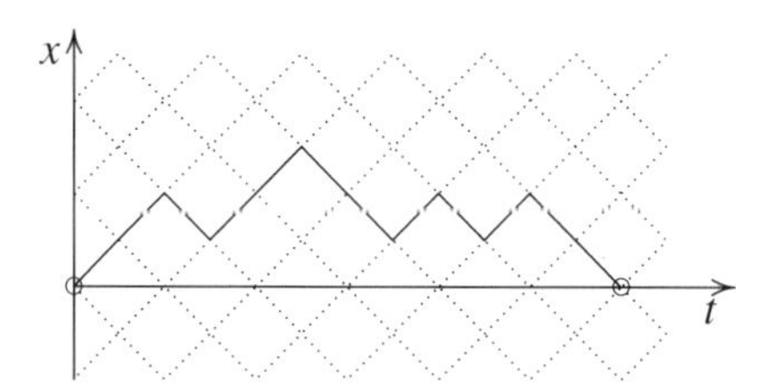

Fig. 4.17: If this staircase walk is projected on the x-axis, then a random walk is obtain. This staircase walk is also an excursion: it starts and terminates on the t-axis, and is otherwise constrained to be above it.

the leading U, and the trailing R in w to obtain a new word w'. This word has the following properties. The number of letters in it satisfies the constraints $|w'|_U = |w'|_R$ and for any prefix u of w', $|u|_U \geq |u|_R$. Thus, excursions without their first and last edges correspond to the words in the Dyck algebraic language. If the first and last edges in an excursion are removed and the resulting walk is translated on to the main diagonal, then a staircase walk which starts and terminates in the main diagonal, and which is constrained not to step below the main diagonal, is obtained. These are Dyck walks [134, 270], and they are in one-to-one correspondence with random walks on the non-negative integers which start and terminate at zero. The generating function of Dyck walks is given by eqn (4.68):

$$G_D(x) = \frac{(1-\sqrt{1-4x^2})}{2x^2} = \frac{2}{1+\sqrt{1-4x^2}}, \tag{4.121}$$

while the generaing function of an excursion is

$$G_e(x) = \frac{(1-\sqrt{1-4x^2})}{2} = \frac{2x^2}{1+\sqrt{1-4x^2}}. \tag{4.122}$$

Notice that this generating function is also the generating function of a simple random walk on the non-negative integers, which starts and terminates at the origin.[5] Excursions are also closely related to the gambler's ruin problem [110], where a gambler bets coins on the roll of a fair two-sided dice. If the gambler goes bankrupt when he runs out of coins, then this event corresponds to the return of a random walk on the non-negative integers to the origin; see Fig. 4.17.

[5] Staircase walks above the line $y = 2x$ were counted by I.M. Gessel [134]. These arguments can be generalized to count staircase walks above or on the line $y = x/p$ with both end-points in the line. Notice that the length of such a walk is always $m(p+1)$ for $m \geq 0$. If an algebraic language is used to enumerate these walks, then any such walk corresponds either to the empty word e, or it first steps in the U direction and then is a staircase walk above the line $y = x/p+1$ until it visits a vertex a_1 in the line $y = x/p+(p-1)/p$ for the first time. It then becomes a staircase walk from from a_1 until it hits a vertex a_2 in the line $y = x/p+(p-2)/p$, and so on, until it finally finds its end-point in the line $y = x/p$. This gives words of the form $w = e + Uw(Rw)^p$, and the generating function satisfies the equation $G(x) = 1 + x^{p+1}[G(x)]^{p+1}$. A real root of this equation will give the generating function. This argument becomes more complicated for staircase walks above the line $y = qx/p$ where q and p are relative primes.

It is possible to find the generating functions in eqn (4.121) from a completely different perspective. If a staircase walk from the origin to the point (n, n) has l left turns, then the coordinates of these left turns can be chosen in $\binom{n}{l}\binom{n}{l}$ ways (choose the X-coordinates in $\binom{n}{l}$ ways, and the Y-coordinate in $\binom{n}{l}$ ways). Let q be a staircase walk which visits a vertex in the subdiagonal $y = x - 1$, and let v be the last vertex in this subdiagonal that it visits. A reflection of the last segment of the staircase walk (from vertex v to (n, n)) through the line $y = x - 1$ gives a staircase walk which ends in the vertex with coordinates $(n + 1, n - 1)$. Conversely, a staircase walk which ends in the vertex with coordinates $(n+1, n-1)$ must visit a vertex v in the subdiagonal $y = x - 1$ a last time, and so can be reflected (through the line $y = x - 1$) to a staircase walk which ends in the vertex (n, n) and which contains a vertex below the main diagonal. The number of staircase polygons with l left turns which ends in the point with coordinates $(n+1, n-1)$ is $\binom{n+1}{l}\binom{n-1}{l}$, and so the number of staircase walks from the origin to the point (n, n), and which remains on or above the main diagonal (these are Dyck walks) is

$$C_n = \sum_{l=0}^{n} \binom{n}{l}\binom{n}{l} - \sum_{l=0}^{n-1} \binom{n+1}{l}\binom{n-1}{l} = \frac{1}{n+1}\binom{2n}{n}. \tag{4.123}$$

C_n is Catalan's number. But notice now that

$$G_D(x) = \sum_{n=0}^{\infty} C_n x^{2n} = \left(1 - \sqrt{1 - 4x^2}\right) \Big/ 2x^2, \tag{4.124}$$

as expected (see eqn (4.68)).

The generating function of staircase walks on or above the main diagonal with a free end-point (see Fig. 4.18) can also be found. Any such walk will visit the main diagonal a last time at a vertex v. The section of the walk between the origin and the vertex v is a Dyck walk while the remaining part of the walk is called its *tail*. If the first edge in the tail is removed, then a staircase walk above or on the superdiagonal $y = x + 1$ with

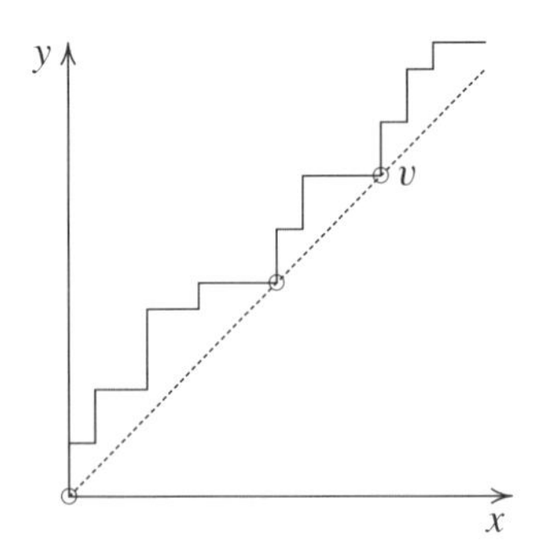

Fig. 4.18: A staircase walk above the diagonal with a free end-point. Such a walk visits the diagonal a last time at vertex v, and then wanders off into the space above the diagonal. The subwalk from the origin to the last visit v on the diagonal is a Dyck walk.

a free end-point is obtained. This can be translated one step in the negative Y-direction to obtain (again) a staircase walk on or above the main diagonal with a free end-point. Thus, the generating function of a staircase walk on or above the main diagonal with a free end-point can be decomposed as

$$G_t(x) = G_D(x) + G_D(x)xG_t(x), \tag{4.125}$$

since it is either a Dyck walk, or it is a Dyck walk followed by a tail generated by $G_t(x)$. Thus

$$G_t(x) = \frac{1 - 2x - \sqrt{1 - 4x^2}}{2x(2x - 1)}. \tag{4.126}$$

4.5.2 Staircase walks adsorbing on the main diagonal

The generating functions obtained in the first two sections can be used to construct a model of staircase walks adsorbing on to the main diagonal. In these models a second variable z is introduced, which will be the activity conjugate to the number of visits between the walk and the main diagonal. An excursion (see Fig. 4.17) has exactly two visits in the main diagonal (at its end-points). It is sufficient to first consider the adsorption of a Dyck walk on the diagonal. A Dyck walk is either a single visit of weight z, or an excursion followed by a Dyck walk. Counting visits gives

$$G_D(x, z) = z + zG_e(x)G_D(x, z). \tag{4.127}$$

Use eqn (4.122) and solve for $G_D(x, z)$. Then

$$G_D(x, z) = \frac{z}{1 - z(1 - \sqrt{1 - 4x^2})/2}. \tag{4.128}$$

Expanding this generating function in x gives

$$\begin{aligned} G_D(x, z) = {} & z + z^2x^2 + (z^2 + z^3)x^4 + (2z^2 + 2z^3 + z^4)x^6 \\ & + (5z^2 + 5z^3 + 3z^4 + z^5)x^8 + (14z^2 + 14z^3 + 9z^4 + 4z^5 + z^6)x^{10} + \cdots \end{aligned} \tag{4.129}$$

exactly as expected [29]. The radius of convergence $x_c(z)$ of $G_D(x, z)$ is given by either $x = 1/2$ if $1 > z(1 - \sqrt{1 - 4x^2})/2$, or by $1 = z(1 - \sqrt{1 - 4x^2})/2$, which implies that

$$x_c(z) = \begin{cases} \frac{1}{2}, & \text{if } z \leq 2; \\ \sqrt{z - 1}/z, & \text{if } z > 2. \end{cases} \tag{4.130}$$

This defines the free energy $\mathcal{F}_D(z) = -\log x_c(z)$ for this model, and note that there is a non-analyticity at $z = 2$, where two curves of branch points in $G_D(x, z)$ meet. This

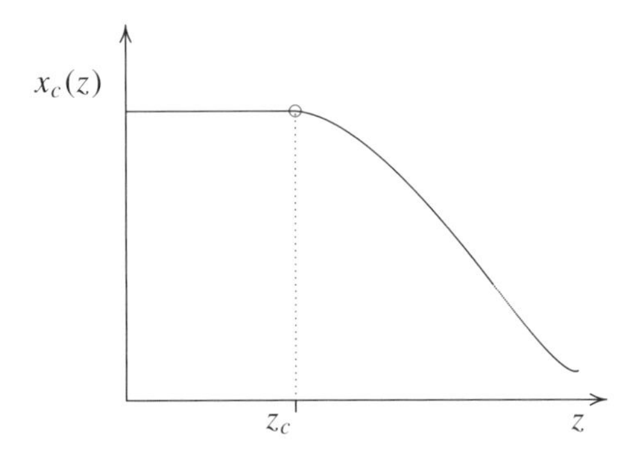

Fig. 4.19: The radius of convergence of the generating function of adsorbing staircase walks. The critical poim corresponds to a non- analyticity in the free energy which is a multicritical point. If $z < z_c$, then the free energy is a constant, and its first derivative (which is the density of visits) is zero. If $z > z_c$, then the density of visits is positive. This derivative is a continuous function, which implies that we have a second-order phase transition. The shape of the critical curve is typical of asymmetric models.

is a multicritical point; a plot of $x_c(z)$ is in Fig. 4.19. The density of visits is the first derivative of the free energy to $\log z$; direct computation gives

$$\langle v \rangle_z = \begin{cases} 0, & \text{if } z < 2; \\ (z-2)/(2z-2), & \text{if } z > 2. \end{cases} \tag{4.131}$$

The density function of adsorbing staircase walks can be explicitly computed from eqn (4.130), using Theorem 3.19:

$$\mathcal{P}_D(\epsilon) = \sqrt{1-2\epsilon}\left[\frac{2-2\epsilon}{1-2\epsilon}\right]^{1-\epsilon}, \qquad \text{where } \epsilon \in (0, 1/2). \tag{4.132}$$

The right-derivative at $\epsilon = 0^+$ is finite, and there is a second-order transition from a desorbed to an adsorbed phase (see Lemma 3.20). Observe that the critical behaviour of adsorption is not tricritical, but may instead be interpreted as a meeting point of two λ-lines, with cross-over behaviour on the adsorbed side.

The gt-plane is defined by putting the origin at the multicritical point in Fig. 4.19, and by choosing the t-axis tangent to the critical curve at this point (and thus parallel to the z-axis). The g-axis is chosen normal to the critical curve at the origin, and thus parallel to the x-axis in Fig. 4.19. The cross-over exponent is most conveniently found by computing the shift exponent, which describes the shape of the critical curve in Fig. 4.19. From eqn (4.130) it follows that

$$g = \frac{1}{2} - x_c(z) = \frac{1}{2} - \frac{\sqrt{z-1}}{z} = \frac{(\sqrt{t+1}-1)^2}{2(t+2)} \simeq \frac{t^2}{8(t+2)} \tag{4.133}$$

as $t \to 0$, which implies that $\psi = 2$. The cross-over exponent is therefore

$$\phi = 1/2. \tag{4.134}$$

Table 4.2: Exponents of adsorbing staircase walks

ϕ	α	γ_t	γ_u	ν_t	y_t	γ_+	γ_-
$\frac{1}{2}$	0	$\frac{1}{2}$	1	$\frac{1}{4}$	2	1	$-\frac{1}{2}$

If $t = 0$ (or $z = 2$), then the exponent γ_t can be found; in this case

$$G_D(x, 2) = 2/\sqrt{1 - 4x^2} \sim g^{-1/2}, \tag{4.135}$$

from eqn (4.128), and by comparison to eqn (2.16) the result is that $\gamma_t = 1/2$. In this model the scaling assumption in eqn (2.16) is entirely appropriate, in contrast with the results obtained for the convex vesicle models in Sections 4.2, 4.3 and 4.4. On the other hand, even though I have argued that this is not tricritical, it is also the case that $G_D(x, z)$ is finite along the line $x = 1/2$, $z < 2$. Thus, the scaling assumption in eqn (2.15) is appropriate, and from eqn (4.128) it may be computed that $\gamma_u = 1$.[6] Observe how the relation $\phi = \gamma_t/\gamma_u$ (Chapter 2, footnote 7) is obeyed by the values of these exponents. The rest of the critical exponents can be found from eqns (2.26), (2.36) and (2.37): these are $\nu_t = 1/4$, $y_g = 4$, $y_t = 2$ and $\alpha = 0$, and they are listed in Table 4.2.

It is possible to modify this model to staircase walks with a free end-point adsorbing on the lattice: simply multiply eqn (4.128) by $(1 + xG_t(x))$ in eqn (2.13) to find the generating function, or change eqn (4.127) to include the interaction with the diagonal (note that there are no visits between the tail and the diagonal):

$$G_t(x, z) = G_D(x, z) + G_D(x, z)xG_t(x). \tag{4.136}$$

This change does not change the shape of the critical curve, so that the same value of the exponent ϕ is obtained.

4.5.3 *Staircase walks adsorbing on to a penetrable diagonal*

In this model the staircase walk interacts with a main diagonal which they may penetrate; this is for example a model of a polymer interacting with an interface between two fluids. The generating function of staircase walks starting and terminating on the diagonal is given by eqn (4.62). To find that version of this generating function which includes an activity conjugate to the number of visits, the starting point is the generating function of an excursion in eqn (4.122). In this model an excursion can be above, or below the diagonal, so that eqn (4.122) should be multiplied by 2. The renewal equation for these walks is then given by the same argument which led to eqn (4.127). The generating function of staircase walks adsorbing on to a diagonal which it can penetrate is defined by

$$G_\delta(x, z) = z + 2zG_e(x)G_\delta(x, z). \tag{4.137}$$

Thus

$$G_\delta(x, z) = \frac{z}{1 - z(1 - \sqrt{1 - 4x^2})}. \tag{4.138}$$

[6] Put $x = 1/2$ in eqn (4.128) above; this gives $G(1/2, z) = z/(1 - z/2)$, and thus as $z \to 2^-$ the divergence in $G(1/2, z)$ indicates that $\gamma_u = 1$.

Equation (4.62) is recovered by putting $z = 1$. The free energy can be computed from the radius of convergence of $G_\delta(x, z)$:

$$x_d(z) = \begin{cases} 1/2, & \text{if} z \leq 1; \\ \sqrt{2z-1}/2z, & \text{if} z > 1. \end{cases} \tag{4.139}$$

The phase diagram is similar to Fig. 4.19, but with the critical point at $z = 1$. Calculations similar to the above indicate that $\phi = 1/2$ in this model as well, with $\gamma_t = 1/2$ and $\gamma_u = 1$. The rest of the exponents are listed in Table 4.2. This transition is in the same universality class as for the impenetrable model. An interesting difference is that the critical point is at $z = 1$, which means that there is no interaction between the walk and the diagonal at the critical point. The adsorbed phase is found for any attractive value of the activity.

4.5.4 Adsorbing staircase walks with an area activity

A Dyck walk has both end-points fixed on the diagonal, and encloses an area above the diagonal and under the walk, as seen in Fig. 4.17. The minimum area is obtained when every second vertex in the walk is adsorbed in the diagonal. Discount this area, and define this walk to enclose zero area. The largest area that can be enclosed is $n^2/8 - n/4$. If the Dyck walk is translated one step away from the diagonal, while two new edges are inserted to keep its end-points in the diagonal, then an excursion (such as in Fig. 4.17) is obtained. The area enclosed by this excursion depends on its length: each edge in the Dyck walk generates the equivalent of one-half square in area, and the two new edges generate a triangle (which are not considered part of the area, as explained above). Thus, if x is the edge activity, and y^2 the area activity (y is conjugate to half a unit area), then $G_e(x, y) = x^2 G_D(xy, y)$ is the generating function of an excursion, where $G_D(x, y)$ is the generating function of a Dyck walk with an edge and an area activity. Since each Dyck walk is either a vertex, or consists of an excursion followed by a Dyck walk, the following functional equation is obtained:

$$G_D(x, y) = 1 + x^2 G_D(xy, y) G_D(x, y). \tag{4.140}$$

Equation (4.140) is a functional recursion for the area–length generating function of Dyck walks. By noting that $G_D(x, y) = 1/(1 - x^2 G_D(xy, y))$, an iteration of this equation gives a continued fraction as an expression for $G_D(x, y)$ (see also [338]).

$$G_D(x, y) = \frac{1}{1 - x^2 \left(\frac{1}{1 - x^2 y^2 \left(\frac{1}{1 - x^2 y^4 (\cdots)} \right)} \right)}. \tag{4.141}$$

Notice the implied identity with $G_D(x)$ in eqn (4.124) if $y = 1$. By Worpitsky's theorem [356] the radius of convergence of $G_D(x, y)$ is found if $x^2 y^{2p} = 1/4$ for all $p \geq 0$. In other words, if $y = 1$, then $x \leq 1/2$, and if $x > 1/2$, then $x^2 y^2 \leq 1/4$ which implies that $x \leq 1/2y$. It is possible to show that the phase boundary described by the line $y = 1$,

$x \leq 1/2$ is a line of essential singularities in $G_D(x, y)$ which corresponds to a phase of "inflated" staircase walks enclosing a large area. The phase boundary described by $x = 1/2y$ if $y < 1$ gives a phase of "deflated" walks. These phase boundaries meet at a tricritical point at $x = 1/2$, $y = 1$, with an associated cross over exponent $\phi = 1$ ($g = x + 2y - 5/2$, $t = \sqrt{5}(y - 1)$ and $g \sim t/\sqrt{5}$ along the critical curve).

By using $G_e(x, y)$, this can be changed into a model of adsorbing staircase walks with an area activity. Argue as in eqn (4.127) to obtain (where z is the visit activity):

$$G_D(x, y, z) = z + zG_e(x, y)G_D(x, y, z). \tag{4.142}$$

Solve for $G_D(x, y, z)$ and use $G_e(x, y) = x^2 G_D(xy, y)$ to obtain

$$G_D(x, y, z) = \frac{z}{1 - zx^2 G_D(xy, y)}, \tag{4.143}$$

from which an infinite fraction can be developed by using eqn (4.141). This relation gives (in principle at least) the free energy of adsorbing staircase walks with an area activity. In particular, note that there is a line of essential singularities at $y = 1$, which corresponds to a first-order transition to inflated staircase walks (regardless of the value of z). If $y = 1$, then the case discussed in Section 4.5.2 is again obtained; there is an adsorption at $z = 2$. If $y = 0$, then no walks with area are generated, but all walks which are generated have every second vertex a visit. This shows that the adsorption occurs at $z = 0$. Thus, the phase diagram has the general appearance presented in Fig. 4.20. A model of absorbing directed walks with an area activity was studied by A.L. Owczarek and T. Prellberg [292]. Scaling functions can also be found from the non-linear functional equation (4.140), see reference [35].

4.5.5 *The constant term formulation and adsorbing staircase walks*

The correspondence between staircase walks above the main diagonal, and which interact with the main diagonal, and random walks on the positive integers which adsorb on the

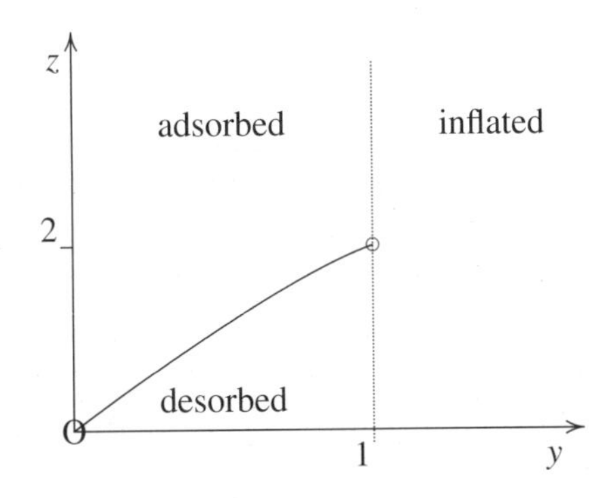

Fig. 4.20: The phase diagram of adsorbing staircase walks with an area fugacity. the transition to inflated walks is an essential singularity in the generating function, corresponding to a first-order transition. The transition between the desorbed and adsorbed phases is second order. The meeting point between these phase boundaries is a critical end-point of the line of adsorption transitions in the deflated phase.

origin has already been pointed out. The problem of staircase walks adsorbing on the (impenetrable) main diagonal can be formulated as a difference equation with a boundary condition: let $v_n(j)$ be the number of random walks of length n steps which terminates at site j on the positive integer line (suppose that the walks started at site j_0), and weighted by z^m where m is the number of visits each walk has made to the origin.[7] Then $v_n(j)$ satisfies the difference equation

$$v_n(j) = v_{n-1}(j-1) + v_{n-1}(j+1), \qquad \forall\, j \geq 1, \quad \forall\, n \geq 1, \tag{4.144}$$

$$v_n(0) = z v_{n-1}(1), \tag{4.145}$$

where the boundary condition will generate new factors of z, each time the walk visits the origin. Notice that the function

$$q_n(j) = \Lambda^n (A_1 e^{ijk} + A_2 e^{-ijk}), \tag{4.146}$$

is a solution of the difference equation (4.144), provided that

$$\Lambda = e^{ik} + e^{-ik}, \tag{4.147}$$

for *any* real value of k. In other words, integrate $q_n(j)$ over k to find a solution of the boundary value problem (note that $k \in (-\pi, \pi]$, due to the periodicity of the complex exponentials). A relationship between A_1 and A_2 can be found by using the boundary condition in eqn (4.145): substitution of eqn (4.146) into eqn (4.145) shows that

$$\frac{A_1}{A_2} = -\frac{\Lambda - z e^{-ik}}{\Lambda - z e^{ik}} = \mathcal{S}(k), \tag{4.148}$$

where $\mathcal{S}(k)$ is called a *scattering function*, since it describes the scattering of the walk from the origin. In other words, a solution of the form

$$v_n(j) = \int_{-\pi}^{\pi} A_2 \Lambda^n (e^{-ijk} + \mathcal{S}(k) e^{ijk})\, dk \tag{4.149}$$

should be expected. Notice that A_2 is still an arbitrary function of k, and that it should be fixed to take into account the fact that the walk started at site j_0 (that is, A_2 will be fixed by the boundary condition). Examination of eqn (4.146) suggests that the reversal of the walk occurs if the sign of k is changed (or take the complex conjugate). Thus, a good choice for A_2 would be

$$A_2 = C_0 [e^{ij_0k} + \mathcal{S}(-k) e^{-ij_0k}]. \tag{4.150}$$

C_0 is a constant which will be fixed later (it may be a function of z). Substitution into eqn (4.149), followed by simplification, and where symmetry of the integral is used, as

[7] In other words, $v_n(j)$ is a partition function. If $v_n(j, m)$ is the number of walks which arrive at j, having visited the origin m times, then $v_n(j) = \sum_{m \geq 0} v_n(j, m) z^m$. The argument z in $v_n(j)$ is suppressed to keep the notation simple.

well as the fact that $\mathcal{S}(-k) = 1/\mathcal{S}(k)$, gives

$$v_n(j;\, j_0) = C_0 \int_{\pi}^{\pi} \Lambda^n (e^{i(j-j_0)k} + \mathcal{S}(k) e^{i(j+j_0)k}) dk. \tag{4.151}$$

This integral becomes a contour integral around the boundary C of the unit disk in the complex plane if the substitution $\zeta = e^{ik}$ is made. One finds that (define $\bar{\zeta} = 1/\zeta$)

$$v_n(j;\, j_0) = C_0 \oint_C \left[\Lambda^n \left(\zeta^{j-j_0} - \frac{\zeta + \bar{\zeta} - z\bar{\zeta}}{\zeta + \bar{\zeta} - z\zeta} \zeta^{j+j_0} \right) \right] \frac{d\zeta}{i\zeta}. \tag{4.152}$$

Observe that this integral is the constant term of the factor in square brackets, multiplied by $2\pi C_0$. Absorb the constants into the C_0. Then

$$v_n(j;\, j_0) = C_0 CT \left[\Lambda^n \left(\zeta^{j-j_0} - \frac{\zeta + \bar{\zeta} - z\bar{\zeta}}{\zeta + \bar{\zeta} - z\zeta} \zeta^{j+j_0} \right) \right], \tag{4.153}$$

where $CT[\cdot]$ selects the constant term from the terms in the square bracket in a Laurent series about $\zeta = 0$. In the case that $j = j_0 = 0$ then it should be possible to recover $G_D(x, z)$ in eqn (4.128) from eqn (4.153): Expand $\Lambda = (\zeta + \bar{\zeta})^n$. Then

$$\begin{aligned} v_{2n}(0;\, 0) &= C_0 CT \left\{ \sum_{l=0}^{2n} \binom{2n}{l} \zeta^{2n-2l} \left[1 - \frac{\zeta^2 + 1 - z}{1 + (1-z)\zeta^2} \right] \right\} \\ &= C_0 CT \left\{ \sum_{l=0}^{2n} \binom{2n}{l} \zeta^{2n-2l} \left[\frac{z(1-\zeta^2)}{1 + (1-z)\zeta^2} \right] \right\} \\ &= z C_0 \sum_{m=0}^{n} \frac{2m+1}{n+m+1} \binom{2n}{n+m} (z-1)^m \end{aligned} \tag{4.154}$$

where the denominator in the term in square brackets has been expanded to find the constant term. Notice that only staircase walks of even length were considered. Multiply eqn (4.154) by x^{2n}, and sum over n to find the generating function. With the help of a symbolic computations program, it is found that

$$\begin{aligned} G_D(x, z) &= C_0 z \sum_{n=0}^{\infty} \sum_{m=0}^{n} \frac{2m+1}{n+m+1} \binom{2n}{n+m} (z-1)^m x^{2n} \\ &= \frac{C_0 z}{1 - z(1 - \sqrt{1-4x^2})/2}, \end{aligned} \tag{4.155}$$

exactly as expected, and this fixes $C_0 = 1$. If j and j_0 have arbitrary values, then

$$\begin{aligned} v_n(j;\, j_0) = &\binom{n}{\frac{n+j-j_0}{2}} - \binom{n}{\frac{n+j+j_0}{2}} \\ &+ z \sum_{m=0}^{(n-j-j_0)/2} \left[\binom{n}{\frac{n+j+j_0}{2} + m} - \binom{n}{\frac{n+j+j_0}{2} + m + 1} \right] (z-1)^m. \end{aligned} \tag{4.156}$$

The first two terms count staircase walks without a factor in z, and so generate all those staircase walks which are disjoint with the main diagonal. The third term is therefore the most interesting. If it is assumed that $j + j_0$ is even, then the generating function found from this term is

$$G_D(x, z; j; j_0) = \frac{(2x)^{j+j_0}}{(1+\sqrt{1-4x^2})^{j+j_0}} \left(\frac{z}{1 - z(1-\sqrt{1-4x^2})/2} \right). \qquad (4.157)$$

This reduces to (4.155) if $j + j_0 = 0$. If $j + j_0$ is odd then exactly the same result as in eqn (4.157) is found. Notice that the critical point is still at $z_c = 2$, and that $\phi = 1/2$. If $j_0 = 0$ and eqn (4.157) is summed over j, then the generating function defined in eqn (4.136) is recovered.

4.5.6 A staircase walks model of copolymer adsorption

A copolymer is a polymer which is composed of two or more different monomers, say of types A and B. A directed model of a copolymer could be a staircase walk composed of vertices which are coloured with two colours (A and B). If these colours alternate along the walk, then a model of an *alternating copolymer* is obtained. If the two colours are arranged such that a sequence of m vertices of colour A is followed by a sequence of m vertices of colour B, then it is a *block copolymer*. In some cases the colours may occur randomly along the copolymer, which is called a *random copolymer*. A partially directed model of a copolymer was examined in reference [362].

Suppose first of all that a model of adsorbing staircase walks is considered, where each vertex has a probability p of interacting with the main diagonal. This averaging of the interaction before the free energy is computed is called *annealing*, and this is an annealed model of copolymer adsorption. The generating function of this model can be immediately written down by noting that for every realization of a walk with v visits, there are $\binom{v}{w}$ ways of colouring w of the visits by A. In other words, if there are $r_n^+(v)$ Dyck walks above the diagonal with length n and with v visits, coloured with probability p by A and otherwise by B, and with A interacting with the diagonal with activity z, then the generating function is $G_D(p; x, z) = \sum_{n=0}^{\infty} \sum_{v=0}^{n+1} \sum_{w=0}^{v} \binom{v}{w} r_n^+(v)(pz)^w(1-p)^{v-w}x^n$. Evaluating the sum over w gives $G_D(p; x, z) = \sum_{n=0}^{\infty} \sum_{v=0}^{n+1} r_n^+(v)(pz+1-p)^v x^n = G_D(x, pz+1-p)$, and so $G_D(p; x, z)$ can be read from eqn (4.128). Thus

$$G_D(p; x, z) = \frac{pz+1-p}{1-(pz+1-p)(1-\sqrt{1-4x^2})/2}. \qquad (4.158)$$

The radius of convergence of this generating function is then

$$x_c(z) = \begin{cases} 1/2, & \text{if } z < z_c; \\ [\sqrt{p(z-1)}]/(1+p(z-1)), & \text{if } z \geq z_c, \end{cases} \qquad (4.159)$$

where z_c is the critical point which has value $z_c = 1 + 1/p$. If $p = 1/2$, then $z_c = 3$. The critical exponents remain unchanged in this model, compared to the homopolymer model examined in Section 4.5.2.

A more interesting scenario is encountered if a model of a true copolymer is considered. In this ensemble the monomers are coloured, and their colours are fixed. This is a *quenched* model of copolymer adsorption, and the outcome of the model depends on the order and arrangement of colours (types of monomers) in the copolymer. I shall consider only two cases: an alternating copolymer, and a block copolymer. Consider first the adsorption of a staircase walk model of an alternating copolymer. An important fact in this model is that only every second vertex in the staircase may adsorb in the diagonal, and so it follows that these vertices should be alternating monomers in the model. Thus, label the vertices along the staircase walk by $\{0, 1, 2 \ldots, n\}$, and let the vertices with labels $\{4m\}_{m=0}$ have colour or type A, and those with labels $\{4m+2\}_{m=0}$ be of type B. Assume that only the vertices of type A can interact with the diagonal. Call all those staircase walks of length $4m$ *even*, and those of length $4m+2$ *odd*. I shall simplify matters by only considering Dyck walks in what follows (tails may be added to the generating functions by using arguments similar to those in Section 4.5.1). An even Dyck walk with two types of vertices is illustrated in Fig. 4.21. Assume that the vertex at the origin is always of type A, and so interacts with the main diagonal.

The generating function $G_{ev}(x)$ of even Dyck walks can be directly computed from Catalan numbers; from eqn (4.124) it immediately follows that

$$G_{ev}(x) = \sum_{n=0}^{\infty} C_{2n} x^{4n} = \frac{\sqrt{2}}{\sqrt{1+\sqrt{1-16x^4}}}. \tag{4.160}$$

Every odd Dyck walk starts at the origin and visits the main diagonal a first time, from where it continues as either an odd or an even Dyck walk. If it continues as an odd Dyck walk, then the first excursion is even, and by removing its first and last edges, an odd Dyck walk is obtained. If it continues as an even Dyck walk, then the first excursion is odd, and by removing the first and last edges, an even Dyck walk is obtained. Thus the generating function of odd Dyck walks satisfies the relation

$$G_{od}(x) = x^2(G_{od}^2(x) + G_{ev}^2(x)), \tag{4.161}$$

and if this is solved, then

$$\begin{aligned} G_{od}(x) &= \frac{2x^2 G_{ev}^2(x)}{1+\sqrt{1-4x^4 G_{ev}^2(x)}} \\ &= \frac{4x^2}{(1+\sqrt{1-16x^4})\left(1+\sqrt{(1-\sqrt{1-16x^4})/2}\right)}. \end{aligned} \tag{4.162}$$

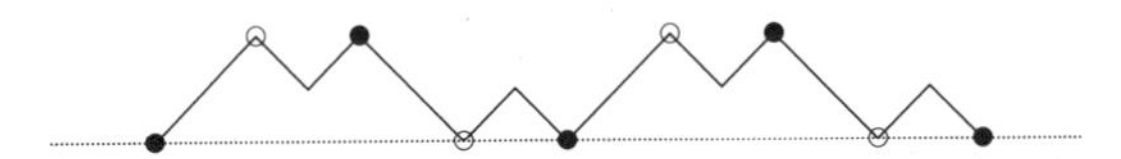

Fig. 4.21: An even staircase walk with vertices of two types as a model of an alternating copolymer. The solid vertices interact with the main diagonal, while those indicated by ∘ do not.

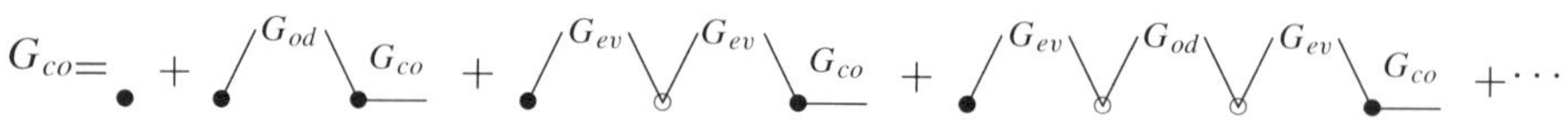

$G_{co}=$ • + G_{od} G_{co} + G_{ev} G_{ev} G_{co} + G_{ev} G_{od} G_{ev} G_{co} $+\cdots$

Fig. 4.22: The generating function of an alternating adsorbing copolymer is either a single vertex, or is an excursion of length $4m$ followed by the copolymer, or two excursions of lengths $4m_1 + 2$ and $4m_2 + 2$ followed by the copolymer, or an excursion of length $4m_1 + 2$ followed by any number of excursions of length $4m_i$, $i = 2, \ldots, N-1$, followed by an excursion of length $4m_N + 2$, and then followed by the copolymer. The vertices indicated by • have an activity z in the adsorbing line, while those indicated by ∘ do not interact with the adsorbing line.

The generating functions of even and odd Dyck walks can now be used to find the generating function of Dyck walks where only the A vertices (with labels $\{4m\}_{m=0}$) interact with an activity z with the main diagonal. The B vertices, with labels $\{4m+2\}_{m=0}$ may visit the main diagonal, but there is no interaction with it. The generating function is derived by arguing as in Fig. 4.22.

From Fig. 4.22 it immediate follows that the generating function of an adsorbing alternating copolymer is given by

$$G_{\text{co}}(x, z) = z + zx^2 G_{od}(x) G_{\text{co}}(x, z) + \frac{zx^4 G_{ev}^2(x)}{1 - x^2 G_{od}(x)} G_{\text{co}}(x, z). \tag{4.163}$$

The generating function $G_{\text{co}}(x, z)$ can be solved in this expression, and expanding it also gives

$$\begin{aligned} G_{\text{co}}(x, z) &= \frac{z}{1 - zx^2 G_{od} - zx^4 G_{ev}^2(x)/(1 - x^2 G_{od}(x))} \\ &= z + 2z^2 x^4 + (10z^2 + 4z^3)x^8 + \cdots \end{aligned} \tag{4.164}$$

The radius of convergence of $G_{\text{co}}(x, z)$ is given by

$$x_c(z) = \begin{cases} \left[\sqrt{\sqrt{8z^3 - 20z^2 + 16z - 4}}\right]\Big/ 2z, & \text{if } z > 2 + \sqrt{2}; \\ 1/2, & \text{if } z \leq 2 + \sqrt{2}. \end{cases} \tag{4.165}$$

The location of the critical point has moved somewhat, but the appearance of the critical curve is still generally given by the curve in Fig. 4.19, where the case of adsorbing homopolymers was considered. There is a second-order adsorption transition at $z_c = 2 + \sqrt{2}$, and the free energy in this model is given by

$$\mathcal{F}_{\text{co}}(z) = \begin{cases} [\log 2]/2 + \log z - [\log((2z-1)(z-1)^2)]/4, & \text{if } z > z_c; \\ \log 2, & \text{if } z < z_c. \end{cases} \tag{4.166}$$

The density of visits is given by

$$z \frac{d}{dz} \mathcal{F}_{\text{co}}(z) = \begin{cases} (z^2 - 4z + 2)/2(2z-1)(z-1), & \text{if } z > z_c; \\ 0, & \text{if } z < z_c. \end{cases} \tag{4.167}$$

The scaling exponents can be found by putting $t = z - z_c$ and then

$$g(t) - 1/2 - x_c(z) \approx \frac{t^2}{4(17+12\sqrt{2})} + O(t^3). \tag{4.168}$$

This shows that the crossover exponent is $\phi = 1/2$ as before. The square-root singularity in eqn (4.164) (this follows really from eqns (4.160) and (4.162)) implies that $\gamma_- = -1/2$ if $z > 2+\sqrt{2}$. If $z < 2+\sqrt{2}$, then $\gamma_+ = 1$, and it can also be checked that $\gamma_t = 1/2$ and $\gamma_u = 1$. In fact, these exponents are exactly those in Table 4.2, and correspond to the adsorption of a staircase walk. This is to be expected; these models should be in the same universality class.

A block copolymer is a copolymer where all the A vertices are arranged as a block in the first half of the copolymer, followed then by a sequence of all the B vertices in the second half of the copolymer. Thus, it has general structure $AAAA\ldots ABBBB\ldots B$. A block copolymer may be alternating and of the form $AA\ldots ABB\ldots BAA\ldots ABB\ldots$, and the number of A and B vertices need not be the same in each block. Consider now an adsorbing block copolymer of length $2n$ consisting of $n+1$ A vertices followed by n B vertices. Suppose that only the A vertices interact with the main diagonal. The result in eqn (4.156) can be used to compute the generating function of this model explicitly. Define $v_n(z; j, j_0) = v_n(j, j_0)$ Then the generating function of a block copolymer of this type, which starts at a height j about the main diagonal, and ends at a height j_0 above the main diagonal, is given by

$$G_B(x, z; j, j_0) = \sum_{n=0}^{\infty}\sum_{k=0}^{\infty} v_n(z; j, k) v_n(1; k, j_0) X^{2n}. \tag{4.169}$$

The first term in eqn (4.156) corresponds to the case that the polymer does not intersect the main diagonal, so consider only the contributions of the second term to G_B. Using a symbolic computational progam, and executing the sums over j and j_0, the result is

$$\begin{aligned} G_B(x,z) = {} & \frac{2(1-2x^2+\sqrt{1-4x^2})^2}{(1-2x+\sqrt{1-4x^2})^2(1-4x^2+\sqrt{1-4x^2})} \\ & \times \left[\frac{2z}{(1+\sqrt{1-4x^2})(1-z(1-\sqrt{1-4x^2})/2)}\right]. \end{aligned} \tag{4.170}$$

In other words, the critical curve is the same as that for an adsorbing homopolymer, given by eqn (4.130). The same critical exponents are also found.

4.5.7 Staircase polygons above the main diagonal

In this section I shall consider staircase polygons with at least one visit in the main diagonal. Imagine a staircase polygon to be composed of two staircase walks with a common first and last vertex, but which are otherwise disjoint. Of these two walks, only the bottom walk will adsorb in the main diagonal; the top walk cannot intersect a vertex in the main diagonal since it is shielded from it by the bottom. I shall consider staircase polygons which are fixed to the main diagonal at their second vertex in the bottom staircase walk. Such staircase polygons are *grafted* to the main diagonal at its beginning. Similarly, a staircase polygon can be grafted to the main diagonal at its end, or both at its beginning and its end. As before, ignore the first two edges, and the last two edges, in the polygons. The orientations of these edges are always fixed, so that they contribute nothing to the combinatorial properties of the polygon. The necessary factors of x will be supplied at the end to account for them.

A suitable starting point is the correspondence between the two-coloured Motzkin language and staircase polygons explored in Section 4.4.2. Then any staircase polygon grafted at its beginning corresponds to words on the four letters $\{O, C, U, D\}$ subject to the following conditions. In any word w it is the case that $|w|_U \geq |w|_D$ and $|w|_O = |w|_C$ and in any prefix u of w it will be the case that $|u|_U \geq |u|_D$, $|u|_O \geq |u|_C$ and $|u|_C + |u|_U \geq |u|_O + |u|_D$. The correspondence of the letters $\{O, C, U, D\}$ to steps in the square lattice gives a random walk from the origin which is confined to the wedge above or on the X-axis, and below or on the main diagonal. A visit is generated each time the random walk visits a site on the main diagonal (this will also correspond to visits the staircase polygon makes to the main diagonal), and the end-point of the walk must be in the X-axis to close the polygon. This wedge will be called the *principal wedge*. In other words, the adsorption of grafted staircase polygons can be studied by considering random walks from the origin in the principal wedge, with an activity z conjugate with the number of visits the random walk makes to the main diagonal, see Fig. 4.23; this is a model of adsorbing random walks in the principal wedge. If the condition that the random walk starts from the origin is relaxed (but starts from the X-axis), then the corresponding staircase polygon can start at any height above the main diagonal; visits will still be generated by visits of the random walk to the main diagonal in the principal wedge. The generating functions of random walks in subgraphs of the square lattice have been considered before [257], and, as will become apparent below, these models are never very simple.

The adsorption of a random walk in the principal wedge on to the main diagonal can be formulated as a system of partial difference equations with boundary conditions. Let $r_n(j, l)$ be the number of walks which arrives at site (j, l), having started at site (j_0, l_0)

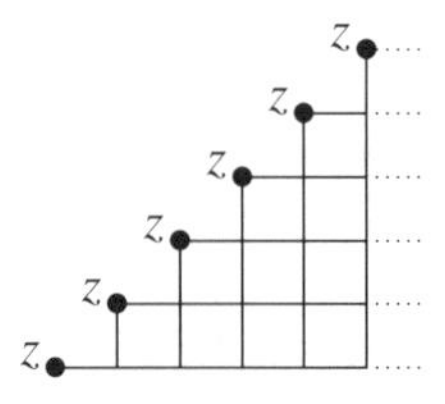

Fig. 4.23: A random walk in this wedge which starts and terminates on the x-axis corresponds to a staircase polygon above the main diagonal and interacting with it.

and after having given n steps. Then

$$\begin{aligned} r_n(j,l) &= r_{n-1}(j-1,l) + r_{n-1}(j+1,l) \\ &\quad + r_{n-1}(j,l-1) + r_{n-1}(j,l+1), && \forall 1 \le l \le j; \\ r_n(j,0) &= r_{n-1}(j-1,0) + r_{n-1}(j+1,0) + r_{n-1}(j,1), && \forall j \ge 1; \\ r_n(j,j) &= z(r_{n-1}(j+1,j) + r_{n-1}(j,j-1)), && \forall j \ge 1. \end{aligned} \tag{4.171}$$

Terms such as $e^{\pm ijk_1} e^{\pm ilk_2}$ seems to be solutions of (4.171); so define

$$\zeta_m = e^{ik_m}, \qquad \bar{\zeta}_m = 1/\zeta_m, \tag{4.172}$$

and exhaust all possible combinations of ζ_1 and ζ_2 in the assumption that the general solution will be of the form

$$\begin{aligned} q_n(j,l) = \Lambda^n \big(& A_1^{11}\zeta_1^j\zeta_2^l + A_1^{21}\bar{\zeta}_1^j\zeta_2^l + A_1^{12}\zeta_1^j\bar{\zeta}_2^l + A_1^{21}\bar{\zeta}_1^j\bar{\zeta}_2^l \\ & + A_2^{11}\zeta_2^j\zeta_1^l + A_2^{21}\bar{\zeta}_2^j\zeta_1^l + A_2^{12}\zeta_2^j\bar{\zeta}_1^l + A_2^{22}\bar{\zeta}_2^j\bar{\zeta}_2^l \big). \end{aligned} \tag{4.173}$$

Substition into eqn (4.171) gives

$$\Lambda = \zeta_1 + \bar{\zeta}_1 + \zeta_2 + \bar{\zeta}_2. \tag{4.174}$$

Substitution into the boundary conditions gives relations amongst the constants A_j^{lm}. The boundary conditions in eqn (4.171) gives the set of relations

$$A_1^{12} = -\bar{\zeta}_2^2 A_1^{11}, \quad A_1^{21} = -\zeta_2^2 A_1^{22}, \quad A_2^{12} = -\bar{\zeta}_1^2 A_2^{11}, \quad A_2^{21} = -\zeta_1^2 A_2^{22}, \tag{4.175}$$

and if the scattering function

$$\mathcal{T}(\alpha,\beta) = -\frac{\alpha + \bar{\alpha} + \beta + \bar{\beta} - z(\alpha+\beta)}{\alpha + \bar{\alpha} + \beta + \bar{\beta} - z(\bar{\alpha}+\bar{\beta})}, \tag{4.176}$$

is defined, then the following relations are found:

$$\begin{aligned} A_2^{11} &= \mathcal{T}(\zeta_1,\bar{\zeta}_2)A_1^{11}, \quad A_2^{12} = \mathcal{T}(\bar{\zeta}_1,\bar{\zeta}_2)A_1^{21}, \\ A_2^{21} &= \mathcal{T}(\zeta_1,\zeta_2)A_1^{12}, \quad A_2^{22} = \mathcal{T}(\bar{\zeta}_1,\zeta_2)A_1^{22}. \end{aligned} \tag{4.177}$$

Simplify the notation by using the shorthand

$$\mathcal{T}_{12} = \mathcal{T}(\zeta_1,\zeta_2), \quad \text{and} \quad \mathcal{T}_{\bar{1}2} = \mathcal{T}(\bar{\zeta}_1,\zeta_2), \tag{4.178}$$

and so on. The relations in eqns (4.175) and (4.177) determine all the constants in $q_n(j,l)$ in terms of (say) A_1^{11}; the result is

$$\begin{aligned} q_n(j,l) = A_1^{11}\Lambda^n \big(& [\zeta_1^j\zeta_2^l + \zeta_2^j\zeta_1^l\mathcal{T}_{1\bar{2}}] - \bar{\zeta}_2^2[\zeta_2\bar{\zeta}_2^l + \bar{\zeta}_2^j\zeta_1^l\mathcal{T}_{12}] \\ & - \bar{\zeta}_1^2\mathcal{T}_{1\bar{2}}[\zeta_2^j\bar{\zeta}_1^l + \bar{\zeta}_1^j\zeta_2^l\mathcal{T}_{12}] + \bar{\zeta}_1^2\bar{\zeta}_2^2\mathcal{T}_{12}[\bar{\zeta}_2^j\bar{\zeta}_1^l + \bar{\zeta}_1^j\bar{\zeta}_2^l\mathcal{T}_{1\bar{2}}] \big). \end{aligned} \tag{4.179}$$

If it is assumed that the random walk starts at the site (j_0, l_0), then the proper choice for A_1^{11} is the complex conjugate of $q_0(j_0, l_0)$, multiplied by a constant C_0, as suggested by the same problem for a walk on the positive integers. Expanding the product gives 64 terms, and integration over k_1 and k_2 implies that symmetries of the integrand can be used to reduce those to just eight terms. The final result is

$$\begin{aligned} r_n(j,l;j_0,l_0) = 8C_0 \int_{-\pi}^{\pi}\int_{-\pi}^{\pi} \Lambda^n \zeta_1^j \zeta_2^l \Big[\big(\bar{\zeta}_1^{j_0} \bar{\zeta}_2^{l_0} + \mathcal{T}_{\bar{1}2} \bar{\zeta}_1^{l_0} \bar{\zeta}_2^{j_0} - \zeta_2^2 \bar{\zeta}_1^{j_0} \zeta_2^{l_0} \\ - \zeta_2^2 \mathcal{T}_{\bar{1}\bar{2}} \bar{\zeta}_1^{l_0} \zeta_2^{j_0} - \zeta_1^2 \mathcal{T}_{\bar{1}2} \mathcal{T}_{\bar{1}\bar{2}} \zeta_1^{j_0} \bar{\zeta}_2^{l_0} - \zeta_1^2 \mathcal{T}_{\bar{1}2} \zeta_1^{l_0} \bar{\zeta}_2^{j_0} \\ + \zeta_1^2 \zeta_2^2 \mathcal{T}_{\bar{1}2} \mathcal{T}_{\bar{1}\bar{2}} \zeta_1^{j_0} \zeta_2^{l_0} + \zeta_1^2 \zeta_2^2 \mathcal{T}_{\bar{1}\bar{2}} \zeta_1^{l_0} \zeta_2^{j_0} \big) \Big] \, dk_1 \, dk_2. \end{aligned} \tag{4.180}$$

As before, the effect of the integrals is to select the constant term in the integrand above, and since the case $j_0 = l_0 = 0$ is particularly interesting (for a random walk starting from the origin in the principal wedge), the constant term formula for these walks is obtained:

$$\begin{aligned} r_n(j,l;0,0) = 8C_0 CT \big[\Lambda^n \zeta_1^j \zeta_2^{l+1} (\bar{\zeta}_2 (1 + \mathcal{T}_{\bar{1}2})(1 + \zeta_1^2 \zeta_2^2 \mathcal{T}_{\bar{1}\bar{2}}) \\ - \zeta_2 (1 + \mathcal{T}_{\bar{1}\bar{2}})(1 + \zeta_1^2 \bar{\zeta}_2^2 \mathcal{T}_{\bar{1}2})) \big]. \end{aligned} \tag{4.181}$$

This equation can be simplified by making the following change of variables. Define η_1 and η_2 by

$$\zeta_1 = \eta_1 \eta_2 \qquad \zeta_2 = \eta_1 \bar{\eta}_2, \tag{4.182}$$

and then the following relations can be checked:

$$\Lambda = (\eta_1 + \bar{\eta}_1)(\eta_2 + \bar{\eta}_2), \qquad \mathcal{T}_{\bar{1}2} = \mathcal{S}(\eta_2), \qquad \mathcal{T}_{\bar{1}\bar{2}} = \mathcal{S}(\eta_1), \tag{4.183}$$

where $\mathcal{S}(\eta)$ is defined in eqn (4.148). The constant term formula for $r_n(j,l;0,0)$ changes to

$$\begin{aligned} r_n(j,l;0,0) = 8C_0 CT \big[\Lambda^n \eta_1^{j+l} \eta_2^{j-l} ((1 + \mathcal{S}(\eta_2))(1 + \eta_1^4 \mathcal{S}(\eta_1)) \\ - \eta_1^2 \bar{\eta}_2^2 (1 + \mathcal{S}(\eta_1))(1 + \eta_2^4 \mathcal{S}(\eta_2))) \big]. \end{aligned} \tag{4.184}$$

Putting $n = 0$ and $j = l = 0$, gives $r_0(0,0;0,0) = 8C_0 z$, from which it is concluded that $C_0 = 1/8$. $r_n(0,0;0,0)$ can be expanded to count staircase polygons grafted at both end-points. Using a symbolic computation program to expand eqn (4.184) gives

$$\begin{aligned} r_0(0,0;0,0) &= z, \\ r_2(0,0;0,0) &= z^2, \\ r_4(0,0;0,0) &= z^2 + 2z^3, \\ r_6(0,0;0,0) &= 3z^2 + 6z^3 + 5z^4, \\ r_8(0,0;0,0) &= 14z^2 + 28z^3 + 28z^4 + 14z^5, \\ r_{10}(0,0;0,0) &= 84z^2 + 168z^3 + 180z^4 + 120z^5 + 42z^6, \\ r_{12}(0,0;0,0) &= 594z^2 + 1188z^3 + 1320z^4 + 990z^5 + 495z^6 + 132z^7. \end{aligned} \tag{4.185}$$

These numbers agree with those obtained in reference [29], and with an enumeration done by computer. If $z = 1$ then the constant term can be found without too much trouble:

$$\begin{aligned} &U(j,l) \\ &\quad = r_n(j,l;0,0)|_{z=1} \\ &\quad = CT\big[(\eta_1+\bar{\eta}_1)^n(\eta_2+\bar{\eta}_2)^n\eta_1^{j+l}\eta_2^{j-l}((1-\eta_1^2)(1-\eta_2)^2(\eta_1^2-\eta_2^2)(\eta_1^2-\bar{\eta}_2^2))\big] \\ &\quad = \frac{(l+1)(j+2)(j-l+1)(j+l+3)}{(n+1)(n+2)(n+3)^2}\binom{n+3}{(n+j-l)/2+2} \\ &\quad\quad \times \binom{n+3}{(n+j+l)/2+2}. \end{aligned} \tag{4.186}$$

$U(j,l)$ is the number of random walks of length n starting from the origin to the point (j,l) in the principal wedge. By expanding denominators in eqn (4.184), and simplifying the resulting expression, it is found that

$$r_n(j,l;0,0) = z\sum_{m_1=0}^{\infty}\sum_{m_2=0}^{\infty}(z-1)^{m_1+m_2}U(j+m_1+m_2, l+m_1-m_2). \tag{4.187}$$

If $z = 1$ then this solution corresponds to random walks in the principal wedge with reflecting boundary conditions; $z = 0$ corresponds to adsorbing boundary conditions. Staircase polygons can only be obtained if $l = 0$, since the two staircase walks comprising the polygon must have end-points close together. The sum over j can be done to obtain a model of staircase polygons which can end at any height above the main diagonal. If n is even then this must be done over even values of j: the result is (with the help of a symbolic computation program)

$$r_{2n}(z) = \frac{z}{n+1}\binom{2n}{n}\sum_{k=0}^{n}\binom{k+1}{n+1}\binom{2n+2}{n-k}(z-1)^k. \tag{4.188}$$

Multiplying this by x^{4n} and summing over n, and inserting an extra factor of x^4, gives the generating function of staircase polygons of perimeter length $4n$ which are grafted at its beginning to the main diagonal at the second vertex in the bottom walk (and which can end anywhere). (Note that each step in the walk generates two edges in the polygon; hence the factor of x^{4n}, an extra factor of x^4 is needed to generate the first two and last two edges in the polygon.) The result is

$$\begin{aligned} G_S^e(x,z) = zx^4 &\sum_{k=0}^{\infty} C_k[x^4(z-1)]^k \\ &\times {}_3F_2([k+1, k+3/2, k+1/2]; [2k+3, k+2]; 16x^4), \end{aligned} \tag{4.189}$$

where C_k is Catalan's number, and ${}_3F_2(\cdot)$ is a generalized hypergeometric function. The generating function for staircase polygons of perimeter length $4n+2$ can be found by summing over odd values of j in eqn (4.187). The result is

$$G_S^o(x,z) = zx^6 \sum_{k=0}^{\infty} \frac{2(2k+1)}{k+2} C_k [x^4(z-1)]^k \times {}_3F_2([k+1, k+3/2, k+1/2]; [2k+3, k+2]; 16x^4). \tag{4.190}$$

The generalized hypergeometric function is convergent on the closed unit disk in the complex plane for any k, with an essential singularity at $16x^4 = 1$, or $x = 1/2$. To find the rest of the radius of convergence of the generating functions in eqns (4.189) and (4.190), examine eqn (4.188). The limiting free energy *per perimeter edge* in the even case is defined by

$$\mathcal{F}_S(z) = \lim_{n\to\infty} \frac{1}{4n} \log r_{2n}(z) = \frac{1}{2} \log[2z/\sqrt{z-1}], \tag{4.191}$$

by direct calculation. The density function can be computed from this through Theorem 3.19:

$$\mathcal{P}_S(\epsilon) = \sqrt{\sqrt{6-16\epsilon}} \left(\frac{2-4\epsilon}{1-4\epsilon}\right)^{1/2-\epsilon}. \tag{4.192}$$

The density ϵ takes values in $[0, 1/4]$, and if the derivative of $\mathcal{P}_S(\epsilon)$ is computed, then $\mathcal{P}_S(0) = 1$ and

$$\left[\frac{d^+}{d\epsilon}\mathcal{P}_S(\epsilon)\right]\Bigg|_{\epsilon=0^+} = -\log 2.$$

Lemma 3.20 then shows that the critical point is at $z_c = 2$. The radius of convergence is

$$x_c(z) = \begin{cases} 1/2, & \text{if } z < 2; \\ \sqrt{\sqrt{z-1}/2z}, & \text{if } z > 2. \end{cases} \tag{4.193}$$

The square of this is similar to eqn (4.130), and the cross-over exponent is thus also $\phi = 1/2$. The continuity of the first derivative of $x_c(z)$ at $z = 2$ indicates a second-order phase transition, as was the case for the staircase walk models considered earlier. Putting $z = 2$ in eqn (4.188) and evaluating eqn (4.189) shows that $G_S^e(x, 2) = {}_3F_2([1, 1/2, 3/2]; [2, 2]; 16x^4)$. But then

$${}_3F_2([1, 1/2, 3/2]; [2, 2]; 16x^4) = \frac{2}{\pi} \sum_{n=0}^{\infty} \frac{\Gamma(n+1/2)\Gamma(n+3/2)}{\Gamma(n+2)\Gamma(n+2)} [16x^4]^n, \tag{4.194}$$

and since

$$\lim_{n\to\infty} \left[n^2 \frac{\Gamma(n+1/2)\Gamma(n+3/2)}{\Gamma(n+2)\Gamma(n+2)}\right] = 1,$$

it is concluded that

$$G_S^e(x, 2) \approx \frac{2}{\pi} \mathcal{L}i_2(16x^4), \tag{4.195}$$

where $\mathcal{L}i_2$ is the dilog function. Taking $x \to 1/2$ with $z = 2$ gives an essential singularity, and expanding $G^e_S(x, 2)$ with $x \to x_c^-$ gives $G^e_S(x, 2) \sim 2(\pi^2/6 - (1 - 16x^4) \times (1 - \log(1 - 16x^4)))/\pi$. Further expansion of $(1 - 16x^4)$ gives a factor of $(1 - 2x)$ and the dominant non-singular term is of the form $\log(1 - 2x)$, from which it may follow that $2 - \alpha_t = 0$. This leaves $2 - \alpha_u$ undefined, and the scaling assumptions in Chapter 2 do not seem to apply strictly in this model.

4.6 Partially directed walks with a contact activity

A walk in the square lattice can step in one of four directions, which will be indicated by $\{N, S, E, W\}$ (these are the four compass directions). In this section a model of walks which start at the origin and give the first step in the E-direction, but which never steps in the W-direction, is examined, and its generating function is found by using a Temperley method, and the presentation here is based on the work of R. Brak *et al.* [31, 32]. Let u_n be the number of such walks of length n. These walks can be concatenated by identifying the first vertex of one walk with the last vertex of a second; thus

$$u_n u_m \leq u_{n+m}, \tag{4.196}$$

so that this is also a supermultiplicative model. An explicit expression for u_n can be found without too much difficulty, and the growth constant can be computed.

Let u_n^r be the number of partially directed walks with the last step in the E-direction, and u_n^u be the number of partially directed walks with the last step in the N- or S-direction. Then $u_n = u_n^r + u_n^u$. If σ is a directed walk of length n counted by u_n^r, then the next step can be given in one of three directions; if it is counted by u_n^u, then the next step is in one of two directions. Therefore, $u_n = 3u_{n-1}^r + 2u_{n-1}^u$. On the other hand, $u_n^r = u_{n-1} = u_{n-1}^r + u_{n-1}^u$, since the last step can be deleted to leave any directed walk of length $n - 1$. Substituting this into the previous expression gives the recursion $u_n - 2u_{n-1} - u_{n-2} = 0$, with $u_1 = 1$ and $u_2 = 4$. The solution is $u_n = \frac{1}{2}(1 + \sqrt{2})^n + \frac{1}{2}(1 - \sqrt{2})^n$, and this implies that

$$\lim_{n\to\infty} \frac{1}{n} \log u_n = \log(1 + \sqrt{2}). \tag{4.197}$$

Let $u_n(k)$ be the number of partially directed walks with m nearest neighbour contacts. In Fig. 4.24(a) such a walk is illustrated. Two such walks of lengths n_1 and n_2,

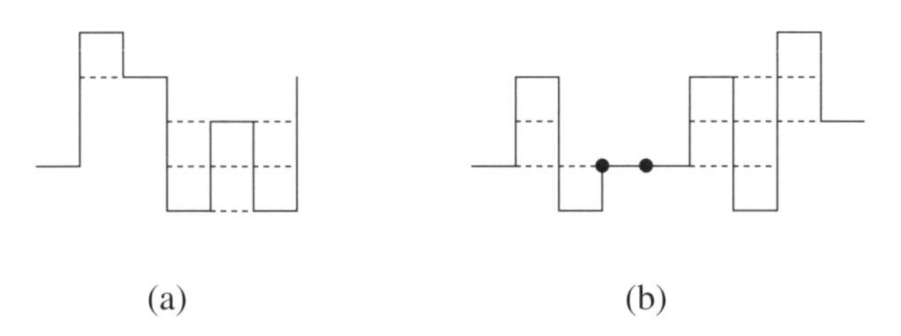

Fig. 4.24: A partially directed walk with seven contacts is illustrated in (a). In (b) two partially directed walks are concatenated by adding an extra edge between them. This avoids the possbility of forming extra contacts.

and numbers of contacts k_1 and k_2 can be concatenated as in Fig. 4.24(b) to give the inequality

$$u_{n_1}(k_1)u_{n_2}(k_2) \leq u_{n_1+n_2}(k_1 + k_2). \tag{4.198}$$

By Assumptions 3.1 and Theorems 3.17 and 3.18 this implies the existence of a finite, convex limiting free energy $\mathcal{F}_\Delta(z)$ and a density function $\mathcal{P}_\Delta(\epsilon)$ (where $\epsilon \in (0, 1/2)$, since the maximum number of contacts in any walk is $\lfloor (n-3)/2 \rfloor$).

The generating function is obtained by deriving a recursion using a Temperley method; let $u_n(k; p)$ be the number of partially directed walks of length n, with k contacts and with the first step in the E-direction followed by p steps in the N- or S-direction. Let $G_p(x, z) = \sum_{n=1}^{\infty} \sum_{k=0}^{n} u_n(k; p)x^n z^k$, so that $G_\Delta(x, z) = \sum_{p \geq 0} G_p(x, z)$ is the generating function. A recursion can be written down for G_p (suppress its arguments to simplify the equations which will follow) by considering Fig. 4.25, which separates G_p into subcases. In particular, G_p can be decomposed into two possible cases with a horizontal step followed by p vertical steps (up or down). Otherwise, there may first be a horizontal step, then p vertical steps, followed by a second horizontal step and then q vertical steps in the opposite direction (with either $p \geq q$, or $p < q$). The last case is a horizontal step, followed by p vertical steps, followed by a horizontal step, followed by q vertical steps in the same direction. Putting together the terms in Fig. 4.25 gives the following expression for G_p:

$$G_p = x^{p+1}\left(2 + \sum_{q=0}^{p}(1 + z^q)G_q + (1 + z^p) \sum_{q=p+1}^{\infty} G_q\right). \tag{4.199}$$

It follows that G_p satisfies the following second order recursion relation:

$$G_{p+2} - (1 + z)xG_{p+1} - (1 - z)z^{p+1}x^{p+3}G_{p+1} + zx^2G_p = 0. \tag{4.200}$$

Notice that solutions of this recursion will be of the form $A(x, z)G_p$, where $A(x, z)$ is not a function of p, and which must be determined separately from initial or boundary conditions. In addition, the solution for G_0 will have an extra factor of 2, since putting $p = 0$ in Fig. 4.25 shows the contributions to G_0 are counted twice. Since walks which are counted by G_0 give two initial steps in the E-direction, it is apparent that

$$G_0 = x + xG_\Delta. \tag{4.201}$$

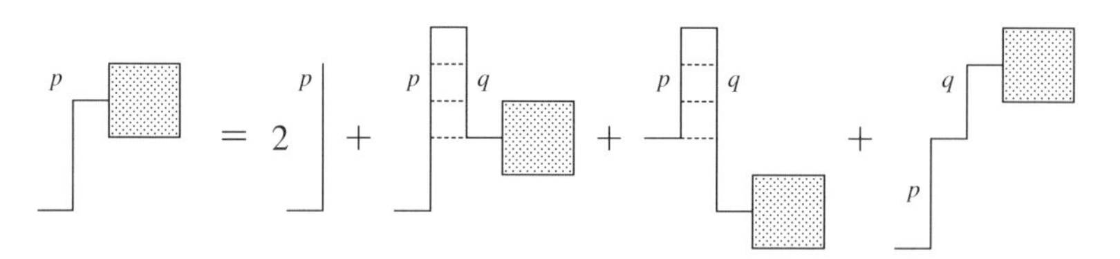

Fig. 4.25: A directed walk counted by the generating function $G_p(x, z)$ is in one of the subcases above.

If instead G_1 is considered in eqn (4.199), then

$$G_1 = x^2(2 + x - xz) + x^2(1 + x + z - xz)G_\Delta = a + bG_\Delta, \tag{4.202}$$

where the functions a and b are defined to simplify the equations. On the other hand, if the solutions of eqn (4.200) are g_p, then $\alpha G_1 = g_1$ and $2\alpha G_0 = g_0$, and from eqns (4.201) and (4.202) one can simultaneously solve for α and G_Δ in terms of g_0 and g_1. This gives

$$G_\Delta = -\frac{2xg_1 - ag_0}{2xg_1 - bg_0}. \tag{4.203}$$

It only remains to determine g_0 and g_1.

Let $y = xz$. Then eqn (4.200) simplifies to

$$g_{p+2} - (2y + (x - y)(1 - y^{p+1}x))g_{p+1} + xyg_p = 0. \tag{4.204}$$

This equation is similar to eqn (4.77), and a solution

$$g_p = \lambda^p \sum_{m=0}^{\infty} r_m(y) y^{mp}, \tag{4.205}$$

with $r_0(y) = 1$ is attempted (see Section 4.4.3). Substitution and simplification leads to the expressions

$$\lambda^2 - (x + y)\lambda + xy = 0, \tag{4.206}$$

and

$$r_m(y) = \frac{\lambda^m (x - y)^m y^{m(m-1)/2}}{(\lambda x^{-1} y; y)_m (\lambda; y)_m}. \tag{4.207}$$

From eqn (4.206) there are two possible choices ($\lambda = x$ or $\lambda = y$) for λ, each producing a solution. Note that $G_1 = 2x^2 + O(x^3)$ which approaches zero as $x \to 0$. If $\lambda = y$, then $g_1 = y(1 + O(x))$ which does not approach zero as $x \to 0$ (with $xz = y$ fixed), and therefore the choice that $\lambda = y$ can be ruled out. Choosing $\lambda = x$ gives the solutions in terms of q-deformed Bessel functions.

$$g_0 = \sum_{m=0}^{\infty} \frac{x^{2m}(1 - z)^m (xz)^{m(m-1)/2}}{(xz; xz)_m (x; xz)_m} = J(x, xz, (z - 1)x^2); \tag{4.208}$$

$$g_1 = x \sum_{m=0}^{\infty} \frac{x^{2m}(1 - z)^m (xz)^{m(m+1)/2}}{(xz; xz)_m (x; xz)_m} = J(x, xz, z(z - 1)x^3). \tag{4.209}$$

From eqn (4.203) it follows that there are singularities in $G_\Delta(x, z)$ at finite values of x if $z = 0$. In this circumstance, $g_0 = 1$ and $g_1 = x$ and from eqn (4.203),

$G_\Delta(x, 0) = x/(1-x)$. This has a singularity if $x = 1$, and since $x_c(z)$ is not increasing with z, so $x_c(z) \le 1$. On the other hand, define

$$h(x, z) = \frac{g_0(x, z)}{g_1(x, z)}, \tag{4.210}$$

which is a ratio of q-deformed Bessel functions, so that

$$G_\Delta(x, z) = -\frac{ah(x, z) - 2x}{bh(x, z) - 2x} = \frac{b - a}{b - 2x/h(x, z)}. \tag{4.211}$$

Then there are singularities in $G_\Delta(x, z)$ at zeros of the denominator. Both g_0 and g_1 have poles when $x^{n+1}z^n = 1$ for $n = 0, 1, 2, \ldots$, and since $g_0 \neq g_1$, there are points between these poles where $bh(x, z) = 2x$. In this case the denominator in eqn (4.211) is zero, and there is a singularity in $G_\Delta(x, z)$. For fixed values of x, the zeros in the denominator accumulate on $z = 1/x$; thus, there is a curve of essential singularities along $xz = 1$ in $G_\Delta(x, z)$. It is more difficult to extract the rest of the singularities in the generating function. Let

$$g_0(t; x, z) = \sum_{m=0}^{\infty} \frac{t^m x^{2m}(1-z)^m (xz)^{m(m-1)/2}}{(xz; xz)_m (x; xz)_m}. \tag{4.212}$$

Then $g_0(1; x, z) = g_0$ in eqn (4.208), and $xg_0(xz; x, z) = g_1$. Moreover, $h(x, z) = g_0(1; x, z)/[xg_0(xz; x, z)]$. It is not difficult to check that $g_0(t; x, z)$ satisfies

$$g_0(t; x, z) - \left(1 + \frac{1}{z} + tx^2(1-z)\right) g_0(xzt; x, z) + \frac{1}{z} g_0((xz)^2 t; x, z) = 0. \tag{4.213}$$

Define $H(t) = g_0(t; x, z)/g_0(xzt; x, z)$ and let $\alpha_p = 1 + 1/z + tx^2(1-z)(xz)^p$, and note that $H(1) = h(x, z)$. Division of equation (4.213) by $g_0(xzt; x, z)$ shows that

$$H(t) = \alpha_0 \left(1 - \frac{1}{z\alpha_0 H(xzt)}\right). \tag{4.214}$$

The critical points on the curve $xz = 1$ can be determined by using eqn (4.214). Put $t = 1$ and $xz = 1$; this gives $\alpha_0 = 1 + 1/z^2$ and

$$H(1) = \alpha_0 - \frac{1}{zH(1)}, \tag{4.215}$$

which can be written as a quadratic and solved:

$$H(1) = (z^2 + 1 \pm \sqrt{(z-1)(z^3 - 3z^2 - z - 1)})/2z^2. \tag{4.216}$$

Since $zH(1) = h(x, z)$ along the curve $xz = 1$ of essential singularities in $G_\Delta(x, z)$, it follows from eqns (4.211) and (4.216) that there is a square root singularity in $G_\Delta(x, z)$ as the tricritical point is approached along the τ-line. By the assumption in eqn (2.15) this implies that $2 - \alpha_u = 1/2$. Furthermore, $b = (z^2 + 1)/z^3$ in eqn (4.211), and since

Table 4.3: Exponents of collapsing partially directed walks

ϕ	α	$2-\alpha_t$	$2-\alpha_u$	ν_t	y_t
$\frac{2}{3}$	$\frac{1}{2}$	$\frac{1}{3}$	$\frac{1}{2}$	$\frac{1}{6}$	4

$h(x,z) = zH(1)$ along the curve $xz = 1$, this gives $bh(x,z) = 2x$ along the curve $xz = 1$ if (after significant simplification)

$$\frac{(z-1)\sqrt{(z-1)(z^3-3z^2-z-1)}}{z(z^2+1)} = 0; \tag{4.217}$$

in other words, when $z = 1$ or when $z = 3.382975\ldots$. It can be checked that the singularity at $z = 1$ cancels in eqn (4.211), so that $z_c = 3.382975\ldots$. If $z > z_c$, then the critical curve $x_c(z) = 1/z$ is a curve of first-order transitions. If $z < z_c$, then the critical curve starts in the point $x_c(0) = 1$, and must finally meet the curve $xz = 1$ at $(z_c, 1/z_c)$, since the free energy is convex and continuous (and therefore, $x_c(z)$ is also continuous). The non-analyticity in $x_c(z)$ at $z = z_c$ is interpreted as the collapse transition.

Lastly, it must also be shown that $G_\Delta(x,z)$ has no singularities if $z > z_c$ and $0 \leq x \leq 1/z$. From eqn (4.214) a continued fraction for $H(t)$ can be written down:

$$H(t) = \alpha_0\left(1 - \cfrac{1}{z\alpha_0\alpha_1\left(1 - \cfrac{1}{z\alpha_1\alpha_2(1-\cdots)}\right)}\right) = \alpha_0(1-C). \tag{4.218}$$

By Worpitsky's theorem [356] this converges if

$$\inf_{p\geq 0} |z\alpha_p\alpha_{p+1}| \geq 4, \tag{4.219}$$

and moreover, $|C - 4/3| \leq 2/3$, from which it is noted that $-z\alpha_0 \leq h(x,z) \leq z\alpha_0/3$. There is a singularity in $G_\Delta(x,z)$ if the denominator in eqn (4.211) is equal to zero. By using monotonicity and calculating the bounds on $h(x,z)$ explicitly at some values of x and z (where $z > z_c$ and $xz \leq 1$), one sees that there are no other singularities in $G_\Delta(x,z)$ if $z > z_c$ and $xz \leq 1$. This model strongly supports the notion that the collapse transition occurs at a tricritical point in the sense explained in Chapter 2.

The functions g_0 and g_1 in eqns (4.208) and (4.209) is related to a q-deformed Bessel function. Derivatives of these can be taken in order to find the derivative of $G_\Delta(x,z)$. Using again a symbolic computational program, this approach gives the gap exponent $\Delta = 3/2$. Thus $\phi = 2/3$ and so $2 - \alpha_t = 1/3$. The rest of the exponents may be computed from the homogeneity hypothesis; it appears that $\nu_t = 1/y_g = 1/6$ and $y_t = 4$. In addition, the specific heat exponent is $\alpha = 1/2$. The exponents are listed in Table 4.3, see also reference [289]. Scaling in the collapsed polymer phase was examined in reference [289], and it was found that $\gamma_- = 1/4$ and $\sigma = 1/2$ in this model.

4.7 Directed animals and directed percolation

It can be shown that there is a collapse transition in a model of self-interacting directed site-animals by using results from a model of directed percolation [108]. The animals will include a contact activity between vertices across the diagonal of unit squares in the square lattice, and at a critical value of the activity a collapse transition to a compact directed animal will be seen. This model is defined in a directed lattice. In the square lattice the positive direction along the main diagonal is the preferred direction, and all edges are oriented such that they have a positive inner product with a unit vector along the main diagonal. If ij is a directed edge from vertex i to vertex j then i is a *predecessor* of j.

Directed percolation is defined as follows. Let the origin be an *unblocked* site, and let every other vertex in the square lattice be unblocked with probability p_s (and *blocked* with probability $q_s = 1 - p_s$). Let any directed edge be *present* with probability p_b and *absent* with probability $q_b = 1 - p_b$. A percolation cluster is now *grown* from the unblocked origin by *occupying* first the unblocked origin, and then occupying recursively only those vertices which are both unblocked, and have a present edge incident *to* it *from* an occupied predecessor. The cluster stops growing when there are no more unblocked vertices adjacent to it via present edges. A typical cluster is illustrated in Fig. 4.26.

A directed percolation cluster is also called a *directed animal*. These site-animals will be counted with respect to number of vertices v, number of edges n, and the number of (next) nearest-neighbour contacts between vertices at sites in the animals with coordinates of the form (i, j) and $(i+1, j-1)$ where $((i, j-1), (i, j))$ and $((i, j-1), (i+1, j-1))$ and/or $((i, j), (i+1, j))$. In any such cluster define v_1 to be the total number of occupied sites with one occupied predecessor in a directed percolation cluster. Let v_2 be the total number of occupied sites with occupied predecessors. Let v_3 be the number of perimeter sites adjacent to the cluster with one occupied predecessor, and let v_4 be the number of perimeter sites with two occupied predecessors. In addition, let the number of contacts be k and the number of edges in the cluster be n. Then counting edges and

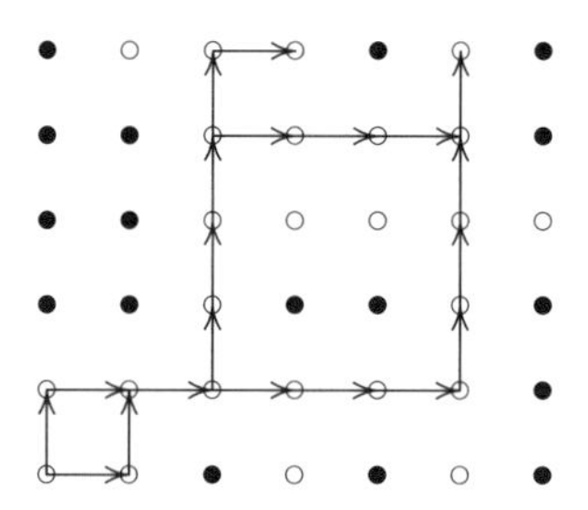

Fig. 4.26: A directed animal. Blocked vertices are indicated by $\bullet$, while unblocked vertices are represented by $\circ$.

vertices in the animal gives the following relationships:

$$\begin{aligned} v_1 &= 2v - n - 2, \qquad & v_2 &= n - v + 1, \\ v_3 &= n - 2k + 2, \qquad & v_4 &= v + k - n - 1. \end{aligned} \tag{4.220}$$

The probability that the origin is in an animal α of v vertices is [77]

$$P_v(\alpha) = p_s^{v-1} p_b^{v_1} (1 - q_b^2)^{v_2} (1 - p_b p_s)^{v_3} (q_s + p_s q_b^2)^{v_4}. \tag{4.221}$$

The relations in eqns (4.220) indicates that this probability can be expressed in terms of $\{v, n, k\}$; substitution and simplification gives

$$\begin{aligned} P_v(\alpha) = &\left[\frac{(1-q_b^2)(1-p_b p_s)^2}{p_s p_b (q_s + p_s q_b^2)}\right]\left[\frac{p_s p_b^2 (q_s + p_s q_b^2)}{(1-q_b^2)}\right]^v \\ &\times \left[\frac{(1-q_b^2)(1-p_b p_s)}{p_b (q_s + p_s q_b^2)}\right]^n \left[\frac{(q_s + p_s q_b^2)}{(1-p_b p_s)^2}\right]^k. \end{aligned} \tag{4.222}$$

This probability is the probability that the origin will be contained in a particular directed animal α.

The partition function of a model of directed animals is

$$a_v^\dagger(y, z) = \sum_{n \geq 0, k \geq 0} a_v^\dagger(n, k) y^n z^k, \tag{4.223}$$

where $a_v^\dagger(n, k)$ is the total number of directed animals (which has the origin as the lexicographically least vertex) with v vertices, n edges and k contacts. Two directed animals can be concatenated as in Fig. 4.27 by adding a single directed edge from the top (lexicographically most) vertex of the first animal to the bottom (lexicographically least) vertex of the second animal. If the first animal had v vertices, n_1 edges, k_1 contacts, and the second had w vertices, $n - n_1$ edges and $k - k_1$ contacts, then this construction shows that

$$\sum_{n_1=0}^{n} \sum_{k_1=0}^{k} a_v^\dagger(n_1, k_1) a_w^\dagger(n - n_1, k - k_1) \leq a_{v+w}^\dagger(n + 1, k). \tag{4.224}$$

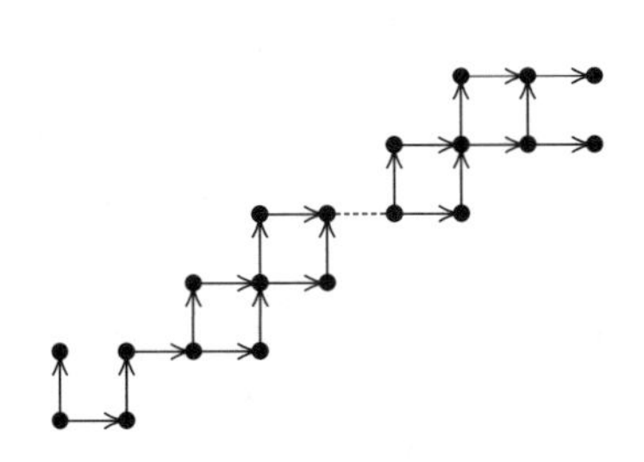

Fig. 4.27: Concatenating two directed animals.

Multiplying by $y^n z^k$ and executing the sums give

$$a_v^\dagger(y,z) a_w^\dagger(y,z) \leq y^{-1} a_{v+w}^\dagger(y,z). \tag{4.225}$$

Thus, by Lemma A.1 there is a limiting free energy in this model defined by

$$\mathcal{F}_\alpha(y,z) = \lim_{v\to\infty} \frac{1}{v} \log a_v^\dagger(y,z), \tag{4.226}$$

and moreover, it is convex in both arguments, and differentiable almost everywhere.

A connection between the limiting free energy of directed animals and directed percolation is found if eqn (4.222) is summed over n and k. This gives

$$P_v = \sum_{n\geq 0, k\geq 0} P_v(\alpha) = \frac{(1-q_b^2)(1-p_b p_s)^2}{p_b p_s (q_s + p_s q_b^2)} \left[\frac{p_s p_b^2 (q_s + p_s q_b^2)}{(1-q_b^2)} \right]^v a_v^\dagger(y_*, z_*), \tag{4.227}$$

where $P_v = \sum_\alpha P_v(\alpha)$ is the probability that the animal at the origin has v vertices, and

$$y_* = \frac{(1-q_b^2)(1-p_b p_s)}{p_b(q_s + p_s q_b^2)}, \qquad z_* = \frac{q_s + p_s q_b^2}{(1-p_b p_s)^2}. \tag{4.228}$$

It is possible to solve explicity for p_b and p_s in these equations in terms of y_* and z_*. Taking the logarithm of the above, dividing by v and letting $v \to \infty$ gives

$$\mathcal{F}_\alpha(y_*, z_*) = -\log \left[\frac{p_s p_b^2 (q_s + p_s q_b^2)}{(1-q_b^2)} \right] + \lim_{v\to\infty} \frac{1}{v} \log P_v, \tag{4.229}$$

If p_b is small enough, then it is known that $P_v \leq e^{-\gamma v}$ for some fixed positive number γ (see Appendix D for a proof of this fact in the case of percolation). On the other hand, if p_b approaches 1, then the percolated phase is found, and in this case $\limsup_{v\to\infty} [\log P_v]/v = 0$ (see eqn (D.4) for a proof). In other words,

$$\mathcal{F}(y_*, z_*) \begin{cases} = -\log \left[\dfrac{p_s p_b^2 (q_s + p_s q_b^2)}{(1-q_b^2)} \right], & \text{if } p_b \text{ is large enough;} \\ \leq -\log \left[\dfrac{p_s p_b^2 (q_s + p_s q_b^2)}{(1-q_b^2)} \right] - \gamma, & \text{if } p_b \text{ is small enough.} \end{cases} \tag{4.230}$$

This is true for each $p_s \in (0,1)$, and since $\gamma > 0$ this result implies the existence of a non-analyticity in $\mathcal{F}(y_*, z_*)$.

It is also the case that a version of directed percolation, called *compact directed percolation* [81], is closely related to staircase polygons. In this model, all lattice sites are unblocked. The directed cluster is generated as follows: if both predecessors of a given site are occupied, then occupy the site with probability one. If only one predessor is occupied, then occupy the site with probability p, and in all other cases the site stays unoccupied. In terms of an algebraic language, an O trapezium (see Fig. 4.7) is next with probability p^2, a U or a D with probability $p(1-p)$ and a C with probability $(1-p)^2$. If a cluster terminates, then the number of Os and Cs are the same, and if the length of the cluster is ℓ, then the weight is $[p(1-p)]^\ell$. The resulting algebraic language is the two-coloured Motzkin language, and the directed clusters are in one-to-one correspondence with staircase polygons, so that the generating function can be read of from eqns (4.64) or (4.73), with z^2 replaced by $p(1-p)$:

$$G_r(p) = \left(1 - 2p(1-p) - \sqrt{1 - 2p(1-p)}\right)\Big/2. \tag{4.231}$$

Thus, the critical value of p is found if $p(1-p) = 1/4$, or $p_c = 1/2$. The expected length of the cluster is found by taking a derivative of $G_r(p)$: if $s(\ell)$ is the number of clusters of length ℓ, then the expected length is

$$\begin{aligned} L(p) &= \sum_{\ell=0}^{\infty} \ell s(\ell)[p(1-p)]^\ell \\ &= \frac{1}{1-2p}\frac{d}{dp}G_r(p) \\ &= \frac{p(1-p)}{\sqrt{1-4p(1-p)}} - p(1-p). \end{aligned} \tag{4.232}$$

Growth of a compact directed cluster near an interface, with a visit activity z, can also be described using the adsorption of staircase polygons [28]. The partition function for even length clusters is given by eqn (4.188), and the generating function of even length clusters is given by eqn (4.189). The critical percolation probability is given by eqn (4.193), where x^2 should be replaced with $p(1-p)$. This shows that the percolation probability is a function of z, in the desorbed phase it remains $p_c = 1/2$, but adsorbed clusters percolate at $p_c = [\sqrt{z} - \sqrt{z - 2\sqrt{z-1}}]/2\sqrt{z}$, where the adsorption occurs at $z_c = 2$.

A review of directed percolation can be found in reference [99], and critical exponents have been estimated. For example, the entropic exponent has been estimated ($\gamma = 2.2783(7)$), and if the correlation length diverges as $(p_c - p)^{-\nu_\perp}$ perpendicular to the prefered direction, and as $(p_c - p)^{-\nu_\parallel}$ parallel to the prefered direction, then $\nu_\perp = 1.0969(3)$ and $\nu_\parallel = 1.7339(3)$ [107, 108]. A connection with self-organized criticality was also noted by Z. Olami *et al.* [277] and P. Grassberger [144]. The collapse of directed animals was examined by D. Dhar [77] and further considered in [181]. Directed column-convex animals were counted in [208] and three-dimensional directed percolation was considered in [305].

5

Interacting models of walks and polygons

5.1 Walks and polygons

There is a close relation between the numbers of lattice polygons and self-avoiding walks. In particular, if c_n is the number of walks of length n,[1] and p_n is the number of polygons of length n, then

$$\lim_{n\to\infty} \frac{1}{n} \log c_n = \lim_{n\to\infty} \frac{1}{n} \log p_n = \log \mu_d, \tag{5.1}$$

a fact which does not have a trivial proof [161, 166] (see Theorems 1.1 and 1.2). This connection can be understood in terms of *unfolded walks*, and in this section I aim to explore this, and other facts, in order to prove a *pattern theorem* [210]. A pattern theorem will generally state that a finite subwalk which can occur three times in a walk (such a subwalk is called a Kesten pattern), will occur with positive density on almost all walks or polygons of a given length. In other words, it is a statement about the density function of the subwalk. A proof of the pattern theorem for walks and polygons is closely related to a proof of eqn (5.1), and in this section I shall develop the ideas which will give eqn (5.1) before I show that they can be used to prove a pattern theorem.

Walks and polygons were defined in Section 1.2, where the existence of the limits in eqn (5.1) was shown (Theorems 1.1 and 1.2). It was noted that the self-avoiding walk is a submultiplicative model, while models of polygons are supermultiplicative; moreover, the proofs that the limits in eqn (5.1) exist rely on the sub- and supermultiplicative nature of the models.

The two most important vertices in a walk or a polygon are the *bottom vertex* and the *top vertex*. These are the first and last vertices in an (increasing) lexicographic ordering of the vertices by their coordinates (first in the X-direction, then in the Y-direction, and so on). The first coordinate of a vertex v_i in a walk or a polygon will be denoted by X_i, the second coordinate by Y_i, and the last coordinate (or d-th coordinate in d dimensions) will always be denoted by Z_i. If two distinct vertices v and w are such that

[1] Walks or polygons can be counted by fixing a vertex at the origin, and then counting the distinct conformations (the walks or polygons are said to be *rooted*). Such walks have a first vertex, and they are called *oriented*. Alternatively, one may consider two walks or polygons to be identical if one can be translated to become identical with the other; these are not oriented. In this chapter, walks and polygons will be counted up to translational equivalence. If there are c_n walks, then there are $2c_n$ oriented walks.

w is lexicographically greater than v, then this is denoted by

$$v \lhd w, \qquad (w \text{ is lexicographically greater than } v). \tag{5.2}$$

The coordinates of the top vertex t will be denoted by $(X_t, Y_t, \dots, Z_t)$ and of the bottom vertex b by $(X_b, Y_b, \dots, Z_b)$. The *X-span* of a walk or a polygon is defined by the difference in the first coordinate between the top vertex and the bottom vertex: $X_t - X_b$ (this is always non-negative). Similarly, the *Y-span* is $Y_t - Y_b$, and so on. The *(total) span* of a polygon or walk A is the sum over all the spans:

$$S(A) = |X_t - X_b| + |Y_t - Y_b| + \cdots + |Z_t - Z_b|. \tag{5.3}$$

A walk is *chiral* if it is not equal to its own reflection, otherwise it is *achiral*.

5.1.1 Unfolded walks

An *fl-walk* is a walk whose end-vertices are also its top and bottom vertices. An f-walk (or an l-walk) is a walk whose bottom vertex (top vertex) is also an end-vertex. An f-walk and an fl-walk are illustrated in Fig. 5.1(a).

Let c_n^{fl}, c_n^f and c_n^l be the numbers of fl-, f- and l-walks respectively. By identifying the top and bottom vertices of two fl-walks A_1 and A_2, they can be concatenated into a new fl-walk, denoted by $A_1 A_2$. In particular, c_n^{fl} is supermultiplicative:

$$c_n^{fl} c_m^{fl} \le c_{n+m}^{fl}, \tag{5.4}$$

and so the limit $\lim_{n\to\infty} \frac{1}{n} \log c_n^{fl}$ exists (see Lemma A.1, and Theorem 1.2, and note that $c_n^{fl} \le (2d)^n$).

The most important construction in the proof of eqn (5.1) is called an *unfolding* of a walk. This construction will change an f-walk into an fl-walk by first cutting it into fl-walks and then reassembling these into an fl-walk. The cutting of an f-walk into fl-walks is called its *partitioning*.

Construction 5.1 (Partitioning an f-walk) Note that the top vertex of an f-walk divides the walk into two parts, one which contains the bottom vertex (denoted by A_1, and which is an fl-walk), and an l-walk (denoted by B_2). The bottom vertex of B_2 divides B_2 into an fl-walk A_2 which contains the top vertex of B_2, and an f-walk (denoted by B_3); see Fig. 5.1(b). Generally, the walk B_{2n-1} is an f-walk which is cut

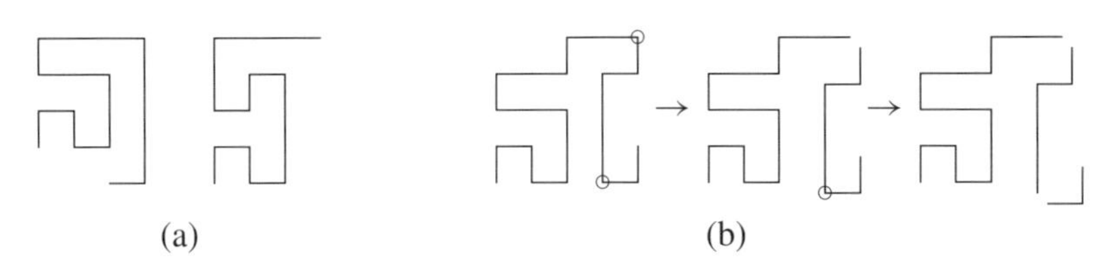

Fig. 5.1: (a) An f-walk and an fl-walk. (b) An f-walk can be cut into fl-walks by successive divisions of the f-walk at top and bottom vertices.

into an fl-walk A_{2n-1} at its top vertex, and an l-walk B_{2n}. In its turn, B_{2n} is cut at its bottom vertex into an fl-walk A_{2n} and an f-walk B_{2n+1}. The process terminates when B_n is an fl-walk, and it partitions a given f-walk A into fl-walks $\{A_1, A_2, \ldots, A_N\}$. An important fact is that the number of fl-walks obtained in a partitioning is at most $C_0 n^{d/(d+1)}$, where C_0 is a fixed constant. This is proven in Lemma 5.2. □

Lemma 5.2 *Let* $\{A_1, A_2, \ldots, A_N\}$ *be the partitioning of an* f*-walk* A *of length* n *into* fl*-walks by alternating cuts of* A *at its top and bottom vertices, as in Construction 5.1. If* $N > [(2d+2)^d]/d!$, *then* $N \leq C_0 n^{d/(d+1)}$, *where* $C_0 = [4/d(d!)^{1/d}]^{d/(d+1)}$.

Proof Let t_i be the top vertices, and b_i be the bottom vertices, of the fl-walks A_i. Denote the length of A_i by $|A_i|$. Note that by Construction 5.1, $t_1 \rhd t_2 \rhd t_3 \rhd \cdots \rhd t_N$, and $b_1 \lhd b_2 \lhd b_3 \lhd \cdots \lhd b_N$, while $b_N \lhd t_N$. Thus

$$t_1 - b_1 \rhd t_2 - b_2 \rhd \cdots \rhd t_N - b_N.$$

Therefore, the d-dimensional *vectors* $t_i - b_i$ are all unequal. In addition $|A_i| \geq S(A_i) = S(t_i - b_i)$. Thus, the number of edges in A_i is minimized if each ξ-span (for $\xi = X, Y, \ldots, Z$) is minimized. Hence, the minimum number of edges is found if all the vectors $t_i - b_i$ are in the first quadrant and with minimal total spans. Consider the points $(i, j, \ldots, k)$ with $(i + j + \cdots + k) \leq m$. The number of such points is

$$\begin{aligned}\sum_{i=0}^{m}\sum_{j=0}^{m}\cdots\sum_{k=0}^{m}\mathcal{I}(i+j+\cdots+k \leq m) &= \sum_{q=0}^{m}\sum_{i=0}^{m}\cdots\sum_{k=0}^{m}\delta(i+j+\cdots+k-q)\\ &= \frac{1}{d!}(m+1)(m+2)\cdots(m+d),\end{aligned}$$

where $\mathcal{I}(P)$ is an indicator function which is equal to one if P is true, and zero otherwise. Each point has span $(i + j + \cdots + k)$, so that the least number of edges counted by fl-walks corresponding to these points is

$$\begin{aligned}&\sum_{i=0}^{m}\sum_{j=0}^{m}\cdots\sum_{k=0}^{m}(i+j+\cdots+k)\mathcal{I}(i+j+\cdots+k \leq m)\\ &\quad= \sum_{q=0}^{m}\sum_{i=0}^{m}\cdots\sum_{k=0}^{m}(i+j+\cdots+k)\delta(i+j+\cdots+k-q)\\ &\quad= \frac{1}{2(d-1)!}m(m+1)(m+1)(m+2)\cdots(m+d-1).\end{aligned}$$

Suppose there are n edges in all. Then this is a lower bound on n:

$$n \geq \frac{1}{2(d-1)!}m(m+1)(m+1)(m+2)\cdots(m+d-1). \qquad (\dagger)$$

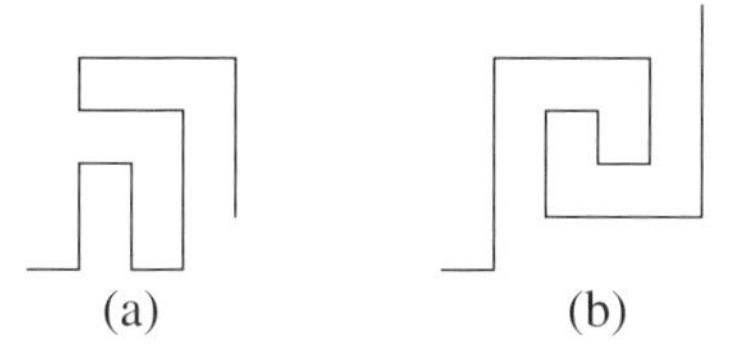

Fig. 5.2: (a) An unfolded walk, and (b) a doubly unfolded walk. A t-walk is obtained by deleting the first and last edges of a doubly unfolded walk.

On the other hand, there are N fl-walks in the partitioning, so choose m such that almost all these are counted. Let m be the largest number such that

$$\frac{1}{d!}(m+1)(m+2)\cdots(m+d) \leq N \leq \frac{1}{d!}(m+2)(m+3)\cdots(m+d+1).$$

Combine this with eqn (†) to obtain

$$n \geq \frac{dm(m+1)}{2(m+d)}N,$$

and since $(m+1)^d \leq d!N \leq (m+d+1)^d$, this becomes

$$n \geq \frac{d}{2}\frac{N\big((d!N)^{1/d}-(d+1)\big)\big((d!N)^{1/d}-d\big)}{\big((d!N)^{1/d}-d+1\big)}.$$

If N is so large that $(d!N)^{1/d} \geq 2(d+1)$, then this can be simplified to

$$n \geq \frac{d(d!)^{1/d}}{4}N^{(d+1)/d}.$$

Thus, if N is greater than $[(2d+2)^d]/d!$, then $N \leq [4n/d(d!)^{1/d}]^{d/(d+1)}$. □

A walk A with vertices $\{A_i\}_{i=0}^n$, labelled sequentially from its least end-point, and with coordinates $\{X_i, Y_i, \ldots, Z_i\}_{i=0}^n$, is an *unfolded walk* if $X_0 < X_i \leq X_n$ for $i = 1, 2, \ldots, n-1$. These walks are also said to be X-unfolded. Y-unfolded walks, Z-unfolded walks, and so on, are similarly defined. Unfolded walks are also called *bridges* [249]. A walk is *doubly unfolded* if both $X_0 < X_i \leq X_n$ and $Z_0 \leq Z_i < Z_n$ for all $i = 1, 2, \ldots, n-1$. Such a walk is also said to be XZ-unfolded. XY-unfolded, or ZX-unfolded, or other combinations, are defined similarly. The number of unfolded walks of length n is denoted by $c_n^\dagger$, and of doubly unfolded walks by $c_n^\ddagger$. The unfolding of an f-walk is a construction which relates an f-walk to an (X)-unfolded walk. Unfolded walks are illustrated in Fig. 5.2. The construction proceeds by partitioning an f-walk into fl-walks, which are joined into an unfolded walk.

Construction 5.3 (Unfolding an f-walk) Let A be an f-walk of length n which is partitioned into the set $\{A_1, A_2, \ldots, A_N\}$ of fl-walks. Let the X-span of A_i be a_i, and let $A_i^\dagger$ be the reflection of A_i through the hyperplane normal to the X-direction and containing the top vertex of A_i. Let i be odd and suppose that $A_i A_{i+1}^{\dagger}$ is the subwalk[2] composed of A_i and the reflected image of A_{i+1}. Then A_i and A_{i+1} can be obtained if the X-span a_i of A_i is known.[3] *Unfold* the walk A now as follows. Suppose that N is even and first construct $A_{N-1}A_N^{\dagger}$. Then $A_{N-2}(A_{N-1}A_N^{\dagger})^{\dagger} = A_{N-2}A_{N-1}^{\dagger}A_N$, and continue in this way until $A_1 A_2^{\dagger} A_3 A_4^{\dagger} \cdots A_N^{(\dagger)}$ is constructed. This walk is an f-walk, with the property that its final vertex has the same X-coordinate as its top vertex.

On the other hand, if an f-walk of length n with its final vertex with a maximal X-coordinate is given and if the X-spans $\{a_i\}$ are all known, then it can be partitioned into $A_1, A_2^{\dagger}, \ldots, A_N^{(\dagger)}$, *provided* that this walk was in the first place obtained by an unfolding. The maximum number of ways of selecting X-spans is $\sum_{k \leq n} \pi_n(k) = P(n)$, where $\pi_n(k)$ is the number of partitions of the integer n into k parts. In other words, at most $P(n)$ f-walks may be mapped to the same walk in the unfolding.

An inequality relating f-walks and fl-walks can be obtained from the unfolding by appending a single edge in the X-direction on the final vertex of the unfolded f-walk (which has a maximal X-coordinate). The result is an fl-walk. If the edge is added in the negative X-direction on the first vertex of the unfolded f-walk, then an unfolded walk is found. In other words, this shows that

$$c_n^f \leq P(n) c_{n+1}^{fl}, \qquad c_n^f \leq P(n) c_{n+1}^{\dagger}.$$

This construction leaves all the coordinates of vertices, except the X-coordinates, unchanged. Notice that the chirality of some subwalks is changed. □

The number of partitions of an integer n is bounded from above by $\pi_n(k) \leq e^{\gamma\sqrt{n}}$ [179, 317], where γ is a positive constant (see footnote 5, Chapter 3). Thus, the inequality obtained in Construction 5.3 gives the following corollary.

Corollary 5.4 *It is possible to partition and reconstitute an f-walk into an fl-walk such that the Z-coordinate of all vertices remains the same. Moreover, there exists a finite positive constant γ such that this construction maps at most $e^{\gamma\sqrt{n}}$ f-walks of length n to a single fl-walk or length $n+1$. Thus*

$$c_n^f = c_n^l \leq e^{\gamma\sqrt{n}} c_{n+1}^{fl}.$$ □

A slightly different outcome can be found if, instead of reflecting fl-walks as in Construction 5.3, they are translated and concatenated bottom to top. Then the walk A is partitioned into fl-walks $\{A_1, A_2, \ldots, A_N\}$ which are then concatenated to form a single fl-walk $A_1 A_2 \cdots A_N$ by identifying the top vertex of A_i with the bottom vertex of A_{i+1}. The resulting inequality in this case is again $c_n^f = c_n^l \leq e^{\gamma\sqrt{n}} c_n^{fl}$, but the

---

[2] Suppose that i is odd. If $A_i A_{i+1}$ is a walk, then $A_i A_{i+1}^{\dagger}$ is also a walk, since A_{i+1} is reflected through that hyperplane normal to the X-direction and containing the top vertex t_i of $A_i A_{i+1}$ (naturally, A_i and A_{i+1} have the same top vertex if i is odd). In other words, every vertex in A_{i+1} is either reflected into the half-space with X-coordinate larger than the X-coordinate of the top vertex of $A_i A_{i+1}$, or is contained in the hyperplane, and remains fixed.

[3] Examine all vertices in the appropriate hyperplane, and choose that vertex which is the top vertex of the first part of the walk, and the top vertex of the reflected image of the second part of the walk. This vertex defines A_i and A_{i+1}. If there is no such vertex, then this walk could not have been obtained by the unfolding, and it contributes to an overcounting of f-walks by cutting them into fl-walks.

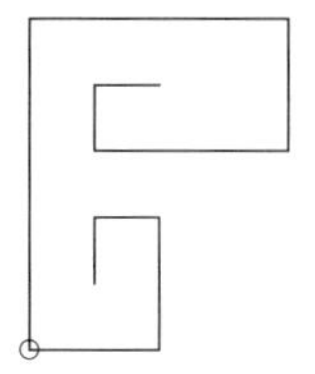

Fig. 5.3: If the walk is divided at its bottom vertex (marked with $\circ$), then two f-walks are obtained.

difference is that the Z-coordinates of vertices are not preserved, as opposed to the case in Construction 5.3.

Any walk can be divided into two f-walks by cutting it at its bottom vertex into two subwalks (see Fig. 5.3). If the walk is cut into an f-walk of length k, and an f-walk of length $n-k$, then

$$c_n \leq \sum_{k=0}^{n} c_k^f c_{n-k}^f. \tag{5.5}$$

The result is that the growth constants of f-walks, fl-walks and walks are all the same.

Theorem 5.5 *The following limits all exist, and are equal:*

$$\lim_{n\to\infty} \frac{1}{n} \log c_n^{fl} = \lim_{n\to\infty} \frac{1}{n} \log c_n^{f} = \lim_{n\to\infty} \frac{1}{n} \log c_n^{l} = \lim_{n\to\infty} \frac{1}{n} \log c_n = \log \mu_d.$$

Moreover, $c_n^{fl} \leq \mu_d^n$.

Proof By eqn (5.5), Corollary 5.4, and eqn (5.4),

$$c_n \leq \sum_{k=0}^{n} e^{\gamma(\sqrt{k}+\sqrt{n-k})} c_{k+1}^{fl} c_{n-k+1}^{fl} \leq \left[\sum_{k=0}^{n} e^{\gamma(\sqrt{k}+\sqrt{n-k})}\right] c_{n+2}^{fl}.$$

On the other hand, $c_n^{fl} \leq c_n^f = c_n^l \leq c_n$, and since $c_n^{fl} c_m^{fl} \leq c_{n+m}^{fl}$ the limit $\lim_{n\to\infty}\left[\log c_n^{fl}\right]/n$ exists (see Lemma A.1). By taking logarithms, dividing by n, and letting n tend to infinity in the above inequalities the existence and equality of all the limits is established. The inequality $c_n^{fl} \leq \mu_d^n$ follows from eqn (5.4) and Lemma A.1. $\square$

A walk with vertices $\{A_i\}_{i=0}^n$, labelled sequentially from its least end-point, and with coordinates $\{X_i, Y_i \ldots, Z_i\}_{i=0}^n$, is a *t-walk* if $X_0 \leq X_i \leq X_n$, $Y_0 \leq Y_i \leq Y_n$, ..., $Z_0 \leq Z_i \leq Z_n$, for each $0 \leq i \leq n$. Thus, a t-walk is an fl-walk which can be put in a d-dimensional rectangular box with its end-vertices at antipodal corners of the box. The t-walk is also said to *cross the box*. The number of t-walks is denoted by c_n^t, and c_n^t will also be shown to grow at the same exponential rate as c_n. In addition, any t-walk is an fl-walk, so that

$$c_n^t \leq c_n^{fl}. \tag{5.6}$$

It is possible to unfold an fl-walk into a t-walk: in d dimensions there are $2d - 2$ unfoldings necessary, as is demonstrated in Lemma 5.6.

Lemma 5.6 *There is a finite positive constant number γ such that in d-dimensions, $c_n^{fl} \leq e^{2d\gamma\sqrt{n}} c_n^t$.*

Proof A proof in two dimensions will be given, and the generalization to higher dimensions will follow easily. Let A be an fl-walk with vertices $\{v_0, v_1, \ldots, v_n\}$, and coordinates $(X(v_i), Y(v_i))$. Then $X(v_0) \leq X(v_i) \leq X(v_n)$ for $0 \leq i \leq n$, since A is an fl-walk. Find the *secondary top vertex* of A by a lexicographic ordering first in the Y-direction, and then the X-direction, and let this vertex be t_y (see Fig. 5.4). Divide A into two subwalks A_1 and A_2 at this vertex, and let A_2^* be the reflected image of A_2 through the hyperplane containing the vertex t_y and normal to the Y-axis. This leaves unperturbed all the X-coordinates of vertices in A. Place t_y at the origin. Note that the image of A_1, reflected through the hyperplane $X = -Y$, is an f-walk with bottom vertex t_y, and it can be unfolded as in Construction 5.3. On reversing the reflection through t_y (after the unfolding), all X-coordinates remained unchanged, but the first vertex now has a minimal Y-coordinate. Let this subwalk be $A_1^\dagger$. Similarly, the walk A_2^* can be reflected through the hyperplane defined by $Y = X$ to obtain an f-walk which can be unfolded. Then t_y is the bottom vertex of the resulting unfolded walk. On reversing the reflection, all X-coordinates remained unchanged, and the final vertex now has a maximal Y-coordinate. Let this be the walk $A_2^\dagger$; see Fig. 5.4. But then the walk $A_1^\dagger A_2^\dagger$ is an fl-walk with vertices $\{v_i\}_{i=0}^n$ such that $Y(v_0) \leq Y(v_i) \leq Y(v_n)$ for $0 \leq i \leq n$, and since all X-coordinates remained unchanged, $X(v_0) \leq X(v_i) \leq X(v_n)$ for $0 \leq i \leq n$. In other words, a t-walk of length n is obtained. Finally, if A_1 has length k and A_2 has length $n - k$, then at most $nP(k)P(n - k)$ walks have the same outcome under

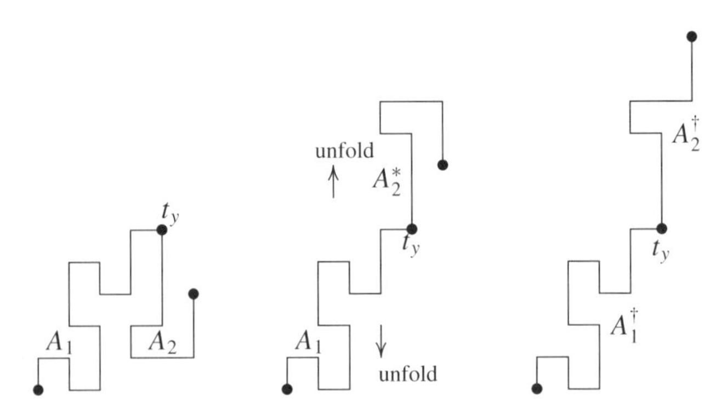

Fig. 5.4: Unfolding an fl-walk into a t-walk.

the unfoldings, where the factor of n is a result of the first reflection. But notice that $\sum_{k=0}^{n} P(k)P(n-k) \leq e^{2\gamma\sqrt{n}}$ for some constant γ. In other words, $c_n^{fl} \leq n\, e^{2\gamma\sqrt{n}} c_n^t$. In more than two dimensions, the construction carries on as above, two unfoldings for each dimension, with the secondary top and bottom vertices appropriately defined. A factor $n\, e^{2\gamma\sqrt{n}}$ is gained for each dimension. The factors of n can be absorbed into the exponential, by increasing γ. □

Every t-walk can be turned into a doubly unfolded walk by appending an edge in the negative X-direction on its bottom vertex, and an edge in the positive Y-direction on its top vertex,

$$c_n^t \leq c_{n+2}^{\ddagger}. \tag{5.7}$$

In addition, by Lemma 5.6,

$$c_n^{fl} e^{-2d\gamma\sqrt{n}} \leq c_n^t \leq c_{n+2}^{\ddagger} \leq c_{n+2}^{\dagger} \leq c_n. \tag{5.8}$$

Combining this with Theorem 5.5 has the important consequence that all these models of walks have the same growth constant.

Theorem 5.7 *The following limits all exist:*

$$\lim_{n\to\infty} \frac{1}{n} \log c_n^t = \lim_{n\to\infty} \frac{1}{n} \log c_n^{\ddagger} = \lim_{n\to\infty} \frac{1}{n} \log c_n^{\dagger} = \lim_{n\to\infty} \frac{1}{n} \log c_n = \log \mu_d.$$

Moreover, there exists a constant $\gamma_0 > 0$ such that the number of walks of length n, c_n, is bounded from above and below by

$$\frac{1}{2}\mu_d^n \leq c_n \leq e^{\gamma_0\sqrt{n}} \mu_d^n.$$

The upper bound on c_n is the Hammersley–Welsh bound on the number of self-avoiding walks [166].

Proof It only remains to show the bounds on c_n. The lower bound is a result of eqn (1.1) and Lemma A.1. The upper bound is derived using the fact that $c_n^{fl} \leq \mu_d^n$ (Theorem 5.5) together with Corollary 5.4 and eqn (5.5). □

Thus, unfolded and doubly unfolded walks grow at the same exponential rate as walks. This is a very important observation: while the self-avoiding walk is a submultiplicative model, unfolded walks are supermultiplicative, and they can be concatenated such that they interact only locally (around the point of concatenation) with one another. The Hammersley–Welsh bound was improved by H. Kesten to $c_n \leq e^{\gamma_0 n^{2/(d+2)}} \mu_d^n$ [211], see reference [249] as well.

5.1.2 Loops and polygons

A walk with vertices $\{v_i\}_{i=0}^{n}$ (where v_i has coordinates $(X_i, Y_i, \ldots, Z_i)$) is an *XY-loop* if $X_0 = X_n < X_i$ and $Y_0 \leq Y_1 \leq Y_n$ for all $1 \leq i \leq n-1$. An XY-loop has X-span h if $h = \max_i |X_0 - X_i|$. XZ-loops, YX-loops, and other variations can be similarly

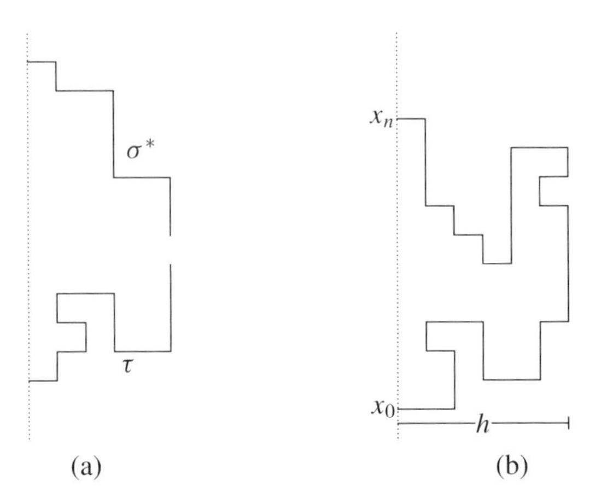

Fig. 5.5: (a) Two doubly unfolded walks can be put together to form a loop. The loop in (b) has X-span h.

defined. Two doubly unfolded walks can be joined into a loop: let A and B be two doubly unfolded walks of X-span h. Let A^* be the image of A after a reflection through the X-axis, or a rotation of π radians about the X-axis. Then A^* and B can be joined into a loop as in Fig. 5.5.

If there are $l_n(h)$ XY-loops of X-span h, and $c_n^\ddagger(h)$ is the number of doubly unfolded walks of X-span h, then the construction in Fig. 5.4(a) shows that for $k \leq n$:

$$c_k^\ddagger(h) c_{n-k}^\ddagger(h) \leq l_n(h). \tag{5.9}$$

Thus, it is possible to show that loops and walks have the same growth constant (this result is due to J.M. Hammersley *et al.* [165]).[4]

Theorem 5.8 $\lim_{n\to\infty}(1/n)\log l_n = \lim_{n\to\infty}(1/n)\log c_n = \mu_d$.

Proof Put $k = \lfloor n/2 \rfloor$ in eqn (5.9). Then

$$c_{2\lfloor n/2\rfloor} \geq l_{2\lfloor n/2\rfloor} \geq \sum_{h\geq 0} c_{\lfloor n/2\rfloor}^\ddagger(h) c_{\lfloor n/2\rfloor}^\ddagger(h).$$

For each n there is an h^* such that $c_n^\ddagger(h^*)$ is a maximum; h^* is the *most popular X-span* of doubly XY-unfolded walks.[5] Then $c_n^\ddagger = \sum_{h\geq 0} c_n^\ddagger(h) \leq n c_n^\ddagger(h^*)$. Substitute this into the equation above. Then

$$\frac{1}{\lfloor n/2\rfloor^2}\left(c_{\lfloor n/2\rfloor}^\ddagger\right)^2 \leq l_{2\lfloor n/2\rfloor} \leq c_{2\lfloor n/2\rfloor}.$$

---

[4] Loops are walks which are confined to a half-space, and are constrained to have their end-vertices in the boundary of the half-space. Let j_n be the number of walks in the half-space $Z \geq 0$ with one end-point in the hyperplane $Z = 0$. Then l_n is the number of walks in the half-space $Z \geq 0$ with both end-points in the hyperplane $Z = 0$. The asymptotic behaviour of both j_n and l_n has received much attention in the literature. In particular, Coulomb gas methods in two dimensions suggest that $j_n \simeq n^{\gamma_1 - 1}\mu_2^n$ and $l_n \simeq n^{\gamma_{11}-1}\mu_2^n$ with $\gamma_1 = 61/64$ and $\gamma_{11} = -3/16$. See for example references [95, 348].

[5] This is called a *most popular* argument.

By Theorem 5.7, the result follows. Strictly speaking, this shows that $\lim_{n\to\infty} \frac{1}{n} \log l_n = \log \mu$ if the limit is taken through even values of n, however, note that by translating the entire loop two steps in the X-direction, five edges can be added (in an obvious way) to join the translated loop with the $X = 0$ hyperplane again. This shows that $l_{n-5} \leq l_n \leq l_{n+5}$, and the same limit is found if $n \to \infty$ through odd numbers. □

Define the numbers $h_Y = |Y_0 - Y_n|, \ldots, h_Z = |Z_0 - Z_n|$; then the YZ-span of a loop is the ordered sequence $(h_Y, \ldots, h_Z)$. Define $l_n(h_Y, \ldots, h_Z)$ to be the number of loops of length n with YZ-span given by $(h_Y, \ldots, h_Z)$. There is a most popular YZ-span $(h_Y^*, \ldots, h_Z^*)$, so that

$$l_n(h_Y^*, \ldots, h_Z^*) \geq l_n(h_Y, \ldots, h_Z), \tag{5.10}$$

and since each of the h's can only take one of n positive values,

$$n^{d-1} l_n(h_Y^*, \ldots, h_Z^*) \geq \sum_{h_Y, \ldots, h_Z} l_n(h_Y, \ldots, h_Z) = l_n. \tag{5.11}$$

Let $l_n^-(h_Y, \ldots, h_Z)$ be the number of "negative loops" with $X_0 = X_n > X_i$ for all $i \in \{1, 2, \ldots, n-1\}$. The reflection symmetry through the $X = 0$ hyperplane implies that $l_n^-(h_Y, \ldots, h_Z) = l_n(h_Y, \ldots, h_Z)$. A negative loop and a loop can be pasted into a polygon if they have the same YZ-span. If a most popular argument is used, then

$$\begin{aligned}(l_{\lfloor n/2 \rfloor})^2 &\leq \lfloor n/2 \rfloor^{2(d-1)} l_{\lfloor n/2 \rfloor}(h_Y^*, \ldots, h_Z^*)\, l_{\lfloor n/2 \rfloor}^-(h_Y^*, \ldots, h_Z^*) \\ &\leq \lfloor n/2 \rfloor^{2(d-1)} p_{2\lfloor n/2 \rfloor} \leq \lfloor n/2 \rfloor^{2(d-1)} c_{2\lfloor n/2 \rfloor - 1},\end{aligned} \tag{5.12}$$

if p_n is the number of polygons of length n, and since a single edge can be removed from the polygon to obtain a walk. If the logarithm of eqn (5.12) is taken, and it is divided by n, and n tends to infinity, then from Theorem 5.8 eqn (5.1) is obtained.

Theorem 5.9 $\lim_{n\to\infty}(1/n) \log p_n = \lim_{n\to\infty}(1/n) \log c_n = \log \mu_d$. □

In other words, the numbers of walks and polygons grow at the same exponential rate, and this completes the proof of eqn (5.1).

5.1.3 *The pattern theorem*

In this section I shall use unfolded walks and fl-walks to prove a pattern theorem for walks and polygons. The proof presented here is different from that in reference [210], and is due to J. Hammersley (unpublished). A *pattern* is any finite self-avoiding walk. A pattern P is said to *occur* in a walk A if it can be translated to coincide with a subwalk of A. The number of times a pattern occurs in a walk A is the total number of distinct subwalks in A which are translations of the pattern. A pattern is called a *Kesten pattern* if it can occur three times (independently) in a walk or a polygon. A little reflection will show that any Kesten pattern can occur an arbitrary number of times in a walk.[6]

[6] This result is called Bellman's theorem by J.M. Hammersley and S.G. Whittington [167], after the quote "What I tell you three times is true" (said by the Bellman) from *The Hunting of the Snark* by Lewis Carroll [50].

Theorem 5.10 *Any Kesten pattern can occur an arbitrary number of times along a walk.*

Proof Orient the walk and let P be a Kesten pattern and let B be the smallest (rectangular) box containing P. Suppose that P_1, P_2 and P_3 are the three occurrences of P in a walk A, and let P_1 occur first along the walk, then P_2, and then lastly P_3. Let the smallest boxes containing these be B_1, B_2 and B_3 respectively, and denote the boundary of a box by ∂B_i. Since the B_i are the smallest boxes to contain the P_i, there are vertices of the walk in ∂B_i for each value of i. In particular, there is a vertex of the walk in ∂B_1 which precedes any vertex of P_2 in the walk (even if B_1 and B_2 are coincident). Let this vertex be v_1. Similarly, there is a vertex in ∂B_3 which is preceded by all the vertices in P_2. Let this vertex be v_3. The existence of v_1 and v_3 guarantees the existence of a last vertex w_1 in ∂B_2 before P_2 is encountered in the walk, and a first vertex w_3 in ∂B_2 after the pattern P_2 has been encountered. Let A' be the subwalk of A from w_1 to w_3. Choose a vertex b (outside of B_2) which is lexicographically less than any vertex in A', and a vertex t (also outside of B_2) which is lexicographically more than any vertex in A'. Add edges (disjoint with B_2) between b and w_1, and between w_3 and t; the resulting walk τ is an fl-walk which contains P at least once (say $k \leq 3$ times). These fl-walks can be concatenated to create walks τ^N which contain P a total of Nk times. But notice that there are f-walks (which are subwalks of τ) which contain P once or twice; since τ is an fl-walk, trace it from its bottom vertex until P has occurred once or twice, and cut it; the resulting walk is an f-walk which contains τ once or twice. By finally concatenating this walk onto τ^N, a walk which contains P an arbitrary number of times can be constructed. These arguments are adapted from J.M. Hammersley and S.G. Whittington [167]. □

The proof of a pattern theorem relies on the use of generating functions of various classes of walks. In particular, define the generating functions $c(x) = \sum_{n\geq 1} c_n x^n$ of walks, and similarly the generating functions $c^f(x)$, $c^{fl}(x)$, and so on, for f-walks, fl-walks, and so on. If an f-walk A is partitioned into fl-walks $\{A_1, A_2, \ldots, A_N\}$ by Construction 5.1, then the fl-walks are lexicographically ordered as $A_1 \rhd A_2 \rhd \cdots \rhd A_N$, and they uniquely determine A (see the proof of Lemma 5.2). Let the lengths of the fl-walks A_i be m_i. Since the f-walk is uniquely determined by a set of fl-walks which can be lexicographically ordered, it follows that any f-walk is either an fl-walk, or is composed of two fl-walks (ordered lexicographically), or three fl-walks which are lexicographically ordered, and so on. Thus

$$\begin{aligned} c_n^f \leq c_n^{fl} &+ \frac{1}{2!} \sum_{m_1,m_2\geq 1}^{n-1} c_{m_1}^{fl} c_{m_2}^{fl} \delta_{m_1+m_2-n} \\ &+ \frac{1}{3!} \sum_{m_1,m_2,m_3\geq 1}^{n-1} c_{m_1}^{fl} c_{m_2}^{fl} c_{m_3}^{fl} \delta_{m_1+m_2+m_3-n} + \cdots, \end{aligned} \tag{5.13}$$

where the factor $1/n!$ accounts for the overcounting in the number of arrangements of the fl-walks (only one arrangement will do). Multiply the above by x^n, and sum over n. This gives the following result.

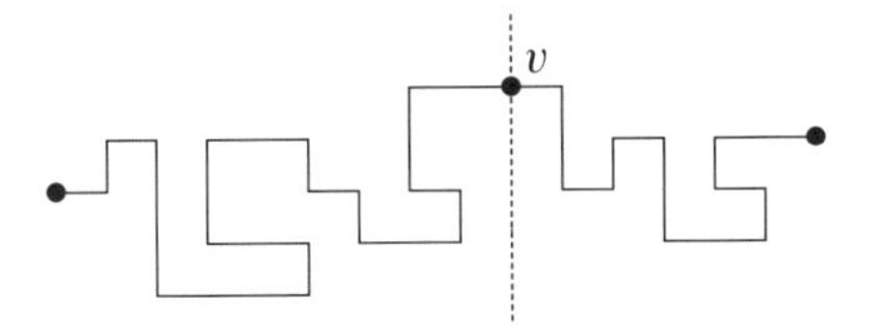

Fig. 5.6: A decomposable pattern walk; by cutting the walk at v, two prime pattern walks are obtained.

Lemma 5.11 *The generating functions of f-walks and fl-walks are related by:*

$$c^{fl}(x) \leq c^f(x) \leq e^{c^{fl}(x)}.$$

Thus, the generating function $c^{fl}(x)$ is finite if and only if the generating function $c^f(x)$ is finite. □

Since any walk is either an f-walk, or can be divided at its bottom vertex into two f-walks,

$$c_n \leq c_n^f + \sum_{k=1}^{n-1} c_k^f c_{n-k}^f. \tag{5.14}$$

Multiply this by x^n and sum over n. Use the convolution theorem to obtain

$$c(x) \leq c^f(x)\big(1 + c^f(x)\big). \tag{5.15}$$

In other words, with Lemma 5.11 this gives the following lemma.

Lemma 5.12 *The generating functions of walks and fl-walks are related by*

$$c^{fl}(x) \leq c(x) \leq e^{c^{fl}(x)}(1 + e^{c^{fl}(x)}).$$

Thus, $c(x) < \infty$ if and only if $c^{fl}(x) < \infty$. □

The important observation is that the bounds relating $c(x)$ and $c^f(x)$ and $c^{fl}(x)$ imply that $c(x)$ is finite if and only if $c^{fl}(x)$ is finite. Now it is known that c_n is submultiplicative, and so by Theorem 5.7, $c_n \geq \mu_d^n/2$. Thus, $c(x) = \sum_{n\geq 1} c_n x^n \geq (1/2)\sum_{n\geq 1}(\mu_d x)^n = (1/2)\mu_d x/(1-\mu_d x)$. Thus, $c(\mu_d^{-1}) = \infty$ and the following corollary is obtained.

Corollary 5.13 *The generating functions $c^f(x)$ and $c^{fl}(x)$ are finite if and only if $c(x)$ is finite. Moreover, $c(\mu_d^{-1}) = c^f(\mu_d^{-1}) = c^{fl}(\mu_d^{-1}) = \infty$.* □

The next important part in the proof is the definition of a *prime pattern-walk*. A *pattern-walk* is an unfolded walk with $X_0 < X_i < X_n$. A pattern-walk is illustrated in Fig. 5.6. A pattern-walk is *decomposable* if there is a dividing hyperplane normal to the X-direction which cuts it in a single vertex, such that each of the half-spaces defined by the hyperplane contains a subwalk which is a pattern-walk. A pattern-walk is a *prime pattern-walk* if it is not decomposable. A prime pattern-walk is illustrated in Fig. 5.7; note that any fl-walk can be turned into a prime pattern-walk.

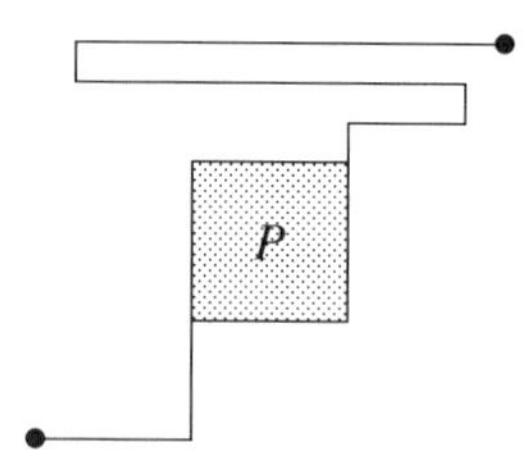

Fig. 5.7: By substituting any fl-walk P in the box, a prime pattern-walk is obtained.

Let the number of pattern-walks of length n be c_n^p, and the number of prime pattern-walks of length n be q_n.[7] Every pattern-walk is either a prime pattern-walk, or can be decomposed into a prime pattern-walk and a pattern-walk by cutting that prime pattern-walk incident with its bottom vertex from it. This gives the renewal equation $c_n^p = q_n + \sum_{m=1}^{n-1} q_m c_{n-m}^p$. Multiplying by x^n and summing over n gives the following relation between generating functions:

$$c^p(x) = \frac{q(x)}{1-q(x)}. \tag{5.16}$$

The number of pattern-walks is related to unfolded and fl-walks as follows. In the first place, any pattern-walk is unfolded, and any unfolded walk can be turned into a pattern-walk by addition of a single edge in the X-direction on its last vertex. Secondly, each fl-walk can be unfolded by adding an edge to its bottom vertex, and each unfolded walk becomes an fl-walk if a single edge in the X-direction is added on the last vertex of the unfolded walk. Thus

$$c_n^p \le c_n^\dagger \le c_{n+1}^p, \qquad c_n^{fl} \le c_{n+1}^\dagger \le c_{n+2}^{fl}. \tag{5.17}$$

Thus, $c^p(x) \le c^\dagger(x) \le x^{-1}c^p(x)$ and $c^{fl}(x) \le x^{-1}c^\dagger(x) \le x^{-2}c^{fl}(x)$. Using these inequalities with eqn (5.16) the following lemma is obtained.

Lemma 5.14 *The generating functions of prime pattern-walks, and of fl-walks, are related by*

$$x^2 c^{fl}(x) \le \frac{q(x)}{1-q(x)} \le x^{-1}c^{fl}(x). \qquad \square$$

Thus, since $c(x)$ is infinite if $x = \mu_d^{-1}$, and finite if $x < \mu_d^{-1}$, it follows that $c^{fl}(x)$ is infinite if $x = \mu_d^{-1}$, and finite if $x < \mu_d^{-1}$, by Corollary 5.13. From Lemma 5.14, the result is that $q(x) < 1$ if $x < \mu_d^{-1}$, and $q(x) = 1$ if $x = \mu_d^{-1}$. This is the key to the proof of a pattern theorem for walks. Define the radius of convergence of $c(x)$ by $x_c = \mu_d^{-1}$.

[7] I have taken a slightly different definition of a prime pattern-walk here [249]. The usual definition would use unfolded walks which are decomposable if they can be cut into subwalks which are unfolded. A prime pattern-walk is then an indecomposable unfolded walk. The reason for the different definition used here will be apparent later, when interacting models of walks are considered. However, notice that there is a one-to-one correspondence between unfolded walks and pattern-walks; since the last edge in a pattern-walk is constrained to point in the X-direction, it can be removed to produce an unfolded walk (with one edge less). On the other hand, any unfolded walk can be changed into a pattern-walk by adding an edge in the X-direction on the last vertex of the unfolded walk. Thus $c_n^\dagger = c_{n+1}^p$.

Theorem 5.15 *Let $c_n(\bar{P})$ be the number of walks which do not contain the prime pattern-walk P. Then*

$$\lim_{n\to\infty}\frac{1}{n}\log c_n(\bar{P}) = \log\mu(\bar{P}) < \log\mu = \lim_{n\to\infty}\frac{1}{n}\log c_n.$$

Proof Orient a walk counted by $c_{n+m}(\bar{P})$, and cut it at its $(n+1)$th vertex into a subwalk of length n, and a subwalk of length m, both oriented. Since there are $2\,c_n(\bar{P})$ such oriented walks of length n, this shows that $c_n(\bar{P})c_m(\bar{P}) \geq [c_{n+m}(\bar{P})]/2$, and by Lemma A.1 there exists a growth constant $\mu_d(\bar{P})$ such that $\lim_{n\to\infty}\frac{1}{n}\log c_n(\bar{P}) = \log\mu_d(\bar{P})$. Let $c(x;\bar{P})$ be the generating function of $c_n(\bar{P})$, and let $c^{fl}(x;\bar{P})$ be the generating function of fl-walks which do not contain the pattern P. Since P is a prime pattern-walk, identical arguments leading to Lemma 5.14 give

$$x^2 c^{fl}(x;\bar{P}) \leq \frac{q(x;\bar{P})}{1-q(x;\bar{P})} \leq x^{-1}c^{fl}(x;\bar{P}), \tag{†}$$

where $q(x;\bar{P})$ is the generating function of prime pattern-walks which do not contain the prime pattern-walk P. Now $c^{fl}(x;\bar{P})$ is infinite if and only if $c(x;\bar{P})$ is infinite; this follows from a sequence of arguments similar to those leading to Corollary 5.13. Let the radius of convergence of $c(x;\bar{P})$ be $x_c(\bar{P}) = [\mu_d(\bar{P})]^{-1}$. By eqn (†) above, $q(x;\bar{P})/(1-q(x;\bar{P}))$ is finite if and only if $c^{fl}(x;\bar{P})$ is finite. Since $c^{fl}(x_c(\bar{P});\bar{P}) = \infty$, it follows that $x_c(\bar{P})$ is a solution of the equation $q(x;\bar{P}) = 1$. Moreover, since $q(x)$ counts at least one walk (the prime pattern-walk P) which $q(x;\bar{P})$ does not count, it follows that $q(x) > q(x;\bar{P})$ (note that $q(x)$ and $q(x;\bar{P})$ are infinite power series with non-negative coefficients). This implies that $1 = q(x_c) > q(x_c;\bar{P})$ (that is, $c^{fl}(x_c;\bar{P}) < \infty$) and this implies that $c(x_c;\bar{P}) < \infty$. Thus $x_c < x_c(\bar{P})$. But $x_c = \mu_d^{-1}$, so this shows that $\mu_d > \mu_d(\bar{P})$. □

The pattern theorem in Theorem 5.15 can be extended to polygons by using doubly unfolded walks and loops as in Sections 5.1.1 and 5.1.2. Some care is needed. In particular, a pattern P may be absent in a walk, and present in its unfolded image (this may happen if the walk contains mirror images of a chiral pattern P, but not P itself). In order to avoid this, let Q_1 be a prime pattern-walk, and let its images under reflections be $\{Q_i\}_{i=2}^{2d}$. Concatenate all the Q_i into a pattern-walk P, and insert it in Fig. 5.7; this creates a prime pattern-walk R which has as subwalks all the Q_i. A model of walks which excludes the walk Q_1 also excludes the prime-pattern walk R and all its reflections, even if reflected images of Q_1 could occur. Unfolding a walk which does not contain R or its reflections cannot create R (since R and any reflected image of R will contain the pattern Q_1). But notice now that $p_n(\bar{Q}_1) \leq p_n(\bar{R})$. And since R is a prime pattern-walk, $\limsup_{n\to\infty}[\log p_n(\bar{Q}_1)]/n \leq \limsup_{n\to\infty}[\log p_n(\bar{R})]/n < \log\mu_d$. Thus, a pattern theorem for prime pattern-walks in polygons is obtained.

Theorem 5.16 *Let $p_n(\bar{P})$ be the number of polygons which do not contain the prime pattern-walk P. Then*

$$\lim_{n\to\infty}\frac{1}{n}\log p_n(\bar{P}) = \log\mu(\bar{P}) < \log\mu = \lim_{n\to\infty}\frac{1}{n}\log p_n.$$

Proof It only remains to show that $\lim_{n\to\infty} \frac{1}{n} \log p_n(\bar{P})$ exists for a prime pattern-walk P. Concatenation of polygons counted by $p_n(\bar{P})$ may result in the creation of copies of the prime pattern-walk P. This can be avoided if more edges than two are added between the polygons in the concatenation in Fig. 1.4(b). This gives $p_n(\bar{P})p_m(\bar{P}) \leq p_{n+m+2k}(\bar{P})$ for some value of $k \geq 0$, where k is chosen to avoid creating copies of P. By Lemma A.1, and since $p_n(\bar{P}) \leq p_n$, the limit $\lim_{n\to\infty} \frac{1}{n} \log p_n(\bar{P})$ exists, provided that $n \to \infty$ through even numbers. □

5.1.4 Density functions and prime patterns

Theorems 5.15 and 5.16 can be strengthened further. In fact, any prime pattern-walk P will occur with positive density along sufficiently long walks or polygons; this gives a natural connection to density functions. To see this, define $c_n(\leq mP)$ to be the number of walks of length n which contains the prime pattern-walk P at most m times, and let $p_n(\leq mP)$ be similarly defined for polygons. Let $c_n(mP)$ and $p_n(mP)$ be the number of walks and polygons which contains the pattern P exactly m times respectively.[8]

Theorem 5.17 *For every prime pattern-walk P there exists a number $\epsilon_0 > 0$ such that for all $\epsilon < \epsilon_0$,*

$$\limsup_{n\to\infty} \frac{1}{n} \log c_n(\lfloor \epsilon n \rfloor P) < \log \mu_d;$$

$$\limsup_{n\to\infty} \frac{1}{n} \log p_n(\lfloor \epsilon n \rfloor P) < \log \mu_d.$$

Proof Consider only the case of $c_n(\lfloor \epsilon n \rfloor P)$; the proof for $p_n(\lfloor \epsilon n \rfloor P)$ is similar. By Theorem 5.16 there exists an $N > 0$ and an $\epsilon > 0$ such that if $n \geq N$, then

$$c_n(\bar{P}) < [\mu_d(1-\epsilon)]^n, \quad c_n < [\mu_d(1+\epsilon)]^n.$$

Let A be a walk counted by $c_n(\leq \lfloor \epsilon n \rfloor P)$. By starting at one end of A, colour subwalks of length m by $0, 1, \ldots$, where $m \geq |P|$, and let $M = \lfloor n/m \rfloor$ be the number of colours used. Then at most $\lfloor \epsilon n \rfloor$ of the monochromatic subwalks will contain the prime pattern-walk P (and each has at most c_m conformations); the remaining subwalks will not contain a copy of P (and each has at most $c_m(\bar{P})$ conformations). The remainder of the walk has $r = n - mM$ edges, and has at most c_r conformations. In other words,

$$\begin{aligned} c_n(\leq \lfloor \epsilon n \rfloor P) &\leq \sum_{j=0}^{\lfloor \epsilon n \rfloor} \binom{M}{j} [c_m]^j [c_m(\bar{P})]^{M-j} c_{n-Mm}, \\ &\leq (\lfloor \epsilon n \rfloor + 1) \binom{M}{\lfloor \epsilon n \rfloor} \mu_d^{Mm} c_{n-Mm} (1+\epsilon)^{m\lfloor \epsilon n \rfloor} (1-\epsilon)^{mM - m\lfloor \epsilon n \rfloor}. \end{aligned}$$

[8] That is, $c_n(mP)$ is the number of walks of length n with energy m, assuming that each occurrence of P contributes a unit of energy.

Take the $1/n$-th power of this, and take the lim sup as $n \to \infty$; this gives

$$\limsup_{n\to\infty}[c_n(\leq\lfloor \epsilon n \rfloor P)]^{1/n} = \frac{\mu_d(1-\epsilon)}{(\epsilon m)^{\epsilon}(1-\epsilon m)^{1/m-\epsilon}}\left(\frac{1+\epsilon}{1-\epsilon}\right)^{\epsilon m} < \mu_d,$$

if ϵ is small enough. Since $c_n(\lfloor \epsilon n \rfloor P) \leq c_n(\leq\lfloor \epsilon n \rfloor P)$, the theorem follows. The proof for polygons is similar. □

This theorem can be further strengthened. In particular, it is possible to show that prime pattern-walks will occur with positive density in "almost all" walks. In particular, for any prime pattern-walk P there is a number $k > 0$ and an $\epsilon > 0$ such that the pattern will occur at least $\lfloor \epsilon n \rfloor$ times in at least $\lfloor (1-e^{-kn})c_n \rfloor$ walks of length n.[9] The same results are true for polygons, and this is proven in Corollary 5.18:

Corollary 5.18 *Let P be any prime pattern-walk and let $\epsilon > 0$ be a small number. Then there exists a $k > 0$ and an $N_0 > 0$ if ϵ is small enough such that*

$$p_n(\lfloor \epsilon n \rfloor) \leq p_n(\leq\lfloor \epsilon n \rfloor) < e^{-kn}p_n,$$

for all $n \geq N_0$. In other words,

$$p_n - p_n(\lfloor \epsilon n \rfloor) \geq p_n - p_n(\leq\lfloor \epsilon n \rfloor) > (1-e^{-kn})p_n.$$

A similar statement is true for walks.

Proof Define $\limsup_{n\to\infty}[\log p_n(\leq\lfloor \epsilon n \rfloor)]/n = \log \mu_d(\epsilon)$. Then

$$\limsup_{n\to\infty} \frac{1}{n}\log[p_n(\leq\lfloor \epsilon n \rfloor)/p_n] = \log(\mu_d(\epsilon)/\mu_d) < 0$$

if $\epsilon < \epsilon_0$ in Theorem 5.17. Let $k = |\log(\mu_d(\epsilon)/\mu_d)|/2 > 0$; then by definition of the lim sup, there is an N_0 such that

$$\frac{1}{n}\log[p_n(\leq\lfloor \epsilon n \rfloor)/p_n] < -k,$$

if $n > N_0$. Thus $p_n(\leq\lfloor \epsilon n \rfloor) < e^{-kn}p_n$. □

The lim sup's in Theorem 5.17 and the proof of Corollary 5.18 can be changed into limits by concatenating polygons (or walks). Since P is a prime pattern-walk it cannot be created or destroyed by concatenation, so that

$$p_{n_1}(m_1 P)p_{n_2}(m_2 P) \leq p_{n_1+n_2+2k}((m_1+m_2)P), \tag{5.18}$$

where $2k$ extra edges were inserted between the polygons to avoid the formation of new copies of P. By Assumptions 3.1 and Theorem 3.4 the existence of a density function can be shown. Similarly, the arguments following Assumptions 3.8 will imply the existence of integrated density functions. If P is not a prime pattern-walk, but is a Kesten pattern,

[9] In other words, the pattern will occur with density at least ϵ in at least $\lfloor (1-e^{-kn})c_n \rfloor$ walks of length n.

then Assumptions 3.8 may be satisfied. If the model is regular, then the density function exists as in Theorem 3.4; otherwise a density function may be shown to exist as in Theorem 3.16. Since any Kesten pattern is a subpattern of a prime pattern-walk (see the proof of Theorem 5.10), Theorem 5.17 can also be relaxed to Kesten patterns. Hence, the pattern theorem for Kesten patterns can be restated in terms of density functions.

Theorem 5.19 *Let $\mathcal{P}_P(\epsilon)$ be the log-concave density function on $(0, \epsilon_M)$ of the Kesten pattern P. Define $\epsilon_c \in (0, \epsilon_M)$ by*

$$\epsilon_c = \inf\{\epsilon \,|\, \mathcal{P}_P(\epsilon) \geq \mathcal{P}_P(\epsilon_1),\ \forall \epsilon_1 \in (0, \epsilon_M)\}.$$

Then $\mathcal{P}_P(\epsilon) < \mathcal{P}_P(\epsilon_c)$ whenever $\epsilon < \epsilon_c$. □

In other words, since it is log-concave, the density function is strictly increasing in an interval $(0, \epsilon_c)$. The theorem is true for any Kesten pattern, since restricting the number of times a Kesten pattern occurs will also restrict the number of times any prime pattern-walk which contains the Kesten pattern will occur.

Two Kesten patterns P and Q are *independent* if their intersection is empty (in other words, P is not a subwalk in Q, and Q is not a subwalk in P). If P and Q are independent patterns, then occurrences of P are not dependent on the occurrences of Q and vice versa. In some circumstances there exists a combined density function for both P and Q. Suppose for example that both patterns are additive under concatenation: $p_n(i_1, j_1)p_m(i_2, j_2) \leq p_{n+m}(i_1 + i_2, j_1 + j_2)$. Then the techniques of Theorem 3.4 can be used to show that the *joint density function*

$$\log \mathcal{P}(\epsilon, \delta) = \lim_{n\to\infty} \frac{1}{n} \log p_n(\lfloor \epsilon n \rfloor P, \lfloor \delta n \rfloor Q) \tag{5.19}$$

exists, is concave in both arguments in $[0, \epsilon_M] \times [0, \delta_M]$ and continuous in $(0, \epsilon_M) \times (0, \delta_M)$ (where $(0, \epsilon_M)$ and $(0, \delta_M)$ are defined as in eqn (3.2)), and differentiable almost everywhere in this rectangle. If P and Q are Kesten patterns, then $p_n(m_1 P, m_2 Q)$ may satisfy assumptions similar to those in Assumptions 3.8, in which case the existence of a density function is shown by using integrated density functions. In addition, the connection to a pattern theorem is seen by fixing ϵ in $(0, \epsilon_M)$, and by replacing c_n by $c_n(\lfloor \epsilon n \rfloor P)$ in the arguments in this section. The result is a pattern theorem for polygons which contains the pattern P with a density of ϵ.

Theorem 5.20 *Let P and Q be two independent Kesten patterns, and let P occur with density ϵ and Q with density δ. Then for every fixed $\epsilon \in (0, \epsilon_M)$ there exists a $\delta_c(\epsilon)$, such that for every $\delta < \delta_c(\epsilon)$,*

$$\mathcal{P}(\epsilon, \delta) < \mathcal{P}(\epsilon, \delta_c(\epsilon)).$$ □

Thus, the pattern Q occurs with a natural density which is not zero. Generalizations of Theorem 5.17 and Corollary 5.18 can similarly be written down. In the next section the connection between these ideas and a pattern theorem for interacting models of walks and polygons will become clear. An alternative approach to the above would be to note that for polygons and independent prime pattern-walks P and Q, concatenation can be done such that $\sum_{i_1=0}^{i} p_n((i - i_1)P, j_1 Q)p_m(i_1 P, j_2 Q) \leq p_{n+m+2k}(iP, (j_1 + j_2)Q)$

where k is chosen to avoid the creation of more copies of P or Q. Multiply this by y^i and sum over i to obtain $p_n(y; j_1 Q) p_m(y; j_2 Q) \leq p_{n+m+2k}(y; (j_1 + j_2)Q)$ where $p_n(y; jQ)$ is the partition function of polygons of length n with j occurrences of the pattern Q and an activity y conjugate to the number of occurrences of the pattern P (with P and Q independent). This partition function satisfies Assumptions 3.1, and so there is a density function of the pattern Q: $\mathcal{P}(y; \epsilon) = \lim_{n\to\infty}[p_n(y; \lfloor \epsilon n \rfloor Q)]^{1/n}$. A Legendre transform gives the free energy $\mathcal{F}(y, z) = \sup_\epsilon \{\log \mathcal{P}(y; \epsilon) - \epsilon \log z\}$. This can be generalized to Kesten patterns by using Assumptions 3.8 instead.

5.2 The pattern theorem and interacting models of walks

In this section the generalization of the results in Section 5.1 to interacting models of walks and polygons is presented. The arguments are similar to those of Section 5.1, but there are some technical difficulties which must be overcome. An immediate and important question is the existence of the free energy in models of interacting walks. While Assumptions 3.1 and 3.8 are also sufficient to give the existence of free energies and density functions in the case of submultiplicative models of walks, there are also other technical difficulties which make it in some cases impossible to prove that a model of walks satisfies a given submultiplicative inequality.

Let $c_n(k)$ be the number of walks of length n and energy k (that is, there are k occurrences of an event which each contributes a unit to the energy of the walks). For example, k could be the number of right angles, or nearest-neighbour contacts, or the number of times a given pattern occurs. Let $p_n(k)$ be the number of polygons similarly counted. I also assume that these events are (almost) additive under concatenation of polygons, in the sense described in Assumptions 3.1 or 3.8 in Chapter 3.[10] Therefore, there is a density function and a limiting free energy $\mathcal{F}_P(z)$ in this model of polygons. An immediate and important question is the following: under what circumstances is there a limiting free energy for the model of walks defined by $c_n(k)$, and if it exists, what is its relation to the limiting free energy of the corresponding model of polygons? This is not a trivial question in some models; for example, if k is the number of nearest-neighbour contacts in models of walks and polygons, then it is known that the limiting free energy exists for both models (and are equal) only if $z \leq 1$. If $z > 1$ then it is known that there is a limiting free energy for the model of polygons, but a proof that there exists a limiting free energy for walks is still outstanding [341]. I shall give some sufficient assumptions

[10] The specific behaviour of the energy depends of course on the particular situation. In most cases concatenating two polygons counted by $p_n(k_1)$ and $p_m(k_2)$ gives an outcome from which the original polygons can be uniquely recovered (decompose the polygon at the first available place which gives n edges in the first, and m in the second; this will fix the values of k_1 and k_2). One can in many cases do better than the supermultiplicative assumptions in Assumptions 3.1 and 3.8. Consider the concatenation of polygons counted by $p_n(k_1)$ and $p_m(k - k_1)$, and sum over all k_1: the observation above then gives the stronger inequality $\sum_{k_1=0}^{k} p_n(k) p_m(k - k_1) \leq \sum_{i=-q}^{q} p_{n+m}(k + i)$, where it is assumed that the energy can change by as much as $q = o(n)$ in concatenation. Multiplication by z^k and summing over k gives $p_n(z) p_m(z) \leq \left[\sum_{i=-q}^{q} z^i\right] p_{n+m}(z)$, where the partition function is defined by $p_n(z) = \sum_{k\geq 0} p_n(k) z^k$. Thus, since $p_n(z)$ is a generalized supermultiplicative function, there is a limiting free energy and thus a density function (note that Theorem A.3 is relevant here (see Theorem 3.19)).

for a limiting free energy to exist; however, it is not known if these are necessary. The partition functions for these models are

$$c_n(z) = \sum_{k\geq 0} c_n(k)z^k, \qquad p_n(z) = \sum_{k\geq 0} p_n(k)z^k. \tag{5.20}$$

The limiting free energy of a model of polygons can be shown directly to exist by the argument in footnote 10, or Theorems 3.16 and 3.18:

$$\mathcal{F}_p(z) = \lim_{n\to\infty} \frac{1}{n} \log p_n(z). \tag{5.21}$$

5.2.1 The free energy of fl-walks

In the previous discussions it appeared that models of polygons have limiting free energies under general assumptions of their behaviour using the construction of concatenation. In particular, the assumption that a model of polygons satisfies Assumptions 3.1, or Assumptions 3.8, is enough to give definitions for both a density function and a limiting free energy. In this section the aim is first to study the limiting free energy of fl-walks, and to relate it to the limiting free energies of walks by unfolding. In particular, sufficient conditions which will give the existence of a limiting free energy in walks (equal to the limiting free energy of fl-walks) will be considered.

A good starting point is some basic assumptions for a model of fl-walks. The supermultiplicative property will be essential, and in its most general form is embodied by Assumptions 3.8(1), (2) and (3′). Under those circumstances, there is a limiting free energy and a density function, as shown in Theorems 3.18 and 3.19. Assume therefore that $c_n^{fl}(k)$ satisfies Assumptions 3.8(1), (2) and (3′), so that the limit

$$\mathcal{F}_{fl}(z) = \lim_{n\to\infty} \frac{1}{n} \log c_n^{fl}(z) \tag{5.22}$$

exists. An important consequence of these assumptions is that the energy of an fl-walk may change by $q = o(n)$ if an edge is appended on its top or its bottom vertex, a fact which must be kept in mind in any construction involving fl-walks. A second crucial issue in relating a model of fl-walks to a model of walks is the process of unfolding. It is in many cases inevitable that this construction will change the energy; and it will be necessary to assume that it does not change it by too much. With these preliminary notions, the assumptions can be stated.

Assumptions 5.21

(1) Assume that $c_n^{fl}(k)$ satisfies all the Assumptions 3.8(1), (2) and (3′).

(2) Assume that if any walk counted by $c_n(k)$ is cut into two pieces at a vertex, and if the two subwalks in the cut are separated by a large distance (so that they are independent), then they have combined energy $k + o(n)$.

(3) Assume that by cutting an f-walk A of energy k at its top vertex, an l-walk B_1 and an fl-walk A_1 are obtained. If $B_1^\dagger$ is the image of B_1 when it is reflected through the hyperplane normal to the first direction, and containing the top vertex, then the new walk $A_1 B_1^\dagger$ has energy $k + o(n^{1/(d+1)})$. □

These assumptions have an immediate consequence when an f-walk is unfolded. Assumption 5.21(3) restricts the change of energy in an unfolding to be $o(n)$.

Lemma 5.22 *Let A be an f-walk of length n and energy k in a model of walks which satisfies Assumptions 5.21. Let A' be an fl-walk obtained from A by the unfolding in Constructions 5.1 and 5.3. Then there is a non-decreasing function $f_1(n) = o(n)$ such that the energy of A' is between $k - f_1(n)$ and $k + f_1(n)$. Moreover,*

$$c_n^f(k) \leq P(n) \sum_{i=-f_1}^{f_1} c_{n+1}^{fl}(k+i),$$

where $P(n)$ is the number of partitions of the integer n.

Proof If an f-walk of energy k is cut at its top vertex into an fl-walk and an l-walk, then the change in energy is $o(n^{1/(d+1)})$ by Assumption 5.21(3). In the unfolding of the f-walk there are at most $C_0 n^{d/(d+1)}$ such cuts by Lemma 5.2, each changes the energy by $o(n^{1/(d+1)})$, so that the maximum change in the energy is $h_1(n) = C_0 n^{d/(d+1)} o(n^{1/(d+1)}) = o(n)$. All the other steps in the unfoldings remain unchanged, so that[11]

$$c_n^f(k) \leq P(n) \sum_{i=-h_1}^{h_1} c_{n+1}^{fl}(k+i),$$

where $h_1(n) = o(n)$. Define $f_1(n) = \max_{0 \leq m \leq n} h_1(m)$. Then $f_1(n)$ is non-decreasing, and can be used to replace $h_1(n)$ in the above. □

The relation between f-walks and walks is seen by cutting a walk into two f-walks at its bottom vertex (see eqn (5.5)). In an interacting model with Assumptions 5.21 this becomes the following:

Lemma 5.23 *Let A be a walk of energy k in a model of walks which satisfy Assumptions 5.21. If A is divided into two f-walks by cutting it at its bottom vertex, then the total energy of the walk changes by at most $o(n)$. Moreover, there is a non-decreasing function $f_2(n) = o(n)$ such that*

$$c_n(k) \leq \sum_{i=-f_2}^{f_2} \sum_{m=0}^{n} \sum_{l=0}^{k} c_m^f(l) c_{n-m}^f(k-l+i).$$

Proof That the energy changes by at most $o(n)$ is seen from Assumption 5.21(2). If walks of energy k are cut at their bottom vertices into f-walks, then

$$c_n(k) \leq \sum_{i=-h_2}^{h_2} \sum_{k=0}^{n} \sum_{l=0}^{k} c_k^f(l) c_{n-k}^f(k-l+i),$$

and by Assumption 5.21(2), $h_2(n) = o(n)$. Define $f_2(n) = \max_{0 \leq m \leq n} h_2(m)$. Then $f_1(n)$ is non-decreasing, and the inequality stays true if $h_2(n)$ is replaced by $f_2(n)$. □

---

[11] An extra edge appears in the unfolding process. This explains the fact that the fl-walks will have length $n+1$. It is also possible that appending this edge may change the energy by $o(n)$ by Assumption 5.21(1); absorb this into $h_1(n)$.

Define the function

$$\phi(z) = z + 1 + z^{-1} > 1. \tag{5.23}$$

Then the following corollary will relate the partition function of walks to that of unfolded walks.

Corollary 5.24 *Let $c_n(z)$ be the partition function of walks in a model which satisfies Assumptions 5.21, and where z is conjugate to the energy of the walks. Then there exists a non-decreasing function $g_1(n) = o(n)$ such that*

$$c_n(z) \leq C(n+1)^2 P(n)^2 [\phi(z)]^{g_1} c_{n+2}^{fl}(z).$$

Proof Replace $c_m^f(l)$ and $c_{n-m}^f(k-l+i)$ in Lemma 5.23 by their respective upper bounds in Lemma 5.22. Since it is assumed that c_n^{fl} satisfies Assumptions 3.8, use Assumption 3.8(3) (where $q = o(n)$) to obtain

$$c_n(k) \leq \sum_{i=-f_2}^{f_2} \sum_{m=0}^{n} \sum_{l=0}^{k} P(m)P(n-m) \sum_{a,b=-f_1}^{f_1} \sum_{j=-q}^{q} c_{n+2}^{fl}(k+i+a+b+j).$$

Multiply this by z^k and sum over k, and use the definition of $\phi(z)$ in eqn (5.23). This gives

$$c_n(z) \leq C(n+1)^2 P(n)^2 [\phi(z)]^{f_1+f_2+q} c_{n+2}^{fl}(z),$$

where the sum over m and l gives $C(n+1)^2$ by Assumption 3.8(2). Define $g_1(n) = \max_{0 \leq m \leq n}\{f_1(m) + f_2(m) + q(m)\} = o(n)$. Then $g_1(n)$ is non-decreasing. □

These results are enough to show the existence of a free energy in an interacting model of walks.

Theorem 5.25 *Let $c_n(k)$ be the number of walks of energy k in a model which satisfies Assumptions 5.21, and let the partition function of the model be $c_n(z)$. Then the limiting free energy*

$$\mathcal{F}_w(z) = \lim_{n \to \infty} \frac{1}{n} \log c_n(z)$$

exists if either

$$c_n(k_1)c_m(k_2) \geq c_{n+m}(k_1 + k_2),$$

or if $c_n^{fl}(k)$ and $c_n(k)$ satisfy Assumptions 5.21.

Proof If $c_n(k_1)c_m(k_2) \geq c_{n+m}(k_1+k_2)$, then $[1/c_n(k)]$ satisfies Assumption 3.1(3), and so there is a density function and the limiting free energy exists. It is possible to relax this assumption, but there will not be much gain if it is done here. Otherwise, note from Corollary 5.24 that $c_n(z) \leq C(n+1)^2 P(n)^2 [\phi(z)]^{g_1} c_n^{fl}(z)$, and that $c_n^{fl}(z) \leq c_n^f(z) \leq c_n(z)$. Since $\lim_{n\to\infty}[\log c_n^{fl}(z)]/n$ exists (by Assumptions 5.21 and Theorem 3.17), so does $\lim_{n\to\infty}[\log c_n(z)]/n$ and $\lim_{n\to\infty}[\log c_n^f(z)]/n$. □

In the proof above it is also shown that if $c_n^{fl}(k)$ and $c_n(k)$ satisfy Assumptions 5.21, then

$$\mathcal{F}_w(z) = \lim_{n\to\infty} \frac{1}{n} \log c_n(z) = \lim_{n\to\infty} \frac{1}{n} \log c_n^f(z) = \lim_{n\to\infty} \frac{1}{n} \log c_n^{fl}(z). \tag{5.24}$$

Theorem 5.25 states that there is a limiting free energy in models of self-interacting walks, provided that unfolding of the walks does not change the energy too much. This situation occurs in many models, some of which will be considered in more detail in subsequent sections.

5.2.2 Interacting models of polygons and walks

It is also possible to show that Assumptions 5.21 are sufficient to prove that a model of interacting walks, and its corresponding model of interacting polygons, have the same free energy. To see this, unfolded walks and loops must be concatenated into polygons, as was done in Section 5.1.2. Define unfolded, doubly unfolded and t-walks as before. The construction in Lemma 5.6 will unfold an fl-walk into a t-walk, and following the arguments in Lemma 5.22 the energy of the fl-walk will change by at most $o(n)$ in this construction. Without repeating the construction, here is the result.

Lemma 5.26 *There exists a finite positive constant γ and a non-decreasing function $f_1(n) = o(n)$ such that*

$$c_n^{fl}(z) \leq e^{2d\gamma\sqrt{n}}[\phi(z)]^{f_1} c_n^t(z). \qquad \square$$

Each t-walk can be turned into a doubly unfolded walk by appending an edge (see eqn (5.7)) at each of the top and bottom vertices. This may change the energy by $o(n)$, therefore

$$c_n^t(z) \leq [\phi(z)]^{f_1} c_{n+2}^{\ddagger}(z), \tag{5.25}$$

where $f_1(n) = o(n)$. In addition,

$$c_n^{\ddagger}(z) \leq c_n^{\dagger}(z) \leq c_n(z). \tag{5.26}$$

By Theorem 5.25, eqn (5.24) and Lemma 5.27:

Lemma 5.27 $\mathcal{F}_w(z) = \lim_{n\to\infty}(1/n)\log c_n^{\ddagger}(z) = \lim_{n\to\infty}(1/n)\log c_n^{\dagger}(z)$. $\square$

In these models of walks, the free energy $\mathcal{F}_w(z)$ is equal to the free energy of the associated model of polygons. This is seen by using a most popular argument, doubly unfolded walks, and loops. These are glued together into polygons. With each such gluing, Assumptions 5.21(2) indicate that the energy changes by at most $o(n)$. This can be absorbed into $f_1(n) = o(n)$. In particular, if $c_n^{\ddagger}(z; h)$ is the partition function of doubly unfolded walks of X-span h, then the construction in Fig. 5.5 can be used to show that

$$\sum_{l=0}^{k} c_{\lfloor n/2\rfloor}^{\ddagger}(l; h) c_{\lfloor n/2\rfloor}^{\ddagger}(k-l; h) \leq \sum_{i=-f_1}^{f_1} l_{2\lfloor n/2\rfloor}(k+i; h), \tag{5.27}$$

where any change in energy was absorbed into f_1. Multiplication by z^k and summing over k gives (with $\phi(z)$ defined in eqn (5.23)):

$$c^{\ddagger}_{\lfloor n/2\rfloor}(z;h)c^{\ddagger}_{\lfloor n/2\rfloor}(z;h) \leq [\phi(z)]^{f_1} l_{2\lfloor n/2\rfloor}(z;h). \tag{5.28}$$

Let $l_n(k; h_y, \ldots, h_z)$ be the number of loops of energy k and with YZ-span $(h_y, \ldots, h_z)$. Reflect one such loop through the plane normal to the first direction to obtain a negative loop, and concatenate its end-points with the end-points of a second loop with the same YZ-span. Absorb any changes in the energy into f_1. This gives $\sum_{l=0}^{k} l_{\lfloor n/2\rfloor}(l; h_y, \ldots, h_z) l_{\lfloor n/2\rfloor}(k-l; h_y, \ldots, h_z) \leq \sum_{i=-f_1}^{f_1} p_{2\lfloor n/2\rfloor}(k+i)$, and multiplication by z^k and summing over k gives

$$l_{\lfloor n/2\rfloor}(z; h_y, \ldots, h_z) l_{\lfloor n/2\rfloor}(z; h_y, \ldots, h_z) \leq [\phi(z)]^{f_1} p_{2\lfloor n/2\rfloor}(z). \tag{5.29}$$

The next step is a most popular argument in the YZ-span of the loops in eqns (5.28) and (5.29). This is done in the proof of Theorem 5.28.

Theorem 5.28 *If a model of walks and polygons satisfies Assumptions 5.21, then*

$$\mathcal{F}_w(z) = \lim_{n\to\infty} \frac{1}{n} \log c_n(z) = \lim_{n\to\infty} \frac{1}{n} \log p_n(z) = \mathcal{F}_p(z).$$

Proof First use a most popular argument in eqn (5.28). Let h^* be that height which maximizes the left-hand side. Then $\lfloor n/2\rfloor c^{\ddagger}_{\lfloor n/2\rfloor}(z, h^*) \geq c^{\ddagger}_{\lfloor n/2\rfloor}(z)$. In addition, sum the right-hand side over h. Then

$$\frac{1}{\lfloor n/2\rfloor^2}[c^{\ddagger}_{\lfloor n/2\rfloor}(z)]^2 \leq [\phi(z)]^{f_1} l_{2\lfloor n/2\rfloor}(z).$$

On the other hand, eqn (5.29) and a most popular argument shows that

$$\frac{1}{\lfloor n/2\rfloor^{2d-2}}[l_{\lfloor n/2\rfloor}(z)]^2 \leq [\phi(z)]^{f_1} p_{2\lfloor n/2\rfloor}(z) \leq [\phi(z)]^{g_1} c_{2\lfloor n/2\rfloor -1}(z),$$

where the second inequality is obtained by deleting a single edge (for example the edge with the least mid-point) in the polygon, and changing its energy by $o(n)$ by Assumptions 5.21. This change is again absorbed into $f_1(n)$ to find $g_1(n)$. By Lemma 5.26, and the last two equations, the conclusion is that

$$\mathcal{F}_w(z) = \lim_{n\to\infty} \frac{1}{n} \log c_n(z) = \lim_{n\to\infty} \frac{1}{n} \log p_n(z).$$

□

5.2.3 The pattern theorem for interacting models

The most important definition in the proof of a pattern theorem of interacting models is that of a prime pattern-walk. In the case of non-interacting walks this was defined with the help of a pattern-walk, which was argued to be in one-to-one correspondence with unfolded walks (see footnote 9). I shall take the same approach here, but there are more technical difficulties. The first is that a decomposition of a pattern-walk into prime pattern-walks may change the energy of the walk; this of course also depends on the specific definition of the energy. The second (and more serious) problem is that the unfolding of an f-walk into an fl-walk may change the energy by $o(n)$, if the assumptions in Assumptions 5.21 are used as a basis for the discussion. This will seriously hinder the derivation leading to Lemma 5.11, which is a crucial result in the proof of a pattern theorem for non-interacting models. It will be possible to overcome this problem if it is assumed that the energy changes by at most a constant (independent of the length of the walk) in each step of an unfolding.

Define a pattern-walk of energy k and length n to be a walk of energy k with vertices whose X-coordinates satisfy the following: $X_0 < X_i < X_n$ for $i = 1, 2, \ldots, n-1$. As observed in footnote 9, there is a one-to-one correspondence to unfolded walks, but in this case the energy may not be preserved if the last edge in the pattern-walk is removed. This must be taken into account in the constructions which lead to a pattern theorem. Thus, a basic assumption about the behaviour of the energy when two pattern-walks are concatenated must be made. Suppose that two t-walks, fl-walks, doubly unfolded walks or pattern-walks are concatenated. In each case the concatenation is done by identifying the top vertex in one walk with the bottom vertex in the second walk. Let the vertex about which the concatenation occurs be v (see Fig. 5.6). I shall assume that the change in the total energy if two t-walks, fl-walks, doubly unfolded walks or pattern-walks are concatenated is *only* dependent on the orientation of edge incident with the top vertex of the first, and the edge incident with the bottom vertex of the second.[12] In other words, only the orientations of the edges incident with the vertex v in Fig. 5.6 will determine the change in energy if two pattern-walks are concatenated (since these edges are always in the X-direction, the energy changes by a constant). An energy with this property is called a *local energy*. Since each edge has d orientations, there are at most d^2 possible changes in the energy when two t-walks or fl-walks are concatenated. The definition of a local energy is made more precise in Definition 5.29.

Definition 5.29 (Local energy) Suppose that A is any walk of length n and energy k. Suppose furthermore that if A is decomposed into two walks of lengths n_1 and n_2, and energies k_1 and k_2, by cutting it at a vertex v into two subwalks, then $k_1+k_2+\alpha(i,j) = k$. If $\alpha(i,j)$ is *only* a function of the relative orientations i and j of the edges in A incident with the vertex v, then the energy is a *local energy*. □

Let the maximum and minimum possible values of $\alpha(i,j)$ be

$$\alpha_M = \max_{i,j}\{\alpha(i,j)\}, \qquad \alpha_m = \min_{i,j}\{\alpha(i,j)\}; \tag{5.30}$$

and define the maximum possible change in the energy in the decomposition in Definition 5.29 by

$$\alpha = \alpha_M - \alpha_m. \tag{5.31}$$

[12] This can in fact be somewhat relaxed, by for example stating that it depends on the subwalks of fixed length N incident with the top and bottom vertices.

As before, a pattern-walk is decomposable if it can be divided by a hyperplane normal to the X-direction into two pattern-walks, with each component confined to a separate half-space defined by the hyperplane. If a pattern-walk is not decomposable, then it is prime. Consider next the effect of Definition 5.29 if two pattern-walks are concatenated. Let the number of pattern-walks of energy k be $c_n^p(k)$.

Lemma 5.30 *Let $c_n^p(k)$ be the number of pattern-walks of length n and energy k, where the energy is local. Then*

$$c_{n_1}^p(k_1)c_{n_2}^p(k_2) \leq c_{n_1+n_2}^p(k_1 + k_2 + \alpha(X, X)).$$

Proof Let A_1 and A_2 be two pattern-walks of energies k_1 and k_2 and lengths n_1 and n_2. The edge incident with the top vertex in A_1, as well as the edge incident with the bottom vertex in A_2, are oriented in the X-direction. If A_1 and A_2 are concatenated, then a pattern-walk of energy $k_1 + k_2 + \alpha(X, X)$ is obtained, by Definition 5.29. Since the concatenated walk has length $n_1 + n_2$, the inequality follows. □

Let the number of prime pattern-walks of energy k be $q_n(k)$. Every decomposable pattern-walk of energy k can be obtained by concatenating a pattern-walk of energy $k - l - \alpha(X, X)$ with a prime pattern-walk of energy l (where $\alpha(X, X)$ compensates for the change in energy as in Lemma 5.30, and where a sum over l is taken). Similarly, any decomposable pattern-walk of energy k can be cut into a prime pattern-walk and a pattern-walk by cutting the prime pattern-walk incident with the bottom vertex from the decomposable pattern-walk. This gives the following:

$$\sum_{m=1}^{n-1} \sum_{l=0}^{k} q_m(l)c_{n-m}^p(k - l - \alpha(X, X)) = c_n^p(k) - q_n(k). \tag{5.32}$$

Multiply eqn (5.32) by z^k and sum over k. This gives

$$c_n^p(z) = q_n(z) + z^{\alpha(X,X)} \sum_{m=1}^{n-1} q_m(z)c_{n-m}^p(z). \tag{5.33}$$

Thus, with the exception of the factor $z^{\alpha(X,X)}$, a renewal equation analogous to the case for non-interacting models is obtained. Define the generating functions $q(x, z) = \sum_{n=1}^{\infty} q_n(z)x^n$ and $c^*(x, z) = \sum_{n=1}^{\infty} c_n^*(z)x^n$, where $* = p, fl, f, l, \dagger, \ddagger$ and so on. Then eqn (5.33) can be cast in terms of generating functions as in the following lemma.

Lemma 5.31 *The generating functions $c^p(x, z)$ and $q(x, z)$ are related by*

$$c^p(x, z) = \frac{q(x, z)}{1 - z^{\alpha(X,X)}q(x, z)}.$$ □

Since every pattern-walk is an fl-walk, it follows that $c^p(x, z) \leq c^{fl}(x, z)$. On the other hand, if edges in the X-direction are concatenated on the first and last vertices of an fl-walk, then a pattern-walk is obtained. But this construction may change the energy of the walk. However, the energy is local, and if the orientations of the first and last edge of the unfolded walk are labelled by i and by j, then the energy of the concatenated walk is $k + \alpha(i, X) + \alpha(j, X)$, as can be seen from Definition 5.29. Thus, $c_n^{fl}(k) \leq \sum_{i,j=X}^{Z} c_{n+2}^{p}(k + \alpha(i, X) + \alpha(j, X))$, where the sums over i and j are over all possible directions that an edge may take (from X to Z). Multiply this by z^k and sum over k. This gives a relation between partition functions: $c_n^{fl}(z) \leq \left[\sum_{i,j=X}^{Z} z^{\alpha(i,X)+\alpha(j,X)}\right] c_{n+2}^{p}(z)$. Define the function $\psi(z)$ by

$$\psi(z) = \sum_{i,j=X}^{Z} z^{\alpha(i,X)+\alpha(j,X)}, \tag{5.34}$$

where the sum is over all orientations $X, Y, \ldots, Z$, and then the above arguments give the following relationships involving $c^p(x, z)$ and $c^{fl}(x, z)$.

Lemma 5.32 *The generating functions $c^p(x, z)$ and $c^{fl}(x, z)$ are related by*

$$c^p(x, z) \leq c^{fl}(x, z) \leq \psi(z) c^p(x, z)/x^2,$$

where $\psi(z)$ is given by eqn (5.34). □

Lemmas 5.31 and 5.32 can be used to relate prime pattern-walks to fl-walks; this will be needed for a pattern theorem.

Corollary 5.33 *Let $\psi(z)$ be defined as in eqn (5.34). Then*

$$[x^2/\psi(z)]\, c^{fl}(x, z) \leq \frac{q(x, z)}{1 - z^{\alpha(X,X)} q(x, z)} \leq c^{fl}(x, z). \qquad \square$$

Thus, if $c^{fl}(x, z)$ is finite, then $z^{\alpha(X,X)} q(x, z) < 1$. The rest of the construction will be to relate the generating function of fl-walks to the generating function of walks. Any walk counted by $c_n^f(k)$ can be partitioned using Construction 5.1 into fl-walks; and Definition 5.29 (of a local energy) states that the energy changes by at most α (defined by eqn (5.31)) for each unfolding of an fl-walk (save the final) in the partitioning. Since the fl-walks in a partitioning are uniquely ordered by a lexicographic ordering, they can be reassembled into the f-walk which was partitioned. Hence

$$\begin{aligned} c_n^f(k) \leq c_n^{fl}(k) &+ \frac{1}{2} \sum_{i=-\alpha_m}^{\alpha_M} \sum_{m_1,m_2=1}^{n-1} \sum_{l=0}^{k+i} c_{m_1}^{fl}(l) c_{m_2}^{fl}(k - l + i) \delta_{m_1+m_2-n} \\ &+ \frac{1}{3!} \sum_{j_1,j_2=-\alpha_m}^{\alpha_M} \sum_{m_1,m_2,m_3=1}^{n-1} \sum_{l_1,l_2,l_3=0}^{k} c_{m_1}^{fl}(l_1) c_{m_2}^{fl}(l_2) c_{m_3}^{fl}(l_3) \delta_{n-\sum m_i} \delta_{k-\sum l_i - \sum j_i} \\ &+ \cdots \end{aligned} \tag{5.35}$$

Multiply this equation by z^k and sum over k. Define $\phi(z) = z + 1 + z^{-1}$ (see eqn (5.23)), and simplify the resulting equation. This gives

$$c_n^f(z) \leq c_n^{fl}(z) + \frac{1}{2}[\phi(z)]^\alpha \sum_{m_1,m_2=1}^{n-1} c_{m_1}^{fl}(z)c_{m_2}^{fl}(z)\delta_{m_1+m_2-n}$$

$$+ \frac{1}{3!}[\phi(z)]^{2\alpha} \sum_{m_1,m_2,m_3=1}^{n-1} c_{m_1}^{fl}(z)c_{m_2}^{fl}(z)c_{m_3}^{fl}(z)\delta_{n-\sum m_i} + \cdots \tag{5.36}$$

Multiply this by x^n and sum over n. Then the N-th term in the above is given by and bounded from above by (since $\phi(z) \geq 1$)

$$\begin{aligned}
&\frac{1}{N!}\sum_{n=1}^{\infty}[\phi(z)]^{(N-1)\alpha} \sum_{m_1=1}^{n-1} \cdots \sum_{m_N=1}^{n-1} \prod_{i=1}^{N} [c_{m_i}^{fl}(z)x^{m_i}]\delta_{\sum_i m_i - n} \\
&\leq \frac{1}{N!}\left(\sum_{n=1}^{\infty}[\phi(z)]^\alpha c_n^{fl}(z)x^n\right)^N \\
&= \frac{1}{N!}\left([\phi(z)]^\alpha c^{fl}(x,z)\right)^N.
\end{aligned} \tag{5.37}$$

Thus, $c^f(x,z)$ can be bounded in terms of $c^{fl}(z)$ as in Lemma 5.34.

Lemma 5.34 *The generating functions $c^f(x,z)$ and $c^{fl}(x,z)$ are related by*

$$c^{fl}(x,z) \leq c^f(x,z) \leq e^{[\phi(z)]^\alpha c^{fl}(x,z)}.$$

Thus, $c^f(x,z)$ is finite if and only if $c^{fl}(x,z)$ is finite. □

Any walk counted by $c_n(k)$ can be cut into a pair of f-walks, but the total energy may change by at most α, since the energy is local. Therefore, the following relation is obtained between walks and f-walks in an interacting model:

$$c_n(k) \leq c_n^f(k) + \sum_{i=-\alpha}^{\alpha} \sum_{m=1}^{n-1} \sum_{l=0}^{k} c_m^f(l)c_{n-m}^f(k-l+i), \tag{5.38}$$

so that after multiplication by $x^n z^k$, and summing over k and n, the result in Lemma 5.35 is obtained.

Lemma 5.35 *The generating functions $c(x,z)$ and $c^f(x,z)$ are related by*

$$c^f(x,z) \leq c(x,z) \leq c^f(x,z)\big(1 + [\phi(z)]^\alpha \left[c^f(x,z)\right]\big).$$

Moreover, $c(x,z)$ is finite if and only if $c^f(x,z)$ is finite. □

These results, together with the definition of a prime pattern-walk and Lemma 5.32, is enough to show that there is a pattern theorem for interacting models of walks with a local energy. The first important observation is the following lemma.

Lemma 5.36 *The radius of convergence of $c(x,z)$ is given by $x_c(z) = e^{-\mathcal{F}_w(z)}$, where $\mathcal{F}_w(z)$ is the limiting free energy of the model. Moreover, $c(x_c(z),z) = \infty$, and $c^{fl}(x,z) < \infty$ if and only if $c(x,z) < \infty$.*

Proof The only issue to address is the claim that $c(x_c(x), z) = \infty$. Consider a walk counted by $c_{n_1+n_2}(k)$ and orient it. Let the vertex v divide it into two walks of lengths n_1 and n_2. In fact, consider all the possible subwalks that can be generated by rotating the two subwalks about v, each exploring $2d$ possible orientations. Alternatively, take two oriented subwalks of lengths n_1 and n_2 and join them into a walk of length $n_1 + n_2$ by rotating them in all possible positions and then by identifying the last vertex of the first with the first vertex of the second. Of the $\binom{d}{2}$ possible relative orientations of the edges around v, let h_s be the number which give a change of s in the total energy. Since each pair of subwalks can give rise (potentially) to $(2d)^2$ connected pairs ($2d$ orientations each), each walk in $c_{n_1+n_2}(k)$ is counted at most $(2d)^2$ times, and many self-intersecting conformations are also produced. Thus

$$4 \sum_{s=\alpha_m}^{\alpha_M} h_s \sum_{k_1=0}^{k} c_{n_1}(k - k_1) c_{n_2}(k_1 - s) \geq 2(2d)^2\, c_{n_1+n_2}(k) \geq 4\, c_{n_1+n_2}(k).$$

Multiply this by z^k and sum over k. Then

$$\left[\sum_{s=\alpha_m}^{\alpha_M} h_s z^s \right] c_{n_1}(z) c_{n_2}(z) \geq c_{n_1+n_2}(z).$$

Define the function $\psi_1(z) = \sum_{s=\alpha_m}^{\alpha_M} h_s z^s$. Then $\psi_1(z) c_n(z)$ is submultiplicative. Thus, by Lemma A.1 the limit $\mu_z = \lim_{n\to\infty} [c_n(z)]^{1/n}$ exists, and $\psi_1(z) c_n(z) \geq \mu_z^n$. Thus $\psi_1(z) c(x, z) \geq 1/(1 - x\mu_z)$, and so $c(\mu_z^{-1}, z) = \infty$. Furthermore, the limiting free energy of the model also exists, and is defined by $-\log \mu_z$. Since the radius of convergence of $c(x, z)$ is also given by $x_c(z) = e^{-\mathcal{F}_w(z)}$, conclude by noting that $\mu_z^{-1} = x_c(z)$, and by noting from Lemmas 5.34 and 5.35 that $c^{fl}(x_c(z), z) = \infty$. □

Lemma 5.36 gives the following important corollary, which will be a key observation in this proof.

Corollary 5.37 *From Corollary 5.33 and Lemma 5.36 if follows that*

$$z^{\alpha(X,X)} q(x, z) \quad \begin{cases} = 1, & \text{if } x = x_c(z); \\ < 1, & \text{if } x < x_c(z). \end{cases}$$
□

The next step in the proof of a pattern theorem would be to repeat the above with walks which do not contain a given prime pattern-walk. Since the first and last edges in every prime pattern-walk are in the X-direction, prohibiting any one of them will not change the outcome in Lemma 5.35, but with the generating functions now defined for a model which do not contain terms corresponding to decomposable pattern-walks which contain the given prime pattern-walk. Let $c_n^*(x, z; \bar{P})$ be the generating function of models of walks which may not contain the prime pattern-walk P, where $*$ is any of f, fl, p, †, ‡ and so on. Let $q(x, z; \bar{P})$ be the generating function of prime pattern-walks, which do not contain any prime pattern-walks P as a factor.

Corollary 5.38 *Let $\psi(z)$ be defined as in eqn (5.34). Then*

$$[x^2/\psi(z)]\,c^{fl}(x,z;\bar{P}) \leq \frac{q(x,z;\bar{P})}{1-z^{\alpha(X,X)}q(x,z;\bar{P})} \leq c^{fl}(x,z;\bar{P}). \qquad \square$$

Arguments identical to those leading to Lemma 5.36 show again that there is a radius of convergence $x_c(z;\bar{P})$ for the generating functions $c_n^*(x,z;\bar{P})$, and moreover that $c_n(x_c(z;\bar{P}),z;\bar{P}) = c_n^{fl}(x_c(z;\bar{P}),z;\bar{P}) = \infty$.

Lemma 5.39 *The radius of convergence of $c(x,z;\bar{P})$ is given by $x_c(z;\bar{P}) = e^{-\mathcal{F}_w(z;\bar{P})}$, where $\mathcal{F}_w(z;\bar{P})$ is the limiting free energy of the model. Moreover, $c(x_c(z;\bar{P}),z;\bar{P}) = \infty$, and $c^{fl}(x,z;\bar{P}) < \infty$ if and only if $c(x,z;\bar{P}) < \infty$.* $\square$

Combining Corollary 5.38 and Lemma 5.39 gives Corollary 5.40, but now for a model of walks which does not include the prime pattern-walk P:

Corollary 5.40 *From Corollary 5.38 and Lemma 5.39 it follows that*

$$z^{\alpha(X,X)}q(x,z;\bar{P}) \quad \begin{cases} = 1, & \text{if } x = x_c(z;\bar{P}), \\ < 1, & \text{if } x < x_c(z;\bar{P}). \end{cases} \qquad \square$$

But now notice that $q(x,z;\bar{P}) < q(x,z)$, since there is at least one term (the term corresponding to P) which contributes to $q(x,z)$ and which does not contribute to $q(x,z;\bar{P})$. In other words, if $z^{\alpha(X,X)}q(x,z) = 1$, then $z^{\alpha(X,X)}q(x,z;\bar{P}) < 1$. Thus $x_c(z) < x_c(z;\bar{P})$, since $c_n(x_c(z),z;\bar{P}) < \infty$. In terms of the limiting free energies, the inequality between $x_c(z)$ and $x_c(z;\bar{P})$ may be expressed as in the following theorem.

Theorem 5.41 *Let $c_n(z;\bar{P})$ be the partition function of a model of interacting walks which does not contain the prime pattern-walk P. Then*

$$\lim_{n\to\infty} \frac{1}{n} \log c_n(z;\bar{P}) = \mathcal{F}_w^{\bar{P}}(z) < \lim_{n\to\infty} \frac{1}{n} \log c_n(z) = \mathcal{F}_w(z).$$

Proof The free energies exist by Lemma 5.36 and Corollary 5.37. The radius of convergence of $c(x,z)$ is $x_c(z) = e^{-\mathcal{F}_w(z)}$ at activity z, and by Lemmas 5.35 and 5.36 note that (1) $c(x,z)$ is finite if and only if $c^{fl}(x,z)$ is, and (2) that if $x = x_c(z)$, then $z^{\alpha(X,X)}q(x_c(z),z) = 1$ (since $c^{fl}(x_c(z),z) = \infty$ and $c^{fl}(x,z) < \infty$ if $x < x_c(z)$ by Corollary 5.37). On the other hand, $q(x_c(z),z;\bar{P}) < q(x_c(z),z)$, since there is at least one term in $q(x_c(z),z)$ which is not present in $q(x_c(z),z;\bar{P})$ (this is the term corresponding to the prime pattern-walk P itself), thus $z^{\alpha(X,X)}q(x_c(z),z;\bar{P}) < 1$ and thus $c^{fl}(x,z;\bar{P})$ is finite if $x = x_c(z)$. In other words, the radius of convergence of $c^{fl}(x,z;\bar{P})$ is strictly bigger than $x_c(z)$: $x_c(z;\bar{P}) > x_c(z)$. Thus $\mathcal{F}_w^{\bar{P}}(z) < \mathcal{F}_w(z)$. $\square$

A pattern theorem for interacting models of polygons follows immediately from Theorem 5.28, given Assumptions 5.21 in addition to a local energy. In that case the corresponding model of polygons has the same free energy as the model of walks.

Corollary 5.42 *Suppose that $c_n(k)$ is the number of walks of length n and with a local energy k. Suppose furthermore that this model of walks satisfies Assumptions 5.21, and let $p_n(z)$ be the partition function of the corresponding model of polygons, and $p_n(z;\bar{P})$*

be the partition function of the corresponding model of polygons which do not contain a given prime pattern-walk P. Then

$$\lim_{n\to\infty} \frac{1}{n} \log p_n(z; \bar{P}) = \mathcal{F}_p^{\bar{P}}(z) < \lim_{n\to\infty} \frac{1}{n} \log p_n(z) - \mathcal{F}_p(z). \qquad \square$$

It is important to note that while Theorem 5.41 and Corollary 5.42 were only proven for a model of walks which do not contain a prime pattern-walk, the applicability is wider. In particular, let P be any Kesten pattern. Then P is contained in a prime pattern-walk P^K. Construct all the reflections of P^K, and denote them by $\{P_j\}$ for $j = 1, \ldots, 2^d$. Construct the composite pattern $P_1 P_2 \ldots P_{2^d}$, and replace P in Fig. 5.7 by this composite pattern P. Call the resulting prime pattern-walk Q^P. Notice that if the Kesten pattern K cannot occur, then neither may P^K nor Q^P, nor any of its reflections. Since $p_n(k; \bar{P}) \leq p_n(k; \bar{Q}^P)$, this shows that

$$\limsup_{n\to\infty} \frac{1}{n} \log p_n(z; \bar{P}) \leq \lim_{n\to\infty} \frac{1}{n} \log p_n(z; \bar{Q}^P). \tag{5.39}$$

Thus, the pattern theorem applies to Kesten patterns.

Corollary 5.43 *Suppose that $c_n(k)$ is the number of walks of length n and energy k, that Assumptions 5.21 are satisfied, and that the energy is local. For any Kesten pattern P there exists a number $\epsilon_0 > 0$ such that for all $\epsilon < \epsilon_0$,*

$$\limsup_{n\to\infty} \frac{1}{n} \log c_n(z; \lfloor \epsilon n \rfloor P) < \mathcal{F}_w(z);$$

$$\limsup_{n\to\infty} \frac{1}{n} \log p_n(z; \lfloor \epsilon n \rfloor P) < \mathcal{F}_p(z).$$

Proof Consider only the case of $c_n(z; \lfloor \epsilon n \rfloor P)$; the proof for $p_n(z; \lfloor \epsilon n \rfloor P)$ is similar. Suppose that P contains $|P|$ edges. By Theorem 5.41 there exists an $N > 0$ and an $\epsilon > 0$ such that if $n \geq N$, then

$$c_n(z; \bar{P}) < [e^{\mathcal{F}_w(z)}(1-\epsilon)]^n, \quad c_n < [e^{\mathcal{F}_w(z)}(1+\epsilon)]^n.$$

Let A be a walk contributing to the partition function $c_n(z; \leq \lfloor \epsilon n \rfloor P)$. By starting at one end of A, colour subwalks of length m by $0, 1, \ldots,$ where $m \geq |P|$, and let $M = \lfloor n/m \rfloor$ be the number of colours used. Then at most $\lfloor \epsilon n \rfloor$ of the monochromatic subwalks will contain the prime pattern-walk P. The remaining subwalks will not contain a copy of P (and each contributes at most $c_m(z; \bar{P})$). The remainder of the walk has $r = n - mM$ edges, and contributes at most $c_r(z)$. In addition, the $M - 1$ cuts made in this partitioning may change the total energy of the subwalks by at most $\alpha(M-1)$ (by Definition 5.29 and eqn (5.31)). Define $\phi(z) = z + 1 + z^{-1}$. Then

$$\begin{aligned} c_n(z; \leq \lfloor \epsilon n \rfloor P) &\leq \sum_{j=0}^{\lfloor \epsilon n \rfloor} \binom{M}{j} [c_m(z)]^j [c_m(z; \bar{P})]^{M-j} c_{n-Mm}(z) [\phi(z)]^{\alpha(M-1)}, \\ &\leq (\lfloor \epsilon n \rfloor + 1)[\phi(z)]^{\alpha(M-1)} \binom{M}{\lfloor \epsilon n \rfloor} e^{Mm\mathcal{F}_w(z)} c_{n-Mm}(z) \left(\frac{1+\epsilon}{1-\epsilon}\right)^{m\lfloor \epsilon n \rfloor} \\ &\quad \times (1-\epsilon)^{mM}. \end{aligned}$$

Take the $(1/n)$-th power of this, and take the lim sup as $n \to \infty$; this gives

$$\limsup_{n\to\infty}[c_n(z; \leq\lfloor \epsilon n\rfloor P)]^{1/n} \leq [\phi(z)]^{\alpha/m} \frac{e^{\mathcal{F}_w(z)}(1-\epsilon)}{(\epsilon m)^{\epsilon}(1-\epsilon m)^{1/m-\epsilon}} \left(\frac{1+\epsilon}{1-\epsilon}\right)^{\epsilon m} < e^{\mathcal{F}_w(z)},$$

if ϵ is small enough, and m large enough. Since $c_n(z; \lfloor \epsilon n\rfloor P) \leq c_n(z; \leq\lfloor \epsilon n\rfloor P)$, the theorem follows. The proof for polygons is similar. □

In addition, Corollary 5.18 and Theorem 5.19 still apply here, with suitable adjustments to account for the activity z; the proof of Corollary 5.44 is the same as for Corollary 5.18.

Corollary 5.44 *Let P be any Kesten pattern, and let $\epsilon > 0$ be a small number. Then there exists a $k(z) > 0$ and an $N_0 > 0$ (if ϵ is small enough) such that*

$$p_n(z; \lfloor \epsilon n\rfloor) \leq p_n(z; \leq\lfloor \epsilon n\rfloor) < e^{-k(z)n} p_n(z),$$

for all $n \geq N_0$. In other words,

$$p_n(z) - p_n(z; \lfloor \epsilon n\rfloor) \geq p_n(z) - p_n(z; \leq\lfloor \epsilon n\rfloor) > (1 - e^{-k(z)n}) p_n(z).$$

A similar statement is true for walks.

Proof Define $\limsup_{n\to\infty} \frac{1}{n} \log p_n(z; \leq\lfloor \epsilon n\rfloor) = \log \mu_z(\epsilon)$, and $\mu_z = e^{\mathcal{F}_w(z)}$. Then

$$\limsup_{n\to\infty} \frac{1}{n} \log[p_n(z; \leq\lfloor \epsilon n\rfloor)/p_n(z)] = \log(\mu_z(\epsilon)/\mu_z) < 0$$

if $\epsilon < \epsilon_0$ in Corollary 5.43. Let $k(z) = \frac{1}{2}|\log(\mu_z(\epsilon)/\mu_z)| > 0$; then by definition of the lim sup, there is an N_0 such that

$$\frac{1}{n} \log\left(p_n(z; \leq\lfloor \epsilon n\rfloor)/p_n(z)\right) < -k(z),$$

if $n > N_0$. Thus $p_n(z; \leq\lfloor \epsilon n\rfloor) < e^{-k(z)n} p_n(z)$. □

To see that the density function exists if Assumptions 5.21 are satisfied (at least for a model of self-interacting polygons) let $p_n(k; nP)$ be the number of polygons of length n which contains the prime pattern-walk P exactly n times and which has energy k. Then by concatenating two polygons, again inserting $2s$ extra edges to avoid creating extra copies of P, it is found that

$$\sum_{m_1=0}^{m} \sum_{l=0}^{k} p_{n_1}(l; m_1 P) p_{n_2}(k-l; (m-m_1)P) \leq \sum_{i=-f_1}^{f_1} p_{n_1+n_2+2s}(k+i; mP), \quad (5.40)$$

where f_1 is $o(n)$ (if the energy is local, then $f_1 = O(1)$). Multiply the above by $y^m z^k$ and sum over k and define $\phi(z) = z + 1 + z^{-1}$. This gives

$$p_{n_1}(z; y) p_{n_2}(z; y) \leq [\phi(z)]^{f_1} \sum_{j=-q}^{q} p_{n_1+n_2+2s}(z; yP). \quad (5.41)$$

In other words, the limiting free energy $\mathcal{F}_p(z, y) = \lim_{n\to\infty}[\log p_n(z, y)]/n$ exists. By Theorem 3.19 there is also a density function for the pattern P, defined by

$$\mathcal{P}_p(z; \epsilon P) = \lim_{n\to\infty} \left[p_n(z; \lfloor \epsilon n \rfloor P)\right]^{1/n}. \tag{5.42}$$

These arguments can similarly be made (with little change) if P is a Kesten pattern. In terms of density functions Corollary 5.43 can be stated as follows.

Theorem 5.45 *Let $\mathcal{P}_p(z; \epsilon)$ be the density function of a Kesten pattern P on the interval $(0, \epsilon_M)$. Then there exists an $\epsilon_c \in (0, \epsilon_M)$ (defined by $\epsilon_c = \inf\{\epsilon | \mathcal{P}_p(z; \epsilon) \geq \mathcal{P}_p(z; \epsilon_1)\ \forall \epsilon \in (0, \epsilon_M)\}$), such that $\mathcal{P}_p(z; \epsilon) < \mathcal{P}_p(z; \epsilon_c)$ whenever $\epsilon < \epsilon_c$. The free energy is given by $e^{\mathcal{F}_p(z)} = \sup_{\epsilon_m < \epsilon < \epsilon_M} \mathcal{P}_p(z; \epsilon)$.* □

A two-parameter limiting free energy is also defined as

$$\mathcal{F}_p(z, y) = \sup_{0\leq\epsilon\leq\epsilon_M} \{\log \mathcal{P}_p(z; \epsilon) + \epsilon \log y\}. \tag{5.43}$$

This free energy has two activities: z is the original activity conjugate to the energy of the polygons, while y is conjugate to the number of occurrences of a pattern P. Of course, under suitable conditions the joint density function may be defined:

$$\log \mathcal{P}_p(\delta, \epsilon) = \inf_{0\leq z<\infty} \{\log \mathcal{P}_p(z; \epsilon) - z \log \delta\}. \tag{5.44}$$

This makes contact with Theorem 5.20.

5.3 Polygons with curvature

5.3.1 Curvature in polygons

A particularly simple case of an interacting model is a polygon with a stiffness parameter or bending energy. This model has its origin in models of semi flexible chains, see for example references [122, 123]. In this model the energy of a polygon is the number of right angles between adjacent edges (times $\pi/2$, but this factor can be absorbed into the activity). Thus, let $p_n(s)$ be the number of polygons of length n with s right angles between adjacent edges. Note that $p_n(4) = (n-2)/2$, and, by counting partially directed walks which can only step in the East, North or South directions (with North or South following each East step, and East following any North or South step), it follows that[13] $p_n(n) \geq 2^{n/2+2}$.

Concatenation of two polygons as in Fig. 1.4 shows that

$$p_{n_1}(s_1) p_{n_2}(s_2) \leq \sum_{i=-4}^{4} p_{n_1+n_2}(s_1 + s_2 + i), \tag{5.45}$$

since as many as four right angles may be created or destroyed in the concatenation. Thus, $p_n(s)$ satisfies Assumptions 3.8 and integrated density functions exists (Corollary 3.14).

[13] Any partially directed walk is unfolded if its first step is in the East direction. If there are s_n such walks, then concatenation shows that $\lim_{n\to\infty}(1/n) \log s_n \geq \log \sqrt{2}$. Moreover, any one of these walks can be unfolded in the North–South direction to find doubly unfolded partially directed walks; notice that no right angles are created or destroyed in this unfolding, so that the energy does not change. By putting together doubly unfolded partially directed walks into loops and polygons, the result is that $\lim_{n\to\infty}(1/n) \log p_n(n) \geq \log \sqrt{2}$. The bound then follows from the supermultiplicativity of $p_n(n)$; see Theorem 1.2.

Moreover, by Corollary 3.14 and Theorem 3.16, there are integrated density functions of right angles, and a density function of right angles, log-concave in (0, 1), such that

$$\mathcal{P}_s(\epsilon) = \lim_{n\to\infty} \left[p_n(\lfloor \epsilon n \rfloor + \sigma_n)\right]^{1/n} = \min\{\mathcal{P}_s(\leq\epsilon), \mathcal{P}_s(\geq\epsilon)\}, \tag{5.46}$$

where $\sigma_n = o(n)$ is a sequence of integers. There is also a limiting free energy $\mathcal{F}_s(z)$ (Theorem 3.18). Note that $\mathcal{P}_s(0) = 1$, and since $p_n(s) \leq \binom{n}{s}(2d)^s$, it follows that $\mathcal{P}_s(\epsilon) \leq (2d)^\epsilon/\epsilon^\epsilon(1-\epsilon)^{1-\epsilon}$. Taking $\epsilon \to 0^+$ establishes the continuity of the density function at $\epsilon = 0$. In addition, the bounds above give $\sqrt{2} \leq \mathcal{P}_s(1) \leq \mu_d$. Continuity at $\epsilon = 1$ is much more problematic, and is not established.

Theorem 5.46 *Let $\mathcal{P}_s(\epsilon)$ be the density function of polygons counted with respect to their number of right angles. Then $\mathcal{P}_s(\epsilon)$ is continuous in* [0, 1), *and has infinite right-derivative at $\epsilon = 0$. Moreover, the left-derivative of $\mathcal{P}_s(\epsilon)$ diverges as $\epsilon \to 1^-$.*

Proof Continuity in (0, 1) is known from Theorem 3.16, and in the previous paragraph it was shown that the density function is also continuous at $\epsilon = 0$. The number of polygons with s right angles is bounded from below by the number of partition polygons with exactly s right angles. In particular, consider only partition polygons of height l and width $n/2 - l$. Further, consider only strict partitions, where each slab (see Fig. 4.1(a)) is strictly shorter than the previous slab. In such a partition polygon, two right angles are given by the corners of the left-most side, while each vertical step on the right-hand side accounts for two more right angles. In total, there are $2 + 2l$ right angles, and the number of such polygons is $\binom{n/2-l-2}{l-1}$. Let $l = \lfloor \delta n \rfloor$ for some small $\delta > 0$. Then

$$\mathcal{P}_s(2\delta) \geq \lim_{n\to\infty} \binom{n/2 - \lfloor \delta n \rfloor - 2}{\lfloor \delta n \rfloor - 1}^{1/n} = \frac{(1/2-\delta)^{(1/2-\delta)}}{\delta^\delta (1/2 - 2\delta)^{(1/2-2\delta)}}.$$

But the limit of the function on the right is 1 as $\delta \to 0^+$, and its right-derivative is infinite as $\delta \to 0^+$. Since $\mathcal{P}_s(\epsilon)$ has a right-derivative everywhere in [0, 1), its right-derivative at $\epsilon = 0$ is infinite. To see that the left-derivative diverges as $\epsilon \to 1^-$, consider Corollary 5.43. Since $p_n(z; \lfloor \epsilon p \rfloor) \leq e^{-k(z)n} p_n(z)$ for some $k(z) > 0$, and for small enough $\epsilon > 0$, only polygons counted by $p_n(z; \geq \lfloor \epsilon p \rfloor)$ need be considered when computing $\mathcal{F}_s(z)$. Let P be the pattern in Fig. 5.8. Consider all the polygons counted by $p_n(z; \geq \lfloor \epsilon p \rfloor)$ with a density α of right angles. Then $1 \geq \alpha \geq 14\epsilon$, and there is a term in $p_n(z; \geq \lfloor \epsilon p \rfloor)$ which dominates the other terms exponentially, let this be the term corresponding to polygons with a density α of right angles: $p_n(\lfloor \alpha n \rfloor, \geq \lfloor \epsilon p \rfloor) z^{\lfloor \alpha n \rfloor}$. Select $\lfloor \delta n \rfloor$ of the P from $\lfloor \epsilon n \rfloor$, and perform the construction in Fig. 5.8. This shows that

$$\begin{aligned}\binom{\lfloor \epsilon n \rfloor}{\lfloor \delta n \rfloor} p_n(z)(1 - e^{-k(z)n}) &\leq \binom{\lfloor \epsilon n \rfloor}{\lfloor \delta n \rfloor} p_n(\lfloor \alpha n \rfloor, \geq \lfloor \epsilon n \rfloor P) z^{\lfloor \alpha n \rfloor} \\ &\leq p_n(\lfloor \alpha n \rfloor - 2\lfloor \delta n \rfloor) z^{\lfloor \alpha n \rfloor}.\end{aligned}$$

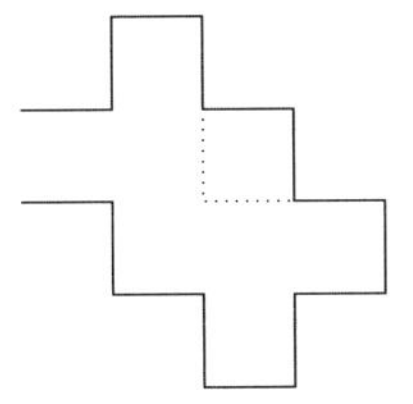

Fig. 5.8: This Kesten pattern occurs with positive density in almost all walks counted by $p_n(n)$. If two edges are replace by the dotted edges, then two right angles disappear.

Take the $(1/n)$-th power and let $n \to \infty$, and note that

$$\begin{aligned}
&\limsup_{n\to\infty} \left[p_n(\lfloor \alpha n \rfloor - 2\lfloor \delta n \rfloor)\right]^{1/n} \\
&\quad \leq \min\{\limsup_{n\to\infty} \left[p_n(\leq \lfloor \alpha n \rfloor - 2\lfloor \delta n \rfloor)\right]^{1/n}, \limsup_{n\to\infty} \left[p_n(\geq \lfloor \alpha n \rfloor - 2\lfloor \delta n \rfloor)\right]^{1/n}\}, \\
&\quad = \min\{\mathcal{P}_s(\leq \alpha - 2\delta), \mathcal{P}_s(\geq \alpha - 2\delta)\} \\
&\quad = \mathcal{P}_s(\alpha - 2\delta)
\end{aligned}$$

and this shows that

$$\frac{\epsilon^\epsilon}{\delta^\delta(\epsilon-\delta)^{\epsilon-\delta}}\, e^{\mathcal{F}_s(z)} \leq \mathcal{P}_s(\alpha - 2\delta) e^{\alpha \log z}.$$

Multiply by $e^{-\alpha \log z}$ and take the supremum with respect to z on the left-hand side, and subtract $\mathcal{P}_s(\alpha)$ from both sides. Then

$$\left[\frac{\epsilon^\epsilon}{\delta^\delta(\epsilon-\delta)^{\epsilon-\delta}} - 1\right] \mathcal{P}_s(\alpha) \leq \mathcal{P}_s(\alpha - 2\delta) - \mathcal{P}_s(\alpha).$$

Now divide this by 2δ and take $\alpha \to 1^-$. Then the left-hand side diverges as $\delta \to 0^+$. □

The infinite right-derivative at zero density of right angles rules out a first-order transition at zero temperature (see Section 3.3.4). There is a phase transition in this model, presumably at zero temperature (infinite stiffness), and presumably continuous. This is seen by considering for example the critical exponents of the model. In particular, for finite values of the activity z the exponents of a polymer in a dilute solution should be expected (see Table 1.1). At infinite stiffness, the polygon is a rectangle, and $\nu = 1$ while $\alpha = 0$.

An important observation in the proof of Theorem 5.46 is that there is a pattern theorem for polygons in this model. In particular, it can be checked that the curvature energy is a local energy, and that the models $c_n^{fl}(s)$, $c_n^f(s)$ and $c_n(s)$ satisfy Assumptions 5.21. Thus, the pattern theorems in Corollary 5.43 and Theorem 5.46 and Theorem 5.45 are all applicable in this model. Some consequences of this are explored in the next sections.

5.3.2 Curvature and knotted polygons

Let P be the pattern-walk in Fig. 5.9. The union of the dual 3-cells centred at the vertices of P is a (topological) ball C, and the pair (C, P) is a knotted ball-pair (if two edges are added in the obvious way to the end-points of P, then the intersection of this augmented P with C is a knotted arc). Let $p_n(z; k)$ be the partition function of polygons of length n with an activity z conjugate to curvature, and which contains the knotted ball-pair (C, P) exactly k times. Since a model of polygons with a curvature energy satisfies Assumptions 5.21 and since the energy is local, Corollary 5.44 implies that there exists a small $\epsilon > 0$ and a $k(z) > 0$ such that $p_n(z; \geq \lfloor \epsilon \rfloor) \geq (1 - e^{-k(z)n}) p_n(z)$, if n is large enough. In other words, the partition function $p_n(z)$ of polygons is dominated in the limit $n \to \infty$ by conformations of polygons which contain a density of the pattern P. This observation gives the following result.

Theorem 5.47 *There exists an $\epsilon_c(z) > 0$ such that for all $0 \leq \epsilon < \epsilon_c(z)$ and $z > 0$,*

$$\lim_{n \to \infty} \frac{p_n(z; \lfloor \epsilon n \rfloor)}{p_n(z)} = 0.$$

Moreover, the rate of approach to zero is exponential.

Proof Notice that P is a Kesten pattern. The result then follows immediately from Corollary 5.44. □

In particular, the partition function $p_n(z)$ is dominated by polygons which contain P with a density of at least $\epsilon_c(z)$. The limiting free energy can be computed by

$$\mathcal{F}_s(z) = \lim_{n \to \infty} \frac{1}{n} \log p_n(z) = \lim_{n \to \infty} \frac{1}{n} \log p_n(z, \geq \lfloor \epsilon_c(z) n \rfloor). \tag{5.47}$$

All the thermodynamic properties of this model may be determined by only considering conformations of polygons which contain P at least $\lfloor \epsilon_c(z) n \rfloor$ times. An important corollary of the above is the Frisch–Wasserman–Delbruck conjecture [67, 129], which (in the context here) states that the probability that a polygon is knotted approaches one as the length of the polygon increases. This conjecture was proven for uniformly weighted polygons by D.W. Sumners and S.G. Whittington [336] and N. Pippenger [292]. In the case of interacting polygons a similar result may be stated; this has been done in the case of polygons with a curvature energy [280].

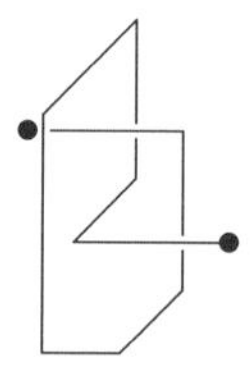

Fig. 5.9: A tight knot; the union of the dual 3-cells of the vertices in this Kesten pattern is a ball, and if the end-points of the arc are extended in straight lines to intersect the boundary of the ball, then a knotted ball-pair is obtained.

Theorem 5.48 *The limiting free energy $\mathcal{F}_s(z)$ of a model of polygons with z conjugate to the curvature of the polygons is completely determined by knotted polygons.* □

Observe that eqn (5.47) is a stronger statement than Theorem 5.48. The limit there is taken over a class of polygons which has knot type determined by compound knots with at least $\lfloor \epsilon_c(z)n \rfloor$ factors. In the limit, the knot complexity (as measured by the number of prime knot factors in any polygon) of this class of polygons becomes arbitrarily large. Obviously, the above arguments can be repeated for any model of interacting polygons which satisfies Assumptions 5.21 and which has a local energy, for example, a model of a polygon adsorbing in a plane in three dimensions can be shown to satisfy these assumptions (and has a local energy) and so Theorem 5.48 also applies in that case [347].

Application of Theorem 5.48 to the model above implies the existence of a free energy

$$\mathcal{F}_s(z; \epsilon) = \lim_{n\to\infty} \frac{1}{n} \log p_n(z; \lfloor \epsilon n \rfloor), \tag{5.48}$$

in a model of polygons which contains the Kesten pattern P in Fig. 5.9 with a density of ϵ. In particular, the function $\mathcal{P}_s(z; \epsilon)$ defined by $\log \mathcal{P}_s(z; \epsilon) = \mathcal{F}_s(z; \epsilon)$ is the density function of P at activity z. Notice that Theorem 5.45 indicates that $\mathcal{F}_s(z; \epsilon) < \mathcal{F}_s(z)$ if $\epsilon < \epsilon_c(z)$, and that there is a joint density function (see eqn (5.44)) defined by

$$\log \mathcal{P}_s(\delta, \epsilon) = \inf_{0\le z\le\infty} \{\log \mathcal{P}_s(z; \epsilon) - z \log \delta\}. \tag{5.49}$$

Thus, the pattern theorem for the tight knots in Fig. 5.9 becomes $\mathcal{P}_s(\delta, \epsilon) < \mathcal{P}_s(\delta, \epsilon_c(\delta))$ for any $\epsilon < \epsilon_c(\delta)$.

There is a tremendous volume of literature devoted to knotted walks and polygons as models of knots in polymers and in DNA. Monte Carlo simulations focused primarily on algorithms for detecting knots efficiently [66, 264, 351] and for computing the incidence of knots as a function of length and other parameters [201, 220]. Theorem 5.47 implies that $[p_n(1; 0)/p_n] \approx e^{-\alpha_0 n + o(n)}$, and numerical simulations indicate that $\alpha_0 = (7.6 \pm 0.9) \times 10^{-6}$ [202]. The issue of knotted arcs in self-avoiding walks was addressed in reference [198]. Theoretical work on the effect of knots on the physical properties of polymers was done by S.F. Edwards [101, 102] and P.-G. de Gennes [65]. The incidence of knots in a model of self-avoiding walks with a bending energy was considered in reference [280]. Experimental work on the occurrence of knots as a function of the degree of polymerization can be found in reference [314].

5.3.3 Curvature and writhe

Let α be any piecewise linear curve in $\mathbb{R}^3$. A projection of α into a plane with normal unit vector $\hat{\xi}$ may in general have crossings (or points where two points in the walk project to the same point) where the projected arcs of the knot pass over one another transversely.[14] The projection is turned into a knot projection by orienting the curve

[14] In fact, it can be shown that the set of vectors $\hat{\xi}$ where this is not the case has zero measure [38].

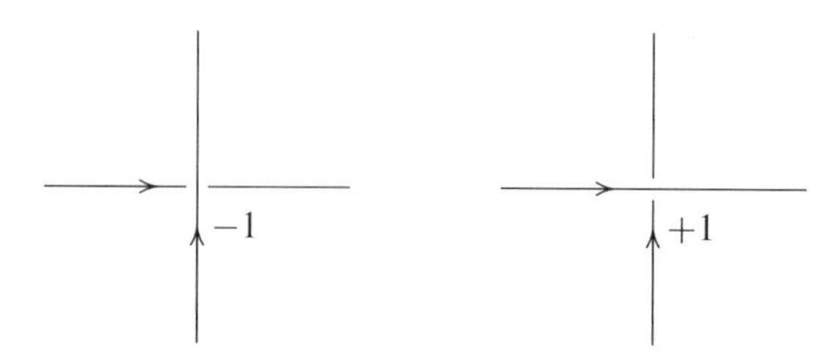

Fig. 5.10: A right-hand rule determines the sign of a crossing in a regular knot diagram.

and by indicating which arc in the projection passes over, and which passes under, at every crossing. This is done by removing a small arc from the underpassing arc about the crossing. The resulting crossing then has the appearance of one of the two cases in Fig. 5.10. A projection which has only transversal crossings and with a small arc removed in the underpassing arc in a crossing is called a *regular knot diagram*. The left-handed crossing (determine this by using a right-hand rule: take the overpassing arc in your right hand; if your fingers curl against the arrow on the underpassing arc, then the crossing is left-handed) is given a negative sign, and the right-handed crossing is given a positive sign. Note that the sign of a crossing is independent of the arrow (orientation) in the knot.

The writhe of a simple and closed curve [130] is defined as the average of the sum of signed crossings over all directions $\hat{\xi}$. This definition defines the writhe of a polygon in $\mathbb{Z}^3$ as well. For example, let B be the polygon with steps $\{\hat{\imath}, \hat{\imath}, -\hat{k}, -\hat{\jmath}, -\hat{\imath}, \hat{\jmath}, \hat{\jmath}, \hat{k}, -\hat{\imath}, -\hat{\jmath}\}$, which is illustrated in Fig. 5.11. Denote the writhe of B by $W(B)$. $W(B)$ can be efficiently computed by using the Lacher–Sumners theorem [223]: the writhe of a lattice polygon in the cubic lattice is the average of the four linking numbers of the polygon with a push-off (of itself) into four non-antipodal octants. The following theorem was proven in reference [196].

Lemma 5.49 $W(B) = +\frac{1}{2}$.

Proof By the Lacher–Sumners theorem, only the linking numbers of B with its four push-offs must be computed. Construct the push-offs by adding the following four vectors to the coordinates of the vertices of B: $\mathbf{v}_1 = (\hat{\imath} + \hat{\jmath} + \hat{k})/2$, $\mathbf{v}_2 = (-\hat{\imath} + \hat{\jmath} + \hat{k})/2$, $\mathbf{v}_3 = (-\hat{\imath} - \hat{\jmath} + \hat{k})/2$ and $\mathbf{v}_4 = (\hat{\imath} - \hat{\jmath} + \hat{k})/2$. These vectors lie in the interiors of unit cubes in four mutually non-antipodal octants. Let B_i be the curve defined by $B + \mathbf{v}_i$. Then the linking number can be computed explicitly: $L(B, B_1) = L(B, B_3) = L(B, B_4) = 1$ and $L(B, B_2) = -1$. Thus, $W(B) = +\frac{1}{2}$. □

Notice that the polygon B and its dual 3-cells form a ball-pair (C, B) where C is the union of the 3-cells. A push-off of B (say B_1) is still contained in C so that the triplet (C, B, B_1) should be considered when the linking numbers above are computed. Let A be a polygon which intersects the polygon B in a self-avoiding walk B' which starts at P and ends at Q in Fig. 5.9. In other words, A traverses all the edges in B, except for the single edge joining P and Q. B' is a Kesten pattern in A, and it can be truncated from A by deleting it between the vertices P and Q, and then reconnecting A by adding the single edge from P to Q. Suppose this truncation leaves the polygon A'.

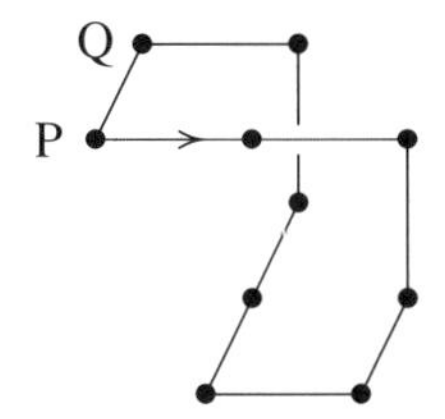

Fig. 5.11: This polygon has writhe equal to $+\frac{1}{2}$.

Lemma 5.50 $W(A) = W(A') + W(B)$.

Proof Consider A and its push-off A_1. Assume that A contains the subwalk $\{\hat{\imath}, \hat{\imath}, \hat{\imath}, -\hat{k}, -\hat{\jmath}, -\hat{\imath}, \hat{\jmath}, \hat{\jmath}, \hat{k}, -\hat{\imath}, -\hat{\imath}\}$ which intersects B in B'. In Fig. 5.11 a projection of this subwalk in the XY-plane is plotted, with a projection of the push-off A_1. If B' and B'_1 are truncated along the dotted lines in Fig. 5.12, then the polygons A' and A'_1 are obtained. Consider now the triplet (C, B', B'_1) of the ball C (the union of 3-cells dual to B') and the two arcs B' and B'_1. By a small deformation of B' inside C, the $(+)$-crossing in the circle in Fig. 5.9 can be changed into a $(-)$-crossing. The resulting pair of curves is isotopic (by an isotopy in C) to the pair (A', A'_1). (This is most easily seen by using a Reidemeister move I inside C to convert A to A' and A_1 to A'_1 [38].) The change in the sign of a single crossing between A and A_1 shows that $L(A, A_1) = L(A', A'_1) + 1$. But $L(B, B_1)$=1. Therefore,

$$L(A, A_1) = L(A', A'_1) + L(B, B_1).$$

Similar calculations show that

$$\begin{aligned} L(A, A_2) &= L(A', A'_2) + L(B, B_2) \\ L(A, A_3) &= L(A', A'_3) + L(B, B_3) \\ L(A, A_4) &= L(A', A'_4) + L(B, B_4). \end{aligned}$$

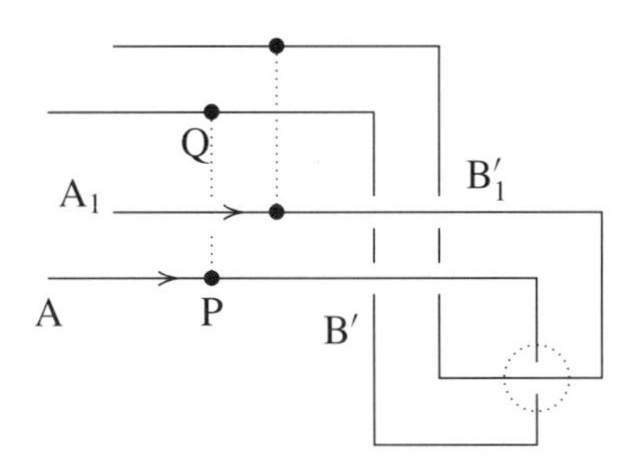

Fig. 5.12: The projections of a subwalk and its push-off in the XY- plane. The subwalks B' and B'_1 can be truncated by adding edges along the dotted lines and removing the resulting polygons. Alternatively, the same can be achieved by reversing the positive crossing in the dotted circle and then by using Reidemeister moves in the knot diagram.

The Lacher–Sumners theorem now states that the average of these four equations gives the writhe of the polygon. □

With the result in Lemma 5.50, it can now be proven that irrespective of the value of the curvature activity z, the expected value of the absolute value of the writhe should increase (almost surely) at least as fast as $\sqrt{n}$, where n is the length of the polygon. The ingredients in the proof are first of all the pattern theorem for interacting models of polygons (Corollary 5.44), and Lemma 5.50. Let $p_n(s)$ be the number of polygons with s right angles, and let $p_n(z)$ be the partition function of this model. This endows the model with a probability distribution $\Pi_n(z, s)$:

$$\Pi_n(z, s) = \frac{p_n(s) z^s}{p_n(z)}. \tag{5.50}$$

Polygons can be sampled from this distribution. For small values of z, polygons with low curvature are likely to be sampled, and at higher values of z polygons with large curvature have more weight.

Theorem 5.51 *Let A be a polygon sampled from $\Pi_n(z, s)$ for any $z > 0$. Then for every function $f(n) = o(\sqrt{n})$, the probability that the absolute value of the expected writhe of A is less than f(n) approaches zero as $n \to \infty$.*

Proof Let $z > 0$, and let $P = (C, B')$. The mirror image (reflected through the XY-plane) of P is $P^* = (C, B^{*\prime})$. Notice that P and P^* are Kesten patterns. Let $p_n(z; \leq l)$ be the partition function of polygons which contain P or P^* at most l times. By Corollary 5.44, if $\epsilon > 0$ is a small enough number, there exists a $k(z) > 0$ such that $p_n(z; \leq \lfloor \epsilon n \rfloor) < e^{-k(z)n} p_n(z)$. In other words, the probability that A contains the patterns P or P^* at least $\lfloor \epsilon n \rfloor$ times is greater than $(1 - e^{-k(z)n})$, for large enough values of n. The distribution of the patterns P and P^* is binomial along A. Consider $\lfloor \epsilon n \rfloor$ occurrences of P and P' along A. The probability that B occurs exactly k times amongst these $\lfloor \epsilon n \rfloor$ occurences is less than $1/\sqrt{\lfloor \epsilon n \rfloor}$ (use Stirling's formula to see this). By Lemma 5.50 the writhe of A is the sum over two terms; the first term is obtained from the polygon found by truncating the $\lfloor \epsilon n \rfloor$ occurrences of P or P^* from A, and the second is from the $\lfloor \epsilon n \rfloor$ copies of B or B^* obtained from these truncations. Suppose that the absolute writhe of A is less than $f(n)$; then the contribution to the writhe of A from the copies of B or B' is at most one of $\lceil 2f(n) + 1 \rceil$ different values (whatever the writhe of the truncated polygon, if the writhe due to B or B^* is added to it, then a result with absolute value less than f(n) must be obtained). In other words,

$$\mathrm{Prob}(|W(A)| < f(n)) \leq Q_n(z) \frac{\lceil 2f(n) + 1 \rceil}{\sqrt{\lfloor \epsilon n \rfloor}} + (1 - Q_n(z)) R_n,$$

where R_n is the contribution from polygons which contain fewer than $\lfloor \epsilon n \rfloor$ occurrences of B or B^*, and where $Q_n(z)$ is the probability that a polygon of length n will contain at least $\lfloor \epsilon n \rfloor$ occurrences of B or B^*. But the probability that a polygon has at least $\lfloor \epsilon n \rfloor$

occurrences of B or B^* is at least $(1 - e^{-k(z)n})$, so that $Q_n(z) \geq (1 - e^{-k(z)n})$. Thus, $(1 - Q_n(z)) \leq e^{-k(z)n}$, and this gives

$$\text{Prob}(|W(A)| < f(n)) \leq Q_n(z)\frac{\lceil 2f(n)+1 \rceil}{\sqrt{\lfloor \epsilon n \rfloor}} + e^{-k(z)n} R_n,$$

and since $Q_n(z) \to 1$ as $n \to \infty$, it follows that $\text{Prob}(|W(A)| < f(n)) \to 0$ as $n \to \infty$ if $f(n) = o(\sqrt{n})$. □

The above is a generalization of the work in reference [196]; see also references [212, 283]. The square-root bound in Theorem 5.51 was tested numerically in reference [196]. If it is assumed that the absolute writhe of a polygon satisfies a power-law, $\langle |W| \rangle \simeq n^\omega$, then it is found that $\omega = 0.522 \pm 0.004$; see also references [283, 348]. The topology and geometry of polymers are discussed in references [212, 213].

5.3.4 Curvature and contacts

Let $p_n(k, s)$ be the number of polygons with curvature s and with k nearest-neighbour contacts between vertices which are nearest neighbours in the lattice, but not adjacent in the polygon. A polygon counted by $p_n(k, s)$ is said to have *curvature energy* s and *contact energy* k. This is an important model of a self-interacting (ring) polymer which is thought to undergo a "collapse transition" at a critical temperature (called the θ-temperature).

The partition function of this model is found by introducing an activity y:

$$p_n(y; s) = \sum_{k \geq 0} p_n(k, s) y^k, \tag{5.51}$$

and a curvature partition function can be defined by introducing an activity z in the same way. Consider the situation when there is a density ϵ of right angles. Concatenation (use two extra edges to separate the polygons in the X-direction to avoid extra contacts, see Fig. 1.4(b)) gives

$$\sum_{k_0=0}^{k} p_{n_1}(k_0, s_1) p_{n_2}(k - k_0, s_2) \leq \sum_{i=-4}^{4} \sum_{j=-2}^{2} p_{n_1+n_2+2}(k + j, s_1 + s_2 + i), \tag{5.52}$$

since up to four extra right angles may be lost or created, and as many as two extra contacts may be lost or created, in this process. Multiplication by y^k and summing over k gives the generalized supermultiplicative inequality

$$p_{n_1}(y; s_1) p_{n_2}(y; s_2) \leq [\phi(y)]^4 \sum_{j=-2}^{2} p_{n_1+n_2+2}(y; s_1 + s_2 + i), \tag{5.53}$$

where $\phi(y) = y + 1 + y^{-1}$, see eqn (5.23). In other words, this model, with a contact activity y, satisfies Assumptions 3.8. Thus, by Corollary 3.14 and Theorem 3.16 there exist integrated density functions and a density function as well as a corresponding free energy (Theorem 3.17) for models of polygons with a fixed density ϵ of right angles (where $\sigma_n = o(n)$ is a sequence of integers):

$$\log \mathcal{P}_k(y;\epsilon) = \lim_{n\to\infty} \frac{1}{n} \log p_n(y; \lfloor \epsilon n \rfloor + \sigma_n) = \min\{\log \mathcal{P}_k(y; \leq\epsilon), \log \mathcal{P}_k(y; \geq\epsilon)\},$$
$$\mathcal{F}_k(y,z) = \sup_{0<\epsilon<1} \{\log \mathcal{P}(y;\epsilon) + \epsilon \log z\}. \tag{5.54}$$

Notice the introduction of the activity z conjugate to the number of right angles in the polygon. The density function and free energy of Section 5.3.1 is recovered if $y = 1$ in eqn (5.54). In contrast with previous models, the contact energy is not local, and the pattern theorem for interacting models does not apply here; thus, Theorem 5.45 is not known to be true for $\mathcal{P}_k(y;\epsilon)$.

On the other hand, there is also a density function of contacts, defined by

$$\log \mathcal{P}_s(\delta; z) = \inf_{0\leq y<\infty} \{\mathcal{F}_k(y,z) - \delta \log y\}, \tag{5.55}$$

where z is the curvature activity, and the density of contacts is δ. Since contacts may occur in Kesten patterns, and since the curvature is a local energy, Theorem 5.45 implies that there is a $\delta_c(z)$ such that $\mathcal{P}_s(\delta; z) < \mathcal{P}_s(\delta_c(z); z)$ whenever $\delta < \delta_c(z)$. In other words, there is a pattern theorem for nearest-neighbour contacts in this model.

If $z = 1$ then the standard model of a polygon (or walk) with a nearest-neighbour interaction is obtained. It is generally accepted that this model has a collapse transition from an expanded polygon (often called "coiled", and in the universality class of a linear polymer in a good solvent) to a "collapsed" or "globule" phase which is a compact object with metric exponent $\nu = 1/d$ in d dimensions. The critical point is called a θ-point, and its tricritical nature was argued by P.-G. de Gennes [62], see also [111]. The expanded (coiled) phase of the polygon has critical exponents listed in Table 1.1. At the θ-point a new set of tricritical exponents is found. There is a large body of work devoted to the evaluation of these exponents, both numerically and by Coulomb gas and conformal invariance methods in two dimensions. In two dimensions the exponent γ_t describes the singularity in the generating function (footnote 5, Chapter 2), and it is generally accepted that $\gamma_t = 8/7$, and the tricritical metric exponent is $\nu_t = 4/7$ [271, 272], see footnote 15, chapter 2. The cross-over exponent has value $\phi = 3/7$ [93, 94, 95, 229, 271, 272, 348]. From this one may compute $\gamma_u = 8/3$. These exponents are listed in Table 5.1. In the collapsed phase the scaling assumption in eqn (2.23) was tested in reference [269], and the data are consistent with a value of $\gamma_- \approx 1$. These results were also checked by ϵ-expansions in $3-\epsilon$ dimensions [88, 89, 333].

Numerical work in two dimensions includes the simulations of H. Meirovitch and H.A. Lim [259, 260, 261]. Many more studies exist in the literature including those in references [15, 51, 145, 252, 287]. The critical point in two dimensions is believed to be $y_c = 1.941 \pm 0.047$ [312]. Monte Carlo simulations were also done in references [185, 301]. Recent work includes the exact enumeration study of collapsing walks in reference [16], and the partition function zeros were studied in reference [303]. For

Table 5.1: θ-point exponents for linear polymers

Dimension	θ	α	γ_t	γ_u	ν_t	y_t
2	$\frac{3}{7}$	$-\frac{1}{3}$	$\frac{8}{7}$	$\frac{8}{3}$	$\frac{4}{7}$	$\frac{3}{4}$
3	$\frac{1}{2}$	0	1	2	$\frac{1}{2}$	$\frac{3}{2}$

more work see references [54, 85]. Three dimensions is the upper critical dimension of the tricritical behaviour [62, 89, 92, 333]. In this case the mean field values $\gamma_t = 1$, $\nu_t = 1/2$ and $\phi = 1/2$ should be found, but with logarithmic corrections to the scaling laws. Computer simulations of collapsing walks and polygons in three dimensions are consistent with this [146, 259, 312, 341, 342]. The critical value of the activity is found to be $y_c = 1.319 \pm 0.043$ [259]. The physical nature of the collapse transition was also discussed by J.F. Douglas and K.F. Freed [86], while experimental work can be found in reference [337]. Series enumeration suggests that the relationship of amplitude ratios $246A_\infty + 91/2 = 182B_\infty$ applies at the θ-point in two dimensions as well (see Section 1.2.3) [287]. If $z \neq 1$, then the collapse transition occurs in the presence of a curvature-fugacity. There are also suggestions of a curvature-induced freezing transition separating two classes of collapsed phases [10, 87].

5.4 Polygons interacting with a surface: adsorption

In this section I shall discuss aspects of the adsorption problem for polygons and walks, a problem which was considered in references [165, 190, 352, 353, 354]. Polygons will be confined to the half-space $Z \geq 0$, and they will interact with the hyperplane $Z = 0$ via the introduction of an activity conjugate to the number of *visits* of the polygon to the hyperplane. The directed version of this problem was considered in Section 4.5, where it was seen that the adsorption transition is continuous. I shall show that there is an adsorption transition here as well, but it will not be possible to characterize it as well as in the directed case.

A polygon or a walk is *attached* to the $Z = 0$ hyperplane if it has a vertex with Z-coordinate in the set $\{-1, 0, 1\}$. An attached polygon or walk is a *positive* polygon or walk if all its vertices have non-negative Z-coordinates. The number of attached or positive polygons will be counted up to a translation parallel to the $Z = 0$ hyperplane. The number of attached polygons will be denoted by $p_n^>$, and the number of attached walks by $c_n^>$. Positive polygons and walks will be denoted by p_n^+ and c_n^+ respectively. The energy of an attached polygon or walk will be the number of vertices with zero Z-coordinate; such vertices are called *visits*. Attached polygons are said to interact with a *defect plane*, while positive polygons interact with a *wall*. Observe that

$$p_n \leq p_n^> \leq (n+2)\, p_n \qquad c_n \leq c_n^> \leq (n+2)\, c_n, \tag{5.56}$$

since a polygon can be translated to become an attached polygon, and at most $(n+2)$ distinct attached polygons can be translated to the same polygon. A similar argument shows that

$$p_n \le p_n^+ \le 2\, p_n \qquad c_n \le c_n^+ \le 2\, c_n, \tag{5.57}$$

since at most two positive polygons can be translated to the same polygon. This shows that attached and positive polygons and walks have the same growth constant μ_d as polygons and walks.

It is first interesting to note that a model of walks which interact with a hyperplane satisfies Assumptions 5.21(2) and (3), and has a local energy. fl-walks which interact with a hyperplane do not satisfy Assumptions 5.21(1), and therefore do not satisfy Assumptions 3.1 or 3.8. This may be overcome by using a most popular argument. Consider positive fl-walks with v visits counted by $c_n^{+,fl}(v; [h_b], [h_t])$ with first (bottom) vertex a distance h_b from the adsorbing hyperplane ($Z = 0$), and last (top) vertex a distance h_t from the adsorbing hyperplane. Let h_b^* and h_t^* be the most popular values of these distances, and notice that by concatenating two walks counted by $c_n^{+,fl}(v; [h_b^*], [h_t^*])$ by first reflecting one through the hyperplane normal to the X-direction and joining them by an edge, $[c_n^{fl}(v; [h_b^*], [h_t^*])]^2 \le c_{2n+1}^{fl}(2v; [h_b^*], [h_b^*])$. Thus, one may choose to work with walks in $c_n^{+,fl}(v, [h_b^*], [h_b^*])$ only, and these can be manipulated by translation and reflection parallel with the adsorbing hyperplane as fl-walks. Techniques such as this will be used in Theorem 5.52, and the outcome here is that a limiting free energy can be shown to exist, and in three and higher dimensions it will also be the case that adsorbing polygons and walks have the same limiting free energies. Moreover, the pattern theorem for interacting models applies, and results very similar to those in the last section can be obtained here. I shall not pursue these matters further here; instead the discussion below will focus on polygons, and their interaction with an adsorbing plane.

The scaling theory of polymer adsorption was developed by P.-G. de Gennes [63]. A self-avoiding walk near an interface was considered by J.M. Hammersley *et al.* [165], see also [155]. This problem is also related to the study of walks in confined geometries, such as a wedge [167]. The entropic exponent of a walk changes near an interface. For example, if a walk in the half-space $Z \ge 0$ has one end-point confined to the line $Z = 0$ in two dimensions, then the value of γ in eqn (1.7) is believed to be $\gamma_1 = 61/64$, and if both end-points are confined to the line $Z = 0$, then $\gamma_{11} = -3/16$ [95]. These numerical values are valid for non-interacting walks, or polymers in a good solvent. Numerical measurements of γ_1 and γ_{11} were carried out in three dimensions in references [240, 241], where it is found that $\gamma_1 = 0.687 \pm 0.005$ and $\gamma_{11} = -0.38 \pm 0.02$. The mean field values of these *surface exponents* are $\gamma_1 = 1/2$ and $\gamma_{11} = -1/2$ [19,60]. The values of the surface exponents change at the critical adsorption point, and also at the critical θ-point.

5.4.1 *Positive polygons*

Let $p_n^+(v)$ be the number of positive polygons with v visits. The partition function of this model is $p_n^+(z) = \sum_{v \ge 0} p_n^+(v) z^v$. The limiting free energy in this model will be shown to exist by using a most popular argument. The argument is somewhat more sophisticated than previous applications, but the principle is still the same. This is explained in the proof of Theorem 5.52. It is also possible to use unfolding and unfolded walks to prove that there is a limiting free energy in a model of an adsorbing polygon in more than two dimensions. The constructions in Construction 5.3 and Corollary 5.4 leave the number of visits in a walk unchanged, and so first loops and then polygons can be formed from doubly unfolded walks.

Theorem 5.52 *There exists a limiting free energy such that*

$$\mathcal{F}_v^+(z) = \lim_{n\to\infty} \frac{1}{n} \log p_n^+(z).$$

Proof The top vertex of a polygon is that vertex with lexicographically most coordinates, and the mid-point of the top edge has lexicographically most coordinates of all the mid-points of edges. The bottom vertex and bottom edge are similarly defined. Note that the top and bottom edges are always perpendicular to the X-direction, but may otherwise be in one of $(d-1)$ orientations. Let $p_n^+(v; [h_b\hat{\phi}], [h_t\hat{\psi}])$ be the number of positive polygons with v visits, and with bottom edge with mid-point having Z-coordinate h_b and oriented in the $\hat{\phi}$-direction, and with top edge with mid-point having Z-coordinate h_t and oriented in the $\hat{\psi}$-direction. The partition function of polygons with given orientations and heights of the bottom and top edges is defined by

$$p_n^+(z; [h_b\hat{\phi}], [h_t\hat{\psi}]) = \sum_{v\geq 0} p_n^+(v; [h_b\hat{\phi}], [h_t\hat{\psi}])z^v.$$

The partition function of the model is defined by

$$p_n^+(z) = \sum_{[h_b\hat{\phi}]}\sum_{[h_t\hat{\psi}]} p_n^+(z; [h_b\hat{\phi}], [h_t\hat{\psi}]),$$

and contains at most $n^2(d-1)^2$ terms. There is a most popular term in the partition function for every value of z. This term defines the most popular height and orientation of the bottom and the top edges. Indicate these most popular values by $\{[h_b^*\hat{\phi}^*], [h_t^*\hat{\psi}^*]\}$. In terms of these,

$$n^2(d-1)^2 p_n^+(z; [h_b^*\hat{\phi}^*], [h_t^*\hat{\psi}^*]) \geq p_n^+(z) \geq p_n^+(z; [h_b^*\hat{\phi}^*], [h_t^*\hat{\psi}^*]). \qquad (\dagger)$$

This equation shows that for any fixed value of z only polygons in the most popular class need be considered asymptotically. However, since $h_b^*\hat{\phi}^*$ may not be equal to $h_t^*\hat{\psi}^*$ in a given most popular class polygon, it is still not possible to concatenate them. Consider the class of polygons $p_n^+(z; [h\hat{\phi}], [h\hat{\phi}])$ where both the top and bottom edges have the same heights and orientations. Let $[h^*\hat{\phi}^*]$ be the most popular height and orientation in this class. Apparently, two polygons counted by $p_n^+(z; [h_b^*\hat{\phi}^*], [h_t^*\hat{\psi}^*])$ can be concatenated if one polygon is reflected through the hyperplane normal to the X-direction. The top edge of the first and an edge of the second (the reflected top edge) will have the same height and orientation and the polygons can be concatenated as in Fig. 1.4. Thus

$$\sum_{v_1=0}^{v} p_n^+(v-v_1; [h_b^*\hat{\phi}^*], [h_t^*\hat{\psi}^*])p_n^+(v_1; [h_b^*\hat{\phi}^*], [h_t^*\hat{\psi}^*]) \leq p_{2n}^+(v; [h_b^*\hat{\phi}^*], [h_b^*\hat{\psi}^*]).$$

Multiply this by z^v and sum over v (and notice especially that $[h_b^*\hat{\phi}^*]$, and $[h_t^*\hat{\psi}^*]$ are independent of v and v_1, the only dependence is on z):

$$[p_n^+(z; [h_b^*\hat{\phi}^*], [h_t^*\hat{\psi}^*])]^2 \leq p_{2n}^+(z; [h_b^*\hat{\phi}^*], [h_b^*\hat{\phi}^*]) \leq p_{2n}^+(z; [h^*\hat{\phi}^*], [h^*\hat{\phi}^*]).$$

The left-hand side of this equation is made smaller if $[h_b^*\hat{\phi}^*]$ and $[h_t^*\hat{\psi}^*]$ are replaced by $[h^*\hat{\phi}^*]$ (or stays the same, at best). Thus

$$[p_n^+(z; [h^*\hat{\phi}^*], [h^*\hat{\phi}^*])]^2 \leq [p_n^+(z; [h_b^*\hat{\phi}^*], [h_t^*\hat{\psi}^*])]^2 \leq p_{2n}^+(z; [h^*\hat{\phi}^*], [h^*\hat{\phi}^*]). \quad (\ddagger)$$

Equations (†) and (‡) show that it is only necessary to consider polygons with the same height and orientation of their top and bottom edges. In particular, concatenate polygons with height and orientation $[h\hat{\phi}]$ for *both* top and bottom edges:

$$\sum_{v_1=0}^{v} p_n^+(v_1; [h\hat{\phi}], [h\hat{\phi}])\, p_m^+(v - v_1; [h\hat{\phi}], [h\hat{\phi}]) \leq p_{n+m}^+(v; [h\hat{\phi}], [h\hat{\phi}]). \qquad (\P)$$

Multiplication by z^v and summing over v gives

$$p_n^+(z; [h\hat{\phi}], [h\hat{\phi}])\, p_m^+(z; [h\hat{\phi}], [h\hat{\phi}]) \leq p_{n+m}^+(z; [h\hat{\phi}], [h\hat{\phi}]),$$

so that the limit $\lim_{n\to\infty}[\log p_n^+(z; [h\hat{\phi}], [h\hat{\phi}])]/n$ exists, for any legitimate choice of $[h\hat{\phi}]$. This limit also exists if one chooses $[h\hat{\phi}] = [h^*\hat{\phi}^*]$, and by eqn (‡) and the squeeze theorem for limits, the limit

$$\mathcal{F}_v^+(z) = \lim_{n\to\infty} \frac{1}{n} \log p_n^+(z; [h_b^*\hat{\phi}^*], [h_t^*\hat{\psi}^*])$$

also exists. Thus, by the squeeze theorem for limits and eqn (†), the existence of the free energy has been established. □

In three and more dimensions the unfolding of walks shows that the free energies for models of adsorbing polygons and walks are the same, but this is not the case in two dimensions (for example, since the maximum number of visits are different in these models, their free energies have different asymptotes). If the unfoldings in three and more dimensions are done in the X-direction and Y-direction, then the result is that

$$\lim_{n\to\infty} \frac{1}{n} \log c_n^+(z) = \lim_{n\to\infty} \frac{1}{n} \log p_n^+(z), \qquad (5.58)$$

where $c_n^+(z)$ is the partition function of positive adsorbing walks. This should be compared to Theorem 5.28, but notice that the proof of eqn (5.58) does not rely on the arguments as presented in Section 5.2. Rather, most popular arguments and unfolding, similar to the arguments in the proof of Theorem 5.52, must be used.

The existence of a density function in this model follows from the existence of the free energy in eqn (5.58), and by Theorem 3.19. More accurately, the arguments in the proof of Theorem 5.52 can be used explicitly to show that there is a density function. Notice that eqn (¶) in the proof of Theorem 5.52 gives

$$p^+_{n_1}(v_1; [h\hat{\phi}], [h\hat{\phi}])\, p^+_{n_2}(v_2; [h\hat{\phi}], [h\hat{\phi}]) \leq p^+_{n_1+n_2}(v_1 + v_2; [h\hat{\phi}], [h\hat{\phi}]), \tag{5.59}$$

so that $p^+_n(v; [h\hat{\phi}], [h\hat{\phi}])$ satisfies Assumptions 3.1 and there is a density function defined for this subclass of polygons. However, most popular arguments show that there are most popular choices for $[h\hat{\phi}]$ which gives the following theorem, from Theorem 3.4, and the most popular arguments in the proof of Theorem 5.52.

Theorem 5.53 *There exist most popular choices, say $[h^*\hat{\phi}^*]$, for the height and orientations of the top and bottom edges of positive polygons, such that the density function of visits, defined by*

$$\log \mathcal{P}^+_v(\epsilon) = \lim_{n\to\infty} \frac{1}{n} \log p^+_n(\lfloor \epsilon n \rfloor) = \lim_{n\to\infty} \frac{1}{n} \log p^+_n(\lfloor \epsilon n \rfloor; [h^*\hat{\phi}^*], [h^*\hat{\phi}^*]),$$

exists. □

The combination of most popular arguments with the techniques in Section 5.2.3 will also produce a pattern theorem for adsorbing walks and polygons. Since the adsorbing plane breaks the symmetry of the problem, the definitions of unfolded walks, fl-walks, and f-walks must all be changed. The energy is local under concatenation, and a pattern theorem can be proven using the same general arguments as in Section 5.2. The consequences of a pattern theorem are the same as before; for example, it may be shown that there is a density of excursions in the adsorbed polygon (from $Z = 0$) which are knotted arcs in three dimensions (this is a result of C. Vanderzande [347]). I shall not pursue those matters further here.

The growth constant of positive polygons is defined by

$$\mathcal{F}^+_v(1) = \log \mu^+_d \tag{5.60}$$

in d dimensions, and in eqns (5.56) and (5.57) it is seen that $\mu^+_d = \mu_d$. If $d \geq 3$, then the polygons counted by $p^+_n(n)$ are all in the hyperplane $Z = 0$; thus

$$\lim_{n\to\infty} \frac{1}{n} \log p^+_n(n) = \log \mu_{d-1}. \tag{5.61}$$

The maximum number of visits in a model of two dimensional polygons is $n/2$, and $p^+_n(n/2) = 1$. It is therefore convenient to define $\mu_1 = 1$ in eqn (5.61). These growth constants are related as in Lemma 5.54.

Lemma 5.54 $\mu_d = \mu^+_d > \mu_{d-1}$.

Proof Let p_n be the number of polygons (equivalent under translations in the lattice), and let p^+_n be the number of positive polygons, counted by equivalence under translations parallel to the hyperplane Z=0. Any polygon counted by p_n can be translated normal to the hyperplane Z=0 until it is a positive polygon. By eqn (5.57), $p_n \leq p^+_n \leq 2p_n$. This proves the equality. To see the inequality, let $p^{(d-1)}_n$ be the number of polygons in

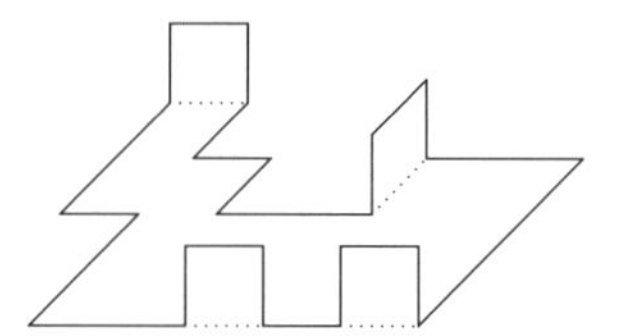

Fig. 5.13: A small excursion can be created in an adsorbed polygon by replacing edges with loops of length three.

$(d-1)$ dimensions. A small excursion in such a polygon can be created by translating any edge in the positive Z-direction, and adding two edges to reconnect it, as illustrated in Fig. 5.13. At most $\lfloor n/2 \rfloor$ such excursions can be created (one on every second edge); choose $\lfloor \epsilon n \rfloor$ edges from the $\lfloor n/2 \rfloor$ and create an excursion each. This gives

$$\binom{\lfloor n/2 \rfloor}{\lfloor \epsilon n \rfloor} p_n^{(d-1)} \leq p_{n+2\lfloor \epsilon n \rfloor}^{+}.$$

Take the $(1/n)$-th power of this, and let $n \to \infty$. Then

$$\left[\frac{\sqrt{1/2}}{\epsilon^{\epsilon}(1/2-\epsilon)^{1/2-\epsilon}}\right] \mu_{d-1} \leq \left(\mu_d^{+}\right)^{1+2\epsilon}.$$

From footnote 3 in Chapter 3 it follows that there is an $\epsilon > 0$ such that

$$\left[\frac{\sqrt{1/2}\,(\mu_d^{+})^{-2\epsilon}}{\epsilon^{\epsilon}(1/2-\epsilon)^{1/2-\epsilon}}\right] > 1.$$

This completes the proof if $d \geq 3$. If $d = 2$, then $\mu_1^{+} = 1$ and this is strictly less than μ_2^{+}. □

It is possible to show that there is an adsorption transition in this model; the polygon undergoes a transition from a desorbed phase with zero density of visits in the wall, to an adsorbed phase with a positive density of visits. This result is due to J.M. Hammersley *et al.* [165].

Theorem 5.55 *For all values of* $z \leq 1$,

$$\mathcal{F}_v^{+}(z) = \log \mu_d.$$

If $z > 1$, *then*

$$\begin{aligned} \log \mu_{d-1} + \log z \leq \mathcal{F}_v^{+}(z) &\leq \log \mu_d + \log z, && \text{if } d > 2; \\ [\log z]/2 \leq \mathcal{F}_v^{+}(z) &\leq \log \mu_2 + [\log z]/2, && \text{if } d = 2. \end{aligned}$$

Proof Consider a polygon counted by $p_n^+(v)$. This polygon can be translated by one step in the Z-direction, giving a polygon with zero visits. Thus, $p_n^+(v) \leq p_n^+(0)$. Consequently, if $z < 1$, then

$$p_n^+(0) \leq p_n^+(z) \leq p_n^+(0) \sum_{v \geq 0} z^v.$$

Take logarithms, divide by n and let $n \to \infty$ to obtain the desired result. If $z > 1$ and $d > 2$ then

$$p_n^+(n) z^n \leq p_n^+(z) \leq p_n^+ z^n.$$

Since $p_n^+(n) = p_n^{(d-1)}$ is the number of polygons in $(d-1)$ dimensions, the first claim follows if logarithms are taken, the result is divided by n, and $n \to \infty$. If d=2 then notice that the maximum number of visits is $n/2$, and that $p_n^+(n/2) = 1$. Thus

$$z^{n/2} \leq p_n^+(z) \leq z^{n/2} p_n^+.$$

Once again, the desired result is found if the logarithm is taken, the result is divided by n and n is taken to infinity. This result is also a direct consequence of the pattern theorem. □

Notice that the asymptotic behaviour of a model of walks will be $\log z \leq \mathcal{F}_v^+(z) \leq \log \mu_2 + \log z$, which is not the same as for polygons. Thus, walks and polygons have different free energies in two dimensions for large enough values of z. Theorem 5.55 also states that the adsorption will occur at a value of z in a finite interval. Let the critical point be at z_c.

Corollary 5.56 *The limiting free energy $\mathcal{F}_v^+(z)$ is a non-analytic function of z. In particular, there is a non-analyticity in $\mathcal{F}_v^+(z)$ at a critical value z_c somewhere in the interval $[1, \mu_d^+/\mu_{d-1}]$ if $d \geq 3$, and in the interval $[1, \mu_2^2]$ if $d = 2$.* □

It is generally accepted that polymer adsorption is not tricritical. Instead, the singularity structure of the generating function is described by Fig. 2.5. Along the desorbed phase ($z < z_c$) the exponents of the model will be those of linear polymers in a good solvent listed in Table 1.1, with the exception that the entropic exponent may have a value which is dependent on the particular model. If a model of walks with one end-point in the interface is studied, then the exponent γ_1 is the entropic exponent; if both end-points are in the interface, then it is γ_{11}. In the case of a polygon the exponent $2-\alpha$ (see eqn (1.6)) should describe the singularity in the generating function. In the adsorbed phase it seems that the model adopts the critical exponents of a linear polymer in a good solvent in one dimension less. At the adsorption critical point the surface exponents γ_s and γ_{ss} are encountered (instead of γ_1 and γ_{11}). Conformal invariance arguments show that $\gamma_s = 93/64$ in two dimensions [46], while in three dimensions Monte Carlo simulations give $\gamma_s = 1.304 \pm 0.016$ and $\gamma_{ss} = 0.806 \pm 0.015$ [241]. The mean field values are believed to be $\gamma_s = -\gamma_{ss} = 1/2$ [19]. It is also known that the surface exponents are related to ν and γ by $\gamma_{11} + \nu = 2\gamma_1 - \gamma$. This can be checked by using the data in Table 1.1 and the values of the surface exponents stated here. This relation is also referred to as Barber's scaling relation [8]. The critical curve in this model is asymmetric in the

Table 5.2: Exponents for adsorbing linear polymers in a good solvent

Dimension	ϕ	γ_1	γ_{11}	γ_s	γ_{ss}
2	$\frac{1}{2}$	$\frac{61}{64}$	$-\frac{3}{16}$	$\frac{93}{64}$	
3	0.5	0.7	−0.4	1.5	0.8

sense that it is constant in the desorbed phase. A cross-over exponent can be computed on the adsorbed side of the critical curve (from the shift exponent). In two dimensions it is $\phi = 1/2$ [11, 17, 40]. The value of the cross-over exponent is also found to be close to $1/2$ in three dimensions [179]. The values of the exponents are listed in Table 5.2; the three-dimensional values are rough estimates based on the Monte Carlo results.

5.4.2 Location of the adsorption transition

The critical activity of the adsorption transition in a model of lattice polygons, denoted by z_c^+, has been shown to be confined to a certain interval in Corollary 5.56. It is still possible for this transition to occur at $z = 1$, and in this section this possibility will be ruled out. The value of the critical point of adsorbing polygons in the hexagonal lattice is known to be $z_c = \sqrt{1+\sqrt{2}} > 1$ [11], but this is not rigorous.

Let the $Z = 0$ plane be a defect plane, and let $p_n^>(v)$ be the number of attached polygons of length n with v visits to the defect plane. The partition function of this model is defined by $p_n^>(z) = \sum_{v=0}^{n+1} p_n^>(v) z^v$, and the limiting free energy $\mathcal{F}_v(z) = \lim_{n\to\infty}[\log p_n^>(z)]/n$ can be shown to exist using the techniques of Theorem 5.52. In addition, Theorem 5.55 and Corollary 5.56 apply unchanged to this model, using similar techniques of proofs.

By Theorem 5.53, there exists a density function of visits $\mathcal{P}_v^+(\epsilon)$ for positive adsorbing polygons (via Assumptions 3.1), where $\epsilon \in (0,1)$ if $d \geq 3$. From Theorem 5.55, it also follows that $\mathcal{P}_v^+(0) = \mu_d$, and that $\lim_{\epsilon\to 0^+} \mathcal{P}_v^+(\epsilon) = \mu_d$, so that the density function is continuous at $\epsilon = 0$. Moreover, the concavity of $\log \mathcal{P}_v^+(\epsilon)$ shows that

$$\log \mathcal{P}_v^+(\epsilon) \leq \log \mu_d - \epsilon \log z_c^+, \tag{5.62}$$

this in fact also follows from Theorem 3.19.

The density functions can be used to prove that $z_c^+ > 1$ in the case of adsorption of positive polygons. The key to the proof is Lemma 3.20,

$$\log z_c^+ = -\left[\frac{d^+}{d\epsilon} \log \mathcal{P}_v^+(\epsilon)\right]\bigg|_{\epsilon=0^+}. \tag{5.63}$$

Similarly, for the defect plane model,

$$\log z_c = -\left[\frac{d^+}{d\epsilon} \log \mathcal{P}_v(\epsilon)\right]\bigg|_{\epsilon=0^+}, \tag{5.64}$$

where $\mathcal{P}_v(\epsilon)$ is the density function of visits of the model of attached polygons interacting with a defect plane. By subtracting the inequalities and using the definition of the right-derivative, Lemma 5.57 is obtained.

Lemma 5.57 *The critical activities z_c and z_c^+ of a model of attached polygons, and a model of positive polygons, are related by*

$$\log z_c^{+} - \log z_c = \mu_d^{-1} \lim_{\epsilon \to 0^+} \frac{1}{\epsilon} \left[\mathcal{P}_v(\epsilon) - \mathcal{P}_v^+(\epsilon) \right].$$

Proof Use the definition of the right-derivative, and the fact that $\mathcal{P}_v(0) = \mathcal{P}_v^+(0) = \mu_d$. □

It is now only necessary to show that the difference in Lemma 5.57 is positive (since $z_c \geq 1$). Observe that each visit in a positive polygon is incident with an edge with both its end-points in the hyperplane $Z = 0$. In other words, if there are v visits, then there are at least $\lfloor v/2 \rfloor$ edges in the hyperplane $Z = 0$. These edges can be used to create polygons in a defect plane model.

Theorem 5.58 *The density functions of positive polygons and attached polygons are related by*

$$(1 + \mu_d^{-2})^{\epsilon/4} \mathcal{P}_v^+(\epsilon) \leq \mathcal{P}_v \left(\frac{2(1 + \mu_d^2)\epsilon}{2(1 + \mu_d^2) + \epsilon} \right).$$

Proof Consider the polygons counted by $p_n^+(\lfloor \epsilon n \rfloor)$. Such polygons have at least $\lfloor \epsilon n/2 \rfloor$ edges in the hyperplane $Z = 0$, not all of them disjoint. Since each edge is adjacent to at most two other edges in the Z=0 hyperplane, there is a subset E of at least $\lfloor \lfloor \epsilon n/2 \rfloor /2 \rfloor$ edges which are disjoint. Choose m of the edges in E, and translate them one step in the negative Z-direction, while new edges are inserted to keep the polygon connected. This is illustrated in Fig. 5.14. The construction creates an attached polygon with v visits and $n + 2m$ edges, so that

$$\binom{\lfloor \lfloor \epsilon n/2 \rfloor /2 \rfloor}{m} p_n^+(v) \leq p_{n+2m}(v).$$

Put $v = \lfloor \epsilon n \rfloor$, and let $m = \lfloor \delta n \rfloor$, where $\delta < \epsilon/4$, take the $(1/n)$-th power of the above, and let $n \to \infty$. The result is

$$\left[\frac{(\epsilon/4)^{\epsilon/4}}{\delta^\delta (\epsilon/4 - \delta)^{\epsilon/4 - \delta}} \right] \mathcal{P}_v^+(\epsilon) \leq \left[\mathcal{P}_v(\epsilon/(1 + 2\delta)) \right]^{1+2\delta}.$$

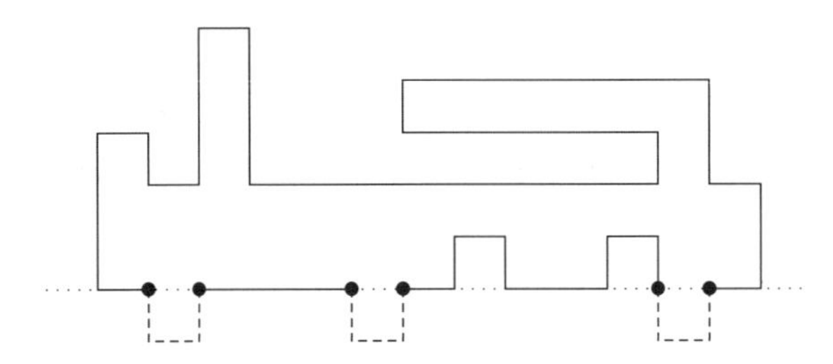

Fig. 5.14: A positive polygon can be changed into an attached polygon by selecting edges in the adsorbing plane, and translating them in the negative z-direction, while inserting new edges to reconnect the polygon.

Now use the fact that $\mathcal{P}_v(\epsilon) \leq \mathcal{P}_v(0) = \mu_d$ (see eqn (5.62)) to find

$$\left[\frac{(\epsilon/4)^{\epsilon/4}\mu_d^{-2\delta}}{\delta^\delta(\epsilon/4-\delta)^{\epsilon/4-\delta}}\right]\mathcal{P}_v^+(\epsilon) \leq \mathcal{P}_v(\epsilon/(1+2\delta)).$$

Lastly, the factor in square brackets is a maximum if

$$\delta = \delta_* = \epsilon/4(1+\mu_d^2),$$

and this gives the claim (see footnote 3 in Chapter 3). □

Combine Lemma 5.57 and Theorem 5.58, and this gives the desired result.

Theorem 5.59

$$\log z_c^+ - \log z_c \geq [\log(1+\mu_d^{-2})]/4 > 0.$$

Proof Use Theorem 5.58 in Lemma 5.57 to eliminate $\mathcal{P}_v^+(\epsilon)$. This gives

$$\log z_c^+ - \log z_c \geq \lim_{\epsilon\to 0^+} \frac{1}{\epsilon}\left[1 - \frac{\mathcal{P}_v\left(\dfrac{2(1+\mu_d^2)\epsilon}{2(1+\mu_d^2)+\epsilon}\right)}{\mathcal{P}_v(\epsilon)}(1+\mu_d^{-2})^{-\epsilon/4}\right]. \qquad (\dagger)$$

Let $\epsilon > 0$ be small. If $\mathcal{P}_v'(0) = 0$, then there is an $\eta > 0$ such that

$$\mathcal{P}_v(0) - \epsilon\eta \leq \mathcal{P}_v(\epsilon) \leq \mathcal{P}_v(0) + \epsilon\eta,$$

and η can be taken to zero if $\epsilon = 0^+$. Thus

$$\begin{aligned}\frac{\mathcal{P}_v\left(\dfrac{2(1+\mu_d^2)\epsilon}{2(1+\mu_d^2)+\epsilon}\right)}{\mathcal{P}_v(\epsilon)} &\leq \frac{\mathcal{P}_v(0) + \eta\left(\dfrac{2(1+\mu_d^2)\epsilon}{2(1+\mu_d^2)+\epsilon}\right)}{\mathcal{P}_v(0)-\epsilon\eta} \\ &= 1 + \frac{\eta\epsilon}{\mu_d}\left[1 + \frac{2(1+\mu_d^2)}{2(1+\mu_d^2)+\epsilon}\right] + O(\epsilon^2).\end{aligned}$$

Substitute this into eqn (†). After some simplification, the result is that

$$\begin{aligned}\log z_c^+ - \log z_c &\geq \lim_{\epsilon\to 0^+}\left[\frac{1}{4}\log(1+\mu_d^{-2}) - \frac{\eta}{\mu_d}\left(1 + \frac{2(1+\mu_d^2)}{2(1+\mu_d^2)+\epsilon}\right)\right] \\ &= \frac{1}{4}\log(1+\mu_d^{-2}) - 2\eta/\mu_d.\end{aligned}$$

Since this is true for any $\eta > 0$, η can be safely taken to zero to obtain the result.

Suppose now that $\mathcal{P}_v'(0) < 0$ (notice that it cannot be positive, since $z_c \geq 0$). Take $\epsilon > 0$ very small so that $\mathcal{P}_v'(\epsilon) \leq 0$, and note that there exists a $\eta > 0$ such that

$$\mathcal{P}_v(0) + \epsilon(1-\eta)\mathcal{P}_v'(0) \geq \mathcal{P}_v(\epsilon) \geq \mathcal{P}_v(0) + \epsilon(1+\eta)\mathcal{P}_v'(0),$$

and if $\epsilon = 0$, then the limit as $\eta \to 0^+$ may be taken. Thus

$$\begin{aligned}\frac{\mathcal{P}_v\left(\dfrac{2(1+\mu_d^2)\epsilon}{2(1+\mu_d^2)+\epsilon}\right)}{\mathcal{P}_v(\epsilon)} &\leq \frac{\mathcal{P}_v(0) + \dfrac{2(1+\mu_d^2)(1-\eta)\epsilon}{2(1+\mu_d^2)+\epsilon}\mathcal{P}_v'(0)}{\mathcal{P}_v(0)+\epsilon(1+\eta)\mathcal{P}_v'(0)} \\ &\leq \left[1 + \left(\frac{2(1+\mu_d^2)(1-\eta)\epsilon}{2(1+\mu_d^2)+\epsilon}\right)\frac{\mathcal{P}_v'(0)}{\mathcal{P}_v(0)}\right] \\ &\quad \times \left[1 - \epsilon(1+\eta)\frac{\mathcal{P}_v'(0)}{\mathcal{P}_v(0)} + \eta_0\epsilon^2\right],\end{aligned}$$

if $\epsilon < -\mathcal{P}_v(0)/(2(1+\eta)\mathcal{P}_v'(0))$, and where $\eta_0 = (1+2(1+\eta)^2)[\mathcal{P}_v'(0)/\mathcal{P}_v(0)]^2$. Further simplification gives

$$\frac{\mathcal{P}_v\left(\dfrac{2(1+\mu_d^2)\epsilon}{2(1+\mu_d^2)+\epsilon}\right)}{\mathcal{P}_v(\epsilon)} \leq 1 - 2\epsilon\eta\frac{\mathcal{P}_v'(0)}{\mathcal{P}_v(0)} + \eta_1\epsilon^2,$$

where η_1 is a finite number. Substitution of this bound into eqn (†) above gives

$$\begin{aligned}\log z_c^+ - \log z_c &\geq \lim_{\epsilon\to 0^+}\frac{1}{\epsilon}\left[1 - \left(1 - 2\epsilon\eta\frac{\mathcal{P}_v'(0)}{\mathcal{P}_v(0)} + \eta_1\epsilon^2\right)(1+\mu_d^{-2})^{\epsilon/4}\right] \\ &= \frac{1}{4}\log(1+\mu_d^{-2}) + 2\eta\frac{\mathcal{P}_v'(0)}{\mathcal{P}_v(0)}.\end{aligned}$$

Notice that $\mathcal{P}_v'(0) \leq 0$. Since the limit $\epsilon \to 0^+$ has been taken, the limit $\eta \to 0^+$ may be taken to finish the proof. □

In other words, the polygons will only adsorb at a strictly attractive value of the visit-activity in the adsorption of positive polygons.

5.4.3 Excursions and the adsorption transition

Any polygon interacting with an adsorbing surface consists of subwalks which lie in the hyperplane $Z = 0$ (these are *incursions*) and loops which lie (except for end-points) in the half-space $Z > 0$ (these are *excursions*). The density of excursions in a model of adsorbing walks is an interesting question, first addressed by J.M. Hammersley *et al.* [165]. I shall study this by defining $p_n^+(w, v)$ to be the number of positive polygons of length n with v visits and w excursions. Consider now the number of polygons with a density δ of excursions: $p_n^+(\lfloor \delta n \rfloor, v)$. The first observation is that there exists a free energy for a model of polygons with a density of excursions: define the partition function $p_n^+(w; z) = \sum_{v\geq 0} p_n^+(w, v)z^v$.

Theorem 5.60 *There exists a limiting free energy in a model of adsorbing positive polygons with a fixed density of excursions:*

$$\lim_{n\to\infty} \frac{1}{n} \log p_n^+(\lfloor \delta n \rfloor; z) = \mathcal{F}_v^+(\delta; z).$$

Moreover, $\mathcal{F}_v^+(\delta; z)$ is a concave function of δ (for fixed values of $z \in (0, \infty)$).

Proof The proof is by a most popular argument. Let $[h\hat{\phi}]$ again be the height and orientation of a top or a bottom edge, and let $p_n^+(m; z; [h\hat{\phi}], [h\hat{\phi}])$ be the partition function of polygons with both bottom and top edges with mid-points at height and orientation $[h\hat{\phi}]$. Concatenating these polygons as in Fig. 1.4 gives

$$p_{n_1}^+(m_1; z; [h\hat{\phi}], [h\hat{\phi}]) p_{n_2}^+(m_2; z; [h\hat{\phi}], [h\hat{\phi}]) \leq p_{n_1+n_2}^+(m_1 + m_2; z; [h\hat{\phi}], [h\hat{\phi}]). \tag{†}$$

Notice that the same outcome is found, even if the top and bottom edges are in the adsorbing plane. Thus, $p_n^+(m; z; [h\hat{\phi}], [h\hat{\phi}])$ satisfies Assumptions 3.1, and the limit

$$\mathcal{F}_{[h\hat{\phi}]}^+(\delta; z) = \lim_{n\to\infty} \frac{1}{n} \log p_n^+(\lfloor \delta n \rfloor; z; [h\hat{\phi}], [h\hat{\phi}]) \tag{‡}$$

exists. But there are most popular choices for h and $\hat{\phi}$, say $[h_b^*\hat{\phi}_b^*]$ for the bottom edge, and $[h_t^*\hat{\phi}_t^*]$ for the top edge, which gives

$$\begin{aligned} p_n^+(\lfloor \delta n \rfloor; z; [h_b^*\hat{\phi}_b^*], [h_t^*\hat{\phi}_t^*]) &\leq p_n^+(\lfloor \delta n \rfloor; z) \\ &\leq n^2(d-1)^2 p_n^+(\lfloor \delta n \rfloor; z; [h_b^*\hat{\phi}_b^*], [h_t^*\hat{\phi}_t^*]). \end{aligned} \tag{¶}$$

Suppose that $p_n^+(\lfloor \delta n \rfloor; z; [h^*\hat{\phi}^*], [h^*\hat{\phi}^*])$ is the most popular class of polygons with mid-points of height h^*, and with parallel top and bottom edges (in the direction $\hat{\phi}^*$) in the partition function $p_n^+(\lfloor \delta n \rfloor; z)$. Then by arguing as in eqn (‡) in the proof of Theorem 5.52,

$$\begin{aligned} \left[p_n^+(\lfloor \delta n \rfloor; z; [h^*\hat{\phi}^*], [h^*\hat{\phi}^*])\right]^2 &\leq \left[p_n^+(\lfloor \delta n \rfloor; z; [h_b^*\hat{\phi}^*], [h_t^*\hat{\phi}_t^*])\right]^2 \\ &\leq p_{2n}^+(2\lfloor \delta n \rfloor; z; [h^*\hat{\phi}^*], [h^*\hat{\phi}^*]). \end{aligned} \tag{ℵ}$$

Thus, from eqn (ℵ),

$$\lim_{n\to\infty} \frac{1}{n} \log p_n^+(\lfloor \delta n \rfloor; z; [h^*\hat{\phi}^*], [h^*\hat{\phi}^*]) = \lim_{n\to\infty} \frac{1}{n} \log p_n^+(\lfloor \delta n \rfloor; z; [h_b^*\hat{\phi}_b^*], [h_t^*\hat{\phi}_t^*]),$$

provided that these limits exist. Let $\mathcal{F}_v^+(\delta; z)$ be the free energy in eqn (‡) defined by taking the most popular class $[h\hat{\phi}] = [h^*\hat{\phi}^*]$. Then the limits above exists, and from eqn (¶),

$$\begin{aligned} \mathcal{F}_v^+(\delta; z) &= \lim_{n\to\infty} \frac{1}{n} \log p_n^+(\lfloor \delta n \rfloor; z; [h_b^*\hat{\phi}^*], [h_t^*\hat{\phi}_t^*]) \\ &= \lim_{n\to\infty} \frac{1}{n} \log p_n^+(\lfloor \delta n \rfloor; z). \end{aligned}$$

Concavity follows directly from eqn (†). □

Thus, the density function of excursions above implies the existence of a limiting free energy (see Theorem 3.18)

$$\mathcal{F}_v^+(y, z) = \lim_{n\to\infty} \frac{1}{n} \log p_n^+(y, z), \tag{5.65}$$

where $p_n^+(y, z) = \sum_{m\geq 0} p_n^+(m; z) y^m$ is the partition function of polygons with activities conjugate to the number of visits (z), and to the number of excursions (y). Indeed, for fixed z, eqn (5.65) is the Legendre transform of the density function $\mathcal{F}_v^+(\delta; z)$. It is a simple exercise to check that the maximum density of excursions is $\delta_M = 1/6$ in two dimensions, and $\delta_M = 1/4$ in more than two dimensions. δ has minimum value $\delta_m = 0$. In addition, the values of the free energy can be computed if $\delta = 0$; since there are no excursions if $\delta = 0$, the polygons lie entirely in the hyperplane $z = 0$. On the other hand, consider the fact that $p_n^+(1, z) \geq z^2 l_{n-1}$, where l_n is the number of loops with n edges. This shows that

$$\mathcal{F}_v^+(0; z) = \mu_{d-1}; \qquad \mathcal{F}_v^+(0^+; z) = \mu_d, \tag{5.66}$$

where Theorem 5.8 was used. There is a jump discontinuity in the density function $\mathcal{F}_v^+(\delta; z)$ at $\delta = 0$. This corresponds to a first-order transition at zero temperature for all values of $0 < z < \infty$, see Section 3.3.4.

The density function of both visits and excursions is defined by

$$\begin{aligned} \log \mathcal{P}_v^+(\delta, \epsilon) &= \inf_{0<z<\infty} \{\mathcal{F}_v^+(\delta; z) - \epsilon \log z\} \\ &= \lim_{n\to\infty} \frac{1}{n} \log p_n^+(\lfloor \delta n \rfloor, \lfloor \epsilon n \rfloor). \end{aligned} \tag{5.67}$$

Naturally,

$$\mathcal{P}_v^+(\epsilon) = \sup_{0\leq\delta\leq\delta_M} \mathcal{P}_v^+(\delta, \epsilon). \tag{5.68}$$

The density of visits takes values $\epsilon \in [0, 1]$ as before, but for every fixed ϵ, note that there is a maximum number of excursions. If there are $\lfloor \delta n \rfloor$ excursions and $\lfloor \epsilon n \rfloor$ visits, then

$$\lfloor \delta n \rfloor \leq \begin{cases} \min\{n/6, \lfloor \epsilon n \rfloor /2\}, & \epsilon \in [0, 1/2]; \quad \text{if } d = 2, \\ \min\{n/4, \lfloor \epsilon n \rfloor /2\}, & \epsilon \in [0, 1]; \quad\;\;\; \text{if } d \geq 3. \end{cases} \tag{5.69}$$

Thus, for $\epsilon \in (0, 1)$, $0 < \delta < \min\{1/4, \epsilon/2\}$ in three and more dimensions, and $0 < \delta < \min\{1/6, \epsilon/2\}$ in two dimensions. This domain of the density function will be denoted by Δ. The density function $\mathcal{P}_v^+(\delta, \epsilon)$ satisfies an important inequality, which will be proven in Theorem 5.61.

Theorem 5.61 *For every $(\delta, \epsilon) \in \Delta$ there exists a $\delta_* > 0$ such that*

$$(1 + \mu_d^{-2})^{(\epsilon-\gamma)/2} \mu_d^{-2\gamma} \mathcal{P}_v^+(\gamma, \epsilon) \leq \mathcal{P}_v^+ \left(\frac{\gamma + \delta_*}{1 + 2\gamma + 2\delta_*}, \frac{\epsilon}{1 + 2\gamma + 2\delta_*} \right).$$

In fact, $\delta_* = (\epsilon - \gamma)/2(1 + \mu_d^2)$ is a sufficient choice for δ_*.

Fig. 5.15: Each excursion in the polygon is translated one step in the positive z-direction, and we reconnected the polygon by adding edges along the dashed lines.

Proof Let $p_n^+(w, v)$ be the number of positive polygons with v visits and w excursions. A new excursion can be created on an edge in such a polygon which is in the $Z = 0$ hyperplane, by translating the edge in the positive Z-direction, and then reconnecting the polygon with two additional edges. This can only be done if the edge is not adjacent to an edge in an excusion, and if there is space available for the translation. To create this space, consider Fig. 5.15. Translate each excursion in the polygon one step in the Z-direction, and reconnect the polygon by adding edges as indicated. The new polygon has $n+2w$ edges, and there are $v-w$ edges which can be used to create new excursions. Note that no more than one excursion on every second edge can be created, so that the maximum number of new excusions is $\lfloor (v-w)/2 \rfloor$. If h locations for new excursions are chosen, then

$$\binom{\lfloor (v-w)/2 \rfloor}{h} p_n^+(w, v) \leq p_{n+2w+2h}^+(w+h, v).$$

Let $v = \lfloor \epsilon n \rfloor$, $w = \lfloor \gamma n \rfloor$. Choose $h = \lfloor \delta n \rfloor$. Substitute these into the above, and take the $(1/n)$-th power, and then let $n \to \infty$. This gives

$$\begin{aligned}\left[\frac{((\epsilon-\gamma)/2)^{(\epsilon-\gamma)/2}}{\delta^\delta((\epsilon-\gamma)/2-\delta)^{(\epsilon-\gamma)/2-\delta}}\right]\mathcal{P}_v^+(\gamma,\epsilon) &\leq \left[\mathcal{P}_v^+\left(\frac{\gamma+\delta}{1+2\gamma+2\delta}, \frac{\epsilon}{1+2\gamma+2\delta}\right)\right]^{1+2\gamma+2\delta} \\ &\leq \mu_d^{2(\gamma+\delta)}\mathcal{P}_v^+\left(\frac{\gamma+\delta}{1+2\gamma+2\delta}, \frac{\epsilon}{1+2\gamma+2\delta}\right).\end{aligned}$$

Thus, since $\mathcal{P}_v^+(\epsilon) \leq \mu_d$,

$$\left[\frac{((\epsilon-\gamma)/2)^{(\epsilon-\gamma)/2}\mu_d^{-2(\gamma+\delta)}}{\delta^\delta((\epsilon-\gamma)/2-\delta)^{(\epsilon-\gamma)/2-\delta}}\right]\mathcal{P}_v^+(\gamma,\epsilon) \leq \mathcal{P}_v^+\left(\frac{\gamma+\delta}{1+2\gamma+2\delta}, \frac{\epsilon}{1+2\gamma+2\delta}\right).$$

The factor in square brackets has a maximum if $\delta_* = (\epsilon-\gamma)/2(1+\mu_d^2)$, and it shows the inequality claimed by the theorem. □

An immediate consequence of Theorem 5.61 is an alternative proof that $z_c^+ > 0$; this circumvents the need to compare positive polygons to attached polygons. Put $\gamma = 0$ in Theorem 5.61. Then

$$(1+\mu_d^{-2})^{\epsilon/2}\mathcal{P}_v^+(0,\epsilon) \leq \mathcal{P}_v^+\left(\frac{\delta_*}{1+2\delta_*}, \frac{\epsilon}{1+2\delta_*}\right), \tag{5.70}$$

where $\delta_* = \epsilon/2(1+\mu_d^2)$. Since the free energy is the Legendre transform of the density function, then by eqn (5.68), it may be supposed that $z > z_c^+$ and that the supremum is achieved at $\epsilon_* > 0$. But suppose that this occurs over a class of polygons with a zero density of excursions:

$$\mathcal{F}_v^+(z) = \log \mathcal{P}_v^+(0, \epsilon_*) + \epsilon_* \log z, \tag{5.71}$$

where $\epsilon_* > 0$. However by eqn (5.70), note that

$$\log \mathcal{P}_v^+(0, \epsilon_*) + \epsilon_* \log z \leq \log \mathcal{P}_v^+\left(\frac{\delta_*}{1+2\delta_*}, \frac{\epsilon_*}{1+2\delta_*}\right) + \frac{\epsilon_*}{1+2\delta_*} \log z$$
$$+ \log \left[\frac{z^{2\delta_*/(1+2\delta_*)}}{\sqrt{1+\mu_d^{-2}}}\right]^{\epsilon_*}, \tag{5.72}$$

which is in contradiction with eqn (5.71) if the factor in square brackets is less than one. However, this is the case if $z^{2\delta_*/(1+2\delta_*)} < \sqrt{1+\mu_d^{-2}}$, unless the desorbed phase is found for these values of z (and thus $\epsilon_* = 0$). Since $\delta_* \leq 1$, it is sufficient to require that $\log z \geq (1+2\delta_*)\left(\log\left[\sqrt{1+\mu_d^{-2}}\right]\right)/2\delta_*$. A lower bound of this is $\log\left[\sqrt{1+\mu_d^{-2}}\right]$, thus

$$z_c^+ \geq \sqrt{1+\mu_d^{-2}}. \tag{5.73}$$

In other words, if $\epsilon_* > 0$ and $\delta_* = 0$ then a contradiction is found. Thus $\epsilon_* > 0$ implies that $\delta_* > 0$, and there is a density of excursions in the adsorbed phase. Lastly, notice that eqn (5.70) implies that

$$\mathcal{P}_v^+(0, \epsilon) < \mathcal{P}_v^+\left(\frac{\delta_*}{1+2\delta_*}, \frac{\epsilon}{1+2\delta_*}\right), \tag{5.74}$$

so that polygons with a zero density of excursions are exponentially rare compared to polygons with a density of excursions.

5.4.4 *A density of excursions*

In this section an explicit proof that there is a density of excursions in the adsorbed phase is given. The discussion will be limited to three dimensions. First, it is shown that new visits can be created in a polygon without adding too many edges. The most natural place to attempt the construction of new visits would be to start around that visit with lexicographically most coordinates; this is the *top-visit*, and it will be indicated by t_v. The construction will easily generalize to higher dimensions, but to keep the discussion simple, only details in three dimensions will be given. The canonical unit vectors are $\{\hat{\imath}, \hat{\jmath}, \hat{k}\}$. Notice that $t_v + \hat{\imath}$ and $t_v + \hat{\jmath}$ are unoccupied lattice sites, but that $t_v + \hat{k}$, may or may not be occupied.

Lemma 5.62 *The density function of excursions and visits, $\mathcal{P}_v^+(\delta, \epsilon)$, satisfies*

$$\mathcal{P}_v^+(\gamma, \epsilon) \leq \left[\mathcal{P}_v^+\left(\frac{\gamma}{1+\alpha}, \frac{\epsilon+\alpha}{1+\alpha}\right)\right]^{1+\alpha}.$$

Proof New visits can be created by adding edges in the vicinity of the top visit. The construction is best followed in Fig. 5.16; the aim is to change a polygon counted by $p_n^+(w, v)$ into a polygon with an edge in the $\hat{j}$-direction incident with t_v (this is illustrated in case (1) in Fig. 5.16). The advantage of these polygons is that this edge can be "slid" in the $\hat{i}$-direction, by replacing it with $(t_v - \hat{j}, t_v + \hat{i} - \hat{j}, t_v + \hat{i}, t_v)$, to create two new visits, while two edges are added to the polygon. In Fig. 5.16 it is shown how to change a polygon so that its top visit is incident with an edge in the $\hat{j}$-direction. Note that the length change in the polygon is either zero, two or four, and that zero, two or three new visits are created. In addition, in at least one case an extra excursion is created in case (2.2). In other words, if $p_n^*(w, v)$ is the number of positive polygons with an edge in the

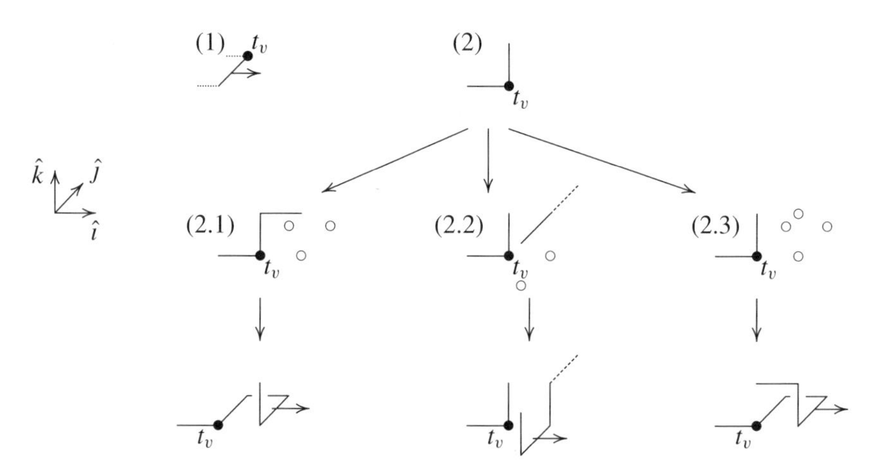

Fig. 5.16: The top visit is either incident with an edge in the $\hat{j}$ direction (case (1)), or not (case (2)). In case (1) there is an edge which can be translated (as indicated by an arrow) in the $\hat{i}$-direction while new edges are inserted to create visists. Every two added edges create two new visits. The other case is illustrated in case (2). The following constructions will change case (2) into case (1). Examine the vertex $t_v + \hat{i} + \hat{k}$; if it is occupied then cases (2.1) and (2.2) are found, if not, then case (2.3) is encountered. In case (2.1) the vertices $t_v + \hat{i}$, $t_v + \hat{j}$ and $t_v + \hat{i} + \hat{j}$ are not occupied, and the two edges $(t_v; t_v + \hat{k}, t_v + \hat{k} + \hat{i})$ can be replaced by $(t_v, t_v + \hat{j}, t_v + \hat{j} + \hat{i}, t_v + \hat{i} \cdot t_v + \hat{i} + \hat{k})$ to find a new polygon which is in case (1). There are two extra edges, and three new visits created in this process. Case (2.2) is similarly dealt with; case (1) is obtained after two new edges and two new visits have been added. Dashed edges may or may not be present. In this case either the solid edge is present, or the dashed edge is present. In case (2.3) the vertices $t_v + \hat{i}$, $t_v + \hat{j}$; $t_v + \hat{i} + \hat{k}$ and $t_v + \hat{i} + \hat{j}$ are not occupied. Four new edges can be added through these, creating three new visits, and giving a polygon which is in case (1).

$\hat{j}$-direction incident on the top-visit (and with v visits and $w \leq v$ excursions), then

$$p_n^+(w, v) \leq \sum_{i=0}^{4} \sum_{j=0}^{3} \sum_{k=0}^{1} p_{n+i}^*(w + k, v + j).$$

On the other hand, $p_n^*(w, v) \leq p_{n+2m}^+(w, v + 2m)$ for any $m \geq 0$, so if $v = \lfloor \epsilon n \rfloor$, $w = \lfloor \gamma n \rfloor$ and $m = \lfloor \alpha n/2 \rfloor$, then

$$p_n^+(\lfloor \gamma n \rfloor, \lfloor \epsilon n \rfloor) \leq \sum_{i=0}^{4} \sum_{j=0}^{3} \sum_{k=0}^{1} p_{n+2\lfloor \alpha n/2 \rfloor + i}^+(\lfloor \gamma n \rfloor + k, \lfloor \epsilon n \rfloor + 2\lfloor \alpha n/2 \rfloor + j).$$

If the $(1/n)$-th power is taken, and $n \to \infty$, then the proof is complete. □

Combine Lemma 5.62 with Theorem 5.61; and pick

$$\alpha = \frac{2\epsilon\delta}{(1 + 2\delta)(1 - \epsilon)}, \tag{5.75}$$

where $\delta = \epsilon/2(1 + \mu_d^2)$. This gives

$$(1 + e^{2\mu_d})^{\epsilon/2} \mathcal{P}_v^+(0, \epsilon) \leq \mu_d^\alpha \mathcal{P}_v^+\left(\frac{\delta}{(1 + 2\delta)(1 + \alpha)}, \epsilon\right). \tag{5.76}$$

The choice for α in eqn (5.75) gives $\alpha \leq \epsilon^2/(1 - \epsilon)$, so that $\left[\sqrt{1 + \mu_d^2}\right]^\epsilon \geq \mu_d^\epsilon > \mu_d^\alpha$ if $\epsilon/(1 - \epsilon) < 1$ or $\epsilon < 1/2$. Thus, if $\epsilon < 1/2$, then $\mathcal{P}_v^+(0, \epsilon) < \mathcal{P}_v^+(\delta/(1 + 2\delta)(1 + \alpha), \epsilon)$, and if the supremum in $\mathcal{F}_v^+(z) = \sup_\epsilon\{\log \mathcal{P}_v^+(\epsilon) + \epsilon \log z\}$ is achieved at some $\epsilon_* > 0$, then there will be a density of excursions, provided that $\epsilon_* < 1/2$.

5.4.5 *Collapsing and adsorbing polygons*

A model of polygons with both a visit activity z and a contact activity y could either collapse at a θ-point, or undergo an adsorption transition. This model is defined by the partition function

$$p_n^+(y, z) = \sum_{v \geq 0} \sum_{k \geq 0} p_n^+(k, v) y^k z^v, \tag{5.77}$$

where $p_n^+(k, v)$ is the number of positive attached polygons with v visits and k contacts. The model for attached and collapsing polygons is similarly defined by the partition function

$$p_n^>(y, z) = \sum_{v \geq 0} \sum_{k \geq 0} p_n^>(k, z) y^k z^v, \tag{5.78}$$

where attached polygons which can pass through the defect plane defined by $Z = 0$ are counted instead. The walk version of this model was studied by T. Vrbová and S.G. Whittington [352], and the discussion here will again be only in three dimensions. There are limiting free energies in these models; this can be shown by appropriately using the constructions in Theorem 5.52 (the theorem does not follow *mutatis mutandis*; instead, care must be taken not to disturbed the number of contacts in concatenation by too much). Thus, there exists limiting free energies in d dimensions defined by

$$\begin{aligned} \mathcal{F}_v^+(y,z) &= \lim_{n\to\infty} \frac{1}{n} \log p_n^+(y,z), \\ \mathcal{F}_v(y,z) &= \lim_{n\to\infty} \frac{1}{n} \log p_n^>(y,z). \end{aligned} \tag{5.79}$$

I shall limit the discussion here to $d \geq 3$ dimensions; the results are slightly different in two dimensions. The first interesting issue is the phase diagram of this model. I shall first show that the free energies are non-analytic functions of z, and that there is an adsorption transition in this model, for any finite value of the contact activity z.

Lemma 5.63 *Let $\mathcal{F}_k(y)$ be the free energy of a model of polygons with contact activity y. Then*

$$\mathcal{F}_v^+(y,z) = \mathcal{F}_v(y,z) = \mathcal{F}_k(y), \quad \text{if } z \leq 1,$$

and if $z > 1$ then

$$\mathcal{F}_k^{(d-1)}(y) + \log z \leq \mathcal{F}_v^+(y,z) \leq \mathcal{F}_v(y,z) \leq \mathcal{F}_k(y) + \log z,$$

where $\mathcal{F}_k^{(d-1)}(y)$ is the limiting free energy of polygons with a contact activity in $d-1$ dimensions.

Proof Let $z \leq 1$. Note that $p_n^+(k,v) \leq p_n^>(k,v)$, and that by translating polygons counted by $p_n^>(k,v)$ in the Z-direction, $p_n^>(k,v) \leq np_n^+(0,k) = np_n(k)$, since $p_n^+(0,k) = p_n(k)$. Thus

$$p_n(y) = p_n^+(y,0) \leq p_n^+(y,z) \leq p_n^>(y,z) \leq n \sum_{k\geq 0} p_n(k) y^k = np_n(y),$$

and by taking logarithms, dividing by n and letting $n \to \infty$, the result claimed for $z \leq 1$ is obtained.

If $z > 1$, then note that

$$\sum_{k\geq 0} p_n^+(k,n) y^k z^n \leq p_n^+(y,z) \leq p_n^>(y,z) \leq z^n \sum_{k\geq 0} p_n^+(k) y^k.$$

Since $p_n^+(k,n)$ is also the number of polygons with k contacts in $d-1$ dimensions, the claimed inequalities are found after taking the logarithm, dividing by n and letting $n \to \infty$. □

An immediate consequence of Lemma 5.63 is that both $\mathcal{F}_v^+(y,z)$ and $\mathcal{F}_v(y,z)$ are non-analytic functions of z, for any finite value of y. The non-analyticities are defined by

$$\begin{aligned} z_c(y) &= \sup\{z \mid \mathcal{F}_v(y,z) = \mathcal{F}_k(y)\}, \\ z_c^+(y) &= \sup\{z \mid \mathcal{F}_v^+(y,z) = \mathcal{F}_k(y)\}, \end{aligned} \tag{5.80}$$

and these values can be bounded. For example, notice that $1 \leq z_c(y) \leq z_c^+(y) \leq e^{\mathcal{F}_k(y)-\mathcal{F}_k^{(d-1)}(y)}$. This can be made more explicit:

Corollary 5.64 *The free energies $\mathcal{F}_v^+(y,z)$ and $\mathcal{F}_v(y,z)$ are non-analytic functions of z for each $y < \infty$ at $z_c^+(y)$ and $z_c(y)$ respectively. Moreover,*

$$1 \leq z_c(y) \leq z_c^+(y) \leq y^{d-1}\frac{\mu_d}{\mu_{d-1}}, \quad \text{if } y \geq 1;$$

$$1 \leq z_c(y) \leq z_c^+(y) \leq \frac{\mu_d}{\sqrt{\mu_{d-1}}}, \quad \text{if } y < 1.$$

where μ_d is the growth constant of walks in d dimensions.

Proof If $y \geq 1$ then it need only be shown that $z_c^+(y) \leq z^{d-1}\mu_d/\mu_{d-1}$. Since $z_c^+(y) \leq e^{\mathcal{F}_k(y)-\mathcal{F}_k^{(d-1)}(y)}$ it is only necessary to find bounds on the free energies. Note that in d dimensions $p_n(y) = \sum_{k\geq 0} p_n(k)y^k \leq y^{(d-1)n}p_n$, since the maximum number of contacts in a polygon in d dimensions is $(d-1)n$. Thus $\mathcal{F}_k^{(d-1)}(y) \leq \log(y^{d-1}\mu_d)$. On the other hand, $p_n(y) \geq p_n$, so that $\mathcal{F}_k^{(d-1)} \geq \log\mu_{d-1}$. Substitution of these bounds gives the results if $y \geq 1$. If $y < 1$ then $p_n \geq p_n(y) \geq p_n(0)$. By subdividing every edge in polygons counted by p_n a polygon is found without contacts, so that $p_{2n}(0) \geq p_n$. Thus $\log\mu_d \geq \mathcal{F}_k(y) \geq \log\sqrt{\mu_d}$. These bounds are enough to find the claimed inequalities if $y < 1$. □

By Corollary 5.64 a desorbed phase is found for small values of z, and an adsorbed phase for larger values of z. It is not known whether there is an expanded phase at small enough values of y and a collapsed phase at large values of y, although there is a tremendous amount of numerical evidence which suggests this; see Section 5.3.4. I shall assume in what follows that there is a collapse transition at a critical value of y where the polygons will go through a θ-point. Let this value of y be $y_c(z)$. I shall show that with a modest assumption, $y_c(z)$ is independent of z for all $z < z_c^+(y_c)$.

Theorem 5.65 *Suppose that $\mathcal{F}_v^+(y,0) = \mathcal{F}_k(y)$ is non-analytic at $y = y_c$. Then $\mathcal{F}_v^+(y,z)$ and $\mathcal{F}_v(y,z)$ are non-analytic functions of y at $y = y_c$ for all $z \leq 0$. Assume furthermore that $z_c^+(y)$ is a continuous function of y at $y = y_c$. Then $\mathcal{F}_v^+(y,z)$ is a non-analytic function of y at $y = y_c$ for all $z \leq z_c^+(y_c)$. A similar statement is true for $\mathcal{F}_v(y,z)$.*

Proof Since $\mathcal{F}_v^+(y,z) = \mathcal{F}_v(y,z) = \mathcal{F}_v(1,y) = \mathcal{F}_k(y)$ if $z \leq 0$ if follows immediately that $y_c(z) = y_c$ if $z \leq 1$, and that the collapse transitions for positive polygons and attached polygons are at the same critical activity $y = y_c$ if $z \leq 1$. Let $\epsilon > 0$ and choose z_d by

$$z_d < \inf\{z_c^+(y) \big| y \in [y_c - \epsilon, y_c + \epsilon]\}.$$

Since $\mathcal{F}_v^+(y,z)$ is analytic for all such z_d if $y \neq y_c$, and $y \in [y_c - \epsilon, y_c + \epsilon]$, the result is that $\mathcal{F}_{v,d}^+(y,z) = \mathcal{F}_k(y)$ for all $y \in [y_c - \epsilon, y_c + \epsilon]$ and $z \leq z_d$ (if not, then there is a phase boundary at (y_d, z_d) for some $y_d \in [y_c - \epsilon, y_c + \epsilon]$, which is a contradiction). If ϵ is small enough, then z_d can be taken arbitrarily close to $z_c^+(y_c)$. The same result is shown for $\mathcal{F}_v(y,z)$ by using these techniques. □

The results in Lemma 5.63, Corollary 5.64 and Theorem 5.65 all justify (but do not prove!) the hypothetical phase diagram in Fig. 5.17. There are four phases: in the first place expanded desorbed polygons (ED-phase) are found, which can collapse to a phase of collapsed and desorbed polygons (CD-phase). Increasing the visit activity gives the expanded adsorbed phase (EA-phase) and the collapsed adsorbed phase (CA-phase). Notice that two triple points are shown in the diagram, rather than one quadruple point, (this is supported by numerical simulations [353, 354]), and that the adsorption transition critical line $z_c^+(y)$ is drawn at values of z greater than one ($z_c^+(y) > 1$). This fact will be proven. In the model of attached polygons the available evidence suggest that $z_c(y) = 1$ for all $y < \infty$, but a proof of this is not known.

That $z_c^+(y) > 1$ can be shown by using the sequence of constructions and arguments in Lemma 5.57 and Theorems 5.58 and 5.59. These generalize to the model here (with some difficulty). Instead of following that route, I shall use a different technique [203], which may have wider applicability. The density function of this model is defined, with the help of Theorem 3.19, by

$$\log \mathcal{P}_v^+(y;\epsilon) = \inf_{z>0}\{\mathcal{F}_v^+(y,z) - \epsilon \log z\}. \tag{5.81}$$

However, the densities of excursions in the model will also play a role, so let $p_n^+(k, w, v)$ be the number of positive polygons with v visits, k contacts and w excursions. Define the partition function

$$p_n^+(y; w, v) = \sum_{k\geq 0} p_n^+(k, w, v) y^k, \tag{5.82}$$

of polygons with v visits and w excursions. Then define a density function

$$\mathcal{P}_v^+(y;\delta;\epsilon) = \lim_{n\to\infty} [p_n^+(y; \lfloor \delta n \rfloor, \lfloor \epsilon n \rfloor)]^{1/n}. \tag{5.83}$$

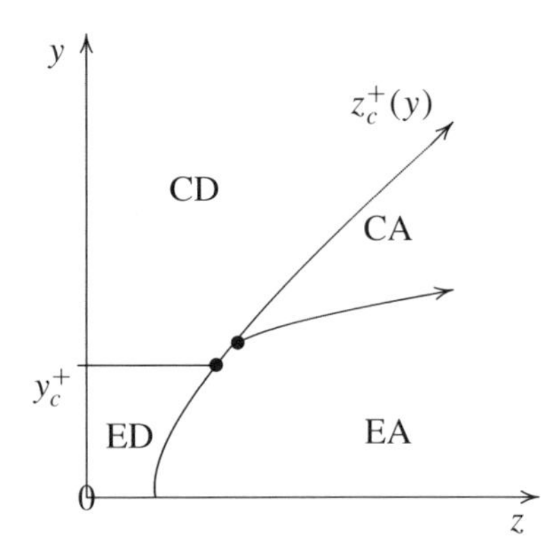

Fig. 5.17: The phase diagram for adsorbing and collapsing positive polygons in more than two dimensions. In two dimensions available evidence from related models suggest that there is no distinct CA phase. Notice that $z_c^+(y) > 1$ in this drawing; in the case of attached polygons it is generally believed that $z_c(y) = 1$.

That this limit exists can be seen by using a most popular argument. Suppose that $p_n^+(y; \lfloor \delta n \rfloor, \lfloor \epsilon n \rfloor | [h\phi h\phi])$ is the partition function of positive polygons with $\lfloor \delta n \rfloor$ excursions and $\lfloor \epsilon n \rfloor$ visits, and with bottom and top edges oriented in the ϕ direction, and mid-points a height h above the adsorbing plane. Concatenation and most popular arguments similar to those in Theorem 5.52 show that the limit in eqn (5.83) exists. Notice the connection to eqn (5.81):

$$\mathcal{P}_v^+(y; \epsilon) = \sup_{0<\delta<\delta_M} \{\mathcal{P}_v^+(y; \delta, \epsilon)\}. \tag{5.84}$$

The domain of (δ, ϵ) is as stated in eqns (5.69), but only the case that $d \geq 3$ will be considered here. Let $\delta_M(\epsilon)$ be the maximum value of δ for a given ϵ. Let A be a polygon counted by $p_n^+(k, w, v)$. Since there are v visits and w excursions in A, there are also $v - w$ edges of A in the plane $Z = 0$. Choose m non-adjacent edges from this set; this can be done in at least $\binom{\lfloor (v-w)/2 \rfloor}{m}$ different ways. Fix these edges in the $Z = 0$ hyperplane, and translate the rest of the polygon one step in the Z-direction to obtain the polygon A'. This is illustrated in Fig. 5.18.

No contacts are broken by this construction, but each of the m chosen edges produces a contact, and each new vertex may have as many as d new contacts; thus, A' may have at least k, and at most $k + (2d + 1)m$ contacts. In addition, there are $2m$ visits and m excursions in A'. Thus

$$\binom{\lfloor (v-w)/2 \rfloor}{m} p_n^+(k, w, v) \leq \sum_{i=0}^{(2d+1)m} p_{n+2m}^+(k + i, m, 2m). \tag{5.85}$$

Multiply this equation by y^c and sum over c. This gives a relation between partition functions:

$$\binom{\lfloor (v-w)/2 \rfloor}{m} p_n^+(y; w, v) \leq (y + 1 + y^{-1})^{(2d+1)m} p_{n+2m}^+(y; m, 2m). \tag{5.86}$$

Define the function $\phi(y) = y + 1 + y^{-1}$ (see eqn (5.23)). The consequence of eqn (5.86) is stated in Theorem 5.66.

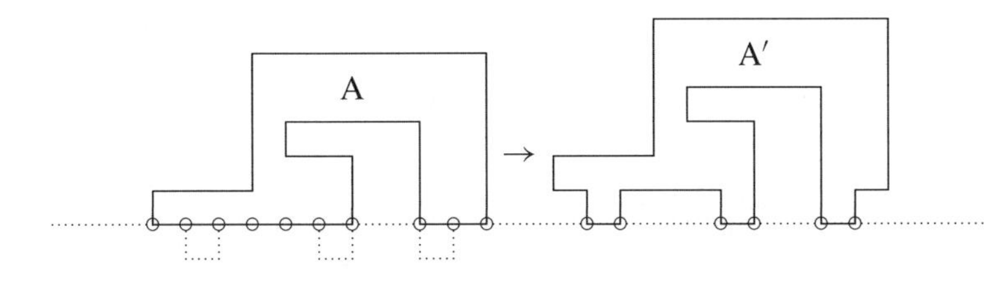

Fig. 5.18: By fixing edges in the $z = 0$ hyperplane and translating the rest of the polygon in the z-direction, new excursions can be created.

Theorem 5.66 *For every* ϵ *in the interval* $(0, 1]$ *there exists a* $\delta_* > 0$ *such that*

$$\left(1 + [\psi(y)e^{2\mathcal{F}_k(y)}]^{-1}\right)^{(\epsilon-\gamma)/2}\mathcal{P}_v^+(y; \gamma, \epsilon) \le \mathcal{P}_v^+(y; \delta_*, 2\delta_*),$$

where $0 < y < \infty$, $2\delta_* < \epsilon - \gamma$ and $\psi(y) = [\phi(y)]^{2d+1}$.

Proof Put $v = \lfloor \epsilon n \rfloor$, $w = \lfloor \gamma n \rfloor$, $m = \lfloor \delta n \rfloor$ in eqn (5.88). Take the $(1/n)$-th power, and take the lim sup as $n \to \infty$ of the left-hand side of the equation. This gives

$$\left[\frac{((\epsilon-\gamma)/2)^{(\epsilon-\gamma)/2}}{\delta^\delta((\epsilon-\gamma)/2-\delta)^{(\epsilon/2-\gamma)-\delta}}\right]\mathcal{P}_v^+(y; \gamma, \epsilon) \le [\psi(y)]^\delta\left[\mathcal{P}_v^+\left(y; \frac{\delta}{1+2\delta}, \frac{2\delta}{1+2\delta}\right)\right]^{1+2\delta}.$$

Since $\mathcal{F}_k(y) \ge \sup_{0<\delta<\delta_M}\{\log \mathcal{P}_v^+(y; \delta)\}$, then, by eqn (5.86), it follows that

$$\left[\frac{((\epsilon-\gamma)/2)^{(\epsilon-\gamma)/2}[\psi(y)e^{2\mathcal{F}_k(y)}]^{-\delta}}{\delta^\delta((\epsilon-\gamma)/2-\delta)^{(\epsilon/2-\gamma)-\delta}}\right]\mathcal{P}_v^+(y; \gamma, \epsilon) \le \mathcal{P}_v^+\left(y; \frac{\delta}{1+2\delta}, \frac{2\delta}{1+2\delta}\right).$$

The left-hand side is a maximum if

$$\delta = \frac{\epsilon-\gamma}{2(1+\psi(y)e^{2\mathcal{F}_k(y)})},$$

in which case

$$(1 + [\psi(y)e^{2\mathcal{F}_k(y)}]^{-1})^{(\epsilon-\gamma)/2}\mathcal{P}_v^+(y; \gamma, \epsilon) \le \mathcal{P}_v^+(y; \delta_*, 2\delta_*),$$

where $\delta_* = \delta/(1+2\delta)$ and where δ is given above. This completes the proof. □

Notice that $\delta_* \le \epsilon/2$ in the proof above, so that the optimal value of δ in Theorem 5.66 is in the interval specified by eqn (5.69). The implication of Theorem 5.66 is that any set of polygons with $\lfloor \epsilon n \rfloor$ visits and no excursions is exponentially rare when compared to a set of polygons with $\lfloor 2\delta_* n \rfloor$ visits and $\lfloor \delta_* n \rfloor$ excursions (and where $2\delta_* = \epsilon/(1+\psi(y)e^{2\mathcal{F}_k(y)})$).

By eqns (5.86) and (5.83), it follows that the free energy is defined by

$$\mathcal{F}_v^+(y, z) = \sup_{0\le\epsilon\le1}\left\{\sup_{0\le\delta\le\delta_M}\{\log(\mathcal{P}_v^+(y; \delta, \epsilon)z^\epsilon)\}\right\} = \log\left[\mathcal{P}_v^+(y; \delta^*, \epsilon^*)z^{\epsilon^*}\right], \quad (5.87)$$

for some ϵ^* and δ^* such that $(\epsilon^*, \delta^*) \in \Delta$. If $z < z_c^+(y)$, then $\epsilon^* = \delta^* = 0$ in eqn (5.87). Theorem 5.66 will give a proof that $z_c^+(y) > 1$. Suppose that $z > z_c^+(y)$ so that $\epsilon^* > 0$

Table 5.3: Exponents for adsorbing linear polymers in a θ-solvent

Dimension	ϕ	$\gamma_{1,\theta}$	$\gamma_{11,\theta}$	$\gamma_{s,\theta}$
2	$\frac{8}{21}$	$\frac{4}{7}$	$-\frac{4}{7}$	$\frac{8}{7}$

in eqn (5.87). Then by Theorem 5.66 it follows that

$$\mathcal{F}_v^+(y,z) = \log\left[\mathcal{P}_v^+(y;\delta^*,\epsilon^*)z^{\epsilon^*}\right] \leq \log\left[\mathcal{P}_v^+(y;\delta^\dagger,2\delta^\dagger)z^{2\delta^\dagger}(z^{\epsilon^*-2\delta^\dagger}/\Delta)\right], \quad (5.88)$$

where $2\delta^\dagger = (\epsilon^* - \delta^*)/\Delta < \epsilon^* - \delta^*$, and where $\Delta = (1 + [\psi(y)e^{2\mathcal{F}_k(y)}]^{-1})^{(\epsilon^*-\delta^*)/2}$. This is a contradiction if both $\epsilon^* > 0$ and

$$\log z < \frac{\log \Delta}{\epsilon^* - 2\delta^*}. \quad (5.89)$$

If it is supposed that $\delta^* = \alpha\epsilon^*$, where $0 \leq \alpha \leq 1/2$, then simplification of eqn (5.89) gives

$$\log z < \frac{(1-\alpha)(1+\psi(y)e^{\mathcal{F}_k(y)})\log(1+[\psi(y)e^{\mathcal{F}_k(y)}]^{-1})}{\alpha + 2\psi(y)e^{\mathcal{F}_k(y)}}. \quad (5.90)$$

This bound is positive for any value of α in $[0, 1/2]$. In other words, if z satisfies the bound in eqn (5.90), then eqn (5.88) is a contradiction (by the definition of $\mathcal{F}_v^+(y,z)$ in eqn (5.87)), *unless* $\epsilon^* = 0$. Thus, if z satisfies the bound in eqn (5.90), then the desorbed phase must be found, and so $z_c^+(y) > 1$.

Numerical support for the phase diagram in Fig. 5.17 was obtained using Monte Carlo simulations of adsorbing and collapsing walks [352, 353]. The surface exponents have also been studied in the vicinity of the θ-point. At the θ-point in the desorbed phase it is found that $\gamma_{1,\theta} = 4/7$ and $\gamma_{11,\theta} = -4/7$ [348], while for θ-walks at the adsorption critical point $\gamma_{s,\theta} = 8/7$ [348] while the cross-over exponent associated with the adsorption at the θ-point is $\phi = 8/21$ [348]. The exponents are listed in Table 5.3.

5.4.6 Copolymer adsorption

A model of copolymer adsorption can be defined by colouring vertices in an adsorbing polygon, and then letting only one colour interact with the adsorbing plane. Random copolymer adsorption is obtained if the assignment of colours of vertices is randomly selected from some distribution. An *annealed* model of random copolymer adsorption is a model where the partition function is averaged. A *quenched* averaged model of random copolymer adsorption is obtained if the free energy is averaged (over all colourings).[15]

The annealed model of random copolymer adsorption can be treated using the same techniques as in Sections 5.4.1–5.4.5. Suppose that vertices are coloured with colours 0 and 1, where colour 0 interacts with the hyperplane $Z = 0$ with activity z, and 1 does

[15] That is, if the free energy is computed for a fixed colouring, and then averaged.

not interact with the hyperplane $Z = 0$. If the colours are assigned from a binomial distribution, and 0 is assigned to a given vertex with probability p, then the model can be dealt with by replacing z in Sections 5.4.1–5.4.5 by $zp + (1 - p)$. This is seen by repeating the argument in Section 4.5.6. Let $c_n^+(v)$ be the number of positive walks with v visits. Of the v visits, w can be coloured by 0 and interact with the adsorbing plane via an activity z, and the remaining $v - w$ visits are coloured by 1, and do not interact with the adsorbing plane. Suppose that a given vertex is coloured by 0 with probability p. Then the generating function of the model is

$$G_v^a(x, z) = \sum_{n=0}^{\infty} \sum_{v=0}^{n+1} c_n^+(v) \sum_{m=0}^{v} \binom{v}{m} (zp)^m (1 - p)^{v-m} x^n. \tag{5.91}$$

The distribution of colours on vertices which are not visits contributed a factor of 1 to the generating function. Evaluating the sum over m shows that $G_v^a(x, z) = G_v^+(x, pz + 1 - p)$, where $G_v^+(x, z)$ is the generating function of the adsorbing homopolymer defined by $G_v^+(x, z) = \sum_{n=0}^{\infty} \sum_{v=0}^{n+1} c_n^+(v) z^v x^n$. However, the critical value of z in the adsorbing homopolymer model is at $z_c > 1$, as shown in Section 5.4.3. Thus the critical value of z in the annealed model of copolymer adsorption is at $z_c^a = (z_c - 1)/p + 1$. Since $z_c > 1$ it follows that $z_c(1 - p) > (1 - p)$ if $p < 1$, and so $(z_c - 1)/p + 1 > z_c$. In other words,

$$z_c^a > z_c > 1, \qquad \forall\, 0 \leq p < 1. \tag{5.92}$$

Thus, the annealed model of copolymer adsorption is similar to homopolymer adsorption.

Therefore, consider the more interesting case of the adsorption of quenched copolymers [283]. To keep the discussion simple, limit the model to three dimensions. In the case of quenched adsorption, the assignment of colours is fixed; cases such as alternating copolymer and block copolymer adsorption can be encountered [148]. Let χ_i be a random variable in the probability space Υ, and defined the random sequence $\chi = (\chi_1, \chi_2, \ldots)$ in $\Omega = \Upsilon \times \Upsilon \times \cdots$. Let $c_n^*(\chi)$ be the numbers of $*$-walks, where $* = \{?, \dagger, \ddagger, l, fl, \ldots\}$, and where it is assumed that each χ_i takes on only one of two colours $\{0, 1\}$, such that the first vertex in a walk has colour χ_1, the second χ_2, and so on. (The model will easily generalize to the case that there are more than two colours.) Observe that χ defines a first vertex in each walk, so that the walks counted by $c_n^*(\chi)$ are oriented; there is an exception if χ is a palindromic sequence. Similarly, $p_n(\chi)$ is the number of polygons with vertices of colours χ_j. The colouring may define a root (or a first vertex) in the polygon, but for some colourings there may be more than one first vertex to start (for example, if χ alternates between 0 and 1). Avoid these cases by rooting each polygon in what follows.

A quenched model is defined by letting $p_n^+(i_1, i_2, \ldots, i_v)$ be the number of positive polygons with their i_1-th, i_2-th, $\ldots$, i_v-th vertices as visits in the hyperplane $Z = 0$. Let χ be defined as above, with two colours selected from the binomial distribution. The first important question is the existence of a limiting free energy. This will be approached with the help of unfolded walks. Let $c_n^{+,\ddagger}(i_1, i_2, \ldots, i_v)$ be the number of XY-unfolded walks with v visits, and with the i_j-th vertices (for $j = 1, \ldots, v$) as visits in the hyperplane $Z = 0$. Any walk counted by $c_n^{+,\ddagger}(i_1, i_2, \ldots, i_v)$ has vertices with X- and Y-components given by $X_0 < X_i \leq X_n$ and $Y_0 \leq Y_i < Y_n$, for $i = 1, \ldots, n-1$. A doubly unfolded walk is *truncated* if its first and last vertices are removed; let $c_n^{+,T}(i_1, i_2, \ldots, i_v)$ be the number of truncated positive walks with vertices $\{i_1, \ldots, i_v\}$ in the hyperplane $Z = 0$. It will be convenient here to work with truncated walks. Consider a positive walk counted by $c^+(i_1, i_2, \ldots, i_v)$. If this walk is unfolded once in the X-direction and then in the Y-direction to find truncated walks (see Construction 5.3), then

$$c_n^{+,T}(i_1, i_2, \ldots, i_v) \leq c_n^+(i_1, i_2, \ldots, i_v) \leq e^{2\gamma\sqrt{n}} c_n^{+,T}(i_1, i_2, \ldots, i_v) \qquad (5.93)$$

is found. This is seen by noting that Constructions 5.1 and 5.3 do not change the Z-component of any vertex. This set of inequalities gives the first lemma in a proof that there is a limiting free energy in the quenched ensemble for adsorbing polygons.

Lemma 5.67 *The partition functions of truncated and positive walks are related by*

$$c^{+,T}(z|\chi) \leq c^+(z|\chi) \leq e^{2\gamma\sqrt{n}} c_n^{+,T}(z|\chi).$$

Proof Choose a colouring χ, and multiply eqn (5.93) by $z^{v-\sum_{j=1}^{v}\chi_{i_j}}$. Sum first over all the i_j, and then over v from 0 to n. □

Two truncated walks can be concatenated by using a most popular argument with respect to the heights of the end-vertices in the truncated walks. This establishes the existence of a limiting free energy.

Theorem 5.68 *Suppose that $\chi \in \Omega$. The quenched free energy*

$$\mathcal{F}_{qu}^+(z) = \lim_{n\to\infty} \frac{1}{n} \langle \log c^+(z|\chi) \rangle_\chi$$

exists, where $\langle \cdots \rangle_\chi$ is the average over all colourings.

Proof Let $c_n^{+,T}(i_1, \ldots, i_v; [h_b h_t])$ be the number of truncated walks with end-vertices a height h_b and h_t above the $Z = 0$ hyperplane. Concatenate these walks by inserting an edge between the last vertex in one walk and the first vertex in the second (if they have the same height). This shows that

$$c_n^{+,T}(i_1, \ldots, i_v; [h_b h_s]) c_m^{+,T}(i_1, \ldots, i_w; [h_s h_t]) \leq c_{n+m+1}^{+,T}(i_1, \ldots, i_{v+w}; [h_b h_t]).$$

Choose random and independent colourings $\chi^{(1)}$ and $\chi^{(2)}$ for the walks above, and multiply by $z^{(v+w)-\sum_{j=1}^{v}\chi_{i_j}^{(1)}-\sum_{k=1}^{w}\chi_{i_k}^{(2)}}$, and define $\chi^{(1)+(2)} = \chi^{(1)}\chi^{(2)}$ to be the sequence obtained if $\chi^{(1)}$ and $\chi^{(2)}$ are concatenated. This shows that

$$c_n^{+,T}(z|\chi^{(1)}; [h_b h_s]) c_m^{+,T}(z|\chi^{(2)}; [h_s h_t]) \leq c_{n+m+1}^{+,T}(z|\chi^{(1)+(2)}; [h_b h_t]). \qquad (\dagger)$$

Observe that by a most popular argument, there are most popular values $[h_b^* h_t^*]$ for the heights. Thus

$$c_n^{+,T}(z|\chi;[h_b^* h_t^*]) \le c_n^{+,T}(z|\chi^{(1)}) \le n^2 c_n^{+,T}(z|\chi;[h_b^* h_t^*]).$$

Apply the last inequalities to eqn (†). This gives, after some simplification,

$$c_n^{+,T}(z|\chi^{(1)})c_m^{+,T}(z|\chi^{(2)}) \le (n+m)^4 c_{n+m+1}^{+,T}(z|\chi^{(1)+(2)}).$$

Now take logarithms, and averages over χ. Then

$$\langle \log c_n^{+,T}(z|\chi)\rangle_\chi + \langle \log c_m^{+,T}(z|\chi)\rangle_\chi \le 4\log(n+m) + \langle \log c_{n+m+1}^{+,T}(z|\chi)\rangle_\chi.$$

Thus, by Theorem A.3 there is a limiting free energy for truncated walks, and by Lemma 5.67 there is a limiting free energy for positive walks in the quenched averaged ensemble. □

Two truncated walks with the same X-span can be concatenated into a YZ-loop, and two YZ-loops can in turn be concatenatated into a polygon. Using most popular arguments, this shows that the limiting free energy of polygons in the quenched averaged ensemble is equal to that of walks given in Theorem 5.68. The next theorem is stated without proof.

Theorem 5.69 *The quenched average free energy of a model of polygons is equal to the corresponding free energy for a model of walks, and is defined by*

$$\mathcal{F}_{qu}^+(z) = \lim_{n\to\infty} \frac{1}{n}\langle \log p_n^+(z|\chi)\rangle_\chi. \qquad \square$$

There is an adsorption transition in these models; this is seen by using techniques similar to that of Section 5.4.1. Note in particular that every positive polygon with v visits can be translated one step in the Z-direction to destroy all the visits. This shows that $p_n^+(z|\chi) \le \max\{1, z^n\} p_n(0|\chi)$. Furthermore, define the growth constant of the quenched copolymer by

$$\lim_{n\to\infty} \frac{1}{n}\langle \log p_n^+(1|\chi)\rangle_\chi = \log \mu_{qu}. \tag{5.94}$$

Since $p_n(0|\chi) \le p_n^+(z|\chi)$, it follows that $\mathcal{F}_{qu}(z)$ is independent of z for all $z \le 1$. Moreover, for $z > 1$ this limiting free energy is a function of z, since there is at least one term in the partition function which has all vertices as visits. By considering the contribution from polygons with n visits to the hyperplane $Z = 0$, it follows that $p_n^+(z|\chi) \ge z^{v(\chi)} q_n(\chi)$, where $v(\chi)$ is the number visits of colour 0 to $Z = 0$ and $q_n(\chi)$ is the number of adsorbed copolymers with n visits. But notice now that $\langle v(\chi)\rangle_\chi = pn$, and so this shows that

$$\mathcal{F}_{qu}^+(z) \ge \lim_{n\to\infty} \frac{1}{n}\langle \log q_n(\chi)\rangle_\chi + \log(pz). \tag{5.95}$$

Thus, for large enough values of z, $\mathcal{F}_{qu}^+(z)$ is dependent on z.

Theorem 5.70 *There exists a critical point $z_c^q \geq 1$ such that $\mathcal{F}_{qu}^+(z) = \log \mu_{qu}$ for all $z < z_c^q$, and $\mathcal{F}_{qu}^+(z) > \log \mu_{qu}$ if $z > z_c^q$. In other words, there is an adsorption transition in this model of quenched averaged random copolymers.* □

Lastly, it is also possible to show that for almost all fixed quenches, there is a limiting free energy, and that it is equal to $\mathcal{F}_{qu}^+(z)$. This result is stated in Theorem 5.71.

Theorem 5.71 *Let χ_0 be an infinite random sequence in Ω. Then*

$$\lim_{n\to\infty} \frac{1}{n} \log c^+(z|\chi_0) = \mathcal{F}_{qu}^+(z),$$

almost surely.

Proof Let $n = Nm + r$ for some fixed value of m and where $0 \leq r < m$. Divide an n-edge positive walk into subwalks of length m vertices by deleting N edges and let the remainder of the walk have length r. Let the first m colours in χ_0 be $\chi^{(1)}$, the second m be $\chi^{(2)}$ and so on, until $\chi^{(N)}$, and the remainder has colours $\chi^{(N+1)}$, where $\chi_0 = \prod_{i=1}^{N+1} \chi^{(i)}$. Then

$$c_n^+(z|\chi_0) \leq (2d)^N \left[\prod_{i=1}^{N} c_{m-1}^+(z|\chi^{(i)})\right] c_{r-1}^+(z|\chi^{(N+1)}).$$

Take logarithms, divide by n and take the lim sup of the left-hand side as $n \to \infty$. If m is kept fixed, then $N \to \infty$, while r varies between 0 and $m - 1$. The result is that

$$\limsup_{n\to\infty} \frac{1}{n} \log c_n^+(z|\chi_0) \leq \liminf_{N\to\infty} \frac{1}{N} \sum_{i=1}^{N} \frac{1}{m} \log c_{m-1}^+(z|\chi^{(i)}),$$

where the lim inf as $N \to \infty$ was taken of the right-hand side. Now use the strong law of large numbers, and let $m \to \infty$ to see that

$$\liminf_{N\to\infty} \frac{1}{N} \sum_{i=1}^{N} \frac{1}{m} \log c_{m-1}^+(z|\chi^{(i)}) = \left\langle \frac{1}{m} \log c_{m-1}^+(z|\chi)\right\rangle_\chi \to \mathcal{F}_{qu}^+(z).$$

Thus, since m is arbitrary,

$$\limsup_{n\to\infty} \frac{1}{n} \log c_n^+(z|\chi_0) \leq \mathcal{F}_{qu}^+(z)$$

almost surely, for almost all colourings χ_0.

An opposite inequality to this may be obtained as follows. Consider again $n = Nm + r$ with $0 \leq r < m$, and construct an n-edge walk of length n from truncated walks with m vertices and with end-points a most popular height above the $Z = 0$ hyperplane. This shows that

$$c_n^+(z|\chi_0) \geq \left[\prod_{i=1}^{N} c_{m-1}^{+,T}(z|\chi^{(i)})\right] c_{r-1}^{+,T}(z|\chi^{(N+1)}).$$

Now take logarithms and argue as above on the right-hand side using the strong law of large numbers. Use Lemma 5.67 to relate the limiting free energy of truncated walks to $\mathcal{F}_{qu}(z)$. This shows that

$$\liminf_{n\to\infty} \frac{1}{n} \log c_n^+(z|\chi_0) \geq \limsup_{N\to\infty} \frac{1}{N} \sum_{i=1}^{N} \frac{1}{m} \log c_{m-1}^{+,T}(z|\chi^{(i)})$$
$$= \left\langle \frac{1}{m} \log c_{m-1}^{+,T}(z|\chi) \right\rangle_\chi .$$

Now take $m \to \infty$ in the last term to complete the theorem. □

From a thermodynamic point of view this model is self-averaging; the limiting quenched average free energy is equal to the limiting free energy for almost all monomer sequences in the copolymer. These ideas were developed in reference [283], see also [363]. The relation between the critical point for quenched adsorption, z_c^q, and the critical values of z for annealed copolymer and homopolymer adsorption can also be stated. Note that the arithmetic geometric mean implies that $\log\langle c_n^+(z|\chi)\rangle_\chi \geq \langle \log c_n^+(z|\chi)\rangle_\chi$. This shows that the annealed model adsorbs before the quenched model. Thus, the critical values of the visit activity are arranged as $z_c^q \geq z_c^a > z_c > 1$.

5.5 Torsion in polygons

In the models considered so far, the density function has always been a function of some substructure (such as contacts, or visits, or excursions). In this section a different situation will be found; the density function will turn out to be a function of the difference between the number of occurrences of two structures in a polygon. The model of polygons in this section will be restricted to three dimensions, although most of the arguments here can be generalized to higher dimensions.

The entanglement complexity of a polygon or a knot has received considerable attention [196, 343, 350]. Torsion is a geometrical measure of entanglement complexity, and it captures local information about the planarity of the polygon. Writhe, in constrast, may be interpreted as a global measure of the supercoiling in the polygon. Observe that a polygon with high plectonemic or solenoidal supercoiling will also have a high torsion, and it is not naive to suspect that polygons with high torsion will also be highly "coiled". In this section the arguments in reference [343] will be followed.

A polygon consists of a sequence of line segments; each line segment consists of a sequence of collinear edges. A sequence of three successive line segments is called a *dihedral angle*. A dihedral angle in the cubic lattice may have size 0 or π radians, in which case the three line segments are coplanar, or it may have size $\pm\pi/2$ (in which case it is not planar). The sign of the dihedral angle is defined by a right-hand rule; see Fig. 5.19. A dihedral angle of size $+\pi/2$ is a *positive dihedral angle*, and a dihedral angle of size $-\pi/2$ is a *negative dihedral angle*. Dihedral angles in a polygon can be labelled by τ_i for $i = 1, 2, \ldots, m$, and their signs can be indicated by $\sigma(\tau_i)$, where $\sigma(\tau) = 0$ if the dihedral angle consists of coplanar line segments, and $\sigma = \pm 1$ if the

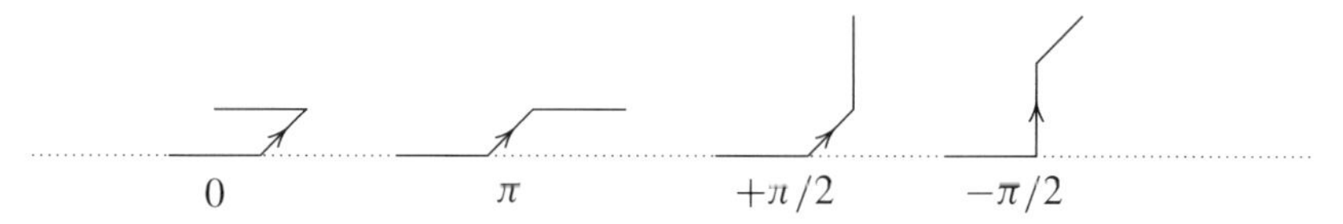

Fig. 5.19: The four possible dihedral angles in a polygon. The sign of the non-planar dihedral angles is defined by a right-hand rule. Note that it is not dependent on the orientation of the polygon (defined by the arrow).

dihedral angle has size $\pm\pi/2$. The *torsion* of a polygon is the sum over the signs of its dihedral angles:

$$t = \sum_{i=1}^{N} \sigma(\tau_i), \tag{5.96}$$

if there are N dihedral angles, while the *excess torsion* of a polygon is the absolute value of t:

$$t_e = \left| \sum_{i=0}^{N} \sigma(\tau_i) \right| . \tag{5.97}$$

With these definitions two models can be defined: in the first model polygons will be counted with respect to torsion, and in the second model polygons will be counted with respect to excess torsion. Let $p_n(t)$ be the number of polygons with torsion t, and $\bar{p}_n(t)$ be the number of polygons with excess torsion t. Then $\bar{p}_n(0) = p_n(0)$ and $\bar{p}_n(t) = 2p_n(t)$ if $t > 0$ (and less than or equal to n). Notice that there are no polygons with negative excess torsion. The partition functions are defined as

$$p_n(z) = \sum_{t=-n}^{n} p_n(t) z^t, \qquad \bar{p}_n(z) = \sum_{t=0}^{n} \bar{p}_n(t) z^t. \tag{5.98}$$

Notice that $p_n(z) = p_n(1/z)$, since $p_n(t) = p_n(-t)$. It can be shown that there are density functions and thermodynamic limits in both these models. Two polygons can be concatenated if the top edge of the first is parallel to the bottom edge of the second. In particular, if the first polygon is chosen in $p_n(t-s)$ ways, then the second polygon can be chosen in only $p_n(s)/2$ ways. Note that the removal of any edge in a polygon can reduce the number of dihedral angles by as many as three, and that the addition of new edges to concatenate the polygons can create as many as three new dihedral angles. So, if the two polygons are concatenated as in Fig. 1.4, then the new polygon may have torsion which is six more or less than the sum of the torsions of the two polygons. Thus

$$\sum_{s=0}^{t} p_n(t-s) p_m(s) \leq 2 \sum_{i=-6}^{6} p_{n+m}(t+i), \tag{5.99}$$

where it is assumed that $n \geq m$ without loss of generality. Thus, $p_n(t)$ satisfies Assumptions 3.8, and there is a density function and a free energy in this model (Theorem 3.18).

Moreover, by Corollary 3.14 and Theorem 3.16 there are integrated density functions and a density function so that the limiting free energy and density function are defined by

$$
\begin{aligned}
\mathcal{F}_t(z) &= \lim_{n\to\infty} \frac{1}{n} \log p_n(z), \\
\mathcal{P}_t(\epsilon) &= \lim_{n\to\infty} [p_n(\lfloor \epsilon n \rfloor + \sigma_n)]^{1/n} = \min\{\mathcal{P}_t(\leq\epsilon), \mathcal{P}_t(\geq\epsilon)\},
\end{aligned}
\tag{5.100}
$$

where $\sigma_n = o(n)$ and are defined as in Theorem 3.16. It is also possible to show that this model is regular (see Remark 3.9), and that σ_n may be taken equal to zero. Notice that the density function does not state the density of a pattern or substructure in the model, but measures the density of the sum of the signed dihedral angles. Somewhat more work is needed to show that there is a limiting free energy and a density function in the model of polygons with excess torsion. The argument has two steps [343]. First, suppose that $z \geq 1$. Then

$$
\sum_{t=-n}^{n} p_n(t) z^t \leq \sum_{t=-n}^{n} p_n(t) z^{|t|} \leq 2 \sum_{t=0}^{n} p_n(t) z^t \leq 2 \sum_{t=-n}^{n} p_n(t) z^t . \tag{5.101}
$$

Take logarithms, divide by n, and let $n \to \infty$. This gives

$$
\bar{\mathcal{F}}_t(z) = \lim_{n\to\infty} \frac{1}{n} \log \bar{p}_n(z) = \mathcal{F}_t(z), \quad \text{if } z \geq 1. \tag{5.102}
$$

On the other hand, if $z < 1$, then eqn (5.99) is used to see that

$$
\sum_{t=-n}^{n} \sum_{s=-m}^{m} p_n(t-s) p_m(s) z^{|t-s|+|s|} \leq 2 \sum_{i=-6}^{6} \sum_{t=-(n+m)}^{n+m} p_{n+m}(t+i) z^{|t|}, \tag{5.103}
$$

where the triangle inequality $|t| \leq |t-s| + |s|$ was used, and if one notes that $||t+i| - |i|| \leq |t|$, then eqn (5.103) becomes

$$
\bar{p}_n(z) \bar{p}_m(z) \leq 2[\phi(z)]^6 \bar{p}_{n+m}(z), \tag{5.104}
$$

where $\phi(z) = z + 1 + z^{-1}$. In other words, the limiting free energy exists by Lemma A.1, and the density function exists by Theorem 3.18. Define these by

$$
\begin{aligned}
\bar{\mathcal{F}}_t(z) &= \lim_{n\to\infty} \frac{1}{n} \log \bar{p}_n(z), \\
\bar{\mathcal{P}}_t(\epsilon) &= \lim_{n\to\infty} [\bar{p}_n(\lfloor \epsilon n \rfloor + \sigma_n)]^{1/n},
\end{aligned}
\tag{5.105}
$$

where $\sigma_n = o(n)$, and defined in Theorem 3.18. Note that $\mathcal{F}_t(0) = \bar{\mathcal{F}}_t(0) = \log \mu_3$, where μ_3 is the growth constant of polygons defined in Theorem 1.1. In Fig. 5.20 an example of a polygon of length 24 edges and torsion 18 is given. By increasing the length of the polygon, a class of polygons with length n and torsion $n - 6$ is found.

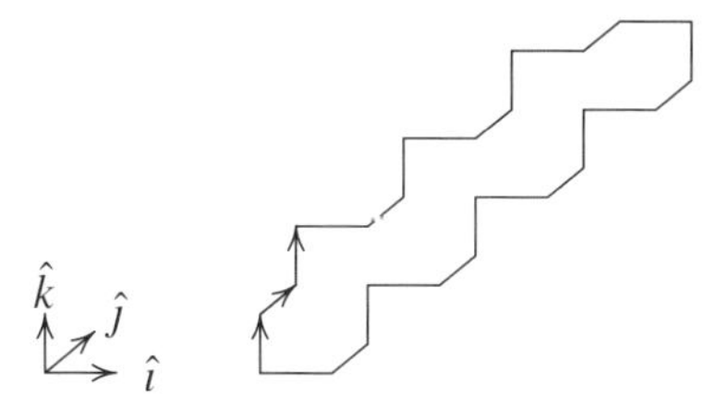

Fig. 5.20: A polygon of length n and torsion $n-6$.

Thus $p_n(z) \geq z^{n-6} p_n(n-6)$. Thus, by taking the logarithm and letting $n \to \infty$, and by noting that $p_n(z)$ is non-decreasing with $z \geq 1$, is follows that

$$\bar{\mathcal{F}}_t(z) = \mathcal{F}_t(z) \geq \max\{\log z, \log \mu_3\}, \quad \text{if } z \geq 1. \tag{5.106}$$

Since $\bar{\mathcal{F}}_t(z) = \mathcal{F}_t(z)$ if $z \geq 1$, and since $\bar{\mathcal{F}}_t(z)$ is non-decreasing for all $z \in [0, \infty)$, there is a non-analyticity in $\bar{\mathcal{F}}_t(z)$ (note that $\mathcal{F}_t(1/z) = \mathcal{F}_t(z)$, and so it follows that $\mathcal{F}_t(z)$ is strictly increasing for large enough values of z; thus $\bar{\mathcal{F}}_t(z) < \mathcal{F}_t(z)$ if $z < 1$ is small enough). In the next theorems it is shown that $\bar{\mathcal{F}}_t(z)$ is independent of z if $z \leq 1$, and that the transition at $z = 1$ is characterized by a change in the density of the excess torsion, which will be shown to be positive for all $z > 1$.

Theorem 5.72 $\bar{\mathcal{F}}_t(z) = \log \mu_3$ *for all* $z \leq 1$.

Proof First note that $\bar{p}_n(0) = p_n(0)$ and $\bar{p}_n(1) = p_n$. Clearly $\bar{p}_n(0) \leq \bar{p}_n(1)$. If two polygons are concatenated, each having n edges, one with torsion k and the other with torsion $-k$, then the inequality

$$\sum_{k=-n}^{n} p_n(k) p_n(-k) \leq 2 \sum_{l=-6}^{6} p_{2n}(l)$$

is obtained (the extra factor of 2 appears because the bottom edge of the second polygon is constrained to have the same orientation as the top edge of the first (see eqn (5.101))). Now use a most popular argument: let the class of polygons with torsion equal to $\pm k^*$ be the most popular:

$$p_n(k^*) \geq p_n(k) \quad \forall |k| \leq n$$

and thus

$$p_n(k^*) \geq p_n/(2n+1).$$

Since $p_n(k^*) = p_n(-k^*)$ this gives

$$\left(\frac{p_n}{2n+1}\right)^2 \leq p_n(k^*) p_n(-k^*) \leq \sum_{k=-n}^{n} p_n(k) p_n(-k) \leq 2 \sum_{l=-6}^{6} p_{2n}(l).$$

Let l^* be the most popular value of l such that $-6 \leq l \leq 6$ and $p_{2n}(l^*) \geq p_{2n}(l)$ for $-6 \leq l \leq 6$. Then

$$\left(\frac{p_n}{2n+1}\right)^2 \leq 26\, p_{2n}(l^*).$$

If $z \leq 1$, then

$$p_{2n} \geq \bar{p}_{2n}(z) \geq p_{2n}(l^*) z^{|l^*|} \geq \left(\frac{p_n}{2n+1}\right)^2 \left[z^{|l^*|}/26\right].$$

Taking logarithms, dividing by $2n$ and letting n go to infinity gives

$$\lim_{n\to\infty} n^{-1} \log \bar{p}_n(z) = \log \mu_3 \quad \forall z \leq 1. \qquad \square$$

$\mathcal{F}_t(z)$ is a convex function of $\log z$, and is differentiable almost everywhere in its domain. Moreover, if $\mathcal{F}_t(z)$ is differentiable at z, then

$$z\frac{d}{dz}\mathcal{F}_t(z) = \lim_{n\to\infty} \frac{1}{n}\left[z\frac{d}{dz} \log p_n(z)\right] = \lim_{n\to\infty} [\langle t\rangle_n/n], \tag{5.107}$$

where $\langle t\rangle_n = \sum_{t=-n}^{n} t p_n(t) z^t / p_n(z)$ is the mean torsion; notice that the limit in eqn (5.107) exists almost everywhere since $\mathcal{F}_t(z)$ is a convex function. The mean torsion per edge is given by $\langle t\rangle_n/n$. The mean excess torsion is similarly defined, and is equal to the mean torsion if $z \geq 1$. In the next theorem, the tossing of a biassed coin is adapted to prove that the limiting value of the mean torsion per edge is positive for any $z > 1$. In particular, this means that $\mathcal{F}_t(z)$ is strictly increasing for all $z > 1$.

Theorem 5.73 *The mean torsion per edge, and the mean excess torsion per edge, are both positive for every $z > 1$, in the limits of infinite n. That is*

$$\lim_{n\to\infty} [\langle t\rangle_n/n] > 0$$

for almost every $z > 1$, and similarly for $\langle t_e\rangle_n/n$.

Proof The mean value of the torsion at finite values of n is

$$\langle t\rangle_n = \frac{\sum_{t=-n}^{n} t p_n(t) z^t}{\sum_{t=-n}^{n} p_n(t) z^t}. \tag{\dagger}$$

Since $\mathcal{F}_t(z) = \bar{\mathcal{F}}_t(z)\ \forall z \geq 1$, only the mean torsion for $z > 1$ has to be considered. Moreover, since $\mathcal{F}_t(z)$ is convex it is also differentiable almost everywhere, and so the limit $\lim_{n\to\infty}[\langle t\rangle_n/n]$ exists almost everywhere. The theorem is now proven by showing that for any $z > 1$ there exists an $\epsilon > 0$ such that

$$\lim_{n\to\infty} [\langle t\rangle_n/n] \geq [2z/(z+z^{-1}) - 1]\epsilon = \left[\frac{z-z^{-1}}{z+z^{-1}}\right]\epsilon$$

whenever this limit exists (almost everywhere). Suppose that $\hat{\imath}$, $\hat{\jmath}$ and $\hat{k}$ are the canonical unit vectors. Let P be the following walk: $[-\hat{\imath}, \hat{k}, -\hat{\imath}, -\hat{\imath}, -\hat{k}, \hat{\imath}, \hat{\jmath}, \hat{k}, -\hat{\imath}, -\hat{k}, -\hat{\imath}, \hat{k}, -\hat{\imath}]$ (see Fig. 5.21). The union of the dual 3-cubes of P is a topological 3-ball C, and P contains two positive and one negative dihedral angle. A reflection of P (to obtain the walk P^*) through its centre of mass leaves its end-points and C unchanged. P^* contains two negative and one positive dihedral angle. P is also a Kesten pattern and since the end-points of P and P^* are the same vertices of C, either P or P^* can be inserted into C in any given polygon containing C. Let α_n be a class of polygons with at least $\lfloor \epsilon n \rfloor$ occurrences of C, each containing either P or P^*, and fixed outside the union of exactly $\lfloor \epsilon n \rfloor$ of the C's. By Corollary 5.18 the number of such classes of polygons is at least $(1 - e^{-kn})2^{-\lfloor \epsilon n \rfloor} p_n$, where $k > 0$ is a small number and p_n is the number of polygons with length n. The factor $2^{-\lfloor \epsilon n \rfloor}$ appears because each class α_n represents $2^{\lfloor \epsilon n \rfloor}$ polygons. (This is a consequence of the binomial choice of either P or P^* in each C.) Suppose that the contribution to the torsion of a polygon in the class α_n from the fixed part outside the union of the C's is t, and let there be t_+ occurrences of the pattern P and t_- occurrences of the pattern P^* in the C's (thus, $t_+ + t_- = \lfloor \epsilon n \rfloor$). In the partition function, the weight of each P is z, and of P^* is z^{-1}. Since the P and P^* are independent and binomially distributed, the normalized contribution of the class α_n to the partition function is

$$z^t \sum_{t_+=0}^{\lfloor \epsilon n \rfloor} \binom{\lfloor \epsilon n \rfloor}{t_+} \left(\frac{z}{z+z^{-1}} \right)^{t^+} \left(\frac{z^{-1}}{z+z^{-1}} \right)^{\lfloor \epsilon n \rfloor - t_+} = z^t.$$

Similarly, the normalized contribution to the numerator in eqn (†) is given by

$$\begin{aligned} z^t \sum_{t_+=0}^{\lfloor \epsilon n \rfloor} (t + t_+ - t_-) \binom{\lfloor \epsilon n \rfloor}{t_+} \left(\frac{z}{z+z^{-1}} \right)^{t_+} \left(\frac{z^{-1}}{z+z^{-1}} \right)^{\lfloor \epsilon n \rfloor - t_+} \\ = z^t \big[t + 2z/(z+z^{-1}) - 1 \big] \lfloor \epsilon n \rfloor. \end{aligned}$$

The contribution from the class α_n^*, the mirror image of α_n, is similarly obtained (by replacing t with $-t$):

$$z^{-t} \big[-t + 2z/(z+z^{-1}) - 1 \big] \lfloor \epsilon n \rfloor.$$

Without loss of generality, assume that $t \geq 0$. In that case, the combined contribution to the numerator in eqn (A) from the class α_n and its mirror image is

$$t(z^t - z^{-t}) + \big[2z/(z+z^{-1}) - 1 \big] \lfloor \epsilon n \rfloor (z^t + z^{-t}).$$

For $z > 1$ the first term is always non-negative, and so a lower bound on the numerator can be obtained by ignoring it. This gives a lower bound for each of the classes α_n.

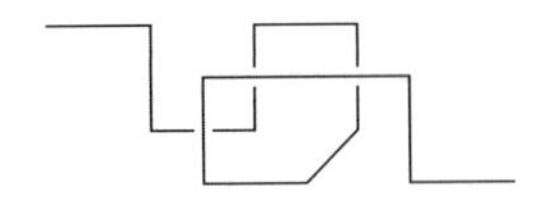

Fig. 5.21: This Kesten pattern contains two positive and one negative dihedral angle.

Let ω_n be the set of polygons with fewer than $\lfloor \epsilon n \rfloor$ occurrences of P or P^* in C, and fixed outside the union of the C's. Let the number of such classes of polygons be q_n. For any $\epsilon > 0$ it can be shown, using standard techniques (see for example [249]), that q_n grows exponentially with n. On the other hand, the pattern theorem for polygons (Corollary 5.18) states that there exists an $\epsilon_0 > 0$ such that for every positive $\epsilon < \epsilon_0$ the number of polygons in ω_n is exponentially small, compared to all polygons. Thus q_n is bounded from above and below by $e^{-k_1 n} p_n \geq q_n \geq e^{-k_2 n} p_n$ if n is large, where $0 < k_1 < k_2$, and for every positive $\epsilon < \epsilon_0$. Assume that $\epsilon < \epsilon_0$ is fixed. Let R be the minimum contribution to the numerator in eqn (†) by a polygon in ω_n, and S be the maximum contribution to the partition function by a polygon in ω_n. Using these bounds, and the contributions from the classes α_n, the mean torsion for n large enough is bounded from below as

$$\frac{\langle t \rangle_n}{n} \geq \frac{(1 - e^{-kn}) 2^{-\lfloor \epsilon n \rfloor} p_n \big[(2z/(z + z^{-1}) - 1)[\lfloor \epsilon n \rfloor / n]\big](z^t + z^{-t}) + e^{-k_2 n} p_n R}{2^{-\lfloor \epsilon n \rfloor} p_n (z^t + z^{-t}) + e^{-k_1 n} p_n S},$$

since each class α_n contains $2^{\lfloor \epsilon n \rfloor}$ polygons. Now take $\epsilon < k_1$; since decreasing ϵ also decreases q_n, this is always possible. Let $n \to \infty$ to obtain

$$\lim_{n \to \infty} [\langle t \rangle_n / n] \geq (2z/(z + z^{-1}) - 1)\epsilon = \left[\frac{z - z^{-1}}{z + z^{-1}}\right]\epsilon. \qquad \square$$

It was noted that $\bar{\mathcal{F}}_t(z) = \log \mu_3$ if $z \leq 1$. In particular, this means that the mean absolute torsion per edge is zero if $z < 1$. The left-derivative of $\bar{\mathcal{F}}_t(z)$ is also zero at $z = 1$, and by Theorem 5.73 the right-derivative at z is either 0 or positive. If it is positive, then there is a jump discontinuity at $z = 1$ in the first derivative of $\bar{\mathcal{F}}_t(z)$, which indicates a *first-order* transition in this model. In addition, since $\mathcal{F}_t(z) = \bar{\mathcal{F}}_t(z)$ if $z \geq 1$, and $\mathcal{F}_t(z) = \mathcal{F}_t(1/z)$, this also indicates a first-order transition in the mean torsion. If the right-derivative at $z = 1$ is 0, on the other hand, then a continuous transition in the excess torsion model is found, while no statement about the mean torsion model can be made. If $z < 1$ then a phase of polygons with the exponents in Table 1.1 can be expected (that is, the polygon is in the universality class of a polymer in a dilute solution). If $z > 1$, then there is a transition to a phase of polygons with excess torsion. It is generally accepted that the conformation of a polygon in these phases will be branched structures called cruciforms composed of supercoiled segments of the polygon [58]. The branched nature of these comformations suggest that branched polymer exponents in a dilute solution could be expected (Table 1.2).

5.6 Dense walks and composite polygons

The θ-transition in linear polymers is usually modelled by a walk with a nearest-neighbour interaction (see Section 5.3.4). Unfortunately, there are not many rigorous results in this model. There is numerical evidence for a collapse transition, and all indications are that the collapse is from a phase of expanded walks to a phase of more densely packed walks (presumably as a solid). In this sense the θ-transition seems to be related to a dilute-dense transition which has been studied in walks confined to a square. The first such model is a model of a walk which crosses a square [364]; it is known that there is a dilute-dense transition in this model [364], and several conjectures (including the location of the critical point) concerning this model were proven in reference [246]. This model also generalizes to higher dimensions, and to models of trees and animals [246].

A second set of models which exhibit a dilute-dense transition are models of composite polygons. These two-dimensional models consist of a containing polygon (which is a model for a vesicle in this context) which contains, in its interior, a polygon, or a tree, or an animal or a disk, or even a walk.[16] In this section I shall explore the phase diagram of these models, starting first with a model of a walk which crosses a square, before considering the more general composite polygon models. I shall show that all these models exhibit a dilute-dense transition, and that there is also an adsorption-like transition in the model of composite polygons.

5.6.1 Walks which cross a square

A walk crosses a square if it is confined to the square (it may visit vertices in the boundary) and if it has end-points on antipodal corners of the square. In addition, note that the walk is a t-walk (see Section 5.1), and that its first (bottom) vertex and its last (top) vertex are located at antipodal corners of the square. Thus, a class of constrained t-walks is found and generalizations to higher dimensions are evident. In this section, only the two-dimensional version of this model will be considered.

Let $\varsigma_n(m)$ be the number of t-walks of length m, with first and last vertices on antipodal corners of a square of side-length n, and area n^2. An interacting model is found by defining the partition function

$$\varsigma_n(z) = \sum_{m=0}^{2\lfloor n(n+2)/2 \rfloor} \varsigma_n(m) z^m. \tag{5.108}$$

The limiting free energy in this model exists, as will be shown below, but it will also be seen to be infinite if $z > 1$. This last fact can also be seen from the construction in Fig. 5.22. Notice that the shortest walk which crosses an $n \times n$-square has length n_{min} and the longest walk which crosses an $n \times n$-square has length n_{max}, where

$$\begin{aligned} n_{min} &= 2n, \\ n_{max} &= 2\lfloor n(n+2)/2 \rfloor. \end{aligned} \tag{5.109}$$

Thus, a walk can explore almost the entire area of a square. Each walk which crosses the shaded squares in Fig. 5.22 can be chosen in two ways, so that $\varsigma_2(8) = 2$. Thus, it follows that $\varsigma_6^c(48) \geq 2^4$. Now suppose that the larger square has side-length $2^k - 2$.

---

[16] This last case cannot be successfully treated with the methods that I shall develop here. If the polygon contains a connected structure in its interior, then it is a *simple composite polygon*; if it is not connected, then it is a *complex composite polygon*.

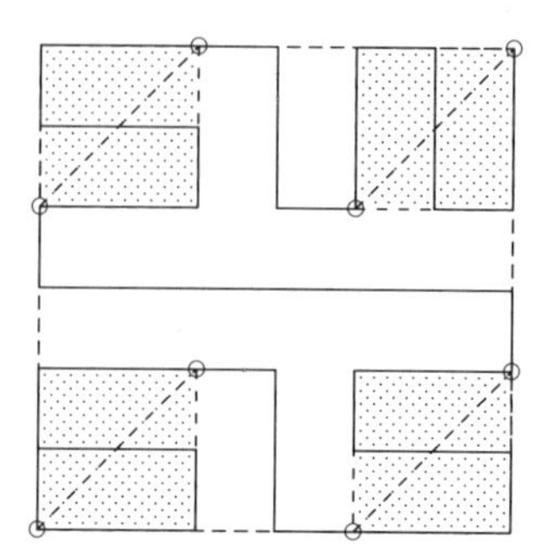

Fig. 5.22: There are at least 2^4 walks which cross a square of side-lenght 6; each of the walks in the smaller shaded squares can be chosen in two ways (by reflection through the diagonal). Thus $c_6^c(48) \geq 2^4$.

Then it follows that $\varsigma_{2^k-2}(2^k(2^k-2)) \geq [2\varsigma_{2^{k-1}-2}(2^{k-1}(2^{k-1}-2))]^4$. In other words, $\varsigma_{2^k-2}(2^k(2^k-2)) \geq 2^{4+4^2+\cdots+4^{k-1}} \geq 2^{4(4^{k-1}-1)/3}$. Thus, $\liminf_{n\to\infty}[\log \varsigma_n(z)]/n \geq \liminf_{k\to\infty}[\log \varsigma_{2^k-2}(2^k(2^k-2))z^{2^k(2^k-2)}]/[2^k-2]$ and so it follows that $\liminf_{n\to\infty}[\log \varsigma_n(z)]/n \geq \liminf_{k\to\infty}[\log 2^{4(4^{k-1}-1)/3}z^{2^k(2^k-2)}]/[2^k-2]$; and this is infinite if $z > 1$.[17]

In the next theorem it is shown that there is a limiting free energy, if it is assumed that it might be infinite. In addition, the limit $\lim_{n\to\infty}[\log \varsigma_n(z)]/n^2$ is also shown to exist. It will later be used to find the critical point in this model.

Theorem 5.74 *The following limits exist for $z \in [0, \infty]$:*

$$\mathcal{F}_\varsigma(z) = \lim_{n\to\infty} \frac{1}{n} \log \varsigma_n(z),$$

$$\mathcal{H}_\varsigma(z) = \lim_{n\to\infty} \frac{1}{n^2} \log \varsigma_n(z).$$

Moreover, $\mathcal{F}_\varsigma(z) = \infty$ if $z \geq 1$, and $\mathcal{H}_\varsigma(z) = 0$ if $\mathcal{F}_\varsigma(z)$ is finite, while $\log z \leq \mathcal{H}_\varsigma(z) \leq \log \mu_2 + \log z$ *if $z \geq 1$.*

Proof If a walk counted by $\varsigma_{n_1}(m - m_1)$ is concatenated with a walk counted by $\varsigma_{n_2}(m_1)$ (by identifying the top vertex of the first with the bottom vertex of the second), then $\sum_{m_1=0}^{m} \varsigma_{n_1}(m-m_1)\varsigma_{n_2}(m_1) \leq \varsigma_{n_1+n_2}(m)$, and multiplication by z^m and summing over m gives

$$\varsigma_{n_1}(z)\varsigma_{n_2}(z) \leq \varsigma_{n_1+n_2}(z).$$

Thus, the model satisfies a supermultiplicative inequality, and the limit exists as claimed (see Lemma A.1), but may be infinite. If $z \geq 1$, then $\mathcal{F}_\varsigma(z) = \infty$, as argued above. Consider now the second limit. If $z \leq 1$ then there is a bound in terms of t-walks: since $\varsigma_n(m) \leq c_m^t$ it follows that $\varsigma_n(z) \leq \sum_{m=n_{min}}^{n_{max}} c_m^t z^m$. If $z \leq 1$, then this

[17] In fact, this argument shows that the free energy is infinite for all values of $z > 2^{-4/3}$.

gives $\varsigma_n(z) \leq (n_{max} - n_{min})c^t_{n_{max}}$, and if $z > 1$ then $\varsigma_n(z) \leq (n_{max} - n_{min})c^t_{n_{max}} z^{n_{max}}$. Thus

$$\limsup_{n\to\infty} \frac{1}{n^2} \log \varsigma_n(z) = \log \mathcal{H}_\varsigma(z) \tag{†}$$

is finite. To see that the limit exist, consider the $n \times n$ square in Fig. 5.23, which contains p^2 squares of size $(M+2) \times (M+2)$, with $p = \lfloor n/(M+2) \rfloor$ and $n = p(M+2)+q$. Each of the smaller $M \times M$ squares (contained in the interior of the $(M+2) \times (M+2)$ squares) can be crossed independently by a walk; label these squares by ij where $1 \leq i, j \leq p$, and let the number of edges in the walk which crosses the ij-th square be n_{ij}. These walks can be made into a longer walk which crosses the $n \times n$ square by adding $2p(p-1)(M+2) + 2q + 4$ edges, as shown in Fig. 5.23. Since the walks cross each of the smaller squares independently, the number of walks in the $n \times n$ square is at least $\prod_{i,j=1}^{p} \varsigma_M(n_{ij})$. Multiplying this with $z^{n_{11}+\cdots+n_{pp}}$, and summing over all the n_{ij}, gives

$$[\varsigma_M(z)]^{p^2} \leq [2M^2 - 2M]^{p^2} z^{-(2p(p-1)(M+2)+2q+4)} \varsigma_n(z), \tag{‡}$$

since each $M \times M$ subsquare contains at least $2M$ and at most $2M^2$ edges. From equation (†) for every $\epsilon > 0$ there exists an infinite set of integers, $S(\epsilon)$, such that

$$\mathcal{H}_\varsigma(z) - \epsilon/2 \leq \frac{1}{n^2} \log \varsigma_n(z) \leq \mathcal{H}_\varsigma(z), \tag{¶}$$

whenever $n \in S(\epsilon)$. From eqn (‡), and whenever $M \in S(\epsilon)$,

$$\begin{aligned}\liminf_{n\to\infty} \frac{1}{n^2} \log \varsigma_n(z) \geq & \frac{p^2 \log \varsigma_M(z)}{(p(M+2)+q)^2} - \frac{p^2 \log(2M(M-1))}{(p(M+2)+q)^2} \\ & + \frac{(2p(p-1)(M+2)+2q+4)|\log z|}{(p(M+2)+q)^2}.\end{aligned}$$

Take $M \to \infty$ in $S(\epsilon)$ in the above, and use eqn (¶) to obtain

$$\liminf_{n\to\infty} \frac{1}{n^2} \log \varsigma_n(z) \geq \mathcal{H}_\varsigma(z) - \epsilon/2 \geq \limsup_{n\to\infty} \frac{1}{n^2} \log \varsigma_n(z) - \epsilon.$$

If $\epsilon \to 0^+$, then this establishes the existence of the limit. If $z \geq 1$, then $\varsigma_n(z) \leq \sum_{m=n_{min}}^{n_{max}} c_m z^{n_{max}}$, where c_m is the number of walks of length m. Thus

$$\log z \leq \lim_{n\to\infty} \frac{1}{n^2} \log \varsigma_n(z) \leq \log \mu_2 + \log z,$$

from eqn (5.109). □

Theorem 5.74 establishes the existence of the free energy, and also shows that it is infinite if $z \geq 1$. The next step would be to show that $\mathcal{F}_\varsigma(z)$ is finite and convex for some values of $z \leq 1$; this will prove that there is a critical point in this model. Notice

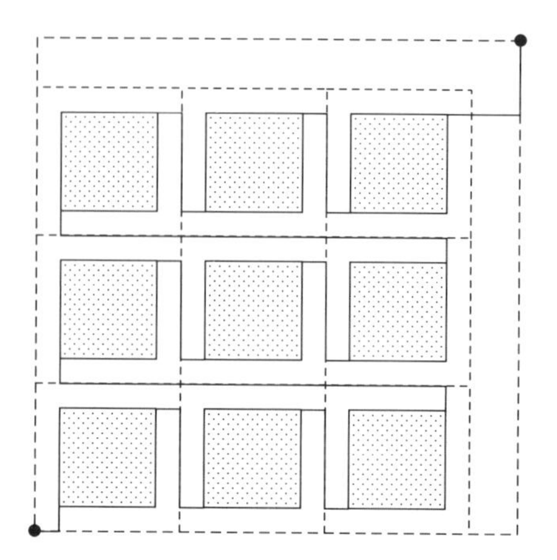

Fig. 5.23: A $n \times n$ square which contains p^2 smaller squares of size $M \times M$, suitably put together into a walk which crosses the larger square.

that Fig. 5.22 implies that $\mathcal{F}_\varsigma(z) > 0$ if $z > 0$. It is also the case that $\sum_{n\geq 0} \varsigma_n(z) \leq \sum_{m\geq 0} c_m z^m = c(z)$, where $c(z)$ is the generating function of walks. By Theorem 1.1, $c(z)$ is finite if $z < \mu_2^{-1}$. In other words, there is a critical point z_c in the interval $[\mu_2^{-1}, 1]$.

It is known that $z_c = \mu_2^{-1}$ [246]. To see this, it is shown that $\mathcal{H}_\varsigma(z) > 0$ if $z > \mu_2^{-1}$ (and thus that $\mathcal{F}_\varsigma(z) = \infty$ if $z > \mu_2^{-1}$). The proof of this fact relies on the use of unfolded walks. Define the number of unfolded walks from the origin to the point (n_1, n_2) by $c_n^\dagger(n_1, n_2)$. Any walk which crosses an $n \times n$ square is not unfolded, but can be made into an unfolded walk by adding a single edge to its bottom vertex in the negative horizontal direction. Thus

$$\varsigma_n(m) \leq c_{m+1}^\dagger(n+1, n) \leq c_{m+1}^\dagger. \tag{5.110}$$

Notice that unfolded walks are supermultiplicative: $c_{n_1}^\dagger c_{n_2}^\dagger \leq c_{n_1+n_2}^\dagger$, so that by Lemma A.1, and theorem 5.7, if $\epsilon > 0$, then there exists a finite fixed number N_0 such that

$$(\mu_2 - \epsilon)^m \leq c_m^\dagger \leq \mu_2^m \tag{5.111}$$

for all $m \geq N_0$. In the next lemma it is seen that eqn (5.111) also applies to a situation where the unfolded walk is forced to terminate with one coordinate of its end-point constrained.

Lemma 5.75 *Let $\epsilon > 0$; then there exists integers m and M, both even, or both odd, such that*

$$c_m^\dagger(M, 0) > (\mu_2 - \epsilon)^{2\lfloor m/2 \rfloor}.$$

Proof By Theorem 5.7 there exists an N_1 such that

$$(2m+1)^{-2} c_m^\dagger > (\mu_2 - \epsilon)^m.$$

There is also a most popular end-point for the unfolded walks, say (k_1, k_2): this means that (since there are at most $(2m+1)^2$ end-points),

$$c_m^\dagger(k_1, k_2) > (\mu_2 - \epsilon)^{m+1}.$$

By symmetry $c_m^\dagger(k_1, k_2) = c_m^\dagger(k_1, \pm k_2)$, and so it may safely be assumed that $k_2 \geq 0$. An unfolded walk counted by $c_m^\dagger(k_1, k_2)$ can be concatenated with one counted by $c_m^\dagger(k_1, -k_2)$ (identify the last vertex in the first with the first vertex in the second). This gives an unfolded walk with last vertex with coordinates $(2k_1, 0)$, so that

$$c_{2m}^\dagger(2k_1, 0) \geq c_m^\dagger(k_1, k_2) c_m^\dagger(k_1, -k_2) \geq [c_m^\dagger(k_1, k_2)]^2 > (\mu_2 - \epsilon)^{2m}.$$

Thus, choose $M = 2k_1$ and replace $2m$ by m to obtain two even integers. Odd integers are obtained if the concatenation above is done by inserting an extra horizontal edge between the two walks. This gives an unfolded walk with last vertex with coordinates $(2k_1 + 1, 0)$, so that

$$c_{2m+1}^\dagger(2k_1 + 1, 0) \geq c_m^\dagger(k_1, k_2) c_m^\dagger(k_1, -k_2) \geq [c_m^\dagger(k_1, k_2)]^2 > (\mu_2 - \epsilon)^{2m}.$$

Put $M = 2k_1 + 1$ and replace $2m + 1$ by m to obtain two odd integers. □

The unfolded walks in Lemma 5.75 can be used to count walks confined to a rectangle, by stacking them vertically. This is shown in Fig. 5.24, and Lemma 5.76 gives a lower bound on the number of these walks.

Lemma 5.76 *Suppose that $\epsilon > 0$ and choose m and M as in Lemma 5.75. Let p be an odd number. Then the number of walks of length $n = 2mp + mpj + (p+1)/2$ which crosses the rectangle $[0, jM + 1] \times [0, 2mp]$ is at least $(\mu_2 - \epsilon)^{2jp\lfloor m/2 \rfloor}$.*

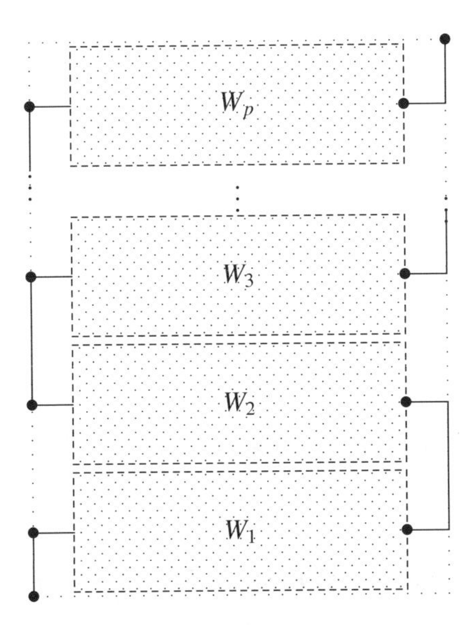

Fig. 5.24: Each shaded rectangle contains an unfolded walk, which is put together into a walk crossing a rectangle of dimensions $[0, jM + 1] \times [0, L]$, where $L = 2(p+1)M$.

Proof Consider the unfolded walks of length jm counted by $c^{\dagger}_{jm}(jM, 0)$ and which are confined to the rectangle $[0, jM] \times (-m, m)$ and with X-span jM, and let the number of these walks be $c^{*}_{jm}(jM, 0)$. Choose p such walks $\{W_i\}$ and arrange them as in Fig. 5.24, with the first vertex of W_1 at the point $(0, m)$, the first point of W_i at $(0, (2i-1)m)$, and add the extra edges as in Fig. 5.24 to join them into a single walk of length $2mp + mpj + (p+1)/2$ which crosses the rectangle $[0, jM+1] \times [0, 2mp]$. Notice that $c^{*}_{jm}(jM, 0) \geq [c^{*}_{m}(M, 0)]^{j}$ by concatenation. Moreover, every walk counted by $c^{\dagger}_{m}(M, 0)$ is in the rectangle $[0, M] \times (-m, m)$ so that $c^{*}_{m}(M, 0) = c^{\dagger}_{m}(M, 0)$. Thus

$$c^{*}_{jm}(jM, 0) \geq [c^{\dagger}_{m}(M, 0)]^{j} \geq (\mu_2 - \epsilon)^{2j\lfloor m/2 \rfloor},$$

by the choices of m and M and from lemma 5.75. But the total number of walks crossing the rectangle $[0, jM+1] \times [0, 2mp]$ is surely larger than $[c^{*}_{jm}(jM, 0)]^{p}$, this gives the result. □

These two lemmas are enough to show that the dilute-dense transition has a critical point $z_c = \mu_2^{-1}$. This is seen as follows. From Lemma 5.76 the number of walks which crosses a rectangle $[0, jM+1] \times [0, 2mp]$, with m, M as in Lemma 5.75, and with p odd, is at least $(\mu_2 - \epsilon)^{2jp\lfloor m/2 \rfloor}$. Let $z > \mu_2^{-1}$ such that $z(\mu_2 - \epsilon) > 1$.

Consider next the generating function of walks which cross a $(jM+1) \times (jM+1)$ square, $\varsigma_{jM+1}(z)$, and in fact, consider only those walks of length $2mp + mjp + (p+1)/2 + 1$ which cross this square. The number of these walks is at least the number of walks of length $2mp + mjp + (p+1)/2$ which crosses a rectangle of dimensions $[0, jM+1] \times [0, jM]$. So choose $2mp = jM$ in Lemma 5.76, so that $p = jM/2m$. Choose m and M odd, and then $j = 2mt$, to give an odd value for p. This implies that $p = Mt$ and $mjp = 2m^2Mt^2$. Therefore, the rectangle has dimensions $[0, jM+1] \times [0, jM]$, and the walks crossing it have length $2mMt + 2m^2Mt^2 + (Mt+1)/2$. The number of these walks is at least $(\mu_2 - \epsilon)^{4mMt^2\lfloor m/2 \rfloor}$, and thence

$$\varsigma_{2mMt+1}(z) \geq (\mu_2 - \epsilon)^{4mMt^2\lfloor m/2 \rfloor} z^{2mMt + 2m^2Mt^2 + (Mt+1)/2 + 1}.$$

Thus

$$\begin{aligned}
\lim_{n\to\infty} \frac{1}{n^2} \log \varsigma_n(z) &\geq \liminf_{n\to\infty} \left[\frac{\log[(\mu_2 - \epsilon)^{4mMt^2\lfloor m/2 \rfloor} z^{2mMt + 2m^2Mt^2 + (Mt+1)/2 + 1}}{(2mMt+1)^2} \right] \\
&\geq \frac{1}{2M} \log((\mu_2 - \epsilon)z) \\
&> 0.
\end{aligned}$$

For every $\epsilon > 0$, a z and a pair (m, M) can be found, thus $\lim_{n\to\infty}(1/n^2)\log \varsigma_n(z) > 0$ if $z > \mu_2^{-1}$. This result proves the following theorem.

Theorem 5.77 $\lim_{n\to\infty}(1/n^2)\log \varsigma_n(z) > 0$ *if* $z > \mu_2^{-1}$. □

The immediate consequence of Theorem 5.77 is that the dilute-dense transition occurs at $z_c = \mu_2^{-1}$.

Corollary 5.78 $\lim_{n\to\infty} \frac{1}{n} \log \varsigma_n(z) = \infty$ *if* $z > \mu_2^{-1}$. *Thus,* $z_c = \mu_2^{-1}$. □

Therefore, the critical point of this model is at μ_2^{-1}. It is also known that $\mathcal{F}_\varsigma(z_c) = 0$, and that this free energy is strictly negative if $z < z_c$, a result which can be proven with the help of the mass of the self-avoiding walk [246]. These results also generalize to higher dimensions.

The dense phase of walks has been considered in references [90, 97]. The appropriate scaling form should be given by eqn (2.23), where σ is a surface exponent dependent on the shape of the containing square (here, one should expect that $\sigma = (d-1)/d$). Moreover, the exponent $\alpha_- - \alpha_t$ is usually replaced by $\gamma_D - 1$, so that the assumption is $\varsigma_n \simeq n^{\gamma_D - 1} \mu_1^{n^\sigma} \mu_2^n$. Thus, γ_D is not the usual entropic exponent, but is related to α_- and α_t. The exponent γ_D is also dominated by the surface term. Studies focused instead on the ratio of walks and polygons ς_n/ρ_n, where ρ_n is the number of polygons in a square. This gives the relation $\varsigma_n/\rho_n \simeq n^{\gamma_D}$, and the value of the exponent is found to be $\gamma_D = 19/16$ [90,97]. In the next sections a model of composite polygons will be studied, where the walk or polygon will be contained in a random polygon itself, and one should expect a different (and unknown) value for σ and γ_D.

The dense phase of a walk is obtained if $z > z_c$, and it characterized by the metric exponent $\nu = 1/2$. The collapsed phase in the θ-collapse of a walk has the same metric exponent, but the relation between these phases is uncertain. For example, Hamiltonian polygons (of a square) have entropic exponent $\gamma_H = 1$ [7], and collapse to a Hamiltonian walk seems to have $\gamma_D = 1$ as well. On the other hand, it was also found that $\gamma_D = 0.92 \pm 0.09$ [15], a result which is not inconsistent with the Hamiltonian walk interpretation. The metric exponent can also be examined from a root mean square radius of gyration, R, point of view. Thus, if $z > z_c$ then $R \simeq \langle n \rangle^2$ where $\langle n \rangle$ is the mean number of edges in the walk. On the other hand, if $z < z_c$, then $R \simeq \langle n \rangle$ [41, 364]. At the critical point $z = z_c$ one should find that $R \simeq \langle n \rangle^{1/y}$, with $y = 1/\nu = 4/3$ and where ν is the metric exponent of a linear polymer in a good solvent. This is supported by a renormalization group argument, which found $y \approx 1.35$ [298].

5.6.2 Complex composite polygons

A composite polygon (see reference [191]) is a generalization of the model of walks which cross a square considered in the previous section. A composite polygon will always be a two-dimensional model; it consists of a lattice polygon (in the square lattice) which contains in its interior a structure which may either be a polygon, or a tree or animal, or a disk composed of unit squares (or a collection of these). The polygon is called a *containing polygon* and it contains an *internal structure*. If the internal structure is connected, then a model of *simple composite polygons* is defined, otherwise the model is called a *complex composite polygon*.

That there exists free energies in the case of complex composite polygons is not too difficult to show. The proof uses concatenation of polygons as in Fig. 1.4. In particular, let $p_n^c(m)$ be the number of complex composite polygons (in a given model), containing an internal structure of size m. Then Fig. 1.4 shows that $\sum_{m_1=0}^{m} p_{n_1}^c(m - m_1) p_{n_2}^c(m_1) \leq p_{n_1+n_2}^c(m)$ and by multiplication with z^m and summing over m, it follows that the partition function satisfies a supermultiplicative inequality:

$$p_{n_1}^c(z) p_{n_2}^c(z) \leq p_{n_1+n_2}^c(z), \tag{5.112}$$

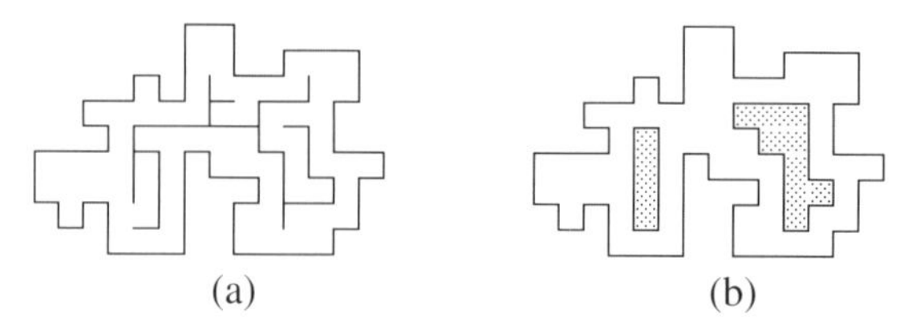

Fig. 5.25: Composite models: (a) is an example of a simple composite model; the model in (b) is a complex composite model containing polygons.

and therefore the limit

$$\mathcal{F}_{co}(z) = \lim_{n\to\infty} \frac{1}{n} \log p_n^c(z) \tag{5.113}$$

exists, but it may be infinite.

Theorem 5.79 *The limiting free energy in models of complex composite polygons exists. Moreover, if $z > 0$, then $\mathcal{F}_{co}(z) = \infty$. In other words, there is a transition at $z = 0$ (zero temperature) to an inflated phase.*

Proof The limiting free energy exists as in eqn (5.113). I shall show that it is infinite if $z > 0$ for internal structures which are polygons; the proofs for other internal structures are simular. Consider a square polygon of side-length l. The maximum number of internal polygons in this square is $\lfloor (l-2)^2/4 \rfloor$, if all these have length four edges, and are packed in the obvious way. Suppose that only $\lfloor \epsilon l^2 \rfloor$ polygons of length four are packed in. Then they can be packed in $\binom{\lfloor (l-2)^2/4 \rfloor}{\lfloor \epsilon l^2 \rfloor}$ ways, so that

$$\begin{aligned}\liminf_{n\to\infty} \frac{1}{n^2} \log p_n^c(z) &> \lim_{l\to\infty} \frac{1}{16l^2} \log \binom{\lfloor (l-2)^2/4 \rfloor}{\lfloor \epsilon l^2 \rfloor} z^{4\lfloor \epsilon l^2 \rfloor} \\ &= \frac{1}{16} \log \left(\frac{(1/4)^{1/4} z^{4\epsilon}}{\epsilon^\epsilon (1/4 - \epsilon)^{1/4-\epsilon}} \right).\end{aligned}$$

By footnote 3, Chapter 3, this is a maximum if $\epsilon = z^4/4(1+z^4)$, in which case

$$\liminf_{n\to\infty} \frac{1}{n^2} \log p_n^c(z) \geq \frac{1}{16} \log(1+z^4)^{1/4}.$$

In other words, if $z > 0$, then $\lim_{n\to\infty} \frac{1}{n^2} \log p_n^c(z) > 0$, so that $\mathcal{F}_{co}(z) = \infty$. □

The result is that models of complex composite polygons exhibit a first-order phase transition at zero "temperature". Notice that the density function of these models has an infinite jump-discontinuity: if ϵ is the density of edges in the internal structure, then $\mathcal{P}_{co}(0) = \mu_2$, and $\mathcal{P}_{co}(\epsilon) = \infty$, if $\epsilon > 0$. From this perspective these models are not interesting, and I shall limit the discussion in the next sections only to simple composite polygons.

5.6.3 *The unfolding of polygons*

In the previous section I have considered complex composite polygons. I have shown that there exists a limiting free energy, but that it is infinite for all values of the activity $z > 0$. In this section I shall consider the existence of a limiting free energy in simple composite models. The proof that it exists relies on a construction which will unfold a polygon through its convex hull, without taking too many steps.

Let A be a polygon and let $C(A)$ be its convex hull (see Fig. 5.26). The interior of $C(A)$ will be denoted by $\tilde{C}(A)$, and the closure of $\tilde{C}(A)$ is $\bar{C}(A) = C(A) \cup \tilde{C}(A)$. The unfolding of the polygon A will be a sequence of steps which will change A into a polygon A^{uu} such that the closure of the convex hull of A is contained entirely in the interior of the convex hull of A^{uu}: $\bar{C}(A) \subset \tilde{C}(A^{uu})$. The basic construction is the reflection of parts of A through the convex hull, but there are many technical details which must be considered in this process. For example, if A is a square, then it is equal to its convex hull, and it cannot be changed into a different polygon using reflections through its convex hull.

$C(A)$ consists of straight line segments joined into a plane polygon (as opposed to a lattice polygon); the first and last vertices of each line segment are vertices in the polygon A, and they are called *pivot points*. Note that exactly two line segments in the convex hull are horizontal (parallel to the X-axis), and two are vertical (parallel to the Y-axis); see Fig. 5.26.

A lexicographic ordering of the vertices in A, first with respect to the X-direction, and then with respect to the Y-direction, will define a unique "lexicographically most" vertex, which is called the *primary top vertex* t_p of A, and a unique "lexicographically least" vertex, which is called the *primary bottom vertex* b_p of A. Similarly, a *secondary top vertex* t_s and a *secondary bottom vertex* b_s can be defined by doing the lexicographic ordering first in the Y-direction, and then in the X-direction. These vertices are indicated in Fig. 5.26. It is possible that the top vertices t_p and t_s are coincident, and that the bottom vertices b_p and b_s are coincident. These top and bottom vertices divide the polygon into four sections, of which two may be empty.

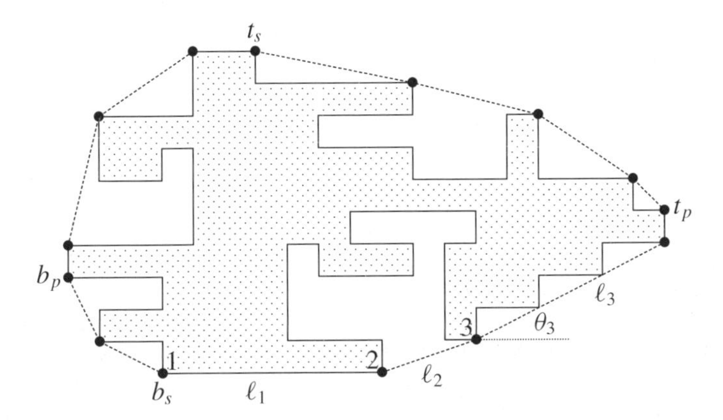

Fig. 5.26: The convex hull of a polygon. Pivot points are indicated by •.

Lemma 5.80 *Let A be a polygon of length n and with convex hull $C(A)$. Then there is a constant $K_0 > 0$ such that $C(A)$ is the union of at most $K_0\lfloor n^{2/3}\rfloor$ line segments. Thus, there are at most $K_0\lfloor n^{2/3}\rfloor$ pivot points in the convex hull.*

Proof Let the primary and secondary top and bottom vertices of A be defined as above. In addition, consider the subwalk of the polygon between b_s and t_p. Label the pivot points in this subwalk by 1, 2, ..., starting at b_s (see Fig. 5.26). The pivot points are lattice points in the square lattice (with integer coordinates), and they are the end-points of the line segments whose union is the segment of $C(A)$ between the vertices b_s and t_p. Let ℓ_i be the line segment between pivot points i and $i+1$. Let θ_i be the angle between ℓ_i and the positive X-direction; then $\tan\theta_i$ is a rational number: define

$$\tan\theta_i = \frac{q_i}{p_i},$$

with (p_i, q_i) relative primes, and suppose that there are M such line segments and angles with $\tan\theta_i \leq 1$. By definition, $\tan\theta_1 = 0$; in this case define $q_1 = 0$ and $p_1 = 1$. Since $C(A)$ is convex,

$$0 = \tan\theta_1 < \tan\theta_2 < \cdots < \tan\theta_M \leq 1,$$

or in other words

$$0 = \frac{q_1}{p_1} < \frac{q_1}{p_2} < \cdots < \frac{q_M}{p_M} \leq 1. \qquad (\dagger)$$

The number of edges in the part of the polygon A which joins the end-points of ℓ_i is at least $p_i + q_i$, so that M is constrained by

$$\sum_{i=1}^{M}(p_i + q_i) \leq n. \qquad (\ddagger)$$

Thus, the M distinct points (p_i, q_i) satisfy the constraints in eqns ($\dagger$) and ($\ddagger$). The largest number of distinct points satisfying these constraints is bounded from above by $C_0\lfloor n^{2/3}\rfloor$; this is seen as follows.

Suppose that the value of M is known, and that a set $\mathcal{S}$ of M points satisfying the constaints above is given. Let the maximum value of p_i+q_i in $\mathcal{S}$ be N. If a point $(p_i, q_i) \in \mathcal{S}$ with $p_i + q_i = N$ is exchanged with a new point $(p_i^*, q_i^*) \notin \mathcal{S}$ but with $p_i^* + q_i^* < N$ (suppose this is possible), then the new set of points $[\mathcal{S} \setminus \{(p_i, q_i)\}] \cup \{(p_i^*, q_i^*)\}$ satisfies the constraints above, the value of M remains unchanged, while the value of the sum $\sum_{i=1}^{M}(p_i + q_i)$ decreases. Thus, the smallest possible values of p_i and q_i should be chosen to find the maximum value of M.

If the first constraint in eqn ($\dagger$) above is relaxed by abandoning the requirements that all pairs (p_i, q_i) are relative primes, then an upper bound on M will be found, since points with smaller values of p_i+q_i can be chosen to satisfy condition ($\ddagger$). Thus, choose the p_i and q_i to be points above or on the p-axis in the pq-plane, but underneath or on the main diagonal $q = p$. Then for each p_i the values of q_i are $\{0, 1, \ldots, p_i\}$, while

$p_i = 1, 2, \ldots$. If $p_i + q_i \leq N$, then the number of points is $1 + 2 + 2 + \cdots + \lfloor N/2 + 1 \rfloor$. Assume that N is odd. Then this sums to at most

$$1 + 2 + 2 + 3 + 3 + \cdots + \lfloor N/2 + 1 \rfloor + \lfloor N/2 + 1 \rfloor = \lfloor N/2 + 1 \rfloor \lfloor N/2 + 2 \rfloor \quad 1, \tag{¶}$$

so that the number of distinct points is bounded from above by $M \leq \lfloor N/2 + 2 \rfloor^2$. On the other hand, for these choices for (p_i, q_i),

$$\begin{aligned} \sum_{i=1}^{M} (p_i + q_i) &\leq \sum_{i=1}^{\lceil N/2 \rceil} i(4i - 3) \\ &= \lceil N/2 + 1 \rceil (\lceil N/2 \rceil + 2)(8\lceil N/2 \rceil + 3)/6 \\ &\leq 4\lceil N/2 + 2 \rceil^3/3. \end{aligned} \tag{ℵ}$$

This is less than n if $\lceil N/2 + 2 \rceil^3 < 3n/4$, or if $\lceil N/2 + 2 \rceil \leq (3n/4)^{1/3}$. But then from eqn (†), $M \leq \lfloor N/2 + 2 \rfloor^2$, thus, $M \leq (3n/4)^{2/3}$, and since M is an integer,

$$M \leq \lfloor (3n/4)^{2/3} \rfloor. \tag{\#}$$

In other words, M cannot grow faster than a $2/3$ power of n. So far, this is only valid if all the points $(p_i + q_i)$ with $p_i + q_i \leq N$ are included in the calculation. If only a subset of these is used (and some which have $p_i + q_i = N$ are discarded to minimize the sum in eqn (‡)), then there are $O(N)$ corrections to eqns (¶) and (ℵ). But these changes will only imply a $O(n^{1/3})$ correction to eqn (#) so that there exists a constant C_0 such that

$$M \leq C_0 \lfloor n^{2/3} \rfloor.$$

By reflecting or rotating the polygon, the number of line segments in the subwalk of the polygon between b_s and t_p which make an angle greater than $\pi/4$ with the positive X-direction can be bounded by $C_0 \lfloor n^{2/3} \rfloor$. Thus there are at most $2C_0 \lfloor n^{2/3} \rfloor$ such line segments in the section of the convex hull between b_s and t_p. In other words, the total number of line segments in the convex hull is at most $8C_0 \lfloor n^{2/3} \rfloor$. □

Lemma 5.80 will be very useful; it states that if a lattice polygon A has length n, and convex hull $C(A)$, then $C(A)$ is an M-gon, where $M \leq K_0 \lfloor n^{2/3} \rfloor$, for some fixed number K_0. The next step is the unfolding of subwalks of A through the sides of $C(A)$; since there are at most $K_0 \lfloor n^{2/3} \rfloor$ such sides, this will limit the number of unfolding subwalks.

Construction 5.81 (Inversion) Let the convex hull of a polygon A of length n be $C(A)$, and suppose that $C(A)$ is an M-gon composed of line segments $\{\ell_i\}_{i=1}^{M}$, where $M \leq K_0 \lfloor n^{2/3} \rfloor$. The end-points of a line segment ℓ_i are pivot points, and they are vertices in A; they are also the end-points of a subwalk A_i of A which is entirely in the closure $\bar{C}(A)$. The basic operation in an unfolding of a polygon is a reflection of A_i through the mid-point of the line segment ℓ_i as illustrated in Fig. 5.27. The reflected image of A_i, denoted by $R(A_i)$, is disjoint with $\tilde{C}(A_i)$. There are potential problems with the

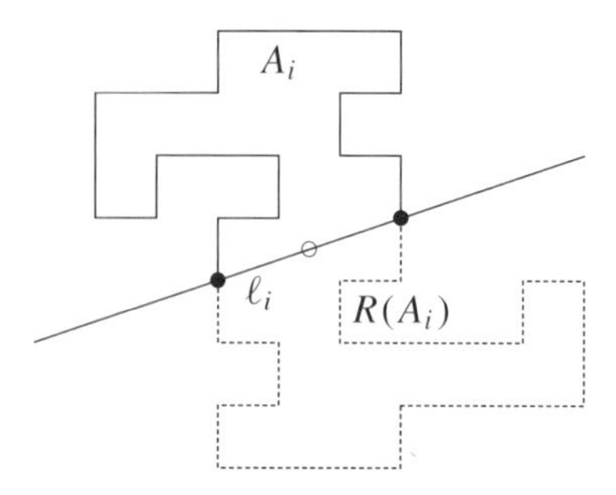

Fig. 5.27: The reflected image of a subwalk.

operation in Fig. 5.27 only if the line ℓ_i is parallel to a lattice axis; in that case the entire subwalk A_i might be contained in ℓ_i (that is, $A_i = \ell_i$), and so $R(A_i) = A_i$. For every other ℓ_i in the convex hull, every edge (except possibly one of its end-points) in A_i is in $\tilde{C}(A)$, and so every edge is moved outside $\tilde{C}(A)$ in the reflection. Only vertices in $R(A_i)$ may still be in $C(A)$. The reflection of A_i to $R(A_i)$ is called an inversion. □

The *primary top and bottom edges* of A are the lexicographically most and least edges with respect to an ordering of their mid-points; first in the X-direction, and then in the Y-direction. Similarly, the *secondary top and bottom edges* of A are the lexicographically most and least edges with respect to an ordering of the mid-points of the edges; first in the Y-direction, and then in the X-direction. To avoid the problems above when ℓ_i is parallel to a lattice axis, I shall slightly change A by adding eight new edges to it.

Construction 5.82 (Augmenting a polygon) A polygon A is *augmented* when two edges are added at each of the primary and secondary top and bottom edges as illustrated in Fig. 5.28. This increases the length of the polygon by eight edges, and the augmented polygon derived from A will be denoted by A^a. An augmented polygon has the important property that *only* its primary and secondary top and bottom edges are contained in its convex hull. Every other edge in A^a is either disjoint with the convex hull, or has at most one end-point in the convex hull. □

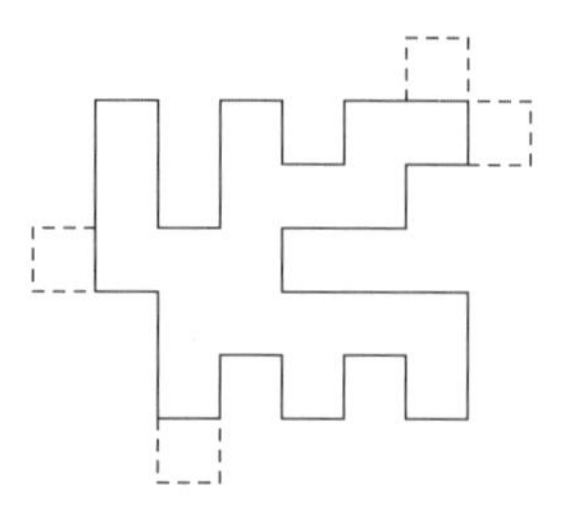

Fig. 5.28: The augmentation of a polygon.

The unfolding of a polygon A always proceeds by operating on the augmented polygon A^a derived from A. The following lemma states the basic construction in an unfolding.

Lemma 5.83 *Let A be an arbitrary polygon, and let $C(A)$ be its convex hull. Then any edge in A^a can be reflected through the convex hull $C(A^a)$ to an image which is disjoint with the interior $\tilde{C}(A)$ of the convex hull of A, and which has at most one end-point in the closure $\bar{C}(A)$ of the convex hull of A.*

Proof Suppose that $C(A) = \cup_{i=1}^{M} \ell_i$ is the union of straight line segments ℓ_i joined at pivot points in A, and let the pivot points cut A into subwalks A_i, where the first and last vertices of A_i are the end-points of ℓ_i (see Fig. 5.29). Augment A to A^a, and similarly let $C(A^a) = \cup_{i=1}^{M_a} \ell_i^a$ be its convex hull, consisting of $M_a \leq M$ line segments ℓ_i^a, with pivot points which cut A^a into subwalks A_i^a. Let e be an arbitrary edge in A^a. There are two possibilities: in the first case, e may be a top or a bottom edge (primary or secondary), or be adjacent to a top or a bottom edge of A^a. By the construction of the augmented polygon, e is then disjoint with $\tilde{C}(A)$, and there is nothing to prove. In the second case, e is contained in some subwalk A_i of A. But A_i is a subwalk of some A_i^a in the augmented polygon, and so e has at most one endpoint in $C(A^a)$. Inverting A_i^a through the mid-point of ℓ_i^a to $R(A_i^a)$ gives $R(e)$ disjoint with $\tilde{C}(A^a)$. But since $\tilde{C}(A) \subset \tilde{C}(A^a)$, $R(e) \cap \tilde{C}(A) = \varnothing$. If one end-point of e is in $\bar{C}(A^a)$, then one end-point of $R(e)$ will also be in $\bar{C}(A^a)$. □

Lemma 5.83 suggests that by reflecting subwalks in an augmented polygon through the convex hull, the polygon can be mapped to an image which is disjoint with the interior of the convex hull of the initial polygon. The result will be a polygon which is *unfolded* with respect to the initial polygon.

Theorem 5.84 *Let A be an arbitrary polygon, and let it be augmented to A^a. Suppose that A has length n. Then by performing at most $K_0\lfloor (n+8)^{2/3} \rfloor$ reflections of subwalks in A^a, a polygon A^u of length $n+8$ will be found, such that $A^u \cap \tilde{C}(A) = \varnothing$, and $\tilde{C}(A^u) \supset A$. Moreover, no edge in A^u is contained in $\bar{C}(A)$ (but some vertices in A^u may be contained in $\bar{C}(A)$).*

Proof Let $C(A^a) = \cup_{i=1}^{M} \ell_i^a$ where the ℓ_i^a are straight line segments. Then $M \leq K_0\lfloor (n+8)^{2/3} \rfloor$ by Lemma 5.80, since A has length n. Order the ℓ_i^a lexicographically with respect to their mid-points, and label the least by 1, the next least by 2 and so on, until the most gets label M. Since A^a is an augmented polygon, exactly four of the ℓ_i^a are parallel to a lattice axis, and they are all of length one. Let A_i^a be that subwalk of A^a with end-points the end-points of ℓ_i^a. By the definition of A^a, ℓ_1^a is a vertical line segment of length one, and it consists of the primary bottom edge of A^a. This is illustrated in Fig. 5.29. Since this edge, and its end-points, is already disjoint with $\bar{C}(A)$, nothing needs to be done here. Invert A_2^a through the mid-point of ℓ_2^a to $R(A_2^a)$, and let $A^{(2)} = (A^a \setminus A_2^a) \cup R(A_2^a)$ be the new polygon. By Lemma 5.83 all the edges in $R(A_2^a)$ are disjoint with $\tilde{C}(A^a)$, and at best have one end-point in $\bar{C}(A^a)$. Continue this process: at the i-th step define $A^i = (A^{(i-1)} \setminus A_i^{(i-1)}) \cup R(A_i^{(i-1)})$ where $A_i^{(i-1)}$ is that subwalk in $A^{(i-1)}$ which contains A_i^a. This is illustrated in Fig. 5.29. Finally,

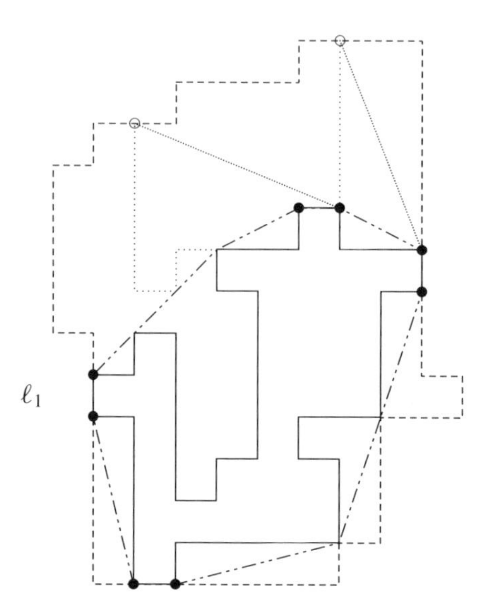

Fig. 5.29: Unfolding a polygon. The dashed lines are images of subwalks in the polygon refelected through the convex hull which is indicated by the dash-dotted lines. The reflections of subwalks take place in increasing lexicographic order of the mid-points of the line segments which makes up the convex hull. Closely dotted lines are images of subwalks which were later reflected again in subsequent reflections, while the two wider spaced dotted lines are parts of the convex hull of a partly unfolded polygon.

the polygon $A^{(M)} = (A^{(M-1)} \setminus A_M^{(M-1)}) \cup R(A_M^{(M-1)})$ is found. Since each $R(A_i^{(i-1)})$ is disjoint with $\tilde{C}(A)$, and since each edge in $R(A_i^{(i)})$ has at most one end-point in $\bar{C}(A)$, the theorem follows. □

This theorem has an important corollary.

Corollary 5.85 *Let A^u be the unfolded image of a polygon A. Then A^u is the image under unfolding of at most $[n^2]^{K_0\lfloor n^{2/3}\rfloor}$ distinct polygons, if the unfolding is done as in Theorem 5.84.*

Proof Each of the polygons to be unfolded to A^u can be reconstructed by choosing pairs of vertices on A^u, and then by reflecting subwalks of A^u between these vertices through the mid-points of the line segment connecting the vertices. A pair of vertices can be chosen in at most $(n+8)^2$ ways, and the reflection for a given pair is done uniquely. Finally, this must be repeated at most $K_0\lfloor n^{2/3}\rfloor$ times, giving rise to at most $(n+8)^2+(n+8)^4+\cdots+[(n+8)^2]^{K_0\lfloor n^{2/3}\rfloor}$ different polygons. Since there are $K_0\lfloor n^{2/3}\rfloor$ terms in this sum, and the last term is the largest, this is at most $K_0\lfloor n^{2/3}\rfloor[(n+8)^2]^{K_0\lfloor n^{2/3}\rfloor}$ polygons. Since the smallest polygon has length 4, increasing K_0 shows that a bound of the form $[n^2]^{K_0\lfloor n^{2/3}\rfloor}$ can be found. □

There is a second important corollary to Theorem 5.84. This corollary will allow the unfolding of a polygon to be disjoint with the closure of its convex hull.

Corollary 5.86 *Let A be a polygon, and let it be augmented to A^a. By unfolding A^a to A^u, and then augmenting and unfolding A^u to find A^{uu}, the following is obtained: $A^{uu} \cap \bar{C}(A) = \varnothing$ and $\tilde{C}(A^{uu}) \supset \bar{C}(A)$.*

Proof Unfold A^a as in the proof of Theorem 5.84 to find A^u. Every edge vw in A^u is either disjoint with $\bar{C}(A)$, in which case it will stay disjoint with $\bar{C}(A)$ if A^u is unfolded again, or has at most one end-point (say v) in $\bar{C}(A)$. Since its other endpoint (w) is not contained in $\bar{C}(A)$, it is the case that v is in $\tilde{C}(A^u)$, and a second unfolding will map v to be disjoint with $\bar{C}(A^u)$, and thus with $\bar{C}(A)$. □

5.6.4 Simple composite polygons

The number of simple composite polygons will be denoted by $p_n^s(m)$ (counted up to translational equivalence). An important assumption about the internal structure of simple composite polygons will be made: let q_m be the number of conformations of the internal structure if it has size m, counted modulo translations, and with the containing polygon disregarded. For example, if the internal structure is a polygon of length m, then q_m is the number of polygons of length m, counted up to translation. I assume that q_m satisfies a generalized supermultiplicative relation of the type

$$q_{m_1} q_{m_2} \leq \sum_{i=-k}^{k} q_{m_1+m_2+i}, \tag{5.114}$$

where k is a constant, and that q_m is bounded from above exponentially in m: $q_m \leq K^m$ for some $K > 1$. This is certainly true for internal structures which are polygons, trees or animals, or disks [161, 200, 219]. An immediate consequence of eqn (5.114) is that the limit

$$\log \xi = \lim_{m \to \infty} \frac{1}{m} \log q_m \tag{5.115}$$

exists (see Lemma A.2), where ξ is the *growth constant* of the objects which are the internal structures in our model. Notice that polygons are also supermultiplicative; the number of polygons of length n will be denoted by p_n, and concatenating them as the containing polygons in Fig. 5.30 gives $p_{n_1} p_{n_2} \leq p_{n_1+n_2}$. In other words, there exists a growth constant for polygons:

$$\log \mu_2 = \lim_{n \to \infty} \frac{1}{n} \log p_n. \tag{5.116}$$

If the internal structure is a polygon, then $\xi = \mu_2$.

In the case of a simple composite model the proof of a supermultiplicative relation is much more complicated. The fact that the concatenation in Fig. 5.30 of two simple composite polygons will give a composite polygon containing two internal structures must be overcome by finding a construction which will concatenate the internal structures as well. It is in this part of the argument that the assumption in eqn (5.114) is important.

The concatenation of simple composite polygons proceeds by the construction in Fig. 5.30. First concatenate the containing polygons. The next step is to concatenate

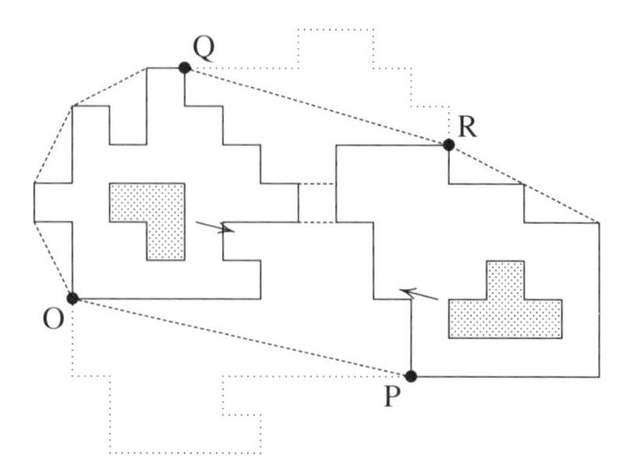

Fig. 5.30: Concatenation of a complex composite model.

the internal structures, and this can only be done with the help of the unfolding of the polygon as in Section 5.6.3. I shall present a proof in the case that the internal structure is a polygon; the other models (trees, animals or disks, and so on) can be handled in a similar way, and the outcome will not differ in an important way from the case of polygons.

Construction 5.87 (Concatenation of simple composite polygons) Let A consist of a polygon containing two internal structures, where the polygon was created in the concatenation of two simple composite polygons. The obstacle to concatenating the two internal structures into one is that one of them, or both, may be entangled with the containing polygon in such a way that it is not possible to translate them into a convenient arrangement which will make the concatenation possible. To disentangle them from the containing polygon, A will be unfolded through its convex hull. The convex hull of the containing polygon A contains line segments which straddle vertices between the constituent polygons from which A was created; these are for example the line segments OP and QR in Fig. 5.30. An important point is that the internal structures in A are both disjoint with OP and QR, and are contained in the wedge or strip formed by elongating OP and QR.

Unfold A twice as in Theorem 5.84 through its convex hull to obtain A^{uu}. Then by Corollary 5.86 there are no vertices in the internal structures B_1 and B_2 adjacent with vertices in A^{uu}, and moreover, $\tilde{C}(A)$ is disjoint with A^{uu}. Since both internal structures are also contained in the convex hull of the concatenated polygons, they are untangled from A^{uu} and they can be translated (as sets in $\mathcal{R}^3$) parallel to a line confined to the wedge made by the lines OP and QR inside the convex hull. Translate them until there are two vertices (one in each) within a unit distance from one another.[18] Since both translated internal structures are still disjoint with OP and QR, they can be pushed

[18] That this can always be done is seen as follows: Translate both B_1 and B_2 normal to OP until they each have a vertex in (say) OP. Then translate both or one of them parallel to OP (one can be translated towards the wider end of the wedge) towards one another until they almost intersect. Finally push them both a short distance off OP on to the lattice. Since the wedge opens in one direction, or is at worst a strip of constant width, this is always possible.

back on to the lattice. The result is that A^{uu} contains two internal structures such that there are two vertices, one in each internal structure, a unit distance apart. The last step is the concatenation of the two internal structures. I shall describe the case for polygons; trees, animals or disks can be handled in a similar way. There are two vertices v and w in the two structures, (one with (say) m_1 edges, and the other with m_2 edges) which are adjacent.

The concatenation of two internal structures which are polygons proceeds by chasing through the diagrams in Fig. 5.31. Either there are two parallel edges (one in each internal structure) a unit distance apart (case (a)), or there are not (case (b)). In case (a) let va and wx be the parallel edges. Remove them and replace them with vw and ax. Then the internal polygons are concatenated and have $m_1 + m_2$ edges. Alternatively, there are no parallel edges. Then there are two vertices v and w a unit distance apart; this situation (or a rotation of it) is in case (b). There are three subcases under case (b). In the first subcase both vertices a and b are unoccupied (case (b1)). Then proceed by deleting pv and wx, and inserting vax and pbw, this gives a polygon with $m_1 + m_2 + 2$ edges. In the second subcase (case (b2)) either a or b are occupied. Without loss of generality, suppose b is occupied, and note that bc cannot be occupied (otherwise there is a pair of parallel edges). Then the only possible case is the one in case (b2). Since a is not occupied,

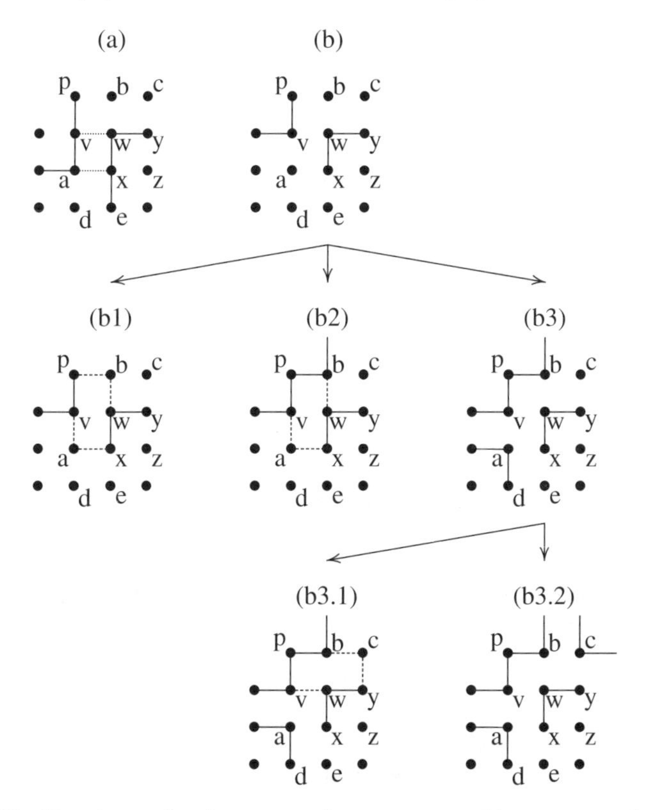

Fig. 5.31: Two internal polygons can be concatenated by a case analysis.

delete wx and vpb, and insert vax and wb to create a polygon of length $m_1 + m_2$. The third subcase is when both a and b are occupied. Since both bc and va are not present, the only possible situation is the one in case (b3). Under case (b3) there are two more subcases. If c is absent, then case (b3.1) is obtained, where c is absent. Delete wy and vpb, and add bcy and vw to obtain a polygon of length $m_1 + m_2$. Otherwise, both c and a are present, and subcase (b3.2) is found. Since xe is absent (there are no parallel edges), xz must be present. Similarly, yz must be present, and so this case can only arise if $m_2 = 4$. Thus, just discard the polygon of length 4 to find a polygon of length $m_1 + m_2 - 4$. This completes the case analysis. Trees, animals and disks can be handled in a similar way. □

Lemma 5.88 *If $p_n^s(m)$ is the number of simple composite polygons consisting of a polygon of length n and containing an internal structure of size m, then there is a constant K_0, and a fixed integer k such that*

$$\sum_{m_1=0}^{m} p_{n_1}^s(m - m_1) p_{n_2}^s(m_1) \le [(n_1 + n_2)^2]^{K_0(n_1+n_2)^{2/3}} \sum_{i=-k}^{k} p_{n_1+n_2+16}^s(m + i).$$

Proof Suppose that a simple composite polygon A_1 of length n_1 and internal structure of size $m - m_1$ is contenated with a simple composite polygon A_2 of length n_2 and internal structure of size m_1. Then the concatenated (and unfolded) polygon has $n_1 + n_2 + 16$ edges, and the concatenated internal structure has at least $m - 4$ and at most $m + 2$ edges, thus choose $k = 4$. By Corollary 5.85 there are at most $[(n_1 + n_2 + 16)^2]^{K_0(n_1+n_2+16)^{2/3}}$ polygons which can be unfolded to the same (augmented) image. The 16 can be omitted since by increasing K_0 we have $n_1 + n_2 \ge 8$. In addition, the internal structures are translated before they are concatenated, so that their top vertices explore at most the entire area of each component polygon. Since the area of A_i is less than n_i^2, another factor $n_1^2 n_2^2$ is sufficient. Now observe that $(n_1 + n_2)^2 \ge n_1 n_2$, and increase K_0 yet again, if necessary, to absorb this factor. □

5.6.5 The free energy of simple composite polygons

The natural definition of the partition function of a model of composite polygons is

$$p_n^s(z) = \sum_{m \ge 0} p_n^s(m) z^m. \tag{5.117}$$

This partition function also satisfies some supermultiplicative inequalities; this follows from Lemma 5.88.

Lemma 5.89 *The partition functions of simple composite polygons satisfy the supermultiplicative relation*

$$p_{n_1}^s(z) p_{n_2}^s(z) \le [(n_1 + n_2)^2]^{K_0(n_1+n_2)^{2/3}} [\phi(z)]^k p_{n_1+n_2+16}^s(z),$$

where $\phi(z) = z + 1 + z^{-1}$.

Proof The inequality is obtained from Lemma 5.88. Multiply the inequality by z^m, and sum over m. Then

$$p^s_{n_1}(z)p^s_{n_2}(z) \leq [(n_1+n_2)^2]^{K_0(n_1+n_2)^{2/3}} \left[\sum_{i=-k}^{k} z^i\right] p^s_{n_1+n_2+16}(z).$$

Observe that $\left[\sum_{i=-k}^{k} z^i\right] \leq [\phi(z)]^k$ to find the result. □

The supermultiplicative relation in Lemma 5.89 suggests the existence of limiting free energies in these models. However, it is also the case that if $z > 1$, then these are infinite. To see this, note that if the containing polygon in a composite model is a square of side-length p, and area p^2, then it may contain an internal structure of size (say) $(p/4)^2$ (the division by 4 gives enough unoccupied vertices to fit the structure into the square). Thus $p_n(z) \geq z^{(p/4)^2}$, and so $\lim_{n\to\infty}[\log p_n(z)]/n \geq \lim_{p\to\infty}[\log z^{(p/4)^2}]/4p = \infty$ if $z > 1$. In other words, the limiting free energy is infinite if $z > 1$. If $z \leq 1$, then the limiting free energy may be finite.

Theorem 5.90 *Suppose that ξ is defined as in eqn (5.115). Then there exists a critical value of z, say z_c, in the interval $[\xi^{-1}, 1]$ such that if $z < z_c$ then there exists a finite limiting free energy in models of composite polygons, with an activity conjugate to the size of the internal structure:*

$$\mathcal{F}_{si}(z) = \lim_{n\to\infty} \frac{1}{n} \log p^s_n(z).$$

Moreover, if $z > z_c$, then this infinite. Lastly, the free energy is a convex function of $\log z$.

Proof That the limit exist is seen from the supermultiplicative inequalities of the partition function in Lemma 5.89 (and Theorem A.3). If $z < z_c$, then

$$p^s_n(z) \leq p_n \sum_{m=0}^{\infty} q_m z^m,$$

and since $q_m = \xi^{m+o(m)}$ by eqn (5.115), the sum above is finite if $z < \xi^{-1}$; thus $\lim_{n\to\infty} \frac{1}{n} \log p_n(z) < \infty$. It has already been shown that the limiting free energy is infinite if $z > 1$. In other words, there exists a critical value of z in the interval $[\xi^{-1}, 1]$. Convexity follows from eqn (2.4). □

A proof that the critical point of a simple composite model containing a polygon is at $z_c = \xi^{-1} = \mu_2^{-1}$ is given in Fig. 5.32, using Corollary 5.78. In fact, arguments of this type also work if the internal structure is a tree, animal or disk.

Theorem 5.91 *The critical value of z is equal to the inverse of the growth constant of the internal structure: $z_c = \xi^{-1}$.* □

In other words, there is a phase transition which corresponds to a divergence in the free energy. This transition occurs when the containing polygon is inflated by the internal

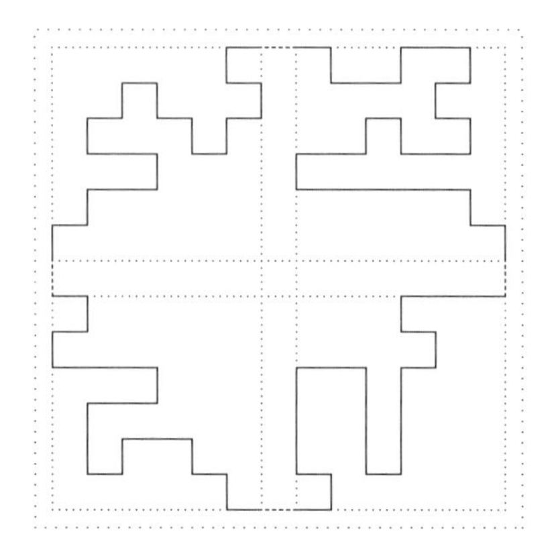

Fig. 5.32: Four walks which cross a square of size $M \times M$ can be arranged as above and put together into a polygon contained in a $(2M + 3) \times (2M + 3)$ square. The result is that $[\varsigma M(z)]^4 \leq p^s_{8M+12}(z)$, and since $\lim_{n\to\infty}[\varsigma M(z)]^{1/M} = \infty$ if $z > \mu_2^{-1}$, $\liminf_{n\to\infty}[p^s_{8M+12}(z)]^{1/M} = \infty$ if $z > \mu_2^{-1}$. In other words, the critical point of a model of simple composite polygons which contains a polygon is at $z_c = \xi^{-1} = \mu_2^{-1}$. This can also be shown in other models of simple composite polygons.

structure, not unlike the transition in inflating vesicles [116] (see Chapter 7). This is best illustrated by introducing the generating function of this model:

$$G_{si}(x, z) = \sum_{n=0}^{\infty} p^s_n(z)x^n. \tag{5.118}$$

Let the radius of convergence of $G_{si}(x, z)$ be $x_c(z)$. The limiting free energy of the model is related to this by

$$\mathcal{F}_{si}(z) = -\log x_z(c). \tag{5.119}$$

The singularity diagram of $G_{si}(x, z)$ is a plot of $x_c(z)$ against z, and a conjectured behaviour of $x_c(z)$ is illustrated in Fig. 5.33. If $z > z_c$, then $x_c(z) = 0$.

If $z > z_c$, then $x_c(z) = 0$. If $z < z_c$, then the internal structure should be small, and its interference with the containing polygon should not be important. In other words, the radius of convergence of the generating function $G_{si}(x, z)$ is given by $x = \mu_2^{-1}$, provided that $z < z_c$. This is seen as follows. Notice that $p_n \leq p^s_n(z) \leq p_n \sum_{m\geq 0} q_m z^m$, so that $\sum_{n\geq 0} p_n x^n \leq G_{si}(x, z) \leq \sum_{n\geq 0} p_n x^n \sum_{m\geq 0} q_m z^m$. For every $z < z_c$ the sum over m is finite, but if $x > \mu_2^{-1}$, then $G_{si}(x, z)$ is infinite.

Theorem 5.92 *The generating function $G_{si}(x, z)$ is finite in a rectangle in the xz-plane, that is, it is finite for all $z < z_c$ and $x < \mu_2^{-1}$. In other words, the radius of convergence of $G_{si}(x, z)$ is equal to μ_2^{-1} if $z < z_c$, and is equal to zero if $z > z_c$.* □

Theorem 5.92 states that the divergence in the generating function as x increases and for $z < z_c$ is due to a divergence in the size of the containing polygon, while the

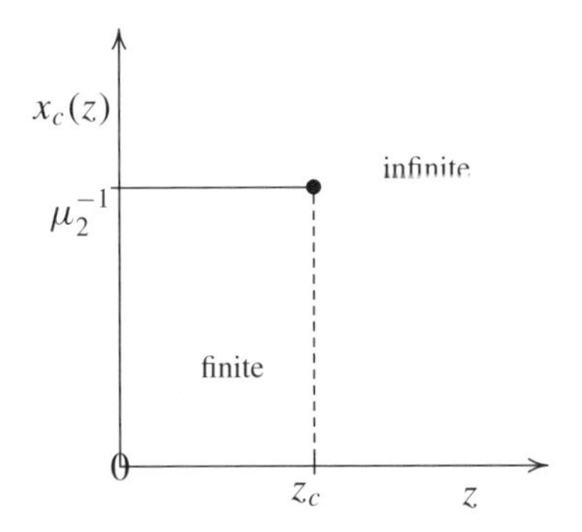

Fig. 5.33: The conjectured singularity diagram of the generating function $G_{si}(x, z)$. The solid line is conjectured to be a line of branch points in $G_{si}(x, z)$, while the dashed line is conjectured to be a line of essential singularities in $G_{si}(x, z)$.

divergence for $x < \mu_2^{-1}$ and increasing z is due to an "inflation" of the polygon by an internal structure of size proportional to the square of the length of the containing polygon. This transition should be first order, and should also not be unlike the transition of a walk crossing a square to its dense phase (where the walk fills the square [246, 364]).

5.6.6 Density functions of simple composite polygons

The existence of the limiting free energy for simple composite models suggests that the growth constant (this will be related to density functions) of composite polygons should be studied. The first result follows directly from the existence of the free energy, and is stated in Theorem 5.93.

Theorem 5.93 *Suppose that the free energy of simple composite polygons, $\mathcal{F}_{si}(z)$, exists and is finite and convex for $z \in [0, z_c)$, but infinite if $z > z_c$. Then there is a function $\sigma_n = o(n)$ such that the function*

$$\mathcal{P}_{si}(\epsilon) = \lim_{n\to\infty} \left[p_n^s(\lfloor \epsilon n \rfloor + \sigma_n)\right]^{1/n},$$

exists for all $0 \leq \epsilon < \infty$. Moreover,

$$\log \mathcal{P}_{si}(\epsilon) = \inf_{z>0}\{\mathcal{F}_{si}(z) - \epsilon \log z\}.$$

$\mathcal{P}_{si}(\epsilon)$ is the density function of this model.

Proof This theorem is a slightly changed version of Theorem 3.19; the exception is that the free energy could be infinite (and allowance must be made for this). Let δ_n be that least value of m (dependent on z) which maximizes $p_n^s(m)z^m$. Then

$$p_n^s(\delta_n)z^{\delta_n} \leq p_n^s(z) \leq n^2 p_n^s(\delta_n)z^{\delta_n}, \tag{†}$$

since there are at most n^2 terms in the partition function. This shows that if the limiting free energy exists and is finite, then the limit

$$\mathcal{F}_{si}(z) = \lim_{n\to\infty} \frac{1}{n} \log p_n^s(\delta_n) z^{\delta_n}, \tag{‡}$$

also exists and is finite. If $\limsup_{n\to\infty} \delta_n/n = \infty$, then eqn (†) shows that $\mathcal{F}_{si}(z) = -\infty$ if $z < 1$; this is a contradiction, so that $\limsup_{n\to\infty} \delta_n/n < \infty$ whenever $z < z_c < 1$. Substract $\epsilon \log z$ from eqn (‡) (where $\epsilon \in [0, \infty)$) to obtain the following

$$\mathcal{F}_{si}(z) - \epsilon \log z = \lim_{n\to\infty} \frac{1}{n} \log p_n^s(\delta_n) + \lim_{n\to\infty} \frac{1}{n}(\delta_n - \lfloor \epsilon n \rfloor) \log z.$$

If the infimum over z is taken, then this shows that $\inf_{z>0}\{\mathcal{F}_{si}(z) - \epsilon \log z\} = \pm\infty$, a contradiction if $z < z_c$, *unless* $\inf_{z>0}\{\lim_{n\to\infty}(1/n)(\delta_n - \lfloor \epsilon n \rfloor)\log z\} = 0$. In other words, $\lim_{n\to\infty} \delta_n/n = \epsilon$, and this limit exists. Thus, $\delta_n = \lfloor \epsilon n \rfloor + o(n)$, and there is a sequence of numbers $\sigma_n = o(n)$ such that

$$\mathcal{P}_{si}(\epsilon) = \lim_{n\to\infty} \left[p_n^s(\lfloor \epsilon n \rfloor + \sigma_n)\right]^{1/n}$$

exists and so that $\log \mathcal{P}_{si}(\epsilon) = \inf_{z>0}\{\mathcal{F}_{si}(z) - \epsilon \log z\}$. □

Observe that the density function $\mathcal{P}_{si}(\epsilon)$ is defined over infimums of z in the interval $[0, z_c]$. Thus, the function $\mathcal{F}_{si}(z) - \epsilon \log z$ for any value of ϵ has an infimum somewhere in the interval $[0, z_c]$; moreover, this infimum is finite whenever it is achieved at a value of $z < z_c$. Thus, the model is only defined on the interval $[0, z_c)$. If it is studied at values of $z > z_c$, then the free energy is infinite, and other methods must be used to describe its thermodynamic behaviour. Therefore, it will be interesting to consider alternative definitions of the density function. For example, consider the density function of composite polygons counted as $p_n^s(\lfloor n^\epsilon \rfloor)$. Suppose that $0 \leq \epsilon < 1$ in the first case, and choose n so large that $n^\epsilon < n/8$. Then every conformation of the internal structure can be fitted into a containing polygon which is a square or almost a square. For these large values of n,

$$p_n \leq p_n^s(\lfloor n^\epsilon \rfloor) \leq n^2 p_n q_{\lfloor n^\epsilon \rfloor}. \tag{5.120}$$

If the $(1/n)$-th power is taken, and $n \to \infty$, then

$$\lim_{n\to\infty} \left[p_n^s(\lfloor n^\epsilon \rfloor)\right]^{1/n} = \mu_2, \tag{5.121}$$

provided that $\epsilon < 1$. If $\epsilon = 1$ then $p_n^s(n) \leq n^2 p_n q_n$, so that

$$\limsup_{n\to\infty} [p_n^s(n)]^{1/n} \leq \mu_2 \xi. \tag{5.122}$$

On the other hand, if $\epsilon > 1$ and a model of composite polygons with internal structures which are polygons, trees or animals is considered, then this limit is infinite. This is seen as follows in the case of spanning polygons. Let s_m be the number of spanning polygons of a square of sidelength $N-1$; then $m = 2N^2$. The construction in Fig. 5.34 shows that $s_{4^k\cdot 16} \geq 4[4[\cdots[s_{16}]^4\cdots]^4]^4$. Thus, $s_{4^k\cdot 16} \geq 4^{1+4+4^2+\cdots+4^k}[s_{16}]^{4^k} = 4^{(4^{k+1}-1)/3}[s_{16}]^{4^k}$. Take logarithms, divide by $4^k\cdot 16$, and taken k to infinity. This shows that $\lim_{k\to\infty}[\log s_{4^k\cdot 16}]/(4^k\cdot 16) \geq [\log 4]/48 + [\log s_16]/16 > 0$. Lastly, if m in s_m is not equal to $4^k\cdot 16$ for some k, then define $k = \lfloor \log(m/16)/\log 4\rfloor$ so that $4^{k+1}\cdot 16 \geq m \geq 4^k\cdot 16$, and notice that $s_m \geq s_{4^k\cdot 16}$. Thus $\liminf_{m\to\infty}(1/m)\log s_m \geq [\log 4]/48 + [\log s_{16}]/16$, and thus $s_m \geq \kappa^m$ for some $\kappa > 1$. In other words, the number of spanning polygons of a square (and thus the number of spanning trees and spanning animals) grows exponentially with its area.

Choose n large enough that $\sqrt{n^\epsilon/2} < n/8$. Then all the spanning polygons of length $\lfloor n^\epsilon\rfloor$ of a square of side-length $\lfloor\sqrt{\lfloor n^\epsilon\rfloor/2}\rfloor$ can be fitted into a (almost) square containing polygon of length n. Thus

$$s_{\lfloor n^\epsilon\rfloor} \leq p_n^s(\lfloor n^\epsilon\rfloor) \leq n^2 p_n q_{\lfloor n^\epsilon\rfloor}, \tag{5.123}$$

and by taking the power $1/n$ and letting $n\to\infty$, it follows that the limit is infinite. If the power $1/n^\epsilon$ is taken instead, then

$$\limsup_{n\to\infty}\left[p_n^s(\lfloor n^\epsilon\rfloor)\right]^{1/n^\epsilon} \leq \xi. \tag{5.124}$$

In other words, this is finite. The existence of this limit is an outstanding issue.

Limits related to the above can be shown to exist; for example, definitions of the integrated density function suggest that one might look at the following:

$$p_n^s(\leq m) = \sum_{j=0}^{m} p_n^s(j). \tag{5.125}$$

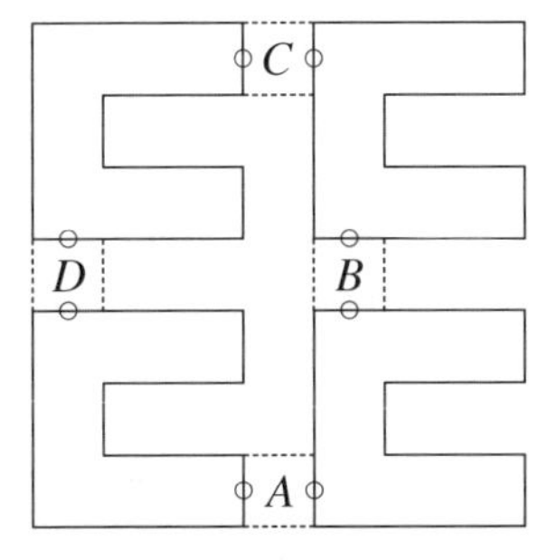

Fig. 5.34: These four spanning polygons of a 3×3 square can be concatenated into a spanning polygon of a 7×7 square by choosing three of the four locations marked by $\{A, B, C, D\}$, deleting the marked edges and replacing them bythe dotted edges. The length of any spanning polygon of a 3×3 square is 16, and the spanning polygon of the 7×7 square has $4\times 16 = 64$ edes. If there are s_M spanning polygons in such a square, then this shows that $s_{64} \geq 4[s_{16}]^4$. Repetition of this construction gives $s_4k._{16} \geq 4[4[\cdots[s_{16}]^4\cdots]^4]^4$.

It follows from arguments similar to those leading to Lemma 5.88 that if $\epsilon \geq 1$, then

$$p^s_{n_1}(\leq \lfloor n_1^\epsilon \rfloor) p^s_{n_2}(\leq \lfloor n_2^\epsilon \rfloor) \leq f(n_1+n_2) p^s_{n_1+n_2+16}(\leq \lfloor (n_1+n_2)^\epsilon \rfloor + 2), \tag{5.126}$$

where $f(n) = 6n^{2K_0 n^{2/3}}$. In addition, the bound $p^s_n(\leq \lfloor n^\epsilon \rfloor) \leq p_n \sum_{m \leq \lfloor n^\epsilon \rfloor} q_m$, is not difficult to derive, so that $p^s_n(\leq \lfloor n^\epsilon \rfloor) \leq K^{\lfloor n^\epsilon \rfloor}$ for some constant K. Thus, the limit

$$\lim_{n\to\infty} \left[p^s_{n-16}(\leq \lfloor n^\epsilon \rfloor - 2) \right]^{1/n^\epsilon} \tag{5.127}$$

exists and is finite, for all $\epsilon \geq 1$ and $\epsilon \leq 2$.

5.6.7 *Interacting models of simple composite polygons*

In this section an interaction between the containing polygon and the internal structure will be introduced. In particular, let $p^s_n(m, k)$ be the number of simple composite polygons with a containing polygon of size n, an internal structure of size m, and with k nearest-neighbour contacts between the internal structure and the containing polygon. The partition function of this model is

$$p^s_n(y, z) = \sum_{m \geq 0, k \geq 0} p^s_n(k, m) y^k z^m. \tag{5.128}$$

Observe that concatenation and unfolding of these simple composite polygons gives, by Lemma 5.88,

$$p^s_{n_1}(0, z) p^s_{n_2}(0, z) \leq f(n_1+n_2) p^s_{n_1+n_2+16}(0, z), \tag{5.129}$$

so that there is a limiting free energy if the interaction between the internal structure and the containing polygon is turned off. The limit

$$\mathcal{F}_{si}(0, z) = \lim_{n\to\infty} \frac{1}{n} \log p^s_n(0, z), \tag{5.130}$$

exists, is finite and convex for all $z \in [0, z_c)$. Arguments similar to those in the previous section show that $z_c = \xi^{-1}$. It is not known whether the free energy exists for all values of y. Instead, if $y < 1$, then it might be argued that if a polygon with an internal structure of size m and with k nearest-neighbour contacts is unfolded twice, then all the nearest-neighbour contacts are destroyed, and by Corollaries 5.85 and 5.86,

$$p^s_n(k, m) \leq [n^2]^{K_0 \lfloor n^{2/3} \rfloor} p^s_{n+16}(0, m). \tag{5.131}$$

Thus, multiplying the above by $y^k z^m$ and summing over k and m,

$$p^s_n(0, z) \leq p^s_n(y, z) \leq [n^2]^{K_0 \lfloor n^{2/3} \rfloor} \left[\frac{p^s_n(0, z)}{1 - y} \right]. \tag{5.132}$$

Take logarithms, divide by n and let $n \to \infty$. Then the following theorem is obtained.

Theorem 5.94 *For all values $z \in [0, \xi^{-1})$ and all $y \in [0, 1]$, the limiting free energy $\mathcal{F}_{si}(y, z)$ exists. Moreover, $\mathcal{F}_{si}(y, z) = \mathcal{F}_{si}(0, z)$ for all $z \in [0, \xi^{-1})$.* □

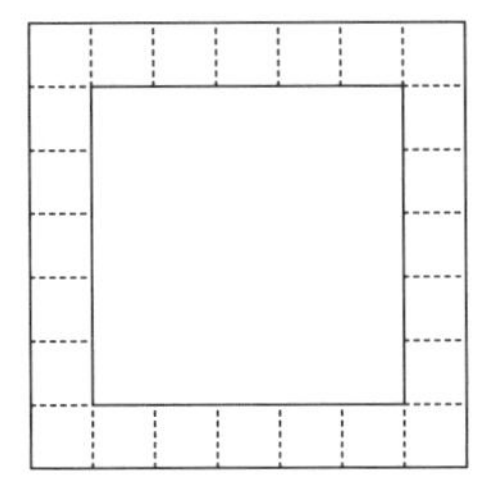

Fig. 5.35: This polygon has length n and $n - 4$ nearest neighbours with an internal structure which is a polygon of length $n - 8$. Contacts are indicated by broken lines.

Theorem 5.94 implies that there exists a critical line $z_c(y)$, with $z_c(y) = \xi^{-1}$ if $y \leq 1$, which corresponds to the inflation of the containing polygon by its internal structure. Assuming that $\mathcal{F}_{si}(y, z)$ exists if $y \geq 1$, it can also be shown that there is a line of non-analyticities $y_c(z) \geq 1$ corresponding to an adsorption transition of the internal structures in the containing polygon. This may be seen by considering Fig. 5.35. The partition function contains a term which corresponds to a square containing a polygon, and an internal structure which consists of at most An edges or unit squares (for some fixed value of A), and with $n - 4$ nearest-neighbour contacts.

It follows from Fig. 5.35 that if $z < z_c$ and $y \geq 1$, then $p_n^s(y, z) \geq y^{n-4} z^{An}$, and thus, if the free energy exists, then

$$\mathcal{F}_{si}(y, z) \geq \log y + A \log z. \tag{5.133}$$

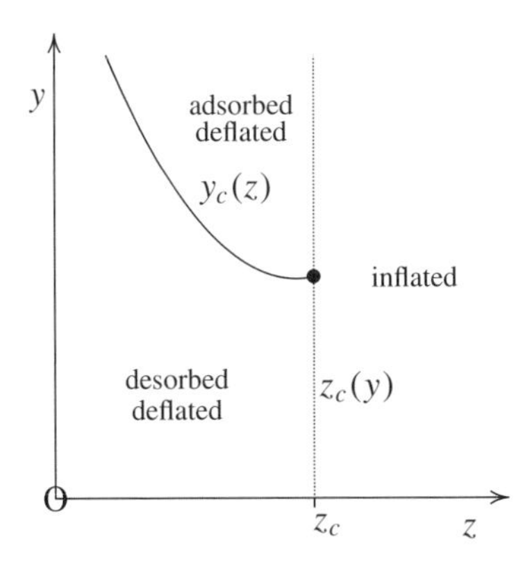

Fig. 5.36: The phase diagram of an interacting model of complex composite polygons. There seems to be three phases; first a desorbed phae of polygons containing internal structures of size no more than the perimerter length of the containing polygon, and an adsorbed phase where the internal structure is adsorbed in the perimeter of the containing polygon. Finally there is an inflated phase, where the internal structure inflates the containing polygon. The critical curve of (presumably continuous) adsorption transitions meet the critical line of (presumably first-order) inflation transitions in a critical end-point.

In other words, for fixed $z < z_c$, there exists a $y_c(z)$ such that $\mathcal{F}_{si}(y, z) > \mathcal{F}_{si}(0, z)$ if $y > y_c(z)$. Thus, there is a second critical line $y_c(z)$ in the phase diagram. Since

$$y \frac{\partial}{\partial y} \mathcal{F}_{si}(y, z) = 0, \quad \text{if } y < y_c(z), \tag{5.134}$$

the critial line $y_c(z)$ corresponds to an adsorption of the internal structure on the containing polygon. Notice also that $y_c(0) = \infty$, since if $z = 0$, then the internal structure is empty, and no adsorption can take place.

The phase diagram of this model is illustrated in Fig. 5.36. A line of first-order transitions separates deflated and inflated composite polygons. The deflated phase is further into adsorbed and desorbed phases by a critical curve of (presumably) continuous adsorption transitions. The line of transitions $z_c(y)$ can be found for all $y \geq 0$, since $\mathcal{F}_{si}(y, z)$ is infinite for any value of y if $z > 1$. If $y < y_c(\xi^{-1})$, then Theorem 5.94 shows that $z_c(y) = \xi^{-1}$, but the argument in Fig. 5.32 shows that $z_c(y) = \xi^{-1}$ for all values of y. This justifies the straight line $z_c(y) = \xi^{-1}$ in Fig. 5.36. The adsorption transition $y_c(z)$ exists, but its representation is hypothetical.

6

Animals and trees

6.1 Lattice animals and trees

In this chapter the limiting free energies and density functions of interacting models of lattice trees and animals will be considered.[1] The first issues are the existence of growth constants, density functions and a pattern theorem. The adsorption of trees will also be considered, and finally the uniform embeddings of graphs, and their relations to walks and polygons in confined geometries will be discussed in Section 6.6.

6.1.1 The growth constant of lattice animals

A *lattice animal* is a connected subgraph of a lattice. In this chapter I shall be interested in *edge-* or *bond-animals* (which have edges as basic building blocks) in the hypercubic lattice. Some of the results in this chapter will also be true for models of *site-animals*, albeit in modified form.[2] I shall also follow the rule of counting animals by edges (rather than by vertices). This will give ensembles of animals weighted differently from ensembles of animals counted by vertices. While all these ensembles are slightly different in their treatment and their properties, one would still expect them to exhibit the same phase behaviour (as a result of universality). The relationship between the various ensembles of animals was studied in references [118, 119, 120, 131, 132]. Let a_n be the number of lattice animals with n edges in the hypercubic lattice, counted up to equivalence under translation. The number a_{n-1} is supermultiplicative, as the concatenation of two animals in Fig. 6.1 demonstrates; this gives the inequality

$$a_n a_m \leq a_{n+m+1}. \tag{6.1}$$

[1] A notational point: free energies of animals will be denoted by $\mathcal{A}(z)$, while density functions will be denoted by $\mathcal{Q}(\epsilon)$. The generating function of a model of animals will be denoted by $A(x)$. In the case of trees, the free energy will be denoted by $\mathcal{T}(z)$, the density function by $\mathcal{O}(\epsilon)$ and the generating function by $T(x)$. This notational convention will avoid confusion with the limiting free energies, density functions and generating functions of polygons and walks considered elsewhere.

[2] A site-animal is a connected section graph of the lattice; two vertices in a site-animal are adjacent (in the animal) if and only if they are adjacent in the lattice; site-animals are said to be *strongly embedded*. Thus, there are no nearest-neighbour contacts in a site-animal, but models of interacting site-animals which have an activity conjugate to the number of pairs of vertices adjacent along a diagonal of a unit square may be defined. Site-trees are acyclic site-animals. Edge-animals are connected subgraphs of the lattice, and trees are acyclic edge-animals. Edge-animals are said to be *weakly embedded*.

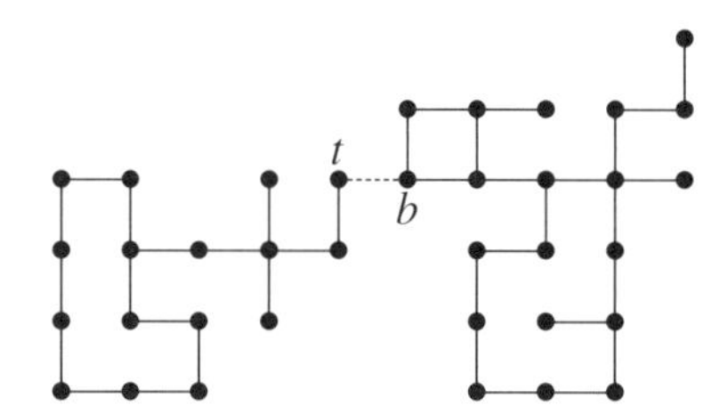

Fig. 6.1: Two animals can be concatenated by translating one animal until its top vertex can be joined with the bottom vertex of the second animal by a single edge. If there are a_n animals on n vertices, then this shows that $a_n a_m \leq a_{n+m+1}$. Notice that no new nearest-neighbour contacts are formed in this construction. If t and b are identified instead, then $a_n a_m \leq a_{n+m}$, but new nearest-neighbour contacts may be formed in the construction.

The concatenation can also be done by identifying the top vertex in one animal with the bottom vertex in the second animal. This gives the inequality

$$a_n a_m \leq a_{n+m}, \tag{6.2}$$

which is slightly stronger than eqn (6.1). In some models it is inappropriate to use the second result; for example, the number of contacts in two animals may change uncontrollably in the concatenation (the concatenation leading to eqn (6.1) preserves the total number of contacts). These inequalities prove the existence of a growth constant for animals, using Lemma A.1, and provided that the function a_n is exponentially bounded in n (there must be a number $K < \infty$ such that $a_n \leq K^n$ for all n). This is proven in Lemma 6.1.

Lemma 6.1 *There exists a finite constant $K > 0$ such that $a_n \leq K^n$ for all $n > 0$.*

Proof Let A be an animal counted by a_n. Label the bottom vertex of A by 1. In the set of vertices nearest neighbour to the vertex labelled by 1, label the lexicographically least vertex by 2. The lexicographically least unlabelled vertex then adjacent with 1 gets labelled by a 3, and so on, until all the nearest neighbours of vertex 1 are labelled. Then consider the unlabelled nearest neighbours of vertex 2, and label those, and so on. At the i-th step, consider the unlabelled nearest neighbours of the vertex labelled by i, and label those in turn with the next available labels. Eventually, the entire animal will be labelled in a unique way, the largest label will be $n+1$ or less, depending on the total number of vertices in the animal. Thus, the animal can be represented by $2dn$ binary digits b_{ij}, where $1 \leq i \leq n$ and $1 \leq j \leq 2d$. The digits b_{ij} correspond to edges incident with the vertex labelled by an i; if there is an edge incident with this vertex in the jth direction, *and the other end of this edge is incident with a vertex with label larger than i*, then define $b_{ij} = 1$, otherwise it is zero.[3] The entire animal can be reconstructed from the b_{ij}: Label a vertex with 1, and consider b_{1j}. If $b_{1j} = 1$, then add an edge in the j-th direction. Label the new vertices created in this way from the lexicographically least by 2, 3, and

[3] Thus, exactly n of these digits will be equal to 1, the rest will be zero.

so on. Consider then the vertex with label 2, and its associated binary digits b_{2j}, and repeat the process. Note that $2dn$ digits are needed, since all the digits associated with the vertex with largest label will be zero. The maximum number of different choices of n 1's from the $2dn$ binary digits is $\binom{2dn}{n}$. But $\lim_{n\to\infty}\binom{2dn}{n}^{1/n} = (2d)^{2d}/(2d-1)^{2d-1}$, so that there is a $K < \infty$ such that $\binom{2dn}{n} < K^n$. □

Together with eqn (6.2) and Lemma A.1 the existence of a growth constant for edge-animals is proven.

Corollary 6.2 *There exists a finite number $\lambda_d > 1$ such that*

$$\lim_{n\to\infty} \frac{1}{n} \log a_n = \sup_{n>0} \frac{1}{n} \log a_n = \log \lambda_d;$$

moreover, $a_n \leq \lambda_d^n$. □

Lattice animals are characterized by several properties. I have already mentioned that there may be *contacts* between adjacent vertices in the lattice which are not adjacent in the animal. In addition, *solvent contacts* are those lattice edges incident with the animal at exactly one end-point. The *perimeter* (also called the "hull") of an animal is the sum of its contacts and solvent contacts: if the animal has k contacts and s solvent contacts, then its perimeter is $\rho = s + k$. In addition, the *cyclomatic index* of an animal is a key property; it is the number of independent cycles in the animal, which is indicated by c (in graph theory it is the dimension of the cycle-space of the animal). The numbers $\{c, k, s\}$ are not independent; in an animal with v vertices and n edges the following relations exist between them (the first is Euler's formula).

$$v - n + c = 1, \qquad 2n + 2k + s = 2dv. \tag{6.3}$$

Animals with cyclomatic index equal to zero are called *lattice trees*. In this model there may still be contacts and solvent contacts, related as in eqn (6.3) with $c = 0$. It is generally believed that lattice animals have the same critical exponents as lattice trees; thus, if a model in which the cyclomatic index does not play a role is considered, then lattice trees are often used instead. Concatenation of lattice trees (Fig. 6.1), by using the bound in Lemma 6.1, also shows that there is a growth constant τ_d. Let t_n be the number of lattice trees with n edges.

Corollary 6.3 *There exists a finite number $\tau_d > 1$ such that*

$$\lim_{n\to\infty} \frac{1}{n} \log t_n = \sup_{n>0} \frac{1}{n} \log t_n = \log \tau_d,$$

and moreover, $t_n \leq \tau_d^n$. □

The generating functions of animals and trees are defined by $A(x) = \sum_{n\geq 0} a_n x^n$ and $T(x) = \sum_{n\geq 0} t_n x^n$. Generally, it is believed that the asymptotic behaviour of t_n is as proposed in eqn (1.32), and animals have the same asymptotic behaviour, but only with τ_d replaced by λ_d. The entropic exponent of both models is θ, given in Table 1.2, and with Flory values in eqn (1.37). The "exact values" for θ in two and three dimensions follows also from a dimensional reduction to an Ising model in an imaginary field in two dimensions less [290]. In Section 6.2 it will be shown that there is a pattern theorem for animals; a consequence of this will be that $\tau_d < \lambda_d$.

6.1.2 A submultiplicative relation for t_n

The upper bound $t_n \le \tau_d^n$ on the number of lattice trees (Corollary 6.3) is due to the supermultiplicative nature of lattice trees. It is also possible to find a lower bound on t_n; this will be done by showing that t_n satisfies a certain generalized submultiplicative inequality [186]. The proof of this fact uses the notion of a "branch" which is defined as follows. A *branch* S of a tree T is a subtree of T which is joined to $T \setminus S$ at a single vertex v. Thus $T \setminus S$ is itself also a tree. The key idea is that it is possible to pick "sizable" (in the sense explained in Lemma 6.4 below) branches from any tree.

Lemma 6.4 *Let T be a tree of size n, and let $m \le n$ be a positive integer. Then there is a branch S in T, of size k, such that $\lceil m/2d \rceil \le k \le m$.*

Proof If $m = n$ then T is itself the branch, so suppose that $m < n$. Choose a vertex p_1 in T. Incident with p_1 are $2d$ branches of T (some of them could be empty). Let these branches be $\{U_i\}_{i=1}^{2d}$, and suppose that U_i has u_i edges. Obviously $\sum_{i=1}^{2d} u_i = n$. If $u_i < m/2d$ for all i, then $\sum_{i=1}^{2d} u_i < m < n$, a contradiction, and so there is at least one branch U_i of size $u_i \ge m/2d$. Since u_i is an integer, this shows that $u_i \ge \lceil m/2d \rceil$. If $u_i \le m$ as well, then the proof is completed, so suppose that $u_i > m$. Without loss of generality, let $i = 1$.

To finish the proof, it will be shown that a branch B of size b can be found, such that $u_1 > b \ge \lceil m/2d \rceil$. If this process is repeated, then eventually a branch of size less than or equal to m and greater than or equal to $\lceil m/2d \rceil$ must be found. Let p_2 be any vertex in U_1, but distinct from p_1. Let the branches incident with p_2 be $\{V_i\}_{i=1}^{2d}$, where V_{2d} is the branch which contains p_1. Let the size of branch V_i be v_i. By construction, each branch V_i is a subtree in U_1, except for V_{2d}. Thus $\sum_{i=1}^{2d-1} v_i < u_1$. The proof now proceeds via a case analysis.

Case (1) Suppose that $\sum_{i=1}^{2d-1} v_i \ge m$. Then there is an i such that $v_i \ge m/(2d-1)$, and thus $v_i \ge \lceil m/2d \rceil$. Let this branch be B of size $b = v_i$. Then $u_1 > b \ge \lceil m/2d \rceil$.

Case (2) Suppose that $\sum_{i=1}^{2d-1} v_i < m$. There are two subcases which must be considered. In case (2a) suppose that $\sum_{i=1}^{2d-1} v_i \ge \lceil m/2d \rceil$. Then the union of all the branches $B = \bigcup_{i=1}^{2d-1} V_i$ is a branch in the desired size range. In case (2b) suppose that $\sum_{i=1}^{2d-1} v_i < \lceil m/2d \rceil$. Let $B' = \cup_{i=1}^{2d-1} V_i$, and now add edges to B' on the path from p_2 to p_1. The first edge to be added to B' is incident with p_2. Let the newly augmented branch be B_1 and let it be joined to the rest of the tree in vertex p_2^1. Incident with p_2^1 there are $2d-2$ branches of T (none of which contains p_1, and some of which may be empty). Augment B_1 by adding these to it one-by-one. There are three possible outcomes: (i) If all these branches are added to B_1, then it is still smaller than $\lceil m/2d \rceil$. Rename this branch B', and repeat case (2b). (ii) A branch may be added which increases the size of B' from below $\lceil m/2d \rceil$ to above m. In this case the added branch has size b at least $m - \lceil m/2d \rceil > \lceil m/2d \rceil$, and so, since it is a subtree in U_1, $u_1 > b \ge \lceil m/2d \rceil$. Thus a branch smaller than U_1 has been found. (iii) The only other possible outcome is that B' will be in the desired size-range once a number of branches has been added. This completes the proof. □

The consequences of this lemma are collected in the next corollary.

Corollary 6.5 *Let T be a tree of size n and let $m \leq n$ be a positive integer. Then k branches $\{B_i\}_{i=1}^k$ (where branch B_i has b_i edges) can be pruned from T such that*

(1) $\sum_{i=1}^k b_i = m$,
(2) $m - \sum_{i=1}^l b_i \geq b_{l+1} \geq \lceil (m - \sum_{i=1}^l b_i)/2d \rceil$,
(3) $k \leq [\log m]/[\log(2d/(2d-1))]$.

Proof The aim is to prune k branches from T such that a total of m edges are removed. The worst possible situation is encountered if the least number edges are pruned from the tree at each step. That is, if p edges must be removed, then only $\lceil p/2d \rceil$ are cut. In these circumstances, suppose that p_j edges must be removed at the j-th step, and that $\lceil p_j/2d \rceil$ are actually removed. At the $(j+1)$-th step at most $p_j - \lceil p_j/2d \rceil = \lfloor (2d-1)p_j/2d \rfloor$ edges remain to be removed. Iterating this shows that after k steps, at most $\lfloor (2d-1)\lfloor (2d-1)\lfloor \cdots \lfloor (2d-1)m/2d \rfloor \cdots \rfloor /2d \rfloor /2d \rfloor$ edges remain to be removed. Thus, m edges will be removed if this expression is equal to 1 after $(k-1)$ iterations, and 0 after k iterations. But after $k-1$ iterations, $m[(2d-1)/2d]^{k-1} \geq \lfloor (2d-1)\lfloor (2d-1)\lfloor \cdots \lfloor (2d-1)m/2d \rfloor \cdots \rfloor /2d \rfloor /2d \rfloor = 1$. Thus it is sufficent that k is large enough that $m[(2d-1)/2d]^k \leq 1$. Thus $k \leq [\log m]/[\log(2d/(2d-1))]$ is sufficient; this shows (3). (1) is seen by noting that it is possible to remove one edge at any stage, and Lemma 6.4 shows that it is never necessary to remove more edges than necessary. (2) is an immediate corollary of Lemma 6.4. □

Corollary 6.5 makes it possible to prove a generalized submultiplicative inequality for t_n.

Theorem 6.6 *The number t_n satisfies the following submultiplicative inequality in any number of dimensions $d \geq 2$:*

$$t_{n+m} \leq [(n+m)m^2]^{\alpha \log m} t_n t_m,$$

so that

$$t_n t_m \geq (n+m)^{-3\alpha \log(n+m)} t_{n+m},$$

where $\alpha = (\log 8)/\log(2d/(2d-1))$.

Proof Let T be a tree of size $(n+m)$. Apply Corollary 6.5 to prune at most $\alpha \log m$ branches from T containing exactly m edges. Any one of these branches can be put back in T in at most $(n+m+1)(m+1)$ ways (select a vertex in each, and identify them). Concatenate the branches as they are pruned into a new tree containing m edges. The total number of ways this tree can be cut into the original branches is at most $(m+1)^{\alpha \log m}$, since the tree must be cut in at most $\alpha \log m$ places, and for each cut, there is a selection of at most $m+1$ vertices to choose for a cut. Thus $t_{n+m} \leq [(n+m+1)(m+1)^2]^{\alpha \log m} t_n t_m$. Now replace $(n+m+1)$ by $2(n+m)$ and $(m+1)$ by $2m$ to obtain the claimed inequality. □

Apply Theorem A.3 to the generalized submultiplicative inequality in Theorem 6.6. Identify the function $g(n)$ in Theorem A.3 as $\log g(n) = 3\alpha[\log n]^2$. Notice furthermore that

$$\sum_{m=2n}^{\infty} \frac{(\log m)^2}{m(m+1)} \leq \int_{2n}^{\infty} \frac{[\log(x-1)]^2}{x(x-1)} dx \leq \int_{n}^{\infty} \left(\frac{\log x}{x}\right)^2 dx.$$

This integral can be computed, and using this bound in Theorem A.3 shows that

$$\frac{\log t_n}{n} \geq \log \tau_d - 9\alpha \frac{[\log n]^2}{n} - 24\alpha \frac{\log n}{n} - \frac{24\alpha}{n}. \tag{6.4}$$

Thus, the number of trees can be bounded as in Theorem 6.7.

Theorem 6.7 *The number of trees t_n in the hypercubic lattice is bounded by*

$$e^{-24\alpha} n^{-24\alpha} e^{-9\alpha[\log n]^2} \tau_d^n \leq t_n \leq \tau_d^n,$$

where $\alpha = (\log 8)/\log(2d/(2d-1))$. □

A bound of this nature can also be derived for animals with a fixed number of cycles [186]. In high dimensions it is known that $t_n \sim C_0 n^{-\theta} \tau_d^n$ [168, 169, 170, 172], but in low dimensions, the best bounds are those given in Theorem 6.7.

6.2 Pattern theorems and interacting models of lattice animals

In this section I present a pattern theorem for lattice animals. The technique of proof is due to N. Madras [248], and the same arguments will also show that there is a pattern theorem for lattice trees. The pattern theorem here will be applicable to interacting models of lattice animals, so define $a_n(k)$ to be the number of lattice animals with n edges and energy k. I shall also assume that $a_{n-1}(k)$ satisfies Assumptions 3.8, and that eqn (3.7) is due to a concatenation of animals as in Fig. 6.1. It will also be necessary to restrict the types of energies in the model of animals here. In particular, a *local energy* should be defined; but in this case, eqn (3.7) already limits the types of energies which may be used. That is, if animals of sizes n_1 and n_2 are concatenated, then the change in energy is at most a constant. In the case of animals or trees, the following definition of a local energy is appropriate.

Definition 6.8 (Local energy) Let A_1 and A_2 be two animals of size n_1 and n_2, and energy k_1 and k_2, respectively. If the energy of any such A_1 changes by at most $O(1)$ if an edge is appended on any vertex of A_1, and if the total energy of A_1 and A_2 changes by at most $O(1)$ under concatenation as in Fig. 6.1, then the energy is local. Thus, if an energy is local, then

$$a_{n_1}(k_1) a_{n_2}(k_2) \leq \sum_{i=-q}^{q} a_{n_1+n_2+1}(k+i),$$

where q is a constant, and

$$a_n(z) \leq [\phi(z)]^K a_{n+1}(z),$$

if $\phi(z) = z + 1 + z^{-1}$ and for some constant K. □

A second important ingredient in the proof of a pattern theorem is the definition of a pattern. This will be done with the help of a *box-animal*, which is illustrated in Fig. 6.2. In two dimensions it is the animal which includes all the edges in the perimeter of a rectangle. In three dimensions it consists of all the edges in the surface of a rectangular box, and so on.

With the definition of a *box-animal*, prime patterns can be defined. A *pattern* in an animal is a connected subanimal. Notice that a pattern is as much the presence as the absence of edges, or both. A pattern P is a *prime pattern* if it cannot be created by concatenating any two animals as in Fig. 6.1 and if there exists a smallest box-animal B such that P can be translated into the interior of B, *and* some edges can be added to P to join it to a vertex of B.

Let P be a prime pattern, and let $a_n(k; mP)$ be the number of animals of size n, local energy k and which contains P exactly m times. The partition function of this model is

$$a_n(z; mP) = \sum_{k \geq 0} a_n(k, mP) z^k. \tag{6.5}$$

Concatenate an animal with energy k_1 and m_1 occurrences of P with an animal of energy $k - k_1$ which contains P a total of $m - m_1$ times. From Assumptions 3.8 and eqn (3.7) this gives

$$\sum_{k_1=0}^{k} \sum_{m_1=0}^{m} a_{n_1}(k_1; m_1 P) a_{n_2}(k - k_1; (m - m_1)P) \leq \sum_{i=-q}^{q} a_{n_1+n_2+1}(k + i; mP). \tag{6.6}$$

Multiply this by z^k and sum over k to obtain the following supermultiplicative inequality:

$$\sum_{m_1=0}^{m} a_{n_1}(z; m_1 P) a_{n_2}(z; (m - m_1)P) \leq \left[\sum_{i=-q}^{q} z^i\right] a_{n_1+n_2+1}(z; mP). \tag{6.7}$$

Let $\phi(z) = z + 1 + 1/z$. Then

$$\sum_{m_1=0}^{m} a_{n_1}(z; m_1 P) a_{n_2}(z; (m - m_1)P) \leq [\phi(z)]^q a_{n_1+n_2+1}(z; mP). \tag{6.8}$$

This result can be used to show that there is a density function for P for every finite value of z; the argument uses Theorem 3.19. Multiply eqn (6.8) by y^m and

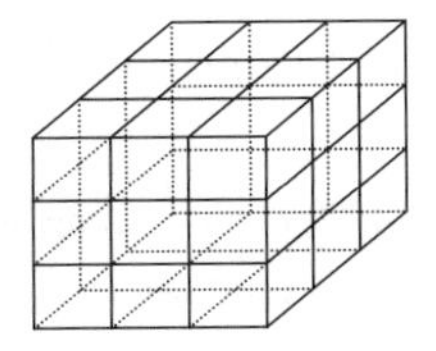

Fig. 6.2: A box-animal.

sum over m. This gives $a_{n_1}(z, y)a_{n_2}(z, y) \leq [\phi(z)]^q a_{n_1+n_2+1}(z, y)$, where $a_n(z, y) = \sum_{m\geq 0} a_n(z; mP)y^m$. By Theorem A.3 there exists a limiting free energy defined by

$$\mathcal{A}_a(z, y) = \lim_{n\to\infty} \frac{1}{n} \log a_n(z, y), \tag{6.9}$$

and so by Theorem 3.19, there is a sequence of integers $\sigma_n = o(n)$ such that the density function is defined by

$$\mathcal{Q}_a(z; \epsilon) = \lim_{n\to\infty} \left[a_n(z; (\lfloor \epsilon n\rfloor + \sigma_n)P)\right]^{1/n}. \tag{6.10}$$

This limit exists for any value of $z \in [0, \infty)$ (this also follows from eqn (6.8) and Theorem 3.16). This result is generally applicable to models of interacting animals (and trees) provided that under concatenation, Assumptions 3.8 are satisfied and provided that the model has a local energy. However, notice that eqn (6.8) also shows that $a_{n_1}(z; m_1 P)a_{n_2}(z; m_2 P) \leq [\phi(z)]^q a_{n_1+n_2+1}(z; (m_1 + m_2)P)$, and so $a_{n-1}(z; mP)/[\phi(z)]^q$ satisfies Assumptions 3.1 and $\sigma_n = 0$ in eqn (6.10).

Theorem 6.9 *Let $a_n(k)$ be the number of animals with a local energy k and n edges, and suppose that $a_n(k)$ satisfies Assumptions 3.8. If P is a prime pattern, and $a_n(z; mP)$ is the partition function of a model of animals containing P exactly m times, and where the activity z is conjugate to the energy k, then there is a density function for P defined by*

$$\mathcal{Q}_a(z; \epsilon) = \lim_{n\to\infty} [a_n(z; \lfloor \epsilon n\rfloor P)]^{1/n}. \qquad \square$$

A density function in a model for trees was examined in reference [238]. A pattern theorem for interacting models of animals with a local energy can be expressed as a property of the density function; in particular, it must be shown that $\mathcal{Q}_a(z; \epsilon)$ is concave and strictly increasing in some interval $(0, \epsilon_c]$, where $\epsilon_c > 0$, and where the minimum density of the prime pattern is assumed to be zero. Since $a_n(k)$ satisfies Assumptions 3.8, it also follows that there is a limiting free energy, which is defined by

$$\mathcal{A}_a(z) = \lim_{n\to\infty} \frac{1}{n} \log a_n(z). \tag{6.11}$$

Comparison of this to eqn (6.10) shows that

$$\mathcal{Q}_a(z; \epsilon) \leq e^{\mathcal{A}_a(z)}. \tag{6.12}$$

A pattern theorem for the prime pattern P may be expressed as a property of the density function $\mathcal{Q}_a(z; \epsilon)$.

Theorem 6.10 *There exists an $\epsilon > 0$ such that $\mathcal{Q}_a(z; 0) < \mathcal{Q}_a(z; \epsilon)$, for all finite values of z.*

Proof Let P be a prime pattern and let A be an animal counted by $a_n(k, 0P)$. Let B be the smallest box-animal which may contain P in its interior, and suppose that the volume (number of vertices) of B is V. If P contains M edges, then $V \geq M/2d$. The animal A contains at least $\lceil n/d\rceil$ vertices, and it can be intersected with at least $\lfloor n/dV\rfloor$

disjoint copies of B. Choose m such copies of B from $\lfloor n/dV \rfloor$, and intersect A with these, deleting all edges of A in the interior of each B, but retaining all the edges (and not just only those in A) in the boundary of B. This leaves a connected animal, which contains m copies of the box-animal B, all with empty interiors. The prime pattern P can be inserted into each of the Bs. Insert P, and join it to the boundary of B by adding extra edges (while care is taken not to violate the restriction on the presence and absence of edges in P). Since there are V vertices in each B, this may change the number of edges in A by as many as mVd. Moreover, since the energy is local, it may be assumed that the energy of A changes by at most a constant K for each added or deleted edge. Since potentially all the edges in the copies of B may be removed, or added, to α, the total change in energy may be as large as $mKVd$. Thus

$$\binom{\lfloor n/dV \rfloor}{m} a_n(k, 0P) \leq \sum_{l=-mVd}^{mVd} \left[\sum_{j=-mKVd}^{mKVd} a_{n+l}(k+j, mP) \right].$$

Multiply this by z^k and sum over k, and define $\phi(z) = z + 1 + z^{-1}$. This gives

$$\binom{\lfloor n/dV \rfloor}{m} a_n(z; 0P) \leq [\phi(z)]^{mKVd} \sum_{l=-mVd}^{mVd} a_{n+l}(z; mP).$$

Let $m = \lfloor \epsilon n \rfloor$ (with $\epsilon < 1/dV$), take the power $1/n$ and let $n \to \infty$. From eqn (6.12), $\mathcal{Q}_a(z; \epsilon) \leq e^{\mathcal{A}_a(z)}$, so that if the fact that the energy is local is used (see Definition 6.8), then

$$\frac{(1/dV)^{1/dV}}{\epsilon^\epsilon (1/dV - \epsilon)^{1/dV-\epsilon}} \mathcal{Q}_a(z; 0) \leq [\phi(z)]^{\epsilon KVd} e^{\epsilon Vd \mathcal{A}_a(z)} \mathcal{Q}_a(z; \epsilon).$$

Define

$$\psi(z) = \left[[\phi(z)]^K e^{\mathcal{A}_a(z)} \right]^{-Vd}. \qquad (\dagger)$$

Then

$$\frac{(1/dV)^{1/dV} [\psi(z)]^\epsilon}{\epsilon^\epsilon (1/dV - \epsilon)^{1/dV-\epsilon}} \mathcal{Q}_a(z; 0) \leq \mathcal{Q}_a(z; \epsilon).$$

Now maximize the left-hand side by choosing $\epsilon_* = \psi(z)/(Vd(1 + \psi(z)))$. This gives

$$(1 + \psi(z))^{1/Vd} \mathcal{Q}_a(z; 0) \leq \mathcal{Q}_a(z; \epsilon_*).$$

Thus, since $0 < \epsilon_* \leq 1/dV$, the proof is completed. □

This theorem is not a "true" pattern theorem yet; a stronger version of it must still be proven, to produce a result similar to Corollary 5.43. Theorem 6.10 can be strengthened by showing that the density function of a prime pattern P, $\mathcal{Q}_a(z; \epsilon)$, is strictly increasing in some interval $[0, \epsilon_c]$. The more important result that "exponentially many" animals contain the pattern P at least $\lfloor \epsilon n \rfloor$ times, if ϵ is small enough, will then follow as a corollary.

In the proof of Theorem 6.10, let A instead be an animal counted by $a_n(k, \lfloor \epsilon n \rfloor P)$. Find again an intersection of A with at least $\lfloor n/dV \rfloor$ disjoint copies of the box B. At most $2^d\lfloor\epsilon n\rfloor$ of these boxes intersect a copy of P,[4] so $\lfloor\delta n\rfloor$ of these boxes (none of which contains a copy of P) may be selected from $(\lfloor n/Vd\rfloor - 2^d\lfloor\epsilon n\rfloor)$ to create new copies of P. Thus

$$\binom{(\lfloor n/Vd\rfloor - 2^d\lfloor\epsilon n\rfloor)}{\lfloor\delta n\rfloor} a_n(k, \lfloor\epsilon n\rfloor P)$$
$$\leq \sum_{i=-\lfloor\delta n\rfloor Vd}^{\lfloor\delta n\rfloor Vd}\left[\sum_{j=-\lfloor\delta n\rfloor KVd}^{\lfloor\delta n\rfloor KVd} a_{n+i}(k+j; (\lfloor\epsilon n\rfloor + \lfloor\delta n\rfloor)P)\right]. \tag{6.13}$$

Multiply this by z^k and sum over k. Take the power $1/n$ and let $n \to \infty$. Define $\psi(z)$ as in Theorem 6.10, eqn (†). This gives

$$\frac{(1/dV - 2^d\epsilon)^{1/dV - 2^d\epsilon}\,[\psi(z)]^\delta}{\delta^\delta(1/dV - 2^d\epsilon - \delta)^{1/dV - 2^d\epsilon - \delta}}\,\mathcal{Q}_a(z;\epsilon) \leq \mathcal{Q}_a(z;\epsilon+\delta). \tag{6.14}$$

Maximize the left-hand side as before. This occurs when $\delta_* = (1/dV - 2^d\epsilon)\psi(z)/(1+\psi(z))$, and then

$$(1+\psi(z))^{1/dV - 2^d\epsilon}\,\mathcal{Q}_a(z;\epsilon) \leq \mathcal{Q}_a(z;\epsilon+\delta_*). \tag{6.15}$$

Thus, if $2^d\epsilon < 1/dV$, then such a δ_* can be found. Since $\log\mathcal{Q}_a(z;\epsilon)$ is a concave function of ϵ, this implies that $\mathcal{Q}_a(z;\epsilon)$ is strictly increasing in some interval $[0,\epsilon_c]$, provided that $\epsilon_c > 0$ is small enough.

Corollary 6.11 *There exists an $\epsilon_c > 0$ such that $\mathcal{Q}_a(z;\epsilon_c) = \sup_{\epsilon\geq 0}\mathcal{Q}_a(z;\epsilon)$, and $\mathcal{Q}_a(z;\epsilon) < \mathcal{Q}_a(z;\epsilon_c)$ whenever $\epsilon < \epsilon_c$. In other words, since $\log\mathcal{Q}_a(z;\epsilon)$ is concave in $[0,\epsilon_c]$, it is also strictly increasing in this interval.* □

The above is only true for prime patterns, but notice that any subanimal can be a subanimal in a prime pattern. Thus, Corollary 6.11 is true for any pattern (which can occur once) in the animal. Moreover, the empty pattern may be defined as the absence of edges in some given (connected) region. This pattern will also occur with a certain density, and thus, given a certain number N, there is a density of cavities in the animal which will contain a hypercube of volume N^d. A consequence is a stronger form of the pattern theorem.

Theorem 6.12 *Let P be any pattern. Then there exists a $k(z) > 0$, an $N_0 > 0$ and a number ϵ_0 such that for all $\epsilon < \epsilon_0$*

$$a_n(z; \lfloor\epsilon n\rfloor P) \leq a_n(z; \leq\lfloor\epsilon n\rfloor P) < e^{-k(z)n}\,a_n(z),$$

for all $n > N_0$.

---

[4] Depending on the arrangement of the boxes, as many as 2^d may intersect a given P.

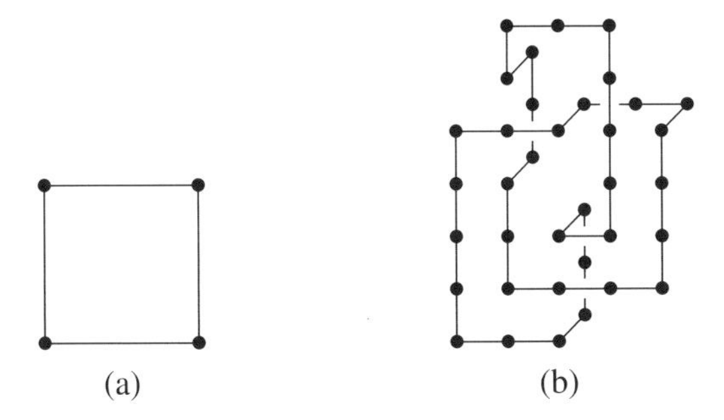

Fig. 6.3: (a) A 4-cycle. (b) A prime pattern which is a knotted cycle.

Proof Since $\mathcal{Q}_a(z;\epsilon) < \mathcal{Q}_a(z;\epsilon_c)$, whenever $\epsilon < \epsilon_c$, note from Corollary 6.11 that

$$\limsup_{n\to\infty} \frac{1}{n} \log \left[\frac{a_n(z; \leq\lfloor \epsilon n \rfloor)}{a_n(z; \lfloor \epsilon_c n \rfloor)} \right] = \log \left[\frac{\mathcal{Q}_a(z; \leq\epsilon)}{\mathcal{Q}_a(z; \epsilon_c)} \right] < 0.$$

Choose the number

$$k(z) = \frac{1}{2} \left| \log \left[\frac{\mathcal{Q}(z; \leq\epsilon)}{\mathcal{Q}(z; \epsilon_c)} \right] \right| > 0.$$

Then there exists an N_0 such that $a_n(z; \leq\lfloor \epsilon n \rfloor) < e^{-kn} a_n(z; \lfloor \epsilon_c n \rfloor) = e^{-kn} a_n(z)$. □

It is an important fact that this pattern theorem applies equally to lattice trees, and with some additional work, similar results can be obtained for other models of lattice animals, such as strongly embedded animals, or animals counted by vertices, and so on; but a proof for site trees may be somewhat harder [248].

The energy of most interesting models of self-interacting models are local in the sense of Definition 6.8. This includes the models of animals with an activity conjugate to the number of nearest-neighbour contacts, or to the number of cycles, in the animal, as well as animals adsorbing in a plane. Thus, the pattern theorem in Corollary 6.11 applies to these models.

There are consequences of the pattern theorem. In particular, choose P to be the prime pattern in Fig. 6.3(a). This pattern is a 4-cycle, and by Theorem 6.12, an exponential fraction of all animals of size n will contain that cycle at least $\lfloor \epsilon n \rfloor$ times, for a small positive value of ϵ. Thus, the mean cyclomatic index of animals will grow as $O(n)$ as the size of the animals is increased. By choosing a pattern such as in Fig. 6.3(b) it also follows that animals in three dimensions will contain cycles which are arbitrarily complex knots. These consequences are true for any finite value of the activity z in an interacting model.

6.3 Collapsing animals

Animals may also be studied with a nearest-neighbour contact energy, or with an energy defined by the number of cycles. In both these models it is thought that there is a collapse transition to a phase of collapsed animals; the critical point is called a θ-point, and it is believed to be tricritical. In this section these models will be considered. It will also turn out that there is a connection to percolation [136, 147].

6.3.1 The cycle model

Let $a_n(c)$ be the number of animals in the hypercubic lattice with c cycles, and define the partition function $a_n(z) = \sum_{c\geq 0} a_n(c)z^c$. Concatenation of animals in this ensemble by the construction in Fig. 6.1 and eqn (6.2) shows that this is a supermultiplicative model:

$$a_{n_1}(c_1)a_{n_2}(c_2) \leq a_{n_1+n_2}(c_1+c_2), \tag{6.16}$$

so that $a_n(c)$ satisfies Assumptions 3.1, and there exists a density function $\mathcal{Q}_c(\epsilon)$ for the number of cycles, defined in Theorem 3.4 by

$$\mathcal{Q}_c(\epsilon) = \lim_{n\to\infty} [a_n(\lfloor \epsilon n \rfloor)]^{1/n}, \tag{6.17}$$

and a limiting free energy as defined in Theorem 3.17. In this section I shall consider the properties of this density function.

Let A be an animal with n edges in the hypercubic lattice. The number of cycles in A is a maximum if it is a hypercube. Since the largest hypercube which can be constructed using at most n edges has at least $\lfloor n/d \rfloor$ vertices, and has side length at least $\lfloor \lfloor n/d \rfloor^{1/d} \rfloor$, and at most $\lceil n/d \rceil$ vertices and side length at most $\lceil \lceil n/d \rceil^{1/d} \rceil$, an animal is said to be *almost a hypercube* if it contains a hypercube of side length $\lfloor \lfloor n/d \rfloor^{1/d} \rfloor$ and is itself contained in a hypercube of side length $\lceil \lceil n/d \rceil^{1/d} \rceil$. Euler's theorem (eqn (6.2)) states that the number of cycles in a graph is $c = n - v + 1$, where n is the number of edges and v is the number of vertices. Thus, since the degree of a vertex is at most $2d$, it follows from handshaking that $2n = \sum(\text{degrees}) \leq 2dv$, or $v \geq \lfloor n/d \rfloor$. Therefore the animal has at most $n - \lfloor n/d \rfloor + 1$ cycles. On the other hand, animals which are almost hypercubes have at most $\lceil n/d \rceil + 0(\lceil n/d \rceil^{(d-1)/d})$ vertices. Define the maximum number of cycles in animals of size n to be $C(n)$, then

$$n - \lceil n/d \rceil - 0(\lceil n/d \rceil^{(d-1)/d}) + 1 \leq C(n) \leq n - \lfloor n/d \rfloor + 1, \tag{6.18}$$

and therefore

$$\lim_{n\to\infty} [C(n)]/n = (d-1)/d. \tag{6.19}$$

Thus, the density function $\mathcal{Q}_c(\epsilon)$ is defined on the interval $(0, (d-1)/d)$. Since $\log \mathcal{Q}_c(\epsilon)$ is concave, it is also continuous on $(0, (d-1)/d)$. It is continuous at $\epsilon = 0$ as well. This fact is proven by using the following lemma [250].

Lemma 6.13 *It is possible to delete k cycles in animals with $c+k$ cycles to find that*

$$\binom{c+k}{k} a_{n+k}(c+k) \leq \binom{C(n)-c}{k} a_n(c).$$

Proof Let $A_n(c)$ be an animal with n edges and c cycles. An animal $A_{n+k}(c+k)$ has at least one set of $c+k$ edges which may be deleted to give a tree. Pick one such set, and delete k edges in $\binom{c+k}{k}$ ways. This gives an animal with n edges and c cycles. However, more than one animal can give the same outcome in this construction, so an upper bound is needed on the possible number of outcomes. Note that if k edges are added to $A_n(c)$, then there are at most $C(n) - c$ contacts where an edge can be added to form a cycle (since there are at most $C(n)$ cycles). Choose k of these possible contacts; this shows that the number of different outcomes if k are deleted in $A_{n+k}(c+k)$ is at most $\binom{C(n)-c}{k}$. Thus, the inequality follows. □

Lemma 6.13 can be used to prove that the density function of cycles in this model is continuous in $[0, (d-1)/d)$.

Theorem 6.14 *The density function $\mathcal{Q}_c(\epsilon)$ of animals counted by cycles is continuous in the interval $[0, (d-1)/d)$. In addition, the right-derivative of $\mathcal{Q}_c(\epsilon)$ at $\epsilon = 0$ is infinite.*

Proof Continuity is known in $(0, (d-1)/d)$, since $\log \mathcal{Q}_c(\epsilon)$ is concave. Corollary 6.11 shows that $\mathcal{Q}_c(0) < \mathcal{Q}_c(\epsilon)$ if ϵ is small enough, but positive. Moreover, let $c = 0$ and $k = \lfloor \epsilon n \rfloor$ in Lemma 6.13. Then $a_{n+\lfloor \epsilon n \rfloor}(\lfloor \epsilon n \rfloor) \leq \binom{n}{\lfloor \epsilon n \rfloor} a_n(0)$ (since $C(n) \leq n$), and by taking the $(1/n)$-th power and letting $n \to \infty$ one finds that

$$\mathcal{Q}_c(0) \leq [\mathcal{Q}_c(\epsilon/(1+\epsilon))]^{1+\epsilon} \leq \frac{\mathcal{Q}_c(0)}{\epsilon^\epsilon (1-\epsilon)^{1-\epsilon}},$$

provided that ϵ is small enough. Taking $\epsilon \to 0^+$ proves that $\mathcal{Q}_c(\epsilon)$ is continuous at $\epsilon = 0$. To see that the right-derivative at $\epsilon = 0$ is infinite, consider the number of trees $t_n = a_n(0)$. Let P be the pattern consisting of three edges in the form ⊓ (that is, a square with the bottom edge absent). This pattern occurs with positive density in almost all trees; by Theorem 6.12, if $\epsilon > 0$ is small enough, there exists a $k > 0$ such that $t_n(\geq \lfloor \epsilon n \rfloor P) = t_n - t_n(< \lfloor \epsilon n \rfloor P) \geq (1 - e^{-kn}) t_n$. Choose $\lfloor \delta n \rfloor$ of the P, and add the fourth edge back to complete a 4-cycle: $\binom{\lfloor \epsilon n \rfloor}{\lfloor \delta n \rfloor} t_n(\geq \lfloor \epsilon n \rfloor P) \leq a_{n+\lfloor \delta n \rfloor}(\lfloor \delta n \rfloor)$. Thus,

$$(1 - e^{-kn}) \binom{\lfloor \epsilon n \rfloor}{\lfloor \delta n \rfloor} a_n(0) \leq a_{n+\lfloor \delta n \rfloor}(\lfloor \delta n \rfloor),$$

and by taking the $(1/n)$-th power and letting $n \to \infty$,

$$\left[\frac{\epsilon^\epsilon}{\delta^\delta (\epsilon - \delta)^{\epsilon - \delta}}\right] \mathcal{Q}_c(0) \leq [\mathcal{Q}_c(\delta/(1+\delta))]^{1+\delta} \leq [\mathcal{Q}_c(\delta/(1+\delta))] \lambda_d^\delta.$$

In other words,

$$\frac{\mathcal{Q}_c(\delta/(1+\delta)) - \mathcal{Q}_c(0)}{[\delta/(1+\delta)]} \geq \frac{\mathcal{Q}_c(0)}{[\delta/(1+\delta)]} \left[\frac{\epsilon^\epsilon \lambda_d^{-\delta}}{\delta^\delta (\epsilon - \delta)^{\epsilon - \delta}} - 1\right]$$

The right-hand side of this equation diverges as $\delta \to 0$. □

It is also the case that $\mathcal{Q}_c(\epsilon)$ is continuous at $\epsilon = (d-1)/d$, and that its left-derivative is infinite at this point. The proofs here are inspired by similar proofs for site-animals discussed in reference [250]. The proof of Theorem 6.15 exploits a connection between percolation and collapsing animals, pointed out by C. Domb [82].

Theorem 6.15 *The density function $\mathcal{Q}_c(\epsilon)$ is continuous at $\epsilon = (d-1)/d$, and $\mathcal{Q}_c((d-1)/d) = 1$. Moreover, the left-derivative at $\epsilon = (d-1)/d$ is infinite.*

Proof It is certainly the case that $\mathcal{Q}_c(\epsilon) \geq 1$. To see continuity, consider animals with n edges, $c = C(n) - \lceil \epsilon n \rceil$ cycles and (therefore) $v = n - C(n) + \lceil \epsilon n \rceil + 1$ vertices. There are at most $b = 2vd - 2n = 2(d-1)n + 2d(\lceil \epsilon n \rceil - C(n)) + 2d$ perimeter edges in such an animal. Let $a_{n,v,b}$ be the number of animals with n edges, v vertices and b perimeter edges.

Next, consider an edge-percolation process. Let p be the probability that an edge is open. Then the probability that the cluster at the origin has exactly n edges is

$$\sum_{v \geq 1} v\, a_{n,v,b} p^n (1-p)^b \leq 1,$$

with exactly $C(n) - \lceil \epsilon n \rceil$ cycles, and b perimeter edges. Take the $(1/n)$-th power and let $n \to \infty$; note also that this model satisfies Assumptions 3.1. Then

$$\mathcal{Q}_c((d-1)/d - \epsilon)\, p(1-p)^{2d\epsilon} \leq 1.$$

This function of p is a minimum if $p = 1/(1 + 2\epsilon d)$ so that

$$\mathcal{Q}_c((d-1)/d - \epsilon) \leq \frac{(1+2d\epsilon)^{1+2d\epsilon}}{(2d\epsilon)^{2d\epsilon}}.$$

As $\epsilon \to 0^+$, the right-hand side approaches 1, and this proves continuity of the density function at $(d-1)/d$.

Next, argue as follows to see that the left-derivative of the density function at $\epsilon = (d-1)/d$ is infinite. Put $c + k = \lfloor \epsilon n \rfloor$ and $c = \lfloor \delta n \rfloor$ in Lemma 6.13, take the $(1/n)$-th power and let $n \to \infty$. After some simplification of the result, the following is obtained:

$$\left[\frac{\epsilon^\epsilon [(d-1)/d - \epsilon]^{(d-1)/d-\epsilon}}{\delta^\delta [(d-1)/d - \delta]^{(d-1)/d-\delta}}\right] \left[\mathcal{Q}_c(\epsilon/(1+\epsilon-\delta))\right]^{1+\epsilon-\delta} \leq \left[\mathcal{Q}_c(\epsilon/(1+\epsilon-\delta))\right]^{1+\epsilon-\delta}.$$

Since $1 \leq \mathcal{Q}_c(\epsilon) \leq \lambda_d$, this may be further simplified to

$$\left[\frac{\lambda_d^{-\epsilon} \epsilon^\epsilon [(d-1)/d - \epsilon]^{(d-1)/d-\epsilon}}{\lambda_d^{-\delta} \delta^\delta [(d-1)/d - \delta]^{(d-1)/d-\delta}}\right] \mathcal{Q}_c(\epsilon/(1+\epsilon-\delta)) \leq \mathcal{Q}_c(\epsilon/(1+\epsilon-\delta)).$$

Now take $\epsilon = (d-1)/d$; then

$$\left[\frac{\lambda_d^{\delta-(d-1)/d}}{\delta^\delta ((d-1)/d - \delta)^{(d-1)/d-\delta}}\right] \mathcal{Q}_c\left(\frac{d-1}{2d-1-d\delta}\right) \leq \mathcal{Q}_c\left(\frac{d\delta}{2d-1-d\delta}\right).$$

Thus

$$\left[1 - \frac{\lambda_d^{\delta-(d-1)/d}}{\delta^\delta ((d-1)/d - \delta)^{(d-1)/d-\delta}}\right] \mathcal{Q}_c\left(\frac{d-1}{2d-1-d\delta}\right)$$
$$\geq \mathcal{Q}_c\left(\frac{d-1}{2d-1-d\delta}\right) - \mathcal{Q}_c\left(\frac{d\delta}{2d-1-d\delta}\right).$$

Divide this by $(d-1)/d - \delta$ and let $\delta \to (d-1)/d$. Then the left-hand side approaches negative infinity, and the right-hand side is the left-derivative at $(d-1)/d$. □

The density function in this model has the appearance displayed in Fig. 3.1, but it is also continuous, and has infinite right- and left-derivatives at the end-points of the interval $[0, (d-1)/d]$. The supremum of the density function is equal to the growth constant for animals: $\tau_d = \mathcal{Q}_c(0) < \sup_\epsilon \mathcal{Q}_c(\epsilon) = \lambda_d$; τ_d is the growth constant for lattice trees, and the inequality is a direct consequence of the pattern theorem in Corollary 6.11 (with $z = 1$). If ϵ_c is defined by $\epsilon_c = \inf\{\epsilon \,|\, \mathcal{Q}(\epsilon) = \lambda_d\}$, then $\epsilon_c > 0$. Notice that the free energy $\mathcal{A}_c(z) = \sup_{\epsilon\in[0,(d-1)/d]}\{\log \mathcal{Q}_c(\epsilon) + \epsilon \log z\}$ is asymptotic to the line $\log \mathcal{Q}_c((d-1)/d) + ((d-1)/d)\log z$, as $z \to \infty$. But $\log \mathcal{Q}_c((d-1)/d) = 0$ by Theorem 6.15, so that the limiting entropy of collapsed animals is equal to zero.

It is strongly suggested by Theorem 6.7, and numerical simulations, that the asymptotic behaviour of trees is $t_n \sim n^{-\theta_0}\tau_d^n$ [27, 44, 137], where θ_0 is the entropic exponent with mean field value $5/2$ [27], and Flory values in eqn (1.37). Notice that Lemma 6.13 shows that $\lim_{n\to\infty}[\log a_n(c)]/n = \log \tau_d$ exists. Subject to the assumption that the exponent θ_0 exists as a limit, it is now possible to show that there is an entropic exponent θ_c for $a_n(c)$, for each fixed value of c, and that $\theta_c = \theta_0 - c$ [326].

Theorem 6.16 *Suppose that the limit* $-\theta_0 = \lim_{n\to\infty}[\log(t_n\tau_d^{-n})]/[\log n]$ *exists. Then the limits*

$$-\theta_c = \lim_{n\to\infty}[\log(a_n(c)\tau_d^{-n})]/[\log n],$$

exists for $c = 1, 2, \ldots,$ *and moreover,* $\theta_c = \theta_0 - c$.

Proof Lemma 6.13 gives

$$a_{n+c}(c) \leq \binom{C(n)}{n} a_n(0).$$

Thus, $\limsup_{n\to\infty}[\log(a_n(c)\tau_d^{-n})]/[\log n] \leq -\theta_0 + c$. On the other hand, choose the pattern $\sqcap$ and use Theorem 6.12 for trees. Let $t_n(\geq\lfloor\epsilon n\rfloor\sqcap)$ be the number of trees of size n which contains the pattern $\sqcap$ at least $\lfloor\epsilon n\rfloor$ times. If $\epsilon > 0$ is small enough, then $\lim_{n\to\infty}[t_n(\geq\lfloor\epsilon n\rfloor\sqcap)]^{1/n} = \tau_d$. Choose c of the $\sqcap$'s in a tree counted by $t_n(\geq\lfloor\epsilon n\rfloor\sqcap)$, and form a cycle by adding an edge. This shows that

$$\binom{\lfloor\epsilon n\rfloor}{c} t_n(\geq\lfloor\epsilon n\rfloor\sqcap) \leq a_{n+c}(c).$$

But $a_n(0) = t_n \geq t_n(\geq\lfloor\epsilon n\rfloor\sqcap) \geq (1 - e^{-kn})a_n(0)$, for some fixed value of k. Thus, $\liminf_{n\to\infty}[\log(a_n(c)\tau_d^{-n})]/[\log n] \geq -\theta_0 + c$. This completes the proof. □

The collapse of animals induced by a cycle fugacity has been studied in two dimensions in references [74, 80, 224, 346]. It is generally believed that the cross-over exponent associated with the collapse has value $2/3$, and that critical θ-animals have $\nu_c = 1/2$, although a Monte Carlo simulation has produced a slightly larger value, $\nu_c = 0.555 \pm 0.005$ [197]. The critical point is believed to be tricritical, and it has been argued to be in the universality class of tricritical zero-state Potts models [346]. If the generating function $A(x, z) = \sum_{n\geq 0}\sum_{c\geq 0} a_n(c)z^c x^n$ is defined and interpreted as in Chapter 2, then some of the tricritical exponents can be assigned values. In particular, along the λ-line the expanded phase of animals is encountered, and by using

Table 6.1: Tricritical exponents for θ-animals (cycle-collapse)

ϕ	α	$2-\alpha_t$	$2-\alpha_u$	ν_t	y_t	$2-\alpha_+$
$\frac{2}{3}$	$\frac{1}{2}$	≈ 1.10	≈ 1.65	≈ 0.55	≈ 1.21	0

$a_n \simeq n^{-\theta}\lambda_2^n$, and eqn (2.12), it seems that $2-\alpha_+ = \theta - 1$. Since $\theta = 1$ in two dimensions (see Table 1.2), it follows that $2-\alpha_+ = 0$. The hyperscaling relation in eqn (2.29) shows that $2-\alpha = 3/2$. Since θ-animals are found at the tricritical point, the Monte Carlo data suggest that $\nu_t \approx 0.55$, in which case eqn (2.36) suggests that $2-\alpha_t \approx 1.10$, and eqn (2.37) shows that $2-\alpha_u \approx 1.65$. These values are tabled in Table 6.1.

6.3.2 The cycle-contact model

A model of collapsing animals is found if a *contact activity* y conjugate to the number of contacts in an animal is introduced in the cycle model. It is unclear whether the tricritical collapse of branched polymers driven by a contact activity is in a different universality class to the collapse driven by a cycle activity. Models of collapsing lattice animals in an ensemble with both cycle and contact activities was considered in references [118, 119, 120], where cross-over exponents associated with contact collapse were estimated.

Consider an animal which is a hypercube of side-length L and with all possible edges present. The number of vertices in such an animal is $v = (L+1)^d$, and the number of edges n is at least dL^d, and at most $d(L+1)^d$. The number of cycles, by eqn (6.3), is at least $c = (d-1)L^d + \sum_{i=1}^{d}(-1)^i \binom{d}{i} L^{d-i}$. Choose $\lfloor \delta n \rfloor$ of the edges, and delete them; this leaves $\lfloor \delta n \rfloor$ contacts and $c - \lfloor \delta n \rfloor$ cycles. Dividing by n, and letting $n \to \infty$, shows that the density of contacts is δ, and the density of cycles is *at most* $(d-1-\delta)/(d-\delta)$. This defines a region Δ over which the joint density function of animals with contacts and cycles are defined, illustrated in Fig. 6.4.

Let $a_n(k,c)$ be the number of animals with n edges, k contacts and c cycles. The cycle partition function of this model is $a_n(k;z) = \sum_{c\geq 0} a_n(k,c)z^c$. Concatenating two

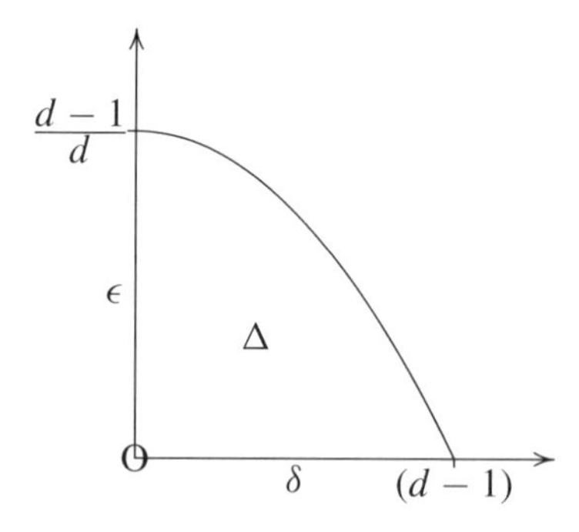

Fig. 6.4: The density function $\mathcal{P}(\epsilon,\delta)$ of animals with a density ϵ of cycles and δ of contacts is defined over the area under the curve $(d-1-\delta)/(d-\delta)$.

animals as in Fig. 6.1, with the first animal with n_1 edges, $c - c_1$ cycles and k_1 contacts, and the second with n_2 edges, c_1 cycles and k_2 contacts, gives (if a sum over c_1 is executed, and after multiplication with z^c, and summing over c),

$$a_{n_1}(k_1; z)a_{n_2}(k_2; z) \leq a_{n_1+n_2+1}(k_1 + k_2; z). \tag{6.20}$$

This shows that a limiting free energy $\mathcal{A}_c(y, z)$ exists in this model. Thus, $a_{n-1}(k; z)$ satisfies Assumptions 3.1, and there is a density function $\mathcal{Q}_c(\delta; z)$ of contacts at a given cycle activity $z \geq 0$. This density function is explicitly defined by the limit

$$\mathcal{Q}_c(\delta; z) = \lim_{n\to\infty} [a_n(\lfloor \delta n \rfloor; z)]^{1/n}; \tag{6.21}$$

see Theorem 3.4. A similar argument shows that the density function

$$\mathcal{Q}_c(y; \epsilon) = \lim_{n\to\infty} [a_n(y; \lfloor \epsilon n \rfloor)]^{1/n}, \tag{6.22}$$

exists; and by Theorem 3.19 there is a joint density function defined by

$$\mathcal{Q}_c(\delta, \epsilon) = \lim_{n\to\infty} [a_n(\lfloor \delta n \rfloor, \lfloor \epsilon n \rfloor)]^{1/n}. \tag{6.23}$$

This joint density function is defined on the domain Δ in Fig. 6.4, as argued in the previous paragraph. The density function $\mathcal{Q}_c(\delta, \epsilon)$ is continuous in the interior of Δ, since $\log \mathcal{Q}_c(\epsilon, \delta)$ is concave in both its arguments. It is also continous on the boundaries of Δ if either ϵ, or δ, is equal to zero. This fact will be proven using Lemma 6.17 below, which is a generalization of Lemma 6.13.

Lemma 6.17 *If $a_n(k, c)$ is the number of lattice animals with n edges, c cycles and k contacts; then*

(1) $a_n(k, c) \leq a_{n+1}(k, c)$,

(2) $\binom{c + k_0}{k_0} a_n(k, c + k_0) \leq \binom{k + k_0}{k_0} a_{n-k_0}(k + k_0, c)$,

(3) $\binom{k + c_0}{c_0} a_n(k + c_0, c) \leq \binom{c + c_0}{c_0} a_{n+c_0}(k, c + c_0)$.

Proof The first inequality is seen by appending a single edge on the lexicographically most vertex in the animal. To find the second inequality, note that there is at least one set of $c + k_0$ edges in an animal counted by $a_n(k, c + k_0)$ such that deleting these edges will leave a tree. Choose k_0 of the edges and remove them from the animal; this removes k_0 cycles and creates k_0 new contacts. But now notice that there are a total of $k + k_0$ contacts, of which k_0 must be replaced with edges to find the original animal. The third inequality is shown by arguing similary to the second. □

Theorem 6.18 *The density function $\mathcal{Q}_c(\delta, \epsilon)$ is continuous in Δ, and on the boundaries of Δ if $\delta = 0$ or $\epsilon = 0$.*

Proof Since $\log \mathcal{Q}_c(\delta, \epsilon)$ is concave in both its arguments, it is also continuous in the interior of Δ. To show continuity if $\epsilon = 0$, let $c = 0$, $k_0 = \lfloor \epsilon n \rfloor$ and $k = \lfloor \delta n \rfloor$ in the second inequality in Lemma 6.17. After some applications of $a_n(k, c) \leq a_{n+1}(k, c)$ (the first inequality in Lemma 6.17), $a_n(\lfloor \delta n \rfloor, \lfloor \epsilon n \rfloor) \leq \binom{\lfloor \epsilon n \rfloor + \lfloor \delta n \rfloor}{\lfloor \epsilon n \rfloor} a_n(\lfloor \epsilon n \rfloor + \lfloor \delta n \rfloor, 0)$. Take the $(1/n)$-th power and let $n \to \infty$. Then

$$\mathcal{Q}_c(\delta, \epsilon) \leq \left[\frac{(\epsilon + \delta)^{\epsilon + \delta}}{\epsilon^\epsilon \delta^\delta} \right] \mathcal{Q}_c(\epsilon + \delta, 0).$$

But $\mathcal{Q}_c(\delta, 0)$ is the density function of a model of trees with a density δ of contacts, and it is concave and continuous for all $\delta \in (0, d-1)$. Take $\epsilon \to 0^+$ to show that $\limsup_{\epsilon \to 0^+} \mathcal{Q}_c(\delta, \epsilon) \leq \mathcal{Q}_c(\delta, 0)$, and by concavity $\liminf_{\epsilon \to 0^+} \mathcal{Q}_c(\delta, \epsilon) \geq \mathcal{Q}_c(\delta, 0)$. Thus $\lim_{\epsilon \to 0^+} \mathcal{Q}_c(\delta, \epsilon) = \mathcal{Q}_c(\delta, 0)$. Continuity along the boundary of $\delta = 0$ is shown by putting $k = 0$, $c = \lfloor \epsilon n \rfloor$ and $c_0 = \lfloor \delta n \rfloor$ in the third equation in Lemma 6.17. This gives $a_n(\lfloor \delta n \rfloor, \lfloor \epsilon n \rfloor) \leq \binom{\lfloor \epsilon n \rfloor + \lfloor \delta n \rfloor}{\lfloor \delta n \rfloor} a_{n + \lfloor \delta n \rfloor}(0, \lfloor \epsilon n \rfloor + \lfloor \delta n \rfloor)$. Taking the $(1/n)$-th power and letting $n \to \infty$,

$$\mathcal{Q}_c(\delta, \epsilon)) \leq \left[\frac{(\epsilon + \delta)^{\epsilon + \delta}}{\epsilon^\epsilon \delta^\delta} \right] [\mathcal{Q}_c(0, (\epsilon + \delta)/(1 + \delta))]^{1+\delta}.$$

Thus, if $\delta \to 0^+$, then $\limsup_{\delta \to 0^+} \mathcal{Q}_c(\delta, \epsilon) \leq \mathcal{Q}_c(0, \epsilon)$, and the result is continuity along the line $\delta = 0$ for $\epsilon \in (0, (d-1)/d)$. Continuity at the point $(\delta, \epsilon) = (0, 0)$ follows from Theorem 6.14. □

The density function $\mathcal{Q}_c(\delta, \epsilon)$ interpolates between a number of different models of animals and trees, all counted with respect to the number of edges. For example, $\mathcal{Q}_c(0, 0)$ is the growth constant of *strongly embedded lattice trees*, while $\sup_{\epsilon \geq 0} \mathcal{Q}_c(0, \epsilon)$ is the growth constant of *strongly embedded animals*. The growth constant of *weakly embedded lattice trees* is obviously $\tau_d = \sup_{\delta \geq 0} \mathcal{Q}_c(\delta, 0)$ (see Corollary 6.3), and the growth constant of *weakly embedded lattice animals* is $\lambda_d = \sup_{\delta \geq 0, \epsilon \geq 0} \mathcal{Q}_c(\delta, \epsilon)$ (see Corollary 6.2). Continuity along the boundary $(d - 1 - \delta)/(d - \delta)$ of Δ is not known. The best bounds are stated in the next theorem.

Theorem 6.19 *For* $0 \leq \delta \leq d - 1$,

$$\mathcal{Q}_c\left(\delta, \frac{d-1-\delta}{d-\delta}\right) \geq \frac{\left(\frac{d-1-\delta}{d-\delta}\right)^{(d-1-\delta)/(d-\delta)}}{\delta^\delta \left(\frac{d-1-\delta}{d-\delta} - \delta\right)^{(d-1-\delta)/(d-\delta)-\delta}}$$

$$\mathcal{Q}_c\delta.\left(\frac{d-1-\delta}{d-\delta}\right) \leq \frac{\left(\frac{2d}{d-\delta} - \delta - 1\right)^{(2d/(d-\delta))-\delta-1}}{\left(\frac{2d}{d-\delta} - \delta - 2\right)^{(2d/(d-\delta))-\delta-2}}.$$

Proof Consider the set of animals with $C(n) - \lfloor \delta n \rfloor$ cycles and $\lfloor \delta n \rfloor$ contacts. Since there exists an animal with $C(n)$ cycles, one may delete $\lfloor \delta n \rfloor$ edges in cycles to find such an animal. In fact, the number of such animals is at least $\binom{C(n)}{\lfloor \delta n \rfloor}$. By taking the $(1/n)$-th power and taking $n \to \infty$, the lower bound is found. To find the upper bound argue as follows. Let $a_{n,v,b}$ be the number of lattice animals with n edges, v vertices, $C(n) - \lfloor \epsilon n \rfloor$ cycles, $\lfloor \delta n \rfloor$ contacts and b perimeter edges. Then the probability that the cluster at the origin has n edges is $\sum_{v \geq 0} v \, a_{n,v,b} p^n (1-p)^b \leq 1$. In an animal with v vertices and n edges, the number of cycles is $c = n - v + 1$ (by eqn (6.3)). Moreover, the size of the perimeter is $b = 2d\lfloor \epsilon n \rfloor + 2d(n+1-C(n)) - 2n - \lfloor \delta n \rfloor$. Taking the $(1/n)$-th power, and taking $n \to \infty$ gives

$$\mathcal{Q}_c\left(\delta, \frac{d-1-\delta}{d-\delta} - \epsilon\right) p(1-p)^{2d\epsilon - \delta - 2 + 2d/(d-\delta)} \leq 1.$$

The left-hand side is a maximum if $p = 1/(2d\epsilon - \delta - 2 + 2d/(d-\delta))$, in which case (if $\epsilon \to 0^+$) the claimed upper bound is obtained. □

The right-derivatives to δ and ϵ of the density function are also infinite along the $\delta = 0$ and $\epsilon = 0$ boundaries of Δ. This fact may be shown by using the pattern theorem in Theorem 6.12.

Theorem 6.20 *The right-derivatives of $\mathcal{Q}_c(\delta, \epsilon)$ to ϵ (if $0 \leq \delta < d-1$) at $\epsilon = 0$, and to δ (if $0 \leq \epsilon < (d-1)/d$) at $\delta = 0$, are infinite.*

Proof Let $a_n(y; 0; \geq \lfloor \gamma n \rfloor P_0)$ be the partition function of animals with $\lfloor \delta n \rfloor$ contacts, no cycles, and which contains the pattern P_0 in Fig. 6.5 at least $\lfloor \gamma n \rfloor$ times. Choose $\lfloor \epsilon n \rfloor$ of the patterns P_0 in each animal (from the $\lfloor \gamma n \rfloor$), and change these patterns as indicated in Fig. 6.5(a). Then

$$\binom{\lfloor \gamma n \rfloor}{\lfloor \epsilon n \rfloor} a_n(y; 0; \geq \lfloor \gamma n \rfloor P_0) \leq a_n(y; \lfloor \epsilon n \rfloor). \qquad (\ddagger)$$

By Theorem 6.12 there is a pattern theorem which states that

$$a_n(y; 0; \geq \lfloor \gamma n \rfloor P_0) \geq (1 - e^{-k(y)n}) a_n(y; 0). \qquad (\dagger)$$

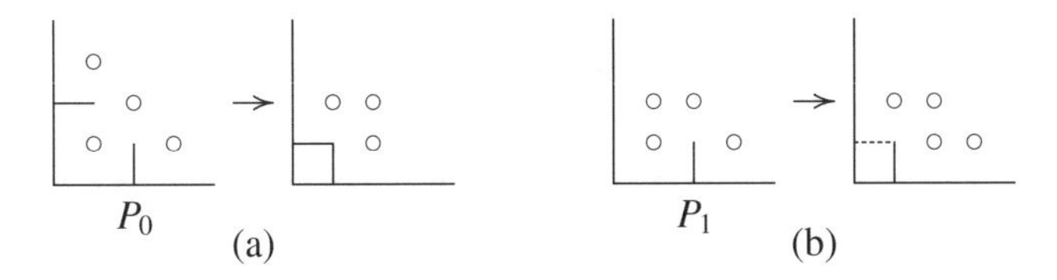

Fig. 6.5: (a) If the two edges in the pattern P_0 are moved as indicated, then a cycle is created, without deleting any existing contacts. (b) If the edge in P_1 is moved one step left, then a contact is created, without deleting any cycles. These patterns and constructions generalizes to higher dimensions (since a pattern is both the presence and the absence of edges).

Define $\mathcal{Q}_c(y; 0; \geq\gamma P_0) = \limsup_{n\to\infty} \left[a_n(y; 0; \geq\lfloor\gamma n\rfloor P_0)\right]^{1/n}$. By taking the $(1/n)$-th power of eqn (†) and taking the lim sup of the left-hand side, then

$$\mathcal{Q}_c(y; 0; \geq\gamma P_0) \geq \mathcal{Q}_c(y; 0),$$

where $\mathcal{Q}_c(y; 0)$ is the density function defined in eqn (6.22). In other words, by taking the $(1/n)$-th power of eqn (‡), and taking $n \to \infty$,

$$\left[\frac{\gamma^\gamma}{\epsilon^\epsilon(\gamma-\epsilon)^{\gamma-\epsilon}}\right]\mathcal{Q}_c(y, 0) \leq \mathcal{Q}_c(y, \epsilon).$$

Multiply this last equation by y^δ, and take the infimum over $y > 0$ of the right-hand side, and use Theorem 3.18. Suppose that this occurs when $y = y_m \geq 0$. Then

$$\left[\frac{\gamma^\gamma}{\epsilon^\epsilon(\gamma-\epsilon)^{\gamma-\epsilon}}\right]\mathcal{Q}_c(y_m, 0)y_m^\delta \leq \mathcal{Q}_c(\delta, \epsilon).$$

But now take the infimum over y_m on the left-hand side. This gives

$$\left[\frac{\gamma^\gamma}{\epsilon^\epsilon(\gamma-\epsilon)^{\gamma-\epsilon}}\right]\mathcal{Q}_c(\delta, 0) \leq \mathcal{Q}_c(\delta, \epsilon).$$

In other words,

$$\frac{1}{\epsilon}\left[\frac{\gamma^\gamma}{\epsilon^\epsilon(\gamma-\epsilon)^{\gamma-\epsilon}} - 1\right]\mathcal{Q}_c(\delta, 0) \leq \frac{1}{\epsilon}\big(\mathcal{Q}_c(\delta, \epsilon) - \mathcal{Q}_c(\delta, 0)\big).$$

If $\epsilon \to 0^+$, then this shows that the right-derivative to ϵ is infinite at $\epsilon = 0^+$, provided that $\mathcal{Q}_c(\delta, 0) > 0$. Since $\mathcal{Q}_c(\delta, 0)$ is the growth constant of trees with a density of contacts, and there are more of these than there are site-trees, it follows that $\mathcal{Q}_c(\delta, 0) > 0$ if $0 \leq \delta < d - 1$. This proves the first claim. That the right-derivative to δ is infinite at $\delta = 0$ follows by a similar argument, where the pattern P_1 in Fig. 6.5(b) is considered instead. □

In addition to these results, it is also known that $\mathcal{Q}_c(\delta, \epsilon)$ is non-analytic. This can be shown by using the connection to edge-percolation as was done in Theorem 6.15. In particular, the probability that the open cluster at the origin contains n edges is

$$\begin{aligned} P_n(p) &= \sum_{k\geq 0, c\geq 0} v a_n(k, c) p^n (1-p)^{s+k} \\ &= p^n(1-p)^{2(d-1)+2d} \sum_{k\geq 0, c\geq 0} (n+1-c) a_n(k, c)(1-p)^{-k}(1-p)^{-2dc} \end{aligned} \tag{6.24}$$

Since $1 \leq (n+1-c) \leq (n+1)$, eqn (6.24) can be turned into a string of inequalities involving the partition function:

$$\begin{aligned} p^n(1-p)^{2(d-1)+2d} a_n((1-p)^{-1}, (1-p)^{-2d}) &\leq P_n(p) \\ &\leq (n+1)p^n(1-p)^{2(d-1)+2d} a_n((1-p)^{-1}, (1-p)^{-2d}), \end{aligned} \tag{6.25}$$

and by taking logarithms, dividing by n and letting $n \to \infty$, the result is that

$$\lim_{n\to\infty} \frac{1}{n} \log P_n(p) = \mathcal{A}_c((1-p)^{-1}, (1-p)^{-2d}) + \log p + 2(d-1)\log(1-p). \tag{6.26}$$

But there exists a non-analytic function $\zeta(p)$, with $\zeta(p) = 0$ if $p > p_c$ and $\zeta(p) > 0$ if $p < p_c$, such that $\lim_{n\to\infty} \frac{1}{n} \log P_n(p) = -\zeta(p)$, and where p_c is the critical percolation probability. This is shown in Appendix D. Thus

$$\mathcal{A}_c((1-p)^{-1}, (1-p)^{-2d}) = -\zeta(p) - \log p - 2(d-1)\log(1-p), \tag{6.27}$$

and so the free energy in this model has a non-analyticity at the point $((1-p_c)^{-1}, (1-p_c)^{-2d})$. To see that there is a non-analyticity in the density function argue as follows. Notice that there is a most popular class in the set of terms $a_n(k,c)(1-p)^{-k}(1-p)^{-2dc}$. In particular, there exists two number ϵ_p and δ_p such that it follows from eqns (6.23) and (6.27) that

$$\begin{aligned}
\lim_{n\to\infty} & \left[\sum_{k\geq 0, c\geq 0} a_n(k,c)(1-p)^{-k}(1-p)^{-2dc} \right]^{1/n} \\
&= \mathcal{Q}_c(\epsilon_p, \delta_p)(1-p)^{-\delta_p}(1-p)^{-2d\epsilon_p} \\
&= e^{-\zeta(p)} p^{-1} (1-p)^{2(1-d)},
\end{aligned} \tag{6.28}$$

where the limit exists since $\mathcal{Q}_c(\epsilon, \delta)$ and $\mathcal{A}_c(y, z)$ exist as in eqn (6.23). Thus, by eqn (6.27), the joint density function is given by

$$\mathcal{Q}_c(\epsilon_p, \delta_p) = p^{-1} e^{-\zeta(p)} (1-p)^{2d(\epsilon_p - 1)+\delta_p+2}. \tag{6.29}$$

In other words, the density function $\mathcal{Q}_c(\epsilon, \delta)$ is non-analytic at the point $(\epsilon_{p_c}, \delta_{p_c})$.

Theorem 6.21 *The limiting free energy of edge-animals in the cycle-contact ensemble is non-analytic at least at one point which corresponds to the critical percolation point:* $(y, z) = ((1-p_c)^{-1}, (1-p_c)^{-2d})$, *where* p_c *is the critical percolation probability. Moreover, there exists a point* $(\epsilon_{p_c}, \delta_{p_c})$ *in* Δ *where the density function of this model is not analytic.* □

The result in Theorem 6.21 can be extended by showing that the limiting free energy is non-analytic at points other than the critical percolation point; these non-analyticities are presumably those corresponding to a collapse transition in this model, and it seems that they are closely related in character to the critical percolation point. These results suggest that the θ-point may be a percolation process.

Theorem 6.22 *There are non-analyticities in the free energy of edge-animals at points other than the critical percolation point.*

Proof Define the function

$$Q_n(\alpha, p) = p^n(1-p)^{2(d-1)n+2d} \sum_{k\geq 0, c\geq 0} (n+1-c)a_n(c,k)\left(\frac{\alpha}{1-p}\right)^k (1-p)^{-2dc}, \tag{†}$$

where $0 \leq \alpha \leq 1$. Then $Q_n(\alpha, p) \leq P_n(p)$. This shows that

$$\limsup_{n\to\infty}[Q_n(\alpha, p)]^{1/n} \leq \lim_{n\to\infty}[P_n(p)]^{1/n} \leq e^{-\zeta(p)} < 1$$

if $p < p_c$, by appendix D. On the other hand, use the relation $2dv = 2n + 2k + s$ to see that

$$Q_n(\alpha, p) = \sum_{k\geq 0, c\geq 0} (n+1-c)a_n(c,k)(p/\alpha^2)^n((1-p)/\alpha)^{s+k}(\alpha^{2d})^v.$$

Observe that $1 - p/\alpha^2 \leq (1-p)/\alpha$, for all values of $p \in [0, 1]$, and long as $\alpha \in (0, 1]$. Thus,

$$Q_n(\alpha, p) \geq \sum_{k\geq 0, c\geq 0} (n+1-c)a_n(c,k)(p/\alpha^2)^n(1-p/\alpha^2)^{s+k}(\alpha^{2d})^v = R_n(\alpha, p).$$

$R_n(\alpha, p)$ is the probability that the open cluster at the origin has size n in a combined edge-site percolation model where edges are open with probability p/α^2, and sites are occupied with probability α^{2d}, provided that $p \leq \alpha^2$ (and where every vertex in the cluster is occupied). If α is close enough to 1, then p can be picked large enough that the open cluster at the origin is infinite with non-zero probability. Thus there are $\alpha_c < 1$ and $p_\alpha < 1$ such that $\lim_{n\to\infty}[Q_n(\alpha, p)]^{1/n} = 1$, for all $\alpha \geq \alpha_c$ and $p \geq p_\alpha$. The free energy $\mathcal{A}_c(\alpha/(1-p), 1/(1-p)^{2d})$ is related to $\lim_{n\to\infty}[Q_n(\alpha, p)]^{1/n}$ (and its derivative). Since $\lim_{n\to\infty}[Q_n(\alpha, p)]^{1/n}$ is singular at a critical value of p, for $\alpha > \alpha_c$, the free energy is non-analytic at (α, p_α) for all $\alpha > \alpha_c$. The critical percolation point is not on the line $(\alpha/(1-p), 1/(1-p)^{2d})$ if $\alpha < 1$ and so this non-analyticity is not the critical percolation point. □

The likely phase diagram of $\mathcal{A}_c(y, z)$ is drawn in Fig. 6.6, based on the notions above and on numerical simulations [197]. The critical curves are thought to be tricritical θ-points, which separate a phase of expanded animals from a phase of collapsed animals. The curve marked by θ in Fig. 6.6 is the locus of tricritical points which corresponds to a collapse of the animal driven by the contact activity; and similarly, the collapse along the θ' curve corresponds to the collapse of the animals driven by a cycle activity. These transitions meet at the critical percolation point, which is also the intersection of a curve in the phase diagram along which the animal is weighted as a percolation cluster; this is directly seen in eqn (6.24), where the percolation probability $P_n(p)$ is seen to be a function of a derivative of the partition function $a_n(y, z)$ of animals in the contact-cycle ensemble. Equation (6.27) explicitly relates the limiting free energy of the animal model to a percolation process. The connection of lattice animals to percolation is also made more interesting by the interpretation of percolation as a one-component Potts model [136, 357, 372, 373]; see also the cluster description of percolation by Fortuin and Kasteleyen [109, 125]. These connections proved useful in studying the

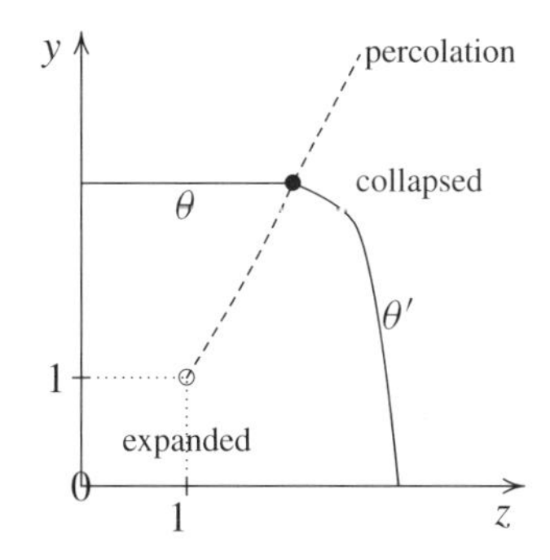

Fig. 6.6: A phase diagram for collapsing animals in the cycle-contact ensemble. The line of θ-transitions are tricritical collapse transitions driven by the contact fugacity y, and the θ'- transitions are tricritical collapse transitions driven by the cycle fugacity z. The dashed curve represents the curve along which percolation clusters are found, and they percolate at the critical percolation point which is on the line of collapse transitions. This curve starts at the point $y = z = 1$.

phases in Fig. 6.6. In particular, there are some arguments and data which suggested that there are two distinct collapsed phases in Fig. 6.6, one phase corresponding to animals undergoing contact collapse, and a second collapse phase for animals undergoing cycle collapse [251]. However, it also seems possible that there is only a single collapsed phase [205, 313, 335]. This single-phase picture is also supported by extended Potts-model calculations on a Bethe lattice [181], and by a Migdal–Kadanoff renormalization group calculation [39, 56] in two dimensions [181].

The usual finite size tricrical assumption suggests that the finite size free energy, defined by $A_n(s,t) = [\log a_n(y,z)]/n$, is a function of two scaling fields s and t. These fields are appropriate combinations of y and z. It is expected that $A_n \sim [\hat{f}(n^{\phi_k} t)]/n$ if the contact collapse is approached (see eqn (2.24), as well as eqn (2.25)), where the cross-over exponent ϕ_k is associated with a collapse transition driven by the contact activity y. Similarly, a cross-over exponent ϕ_c may be associated with the tricritical θ-points driven by the cycle activity, since it is expected that $A_n \sim [\hat{g}(n^{\phi_c} s)]/n$ as the cycle collapse is approached. The specific heat exponent defined in eqn (2.27) is related to ϕ_c and ϕ_k as in eqn (2.28). Numerical simulations in two dimensions show that $\phi_c = 0.62 \pm 0.04$ and $\phi_k = 0.62 \pm 0.03$ [197].

The collapse transitions are also characterized by a change in the metric behaviour of the animal. In particular, the metric exponent ν (see Section 1.2.3) should change discontinuously as the animal is taken through the collapse transition. In two dimensions expanded animals have metric exponent $\nu = 0.644 \pm 0.002$ [195], but $\nu \approx 0.53$ along

Table 6.2: Percolation exponents in two dimensions

ϕ	α	$2-\alpha_t$	$2-\alpha_u$	ν_t	y_t
$\frac{2}{3}$	$\frac{1}{2}$	$\frac{96}{91}$	$\frac{144}{91}$	$\frac{48}{91}$	$\frac{91}{72}$

the critical curve [197]. This value is also the value of the metric exponent for percolation clusters, a fact which may be seen as follows [291]. Use the fact that $P_n(p_c) \simeq n^{-1-1/\delta}$ [147] with $\delta = 91/5$ in eqn (6.25). Thus, the partition function of animals at the critical percolation point has the following finite size scaling: $a_n(1/(1-p_c), 1/(1-p_c)^{2d}) \simeq n^{-2-1/\delta} p_n^{-n}$, from which it appears that the entropic exponent is $\theta_{perc} = 2 + 1/\delta = 187/91$; see eqn (1.3). In other words, by eqn (2.16) it is apparent that $2-\alpha_t = \theta_{perc}-1 = 96/91$, and since $2-\alpha_t = d\nu_t$ in eqn (2.36), the result is that $\nu_t = 48/91 = 0.5275\ldots$. These exponents are the exponent of percolation interpreted as a tricritical θ-point as in Fig. 6.6. If it is also assumed that the cross-over exponent is equal to $2/3$ (as it is along the cycle-collapse critical curve), then the remaining tricritical exponents may be computed.

The θ-transition in a contact-activity model was also argued to be in the Ising universality class with $\phi = 8/15$ and $\nu_t = 8/15$ [56, 313]. The collapse transition in lattice trees with a contact fugacity was discussed in reference [121], and enumeration shows that $\phi = 0.60 \pm 0.03$ in two dimensions and $\phi = 0.82 \pm 0.03$ in three dimensions [132]. Simulations of the contact collapse in lattice animals produced $\phi = 0.619 \pm 0.012$ [335]. In a cycle model, the θ-transition is modelled by a tricritical zero-state Potts model, with $\phi = 2/3$ and $\nu_t = 1/2$ [332]. It was also suggested that three-dimensional vesicles are in the same universality class as animals undergoing a θ-transition driven by a cycle fugacity in three dimensions, and here $\phi = 1$ and $\nu_t = 1/2$ [331]. There is also no evidence for a critical curve separating two distinct collapsed phases in Fig. 6.6 [335]. Taken together, it still seems that there is much uncertainty about the nature of θ-transitions in collapsing animals, and there is scope for more work.

6.4 Adsorbing trees

The interaction between a branched polymer and a surface can be described by a model of a lattice tree adsorbing on a solid wall. A variant of this model is a tree interacting with a penetrable interface or a defect plane. It is appropriate in this case to consider a model of trees, since cycles will play no role in this section.

Let the interface or wall be the $Z = 0$ hyperplane in d dimensions. A tree is *attached* to the wall (adsorbing plane) if it has at least one vertex with Z-coordinate between -1 and 1. The interaction between the attached tree and the adsorbing plane is modelled by counting the number of vertices in the tree with Z-coordinate equal to zero. Such vertices are called *visits*. A tree is a *positive* tree if it is attached and all the Z-coordinates of its vertices are non-negative. A positive tree is a model of a branched polymer interacting with an impenetrable wall, while an attached tree is a model of a branched polymer interacting with a defect plane (such as an interface between two liquids).

The number of attached trees with v visits and k nearest-neighbour contacts will be denoted by $t_n^>(k, v)$ and the number of positive trees with v visits and k contacts will be denoted by $t_n^+(k, v)$. There are two activities in these models: z will be conjugate to the number of visits, while y will be the activity conjugate to the number of contacts. A θ-transition (or collapse transition) is expected at a critical value of y, and the tree should adsorb at a critical value of z. The partition function in the model of attached trees is

$$t_n^>(y,z) = \sum_{v\geq 0}\sum_{k\geq 0} t_n^>(k,v)y^k z^v. \tag{6.30}$$

A model of self-interacting branched polymers interacting with an impenetrable wall is defined by considering positive attached trees instead:

$$t_n^+(y,z) = \sum_{v\geq 0}\sum_{k\geq 0} t_n^+(k,v)y^k z^v. \tag{6.31}$$

At small values of the parameters (y, z) the trees are expected to be desorbed and expanded in a phase which is called the DE-phase. Increasing z should lead to an adsorption into an adsorbed and expanded phase (AE-phase). Similarly, increasing y instead will give a desorbed and collapsed phase (DC-phase). In three and higher dimensions increasing both y and z should give a collapse and adsorbed phase (AC-phase). Studies of directed walk models suggest that the AC-phase is not present in two dimensions [126, 127, 128]. The phase diagram expected in this model is illustrated in Fig. 6.7. It is similar in appearance to the phase diagram for collapsing and adsorbing polygons (see Fig. 5.17) [203].

6.4.1 The free energy of adsorbing and collapsing trees

The existence of free energies follows from concatenation of attached trees. One may use a most popular argument as in Theorem 5.52; but here I shall concatenate two trees directly, as in Fig. 6.8. As before, the bottom and top vertices of a tree are the lexicographically least and most vertices in the tree.

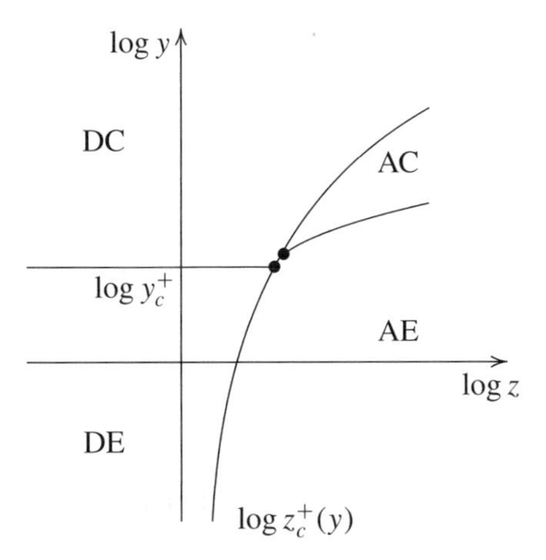

Fig. 6.7: The hypothetical phase diagram for adsorbtion and collapsing positive attached trees in three and more dimensions.

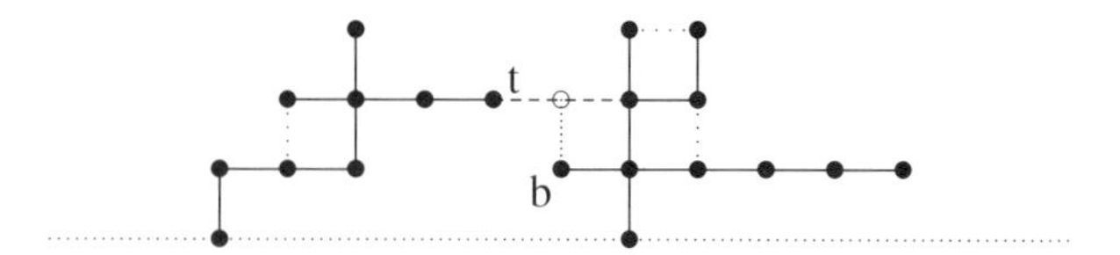

Fig. 6.8: Concatenation of two positive trees. If the tree on the right is shifted two more steps towards the left, then there will be an intersection between them. The trees are concatenated by adding the vertex o and an edge to each tree. This may create as many as $2(d-1)$ new contacts in d dimensions, as well as an extra visit, if o is in the adsorbing surface. The top vertex and bottom vertex are indicated by t and b respectively; they need not be involved in the concatenation.

Theorem 6.23 *There exist functions $\mathcal{T}_v$ and $\mathcal{T}_v^+$ which are the free energies of models of adsorbing and collapsing trees in d dimensions:*

$$\mathcal{T}_v(y,z) = \lim_{n\to\infty} \frac{1}{n} \log t_n^{>}(y,z), \quad \mathcal{T}_v^+(y,z) = \lim_{n\to\infty} \frac{1}{n} \log t_n^+(y,z),$$

for all $y < \infty$ and $z < \infty$. Moreover, $\mathcal{T}_v(y,z)$ and $\mathcal{T}_v^+(y,z)$ are convex functions of both their arguments, and are non-decreasing, continuous, and differentiable almost everywhere.

Proof The proof for positive attached trees will be given; the proof for attached trees is similar. Let T_1 and T_2 be two positive trees with n_1 edges, v_1 visits and k_1 contacts, and n_2 edges, $v - v_1$ visits and $k - k_1$ contacts, respectively (see Fig. 6.8). Translate T_2 parallel to the $Z = 0$ hyperplane until its bottom vertex has the same coordinates as the top vertex of T_1, except for the X- and Z-coordinates. Translate T_2 in the X-direction until the X-coordinate of its bottom vertex is two lattice steps bigger than the X-coordinate of the top vertex of T_1. Next translate T_2 in the negative X-direction, until there is a vertex in T_2 which is within a distance of two steps from a vertex in T_1. At this point, there are no contacts between vertices in T_1 and in T_2. Let w_1 in T_1 and w_2 in T_2 be two vertices with least Z-coordinates and which are exactly two steps apart. Concatenate T_1 and T_2 by adding two new edges and one vertex between w_1 and w_2. The new vertex may be adjacent to at most $2d$ vertices in the two trees, and so as many as $2(d-1)$ new contacts may be created. In addition, if the new vertex is in the adsorbing plane, then there is also a new visit. Thus, a new tree with $n_1 + n_2 + 2$ edges and $v + i$ visits ($i = 0$ or $i = 1$) and $k + j$ contacts ($j \in \{0, 1, \ldots, 2(d-1)\}$) is created. Each distinct pair of trees will give a different outcome, thus

$$\sum_{k_1=0}^{k} \sum_{v_1=0}^{v} t_{n_1}^+(k_1, v_1) t_{n_2}^+(k-k_1, v-v_1) \leq \sum_{i=0}^{1} \sum_{j=0}^{2(d-1)} t_{n_1+n_2+2}^+(k+j, v+i). \qquad (\dagger)$$

In other words, $t_{n-2}^+(k,v)$ is a generalized supermultiplicative function. Multiply this equation by $y^k z^v$, and sum over k and v. Define $\phi(z) = z + 1 + z^{-1}$. Then

$$t_{n_1}^+(y,z) t_{n_2}^+(y,z) \leq [\phi(y)]^{2(d-1)} [\phi(z)] t_{n_1+n_2+2}^+(y,z).$$

Thus, $t^+_{n-2}(y,z)/\left([\phi(y)]^{2(d-1)}[\phi(z)]\right)$ is a supermultiplicative function. In addition, since there are at most $n+1$ visits and $(d-1)n$ contacts in a tree,

$$t^+_n(y,z) \leq \begin{cases} t^+_n, & \text{if } y \leq 1 \text{ and } z \leq 1; \\ t^+_n z^{n+1}, & \text{if } y \leq 1 \text{ and } z > 1; \\ t^+_n y^{(2d-1)n}, & \text{if } y > 1 \text{ and } z \leq 1; \\ t^+_n y^{(2d-1)n} z^{n+1}, & \text{if } y > 1 \text{ and } z > 1. \end{cases}$$

But there is a finite constant $K > 0$ such that $t^+_n < K^n$ by Lemma 6.1, thus $t^+_n(y,z)$ is bounded exponentially in n for all finite values of $y \geq 0$ and $z \geq 0$. Therefore, the claimed limit exists (Lemma A.1). □

The construction in this theorem is also enough to prove that the density functions exists. Notice that eqn (†) in Theorem 6.23 gives

$$\sum_{v_1=0}^{v} t^+_{n_1}(k_1, v_1)t^+_{n_2}(k_2, v-v_1) \leq \sum_{i=0}^{1}\sum_{j=0}^{2(d-1)} t^+_{n_1+n_2+2}(k_1+k_2+j, v+i). \tag{6.32}$$

Multiply this by z^v and sum over v:

$$t^+_{n_1}(k_1;z)t^+_{n_2}(k_2;z) \leq [\phi(z)] \sum_{j=0}^{2(d-1)} t^+_{n_1+n_2+2}(k_1+k_2+j;z). \tag{6.33}$$

Thus, $t^+_{n-2}(k;z)$ satisfies Assumptions 3.8, and by Corollary 3.14 and Theorem 3.16 there are integrated density functions such that the density function may be defined by

$$\mathcal{O}^+_v(\epsilon;z) = \lim_{n\to\infty}[t^+_{n-2}(\lfloor \epsilon n \rfloor + \sigma_n;z)]^{1/n} = \min\{\mathcal{O}^+_v(\leq\epsilon;z), \mathcal{O}^+_v(\geq\epsilon;z)\}, \tag{6.34}$$

where $\sigma_n = o(n)$ is defined as in Theorem 3.16. Similarly, it may be shown that there is a density function of visits at a given contact activity.

Corollary 6.24 *The density functions of contacts at a given visit activity, and of visits at a given contact activity, exist and are defined by*

$$\mathcal{O}^+_v(\epsilon;z) = \lim_{n\to\infty}[t^+_{n-2}(\lfloor \epsilon n \rfloor + \sigma_n;z)]^{1/n},$$
$$\mathcal{O}^+_v(y;\delta) = \lim_{n\to\infty}[t^+_{n-2}(y;\lfloor \delta n \rfloor + \sigma'_n)]^{1/n},$$

where $\sigma_n = o(n)$ and $\sigma'_n = o(n)$ are defined as in Theorem 3.16. □

There also exists a joint density function, which may conveniently be defined by using Theorem 3.19:

$$\log \mathcal{O}^+_v(\epsilon,\delta) = \inf_{z>0}\{\log \mathcal{O}^+_v(\epsilon;z) - \delta \log z\}. \tag{6.35}$$

The growth constant of positive attached trees in d dimensions, τ^+_d, is defined by $\mathcal{T}^+_v(1,1) = \log \tau^+_d$.

Lemma 6.25 $\tau_{d-1} < \tau_d^+ = \tau_d$.

Proof Let $t_n^{(d)}$ be the number of trees, counted up to translation, in the d-dimensional hypercubic lattice. Let T be a tree in the hyperplane $Z = 0$. Then T is a $(d-1)$-dimensional tree, and an edge in the Z-direction may be added to any vertex of T to create a positive tree. If k edges are added to the tree in this way in $\binom{n+1}{k}$ ways, then

$$\binom{n+1}{k} t_n^{(d-1)} \leq t_{n+k}^+.$$

Let $k = \lfloor \epsilon n \rfloor$ in this, take the $(1/n)$-th power of this, and let $n \to \infty$. This gives

$$\left[\frac{(\tau_d^+)^{-\epsilon}}{\epsilon^\epsilon (1-\epsilon)^{(1-\epsilon)}}\right] \tau_{d-1} \leq \tau_d^+.$$

The factor in the square brackets is equal to its maximum $(1 + 1/\tau_d^+)$ when $\epsilon = 1/(1 + \tau_d^+)$. This proves the inequality.

On the other hand, each attached tree counted by t_n can be translated normal to the plane $Z = 0$ until it is a positive tree. Since at most $n + 2$ such trees can be translated to the same positive tree, $t_n^+ \leq t_n \leq (n+2)t_n^+$, and by Corollary 6.3, $\tau_d^+ = \tau_d$. □

Non-analyticities in the free energy will signal thermodynamic phase transitions in the models above. In particular, there should be critical lines in the phase diagram which corresponds to a collapse transition (these are the θ-points), and to an adsorption transition. The θ-points are thought to be tricritical, and if $z = 1$, then the singularity in $\mathcal{T}_v(y, 1)$ is expected to have the general form suggested by eqn (2.24). More generally, it is not unreasonable to expect that the θ-transition will occur for all values of the visit activity z corresponding to a phase of desorbed trees.

The critical exponents for adsorbing trees have been computed using various means. Consider the pure adsorption problem first. Let $t_n^+(v|r)$ be the number of positive trees, with partition function $t_n^+(z|r)$, and which is rooted at one vertex in the hyperplane $Z = 0$. The natural finite size scaling ansatz is $t_n^+(z|r) \sim n^{-\theta_1}\mu_t^n$, where θ_1 is a "surface entropic exponent". The value of θ_1 may be estimated as follows. Notice that $t_n^+(v|r) = vt_n^+(v)$, since any visit can be rooted. Thus, $t_n^+(z|r) = \sum_{v=0}^{n+1} vt_n^+(v)z^v = \langle v \rangle_z t_n^+(z)$. Now if $t_n^+(z|r) \sim n^{-\theta_1}\mu_t^n$, and $t_n^+(z) \sim n^{-\theta}\mu_t^n$, then $[t_n^+(z|r)/t_n^+(z)] \sim n^{\theta-\theta_1}$, and so if $z \leq z_c$ then $\theta = \theta_1$, and $z > z_c$ then $\theta = \theta_1 + 1$. See reference [61] for more on this. The adsorption transition has cross-over exponent $\phi = 1/2$ [70, 204]; this exponent may be "superuniversal" [179] (that is, independent of dimensions). At the collapse and adsorption critical points (the "special point") it has been argued that $\theta_s = [(d-3)/(d-2)](\theta - 1) + (2 - \phi)$ [204, 348]. The generating function $T^+(x, z) = \sum_{n \geq 0} t_n^+(z)x^n$ has a singularity diagram similar to Fig. 2.5, and by using the values of θ_s, it is possible to compute the values of α_+ and α_-. These exponents will be unaffected in the expanded phase if there is also a nearest-neighbour interaction in the trees, but little is known about adsorbing trees at the θ-point or in the collapsed phase. The value of θ_s is not known at the θ-point, but for adsorbing animals with a cycle fugacity at the θ-point it was argued that the adsorption cross-over exponent is $\phi_s = 1/3$ [313]. For trees with a contact fugacity there is no estimate for this exponent.

6.4.2 The phase diagram of adsorbing and collapsing trees

In this section I discuss the adsorption transition in the model of collapsing and adsorbing positive trees. It will turn out that there is a critical value, $z^+(y)$, of the visit activity, where adsorption of the tree occurs in the model of positive trees. Secondly, the line of θ-points separating the DE-phase from the DC-phase in Fig. 6.7 will be shown to be a straight line; if it is there, and given a mild condition on $z^+(y)$.

Let $t_n^+(k) = \sum_{v\geq 0} t_n^+(k, v)$ be the number of positive trees with k contacts, and define the partition function $t_n^+(y) = \sum_{k\geq 0} t_n^+(k)y^k$. The limiting free energy defined by

$$\mathcal{T}_v^+(y) = \lim_{n\to\infty} \frac{1}{n} \log t_n^+(y), \tag{6.36}$$

exists by the same arguments as in Theorem 6.23. Moreover, since any tree with v visits can be translated one step in the Z-direction to find a positive tree with zero visits,

$$t_n^+(k, 0) \leq t_n^+(k) \leq (n+1)t_n^+(k). \tag{6.37}$$

The consequence of this is that the free energies of trees with only a contact-activity is equal to the free energy of adsorbing trees if the visit activity is zero:

$$\mathcal{T}_v^+(y, 0) = \mathcal{T}_k^+(y) = \mathcal{T}_k(y). \tag{6.38}$$

These observations are all that is needed to show that there is an adsorption transition in this model.

Theorem 6.26 *For every finite value of $y > 0$, the limiting free energy $\mathcal{T}_v^+(y, z)$ is independent of z for all $z \leq 1$ (that is, $\mathcal{T}_v^+(y, z) = \mathcal{T}_k(y)$ for all $z \leq 1$).*

Proof Consider any positive tree with v visits and k contacts. Such a tree can be translated one step in the Z-direction to find a positive tree with zero visits, and k contacts. Thus $t_n^+(k, v) \leq t_n^+(k, 0)$. Use this, and the fact that $z \leq 1$, in the following string of inequalities:

$$\sum_{k\geq 0} t_n^+(k,0)y^k \leq t_n^+(y,z) \leq \sum_{k\geq 0}\sum_{v\geq 0} t_n^+(k,0)y^k z^v \leq (n+1)\sum_{k\geq 0} t_n^+(k,0)y^k.$$

Take logarithms, divide by n and let $n \to \infty$. This shows that there is a limiting free energy $\mathcal{T}_k(y) = \mathcal{T}_v^+(y, 0)$ for a model of self-interacting trees, and that $\mathcal{T}_v^+(y, z) = \mathcal{T}_v(y)$ for all $z \leq 1$. □

Suppose now that there is a collapse transition in this model at $y = y_c^+$ for a given $z \leq 1$. Then Theorem 6.26 implies that there is a collapse transition at y_c^+ for all values of $z \leq 1$; the critical curve of θ-points $y_c^+(z)$ is a straight line in Fig. 6.7 for all $z \leq 1$. Theorem 6.26 also suggests that there may be an adsorption transition in this model. Since $\mathcal{T}_v^+(y, z)$ is a constant function of z if $z \leq 1$, it is only needed to show that it is a non-constant function of z for some value of $z > 1$ to prove that it is a non-analytic function of z.

Lemma 6.27 *For all finite $y > 0$ and $z > 1$,*

$$\max\{\mathcal{T}_k(y), \mathcal{T}_k^{(d-1)}(y) + \log z\} \le \mathcal{T}_v^+(y, z) \le \mathcal{T}_k(y) + \log z,$$

where $\mathcal{T}_k^{(d-1)}(y)$ is the limiting free energy of trees with a contact activity in $(d-1)$ dimensions.

Proof Since $t_n^+(y, z)$ is a non-decreasing function of z, $\mathcal{T}_v^+(y, 1) \le \mathcal{T}_v^+(y, z)$, for all $z \ge 1$. By picking out only those terms in the sum of $t_n^+(y, z)$ with $n+1$ visits, completely adsorbed trees are found (there are $t_n^{(d-1)}(k)$ such trees with k contacts in $(d-1)$ dimensions) and

$$\sum_{k \ge 0} t_n^{(d-1)}(k) y^k z^{n+1} \le t_n^+(y, z).$$

Take the logarithm of the above, divide by n and let $n \to \infty$. Then

$$\mathcal{T}_k^{(d-1)}(y) + \log z \le \mathcal{T}_v^+(y, z).$$

The upper bound is obtained by noting that the maximum value of v is $n+1$, and that $z \ge 1$. Thus, put $v = n+1$ in the factor z^v in $t_n^+(y, z)$; then

$$t_n^+(y, z) \le z^{n+1} \sum_{k \ge 0} t_n^+(k) y^k.$$

The bound follows if the logarithm is taken, the result is divided by n and $n \to \infty$. □

The lower bound on $\mathcal{T}_v^+(y, z)$ in Lemma 6.27 suggests that it will become a non-constant function of z (for fixed y) at some value of z in the interval $[0, \mathcal{T}_k(y) - \mathcal{T}_k^{(d-1)}(y)]$. In other words, for every $y \ge 0$ there is a non-analyticity in $\mathcal{T}_v^+(y, z)$ somewhere in this interval.

Theorem 6.28 *The limiting free energy of self-interacting positive trees interacting with a surface, $\mathcal{T}_v^+(y, z)$, is a non-analytic function of z for every finite value of $y \ge 0$. Moreover, the phase boundary $z_c^+(y)$ is in the interval* $[0, \log \tau_d - \frac{1}{2} \log \tau_{d-1}]$ *if $y \le 1$ and in the interval* $[0, \log \tau_d - \log \tau_{d-1} + (d-1) \log y]$ *if $y \ge 1$.*

Proof Since $\mathcal{T}_v^+(1, 1) = \mathcal{T}_v(1, 1) = \log \tau_d$, and from Theorem 6.26 and Lemma 6.27, for every finite $y \ge 0$ there must be a non-analyticity in $\mathcal{T}_v^+(y, z)$ at

$$z_c^+(y) = \sup\{z | \mathcal{T}_v^+(y, z) = \mathcal{T}_k(y)\}.$$

In addition, the location of $z_c^+(y)$ is in the interval $[0, \mathcal{T}_k(y) - \mathcal{T}_k^{(d-1)}(y)]$. If $y \ge 1$, then the maximum number of contacts in the tree is $(d-1)n$, so that $t_n^+(y, 1) \le \sum_{k \ge 0} t_n^+(k) y^{(d-1)n} = t_n^+ y^{(d-1)n}$. Thus, $\mathcal{T}_k(y) \le \log \tau_d + (d-1) \log y$. In addition, $t_n^+(y, 1) \ge \sum_{k \ge 0} t_n^+(k) = t_n^+$. Thus, $\mathcal{T}_k^{(d-1)}(y) \ge \log \tau_{d-1}$, and so $\mathcal{T}_k(y) - \mathcal{T}_k^{(d-1)}(y) \le \log \tau_d - \log \tau_{d-1} + (d-1) \log y$ if $y \ge 1$.

If $y \le 1$, then $t_n^+(y, 1) = \sum_{k \ge 0} t_n^+(k) y^k \le t_n^+$, so that $\mathcal{T}_k(y) \le \log \tau_d$. On the other hand, if every edge in the trees counted by t_n^+ is subdivided, then $t_n^+ \le t_{2n}^+(0)$, since the resulting trees will have no contacts. Thus, $t_n^+(y, 1) \ge t_n^+(0)$ implies that $\mathcal{T}_v^{(d-1)}(y) \ge \frac{1}{2} \log \tau_{d-1}$. This completes the proof if $y \le 1$. □

Theorem 6.28 establishes the existence of the critical curve $z_c^+(y)$ of adsorption transitions in Fig. 6.7. It was already argued that if there is a θ-transition for some $z \leq 1$, then there is a straight line of θ-transitions in the phase diagram for $z \leq 1$. In the next theorem an assumption that the critical curve $z_c^+(y)$ is continuous at $y = y_c^!$ shows that the line of θ-transitions is in fact straight until it meets the curve of adsorption transitions.

Theorem 6.29 *Assume that $\mathcal{T}_v^+(y, 1)$ is singular at $y = y_c^+$, and that the phase boundary $z_c^+(y)$ is continuous at $y = y_c^+$. Then $\mathcal{T}_v^+(y, z)$ is singular at $y = y_c^+$ for every $z \leq z_c^+(y_c^+)$.*

Proof Let $\epsilon > 0$ and choose z_d by

$$z_d < \inf\{z_c^+(y) \big| y \in [y_c^+ - \epsilon, y_c^+ + \epsilon]\}.$$

Since $\mathcal{T}_v^+(y, z)$ is analytic for all such z_d, as long as $y \neq y_c^+$ and $y \in [y_c^+ - \epsilon, y_c^+ + \epsilon]$, it follows that $\mathcal{T}_v^+(y, z_d) = \mathcal{T}_v^+(y, 1)$ for $y \in [y_c^+ - \epsilon, y_c^+ + \epsilon]$ (if this is not so, then there is a phase boundary at (z_d, y_d^+) with y_d^+ some point in $[y_c^+ - \epsilon, y_c^+ + \epsilon]$; this is a contradiction). By taking ϵ small z_d can approach $z_c^+(y_c^+)$, and the result follows. □

By using identical arguments to the above, it is possible to show that all the results obtained for positive attached trees are also true for a model of attached trees (interacting with a defect plane). Notice that since $t_n^+(k, v) \leq t_n(k, v)$, it follows that

$$\mathcal{T}_v^+(y, z) \leq \mathcal{T}_v(y, z), \tag{6.39}$$

and in Theorem 6.30 below it will be seen that if $z \leq 1$, then $\mathcal{T}_v^+(y, z) = \mathcal{T}_v(y, z)$.

Theorem 6.30 *The limiting free energy of self-interacting attached trees is a non-analytic function of z for every finite value of $y \geq 0$, and the phase boundary $z_c(y)$ is bounded by $z_c(y) \in [0, \log \tau_d - \frac{1}{2} \log \tau_{d-1}]$ if $y \leq 1$ and $z_c(y) \in [0, \log \tau_d - \log \tau_{d-1} + (d-1) \log y]$ if $y \geq 1$.*

Proof That $\mathcal{T}_v(y, z) = \mathcal{T}_v(y, 1)$ for all finite $y > 0$ and $z \leq 1$ is seen as follows: by eqn (6.39) and Theorem 6.26, note that $\mathcal{T}_k(y) = \mathcal{T}_v^+(y, z) \leq \mathcal{T}_v(y, z) \leq \mathcal{T}_v(y, 1) = \mathcal{T}_k(y)$. Thus $\mathcal{T}_v^+(y, z) = \mathcal{T}_v(y, z)$ if $z \leq 1$. The rest of the results are obtained by using arguments analogous to those for positive attached trees. □

6.4.3 *The location of the adsorption transition*

In this section it will be shown that positive trees adsorb at a critical activity $z_c^+(y) > z_c(y) \geq 1$ for any finite $y \geq 0$. In particular, there exists a non-increasing function $K(y) > 0$ such that

$$\log z_c^+(y) - \log z_c(y) > K(y) > 0, \tag{6.40}$$

whenever $y \geq 0$. The proof will rely on the use of density functions, and in particular, on the density functions $\mathcal{O}_v^+(y; \delta)$ and $\mathcal{O}_v(y; \delta)$ of visits, as defined in Corollary 6.24.

Notice that

$$\log \mathcal{O}_v^+(y;0) = \log \mathcal{O}_v(y;0) = \mathcal{T}_k(y). \tag{6.41}$$

The free energies are constant functions of z if $z \leq z_c(y)$; this implies that Lemma 3.20 can be used to determine the location of the critical point; in particular,

$$\begin{aligned} \log z_c^+(y) &= -\left[\frac{d^+}{d\epsilon}\log \mathcal{O}^+(y;\epsilon)\right]\Big|_{\epsilon=0^+}; \\ \log z_c(y) &= -\left[\frac{d^+}{d\epsilon}\log \mathcal{O}(y;\epsilon)\right]\Big|_{\epsilon=0^+}. \end{aligned} \tag{6.42}$$

Thus, by using the definition of the right-derivatives, subtracting the second from the first, and using eqn (6.41), the following lemma is proven.

Theorem 6.31 *For every finite value of $y \geq 0$,*

$$\log z_c^+(y) - \log z_c(y) = e^{-\mathcal{T}_k(y)} \lim_{\epsilon\to 0^+} [(\mathcal{O}_v(y;\epsilon) - \mathcal{O}_v^+(y;\epsilon))]/\epsilon. \qquad \square$$

Thus, the critical curves $z_c(y)$ and $z_c^+(y)$ can be compared if more is known about the density functions. By eqn (6.42) it follows that the right-derivatives of the density functions are finite at $\epsilon = 0$. The next step is to find a relationship between the density functions in Theorem 6.31.

Consider a tree counted by $t_n^+(k, \lfloor \epsilon n \rfloor)$ (see Fig. 6.9). These are all positive trees, and they can be changed into attached trees by adding new edges in the negative Z-direction on the visits. However, it will complicate matters if edges are added on adjacent visits; this will increase the number of contacts. Instead, choose from those visits whose coordinates add to either an odd number, or to an even number (whichever is more), in each tree. This means that there will be at least $\lfloor \epsilon n/2 \rfloor$ visits to choose from. If $\lfloor \delta n \rfloor$ visits are chosen from $\lfloor \epsilon n/2 \rfloor$ visits, then

$$\binom{\lfloor \epsilon n/2 \rfloor}{\lfloor \delta n \rfloor} t_n^+(k, \lfloor \epsilon n \rfloor) \leq t_{n+\lfloor \delta n \rfloor}(k, \lfloor \epsilon n \rfloor). \tag{6.43}$$

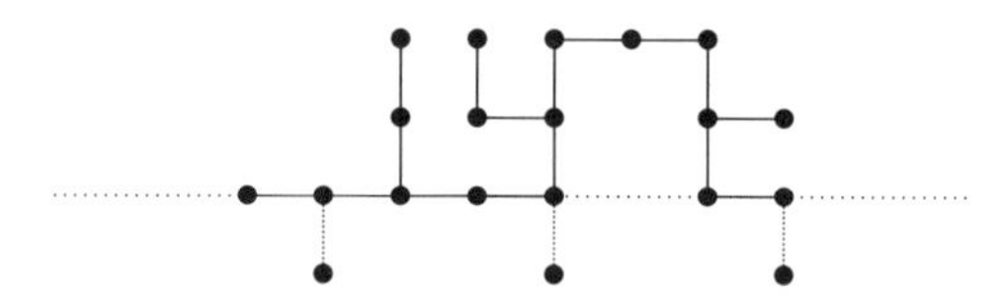

Fig. 6.9: By adding edges in the $-Z$-direction to the visits of this positive attached tree, an attached tree is found.

Multiply this by y^k, sum over k, take the $(1/n)$-th power, and let $n \to \infty$. Then

$$\left[\frac{(\epsilon/2)^{\epsilon/2}}{\delta^\delta(\epsilon/2-\delta)^{\epsilon/2-\delta}}\right] \mathcal{O}_v^+(y;\epsilon) \leq [\mathcal{O}_v(y;\epsilon/(1+\delta))]^{1+\delta}. \tag{6.44}$$

Notice that the free energy of collapsing trees is given by

$$\mathcal{T}_k(y) = \sup_{0\leq\epsilon\leq 1} \log \mathcal{O}_v(y;\epsilon) \geq \log \mathcal{O}_v(y;\epsilon), \tag{6.45}$$

Thus, eqn (6.44) may be written as

$$\left[\frac{(\epsilon/2)^{\epsilon/2}e^{-\delta\mathcal{T}_k(y)}}{\delta^\delta(\epsilon/2-\delta)^{\epsilon/2-\delta}}\right] \mathcal{O}_v^+(y;\epsilon) \leq \mathcal{O}(y;\epsilon/(1+\delta)). \tag{6.46}$$

The factor in square brackets is a maximum if $\delta = \epsilon/2\left(1+e^{\mathcal{T}_k(y)}\right)$ in which case eqn (6.46) becomes the inequality stated in Lemma 6.32.

Lemma 6.32 *The density functions of visits in self-interacting positive attached trees, and in self-interacting attached trees, are related by*

$$\left(1+e^{-\mathcal{T}_k(y)}\right)^{\epsilon/2} \mathcal{O}_v^+(y;\epsilon) \leq \mathcal{O}_v(y;\epsilon/(1+\delta)),$$

where $\mathcal{T}_k(y)$ is the free energy of trees with a contact activity, and where $\delta = \epsilon/2(1+e^{\mathcal{T}_k(y)})$. □

The result in Lemma 6.32 can now be used to show that positive trees adsorb at a strictly positive value of the visit activity.

Theorem 6.33 *For every finite value of $y \geq 0$,*

$$\log z_c^+(y) - \log z_c(y) \geq \frac{1}{2}\log(1+e^{-\mathcal{T}(y)}) > 0.$$

Proof From Theorem 6.31 and Lemma 6.32 note that

$$\log[z_c^+(y)/z_c(y)] \geq \lim_{\epsilon\to 0^+} \frac{1}{\epsilon}\left[1 - \frac{\mathcal{O}_v(y;\epsilon/(1+\delta))}{\mathcal{O}_v(y;\epsilon)}\left[1+e^{-\mathcal{T}_k(y)}\right]\right]^{-\epsilon/2}, \tag{†}$$

where the fact that $e^{-\mathcal{T}_k(y)}\mathcal{O}_v(y;0) = 1$ and eqn (6.41) were used, and where δ is given by $\delta = \epsilon/2(1+e^{\mathcal{T}_k(y)})$. Since $z_c^+(y) \geq z_c(y) \geq 0$, it follows that the right-derivatives of $\log \mathcal{O}_v^+(y;\epsilon)$ and $\log \mathcal{O}_v(y;\epsilon)$ are non-positive. Suppose first that the right-derivative of $\log \mathcal{O}_v(y;\epsilon)$ at $\epsilon = 0$ is zero. In this case $\mathcal{O}_v(y;\epsilon) = \mathcal{O}_v(y;0) + O(\epsilon^2)$ so that

$$\frac{\mathcal{O}_v(y;\epsilon/(1+\delta))}{\mathcal{O}_v(y;\epsilon)} = 1 + O(\epsilon^2).$$

Substitute this into eqn (†) and take the limit to obtain the desired result. On the other hand, if the right-derivative of $\log \mathcal{O}_v(y;\epsilon)$ is negative, then for a givenvalue of $\epsilon > 0$, there is an $\eta > 0$ such that

$$\mathcal{O}_v(y;0) + \epsilon(1+\eta)\left[\frac{d^+}{d\epsilon}\mathcal{O}'_v(y;\epsilon)\right]\bigg|_{\epsilon=0^+} \leq \mathcal{O}_v(y;\epsilon)$$
$$\leq \mathcal{O}_v(y;0) + \epsilon(1-\eta)\left[\frac{d^+}{d\epsilon}\mathcal{O}'_v(y;\epsilon)\right]\bigg|_{\epsilon=0^+},$$

and where one may take $\eta \to 0^+$ if $\epsilon = 0^+$. These bounds can be used to show that there exists a finite number η_2 (possibly dependent on y), such that

$$\frac{\mathcal{O}_v\left(y;\epsilon/(1+\delta)\right)}{\mathcal{O}_v(y;\epsilon)} \leq 1 - \epsilon\left(2\eta + \frac{\delta}{1+\delta}\right)\left[\frac{d^+}{d\epsilon}\log\mathcal{O}_v(y;\epsilon)\right]\bigg|_{\epsilon=0^+} + \eta_2\epsilon^2.$$

Notice that $\mathcal{O}'(y;0^+) \leq 0$, and that $\delta \leq \epsilon/2$. Thus,

$$\frac{\mathcal{O}_v\left(y;\epsilon/(1+\delta)\right)}{\mathcal{O}_v(y;\epsilon)} \leq 1 - 2\epsilon\eta\left[\frac{d^+}{d\epsilon}\log\mathcal{O}_v(y;\epsilon)\right]\bigg|_{\epsilon=0^+}$$
$$+\left(\eta_2 - \frac{1}{2}\left[\frac{d^+}{d\epsilon}\log\mathcal{O}_v(y;\epsilon)\right]\bigg|_{\epsilon=0^+}\right)\epsilon^2.$$

Substitute this last bound in eqn (†), simplify and take the limit $\epsilon \to 0^+$. This gives

$$\log z_c^+(y) - \log z_c(y) \geq \frac{1}{2}\log(1 + e^{-\mathcal{T}_k(y)}) + 2\eta\left[\frac{d^+}{d\epsilon}\log\mathcal{O}_v(y;\epsilon)\right]\bigg|_{\epsilon=0^+}.$$

Since $\epsilon = 0^+$, η can be safely taken to 0^+ to finish the proof. □

Lastly, choose $K(y) = \frac{1}{2}\log(1 + e^{-\mathcal{T}_k(y)})$ to find eqn (6.40).

6.4.4 Excursions and roots in the adsorption transition

Those maximal subtrees with every edge (but not every vertex) in the half-space $Z > 0$ are called *excursions*. A maximal subtree which is completely contained in the hyperplane $Z = 0$ is an *incursion*. The edges of an excursion which are incident with visits are called *roots*.

Let $t_n^+(k, v, r)$ be the number of positive trees with k contacts, v visits and r roots. Then $v \geq r$ and r of the v visits is incident with a root. Let $t_n^+(y; v, r|[hh])$ be the partition function of positive trees with a visit activity y, and with both bottom and top vertices a distance h above the adsorbing plane. Trees counted by $t_n^+(y; \lfloor \epsilon n \rfloor, \lfloor \rho n \rfloor|[hh])$ can be concatenated, and most popular arguments then show that the limit

$$\mathcal{O}_v^+(y;\epsilon,\rho) = \lim_{n\to\infty}\left[t_n^+(y; \lfloor \epsilon n \rfloor, \lfloor \rho n \rfloor)\right]^{1/n}, \tag{6.47}$$

of positive attached trees with a contact activity y, and a density ϵ of visits, and a density ρ of roots, exists. The partition function of this model is defined by

$$t_n^+(y; \lfloor \epsilon n \rfloor, \lfloor \rho n \rfloor) = \sum_{k\geq 0} t_n^+(k, \lfloor \epsilon n \rfloor, \lfloor \rho n \rfloor)y^k. \tag{6.48}$$

A tree counted by $t_n^+(v, c, r)$ is illustrated in Fig. 6.10. Choose m of the visits and append m edges in the $-Z$-direction; after translating the tree one step in the Z-direction, these edges will be roots. Thus

$$\binom{v}{m} t_n^+(k, v, r) \leq \sum_{i=0}^{2(d-1)m} t_{n+m}^+(k+i, m, m), \tag{6.49}$$

since each of the m new vertices may have as many as $2(d-1)$ contacts. Multiply eqn (6.49) by y^k and sum over k; this gives a relation between the partition functions:

$$\binom{v}{m} t_n^+(y; v, r) \leq \left[\sum_{i=0}^{2(d-1)m} y^{-i}\right] t_{n+m}^+(y; m, m). \tag{6.50}$$

Let $v = \lfloor \epsilon n \rfloor$, $r = \lfloor \rho n \rfloor$ and $m = \lfloor \delta n \rfloor$, where $\epsilon \geq \rho$ and $\epsilon \geq \delta$. Define $\phi(y) = y + 1 + y^{-1}$, and take the $(1/n)$-th power and let $n \to \infty$. This gives

$$\left[\frac{\epsilon^\epsilon}{\delta^\delta(\epsilon-\delta)^{\epsilon-\delta}}\right] \mathcal{O}_v^+(y; \epsilon, \rho) \leq [\phi(y)]^\delta \left[\mathcal{O}_v^+\left(y; \frac{\delta}{1+\delta}, \frac{\delta}{1+\delta}\right)\right]^{1+\delta}, \tag{6.51}$$

and since $\mathcal{O}_v^+(y; \epsilon, \delta) \leq e^{\mathcal{T}_k(y)}$,

$$\left[\frac{\epsilon^\epsilon}{\delta^\delta(\epsilon-\delta)^{\epsilon-\delta}}\right] \mathcal{O}_v^+(y; \epsilon, \rho) \leq \phi(y)^\delta e^{\delta \mathcal{T}_k(y)} \mathcal{O}_v^+\left(y; \frac{\delta}{1+\delta}, \frac{\delta}{1+\delta}\right). \tag{6.52}$$

Theorem 6.34 *For every $\epsilon > 0$ in $(0, 1]$ and $\rho \in [0, \epsilon]$*

$$\left(1 + \phi(y)^{-1} e^{-\mathcal{T}_k(y)}\right)^\epsilon \mathcal{O}_v^+(y; \epsilon, \rho) \leq \mathcal{O}_v^+(y; \delta_*, \delta_*),$$

where

$$\delta_* = \frac{\delta}{1+\delta}, \quad \text{and} \quad \delta = \frac{\epsilon}{1+\phi(y)e^{\mathcal{T}_k(y)}} < \epsilon.$$

Proof Write eqns (6.51) and (6.52) in the following form:

$$\left[\frac{\epsilon^\epsilon \phi(y)^{-\delta} e^{-\delta \mathcal{T}_k(y)}}{\delta^\delta(\epsilon-\delta)^{\epsilon-\delta}}\right] \mathcal{O}_v^+(y; \epsilon, \rho) \leq \mathcal{O}_v^+\left(y; \frac{\delta}{1+\delta}, \frac{\delta}{1+\delta}\right).$$

The maximum of the left-hand side of this inequality is obtained when $\delta = \epsilon/(1 + \phi(y)e^{\mathcal{T}_k(y)})$. This proves the theorem. □

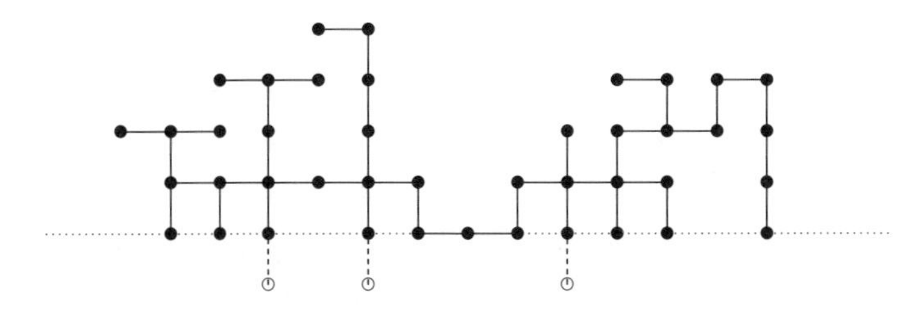

Fig. 6.10: Appending the dashed edges, and translating the entire tree one step in the Z-direction gives a tree with three visits and three roots.

An immediate consequence of Theorem 6.34 is that $z_c^+(y) > 1$, for all finite $y \geq 0$. To see this, suppose that $z > z_c^+(y)$, so that there is a density of visits. Then Theorem 3.18 implies that there is a $\rho_* \geq 0$ and an $\epsilon_* > 0$ such that

$$\mathcal{T}_v^+(y,z) = \log \mathcal{O}_v^+(y;\epsilon_*,\rho_*) + \epsilon_* \log z \geq \log \mathcal{O}_v^+(y;\delta_*,\delta_*) + \delta_* \log z, \tag{6.53}$$

where δ_* is defined in Theorem 6.34. On the other hand, by Theorem 6.34,

$$\log[\mathcal{O}_v^+(y;\epsilon_*,\rho_*) z^{\epsilon_*}] \leq \log\left[\mathcal{O}^+(y;\delta_*,\delta_*) z^{\delta_*}[z^{\epsilon_*-\delta_*}/\Delta]\right], \tag{6.54}$$

where $\Delta = (1+\phi(y)^{-1}e^{-\mathcal{T}_k(y)})^{\epsilon_*}$. This is a contradiction if $z^{\epsilon_*-\delta_*} < \Delta$. Let $\delta_* = \gamma\epsilon_*$ where $\gamma < 1$. Then this implies that the above is a contradiction to eqn (6.53) if

$$\log z < \frac{1}{1-\gamma} \log\left(1+\phi(y)^{-1}e^{-\mathcal{T}_k(y)}\right), \tag{6.55}$$

unless $\epsilon_* = 0$, in which case the desorbed phase is found. But then

$$\log z_c^+(y) \geq \frac{1}{1-\gamma} \log(1+\phi(y)^{-1}e^{-\mathcal{T}_k(y)}) > 0.$$

Thus, by examining the density of roots, a proof that the adsorption occurs at a strictly positive value of $\log z$ is found, for all finite $y \geq 0$.

6.4.5 *Excursions and roots in adsorbing trees*

In this section the density of excursions is examined if the contact activity y is equal to 1. In that case, the sums over the contacts in the partition function may be executed, so that contacts play no role in the model. The basic construction is illustrated in Fig. 6.11, where new excursions are created by first translating existing excursions away from the adsorbing plane.

In a positive tree with v visits and r roots, choose $m \leq v - r$ visits, and construct new excursions (these are also roots) as illustrated in Fig. 6.11. The new tree will still have v visits, but there are $m + r$ roots and $n + m$ edges; the number of contacts is not relevent, since $y = 1$. This gives a relation between the partition functions

$$\binom{v-r}{m} t_n^+(1;v,r) \leq t_{n+m}^+(1;v,m+r). \tag{6.56}$$

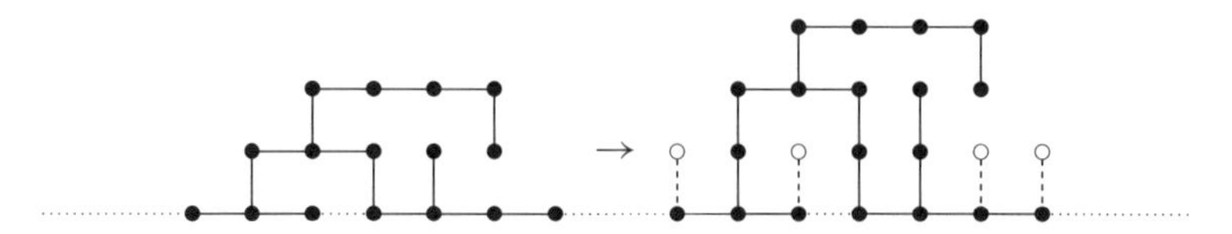

Fig. 6.11: Construction of new roots in an adsorbed tree. The construction has two steps: first subdivide the roots; this creates room for the construction of roots on visits which are not yet incident with a root.

Choose $v = \lfloor \epsilon n \rfloor$, $r = \lfloor \rho n \rfloor$, and $m = \lfloor \eta n \rfloor$, and use the fact that $\mathcal{T}_k^+(1) = \log \tau_d$. Then, after taking the $(1/n)$-th power of eqn (6.56), and taking $n \to \infty$, the result is

$$\left[\frac{(\epsilon - \rho)^{\epsilon-\rho} \tau_d^{-\eta}}{\eta^\eta (\epsilon - \rho - \eta)^{\epsilon-\rho-\eta}} \right] \mathcal{O}_v^+(1; \epsilon, \rho) \le \mathcal{O}_v^+ \left(1; \frac{\epsilon}{1+\eta}, \frac{\rho+\eta}{1+\eta} \right). \tag{6.57}$$

The factor in square brackets is a maximum if $\eta = (\epsilon - \rho)/(1 + \lambda_d)$, in which case

$$\left[\frac{(1 + \tau_d^{-1})^\epsilon}{(1 + \tau_d^{-1})^\rho} \right] \mathcal{O}_v^+(1; \epsilon, \rho) \le \mathcal{O}_v^+ \left(1; \frac{\epsilon}{1+\eta}, \frac{\eta+\rho}{1+\eta} \right). \tag{6.58}$$

This gives the following theorem if $\rho = 0$.

Theorem 6.35 *Put $\eta = \epsilon/(1 + \lambda_d)$ in eqn (6.58) (and $\rho = 0$). Then*

$$(1 + \tau_d^{-1})^\epsilon \, \mathcal{O}_v^+(1; \epsilon, 0) \le \mathcal{O}_v^+ \left(1; \frac{\epsilon}{1+\eta}, \frac{\eta}{1+\eta} \right). \qquad \square$$

Theorem 6.35 cannot be used directly, since the density of visits changed in the construction; this fact makes explicit calculations difficult. The next lemma will help.

Lemma 6.36 *The density function $\mathcal{O}_v^+(1; \epsilon, \rho)$ satisfies the following inequality:*

$$\mathcal{O}_v^+(1; \epsilon, \rho) \le \left[\mathcal{O}_v^+ \left(1; \frac{\epsilon+\delta}{1+\delta}, \frac{\rho}{1+\delta} \right) \right]^{1+\delta}.$$

Proof Let $t_n^+(v, r)$ be the number of positive trees with v visits and r roots. Let t_v be the lexicographically most visit in a tree, and add q edges in the X-direction in the $Z = 0$ hyperplane to t_v one by one. This generates a tree with $v + q$ visits and $n + q$ edges, while there are still r roots. Let $v = \lfloor \epsilon n \rfloor$, $r = \lfloor \rho n \rfloor$ and $q = \lfloor \delta n \rfloor$. Take the $(1/n)$-th power and let $n \to \infty$. This gives the result above. $\square$

If Lemma 6.36 and Theorem 6.35 are combined (and the fact that $\mathcal{O}_v^+(1; \epsilon, \rho) \le \tau_d$ is used), then

$$(1 + \tau_d^{-1})^\epsilon \, \mathcal{O}_v^+(1; \epsilon, 0) \le \tau_d^\delta \, \mathcal{O}_v^+ \left(1; \epsilon, \frac{\eta}{(1+\eta)(1+\delta)} \right), \tag{6.59}$$

where

$$\delta = \frac{\epsilon\eta}{(1-\epsilon)(1+\eta)}. \tag{6.60}$$

But $\eta = \epsilon/(1 + \tau_d)$ in Theorem 6.35, so that

$$\mathcal{O}_v^+(1; \epsilon, 0) \le \left[\frac{\tau_d^{\epsilon^2/((1+\tau_d)(1-\epsilon)(1+\eta))}}{(1 + \tau_d^{-1})^\epsilon} \right] \mathcal{O}_v^+ \left(1; \epsilon, \frac{\epsilon}{(1+\tau_d)(1+\delta)(1+\eta)} \right). \tag{6.61}$$

But for ϵ small enough (but not zero!) this implies that $\mathcal{O}_v^+(1; \epsilon, 0) < \mathcal{O}_v^+(1; \epsilon, \epsilon^\dagger)$ for some $\epsilon^\dagger > 0$. In other words, if the supremum in eqn (6.53) is achieved at a small value of ϵ, then there is a density of roots in the adsorbed phase. Incidentally, since the roots constructed in Fig. 6.11 are also excursions, this implies that there is also a density of excursions. This argument does not work if the contact activity is included; the construction in Fig. 6.11 destroys too many contacts, and Theorem 6.35 cannot be proven.

Lastly, if both sides of eqn (6.59) are multiplied by $z^\epsilon < (1+\tau_d^{-1})^\epsilon$ (and where $z > 1$), then, if ϵ is small enough, $\mathcal{O}_v^+(1;\epsilon,0)z^\epsilon < \mathcal{O}_v^+(1;\epsilon,\epsilon^\dagger) \leq \mathcal{O}_v^+(1;0,0)$. Thus, $\mathcal{T}_v^+(1,z) < \mathcal{T}_v^+(1,0)$. This is a contradiction, unless $\epsilon = 0$, which means that for these values of z the desorbed phase is found. Thus, $z_c^+(1) > (1+\tau_d^{-1})$.

6.4.6 Adsorption of percolation clusters

It is possible to consider the adsorption transition of percolation clusters in a half-space [59] (see also reference [60] and references therein). This is done by defining a model of edge-animals weighted with respect to their perimeters and with respect to the number of visits to a wall. In particular, let $a_n^+(\rho, v)$ be the number of positive attached animals with total perimeter ρ and with v visits to the hyperplane $Z = 0$ (the total perimeter ρ includes all the perimeter edges of the animal, including those which are in the half-space $Z \leq 0$). Note that the number of perimeter edges in the half-space $Z \geq 0$ is $\rho - v$, and define the partition function of these animals by

$$a_n^+(p, z/(1-p)) = \sum_{\rho\geq 0}\sum_{v\geq 0} a_n^+(\rho, v)(1-p)^{\rho-v}z^v. \tag{6.62}$$

By concatenating two animals as in Fig. 6.1, a generalized supermultiplicative inequality

$$\sum_{\rho_1=0}^{\rho}\sum_{v_1=0}^{v} a_{n_1}^+(\rho_1, v_1)a_{n_2}^+(\rho-\rho_1, v-v_1) \leq \sum_{k=0}^{2d-2} a_{n_1+n_2+2}^+(\rho+k, v), \tag{6.63}$$

is obtained. Multiplying by $[z/(1-p)]^v$ and summing over v gives the relation between partition functions

$$\sum_{\rho_1=0}^{\rho} a_{n_1}^+\left(\rho_1; \frac{z}{1-p}\right) a_{n_2}^+\left(\rho-\rho_1; \frac{z}{1-p}\right) \leq \sum_{k=0}^{2d-2} a_{n_1+n_2+2}^+\left(\rho+k; \frac{z}{1-p}\right). \tag{6.64}$$

Thus, $a_{n-2}^+(\rho; z/(1-p))$ satisfies Assumptions 3.8, and there is a limiting free energy defined by

$$\mathcal{A}_\rho^+(p, z/(1-p)) = \lim_{n\to\infty}\frac{1}{n}\log a_n^+(p, z/(1-p)), \tag{6.65}$$

and a density function $\mathcal{Q}_\rho^+(\epsilon; z/(1-p))$ of perimeter edges, as defined in Theorem 3.19. There is an adsorption transition in this model for every value of the percolation probability $p \in (0,1)$; this fact is proven in Theorem 6.37.

Theorem 6.37 *$\mathcal{A}^+_\rho(p, z/(1-p))$ is a non-analytic function of z for every $p \in (0,1)$. Moreover, if the critical curve*

$$z_c^+(p) = \inf\{z \mid \mathcal{A}^+(p, z/(1-p)) \text{ is non analytic at } (p, z/(1-p))\}$$

is defined, then $z_c^+(p) \geq 1-p$.

Proof By translating an animal counted by $a_n^+(\rho, w)$ one step in the positive z-direction, it is seen that

$$a_n^+(\rho, w) \leq a_n^+(\rho, 0) = a_n(\rho), \qquad (\dagger)$$

where $a_n(\rho)$ is the number of animals with n edges and perimeter ρ. By only keeping terms with zero visits, it also follows that if $z < 1-p$, then

$$a_n^+(p, z/(1-p)) \geq \sum_{\rho \geq 0} a_n(\rho)(1-p)^\rho,$$

and from eqn (†),

$$a_n^+(p, z/(1-p)) \leq \left[\sum_{\rho \geq 0} a_n(\rho)(1-p)^\rho\right] \Big/ \left(1 - \frac{z}{1-p}\right).$$

In other words, by taking logarithms of the above inequalities, dividing by n and letting $n \to \infty$,

$$\mathcal{A}^+_\rho(p, z/(1-p)) = \mathcal{A}^+_\rho(p, 0) \qquad \text{if } z < 1-p.$$

Note that $a_n^+(p, z/(1-p)) \geq (1-p)^{2(d-1)n+2} z^{n+1}$ (consider the animal which is a straight line segment (of length $n+1$), and which is completely adsorbed in the hyperplane $Z = 0$ and with total perimeter $(2d-3)n+2$). Thus, the free energy is dependent on z for large enough values of z. In other words, for each p in $(0, 1)$ there is a critical value of z, denoted by $z_c^+(p)$ where adsorption of the cluster occurs. Note also that $z_c^+(p) \geq 1-p$. □

With increasing p a critical percolation point of the *desorbed* cluster might be encountered. That this is so can be demonstrated by considering the probability that the cluster at the origin has exactly n edges:

$$P_n(p) = \sum_{\rho \geq 0} w\, a_n(\rho) p^n (1-p)^\rho, \qquad (6.66)$$

where the number of vertices in the cluster is w, and since $1 \leq w \leq n+1$,

$$\sum_{\rho \geq 0} a_n(\rho) p^n (1-p)^\rho \leq P_n(p) \leq (n+1) \sum_{\rho \geq 0} a_n(\rho) p^n (1-p)^\rho. \qquad (6.67)$$

In other words, if this is compared to the definition of $a_n^+(p, z/(1-p))$ and by using eqn (†) from Theorem 6.37, then

$$\left[\frac{p^{1-n}}{n}\right] P_{n-1}(p) \leq a_n^+(p, z/(1-p)) \leq \left[\frac{p^{-n}}{1 - \frac{z}{1-p}}\right] P_n(p), \qquad (6.68)$$

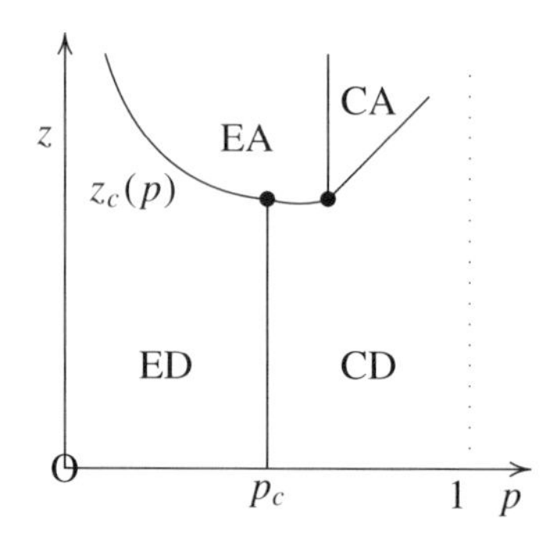

Fig. 6.12: The hypothetical phase diagram of adsorbing percolation clusters. If $z < z_c(p)$ then there are two desorbed phases, one a phase of subcritical clusters (ED), and the other a phase of supercritical clusters (CD). If $z > z_c(p)$ then we are in the adsorbed phase. In this case there is (presumably) also a phase of adsorbed subcritical clusters (EA), and adsorbed supercritical clusters (CA).

as long as $z < 1 - p$. Take the logarithm of the above, divide by n and let $n \to \infty$. By Theorem D.4 (in Appendix D),

$$\mathcal{A}_\rho^+(p, z/(1-p)) = -\log p - \log \zeta(p), \tag{6.69}$$

where $\zeta(p) > 0$ if $p \leq p_c$ and $\zeta(p) = 0$ if $p > p_c$. In other words, if $z < 1 - p$, then there is a non-analyticity in $\mathcal{A}_\rho^+(p, z/(1-p))$ at $p = p_c$.

Theorem 6.38 *$\mathcal{A}_\rho^+(p, z/(1-p))$ is non-analytic at the critical percolation point $p = p_c$ for all $z < 1 - p_c$.* □

If the phase boundary $z_c^+(p)$ is continuous at $p = p_c$, then it is possible to show that there is a critical percolation point at $p = p_c$ for all values of $z < z_c^+(p_c)$.

Theorem 6.39 *Suppose that the phase boundary $z_c^+(p)$ is continuous at $p = p_c$. Then $\mathcal{A}_\rho^+(p, z/(1-p))$ is non-analytic at $p = p_c$ for all values of $z \leq z_c^+(p_c)$.*

Proof Let $\epsilon > 0$ and choose a number z_d by

$$z_d < \inf\{z_c^+(p) \big| \, p \in [p_c - \epsilon, p_c + \epsilon]\}.$$

Notice that $\mathcal{A}_\rho^+(p, z/(1-p))$ is analytic if $p \neq p_c$, $p \in [p_c - \epsilon, p_c + \epsilon]$, and $z < z_d$. By taking ϵ to zero, one may choose z_d arbitrarily close to $z_c^+(p_c)$. □

The point $(p_c, z_c^+(p_c))$ is the meeting point of at least three critical curves in the phase diagram; desorbed subcritical percolation clusters, desorbed supercritical percolation clusters, and adsorbed subcritical percolation clusters are all simultaneously critical at this point. It is unclear what the situation is if $z > z_c^+(p)$. This is the adsorbed phase of the cluster, and there is a non-zero fraction of vertices in the wall $Z = 0$. With increasing p it is certainly possible that the parts of the cluster which are not in the wall may percolate at p_c, but it is unclear whether vertices adsorbed in $Z = 0$ will also percolate at this, or some other, critical point.

6.4.7 Branched copolymer adsorption

Models of branched copolymers can be defined in a variety of ways, depending on a rule for assigning colours to vertices. A simple model of a branched copolymer can be defined by simply colouring vertices in a tree with two colours, say red and blue, alternately. That is, colour a vertex (say the bottom vertex) of a tree, and then continue by colouring all nearest-neighbour vertices of red vertices blue, and all nearest neighbours of blue vertices red. Since the tree is a bipartite graph, this colouring can always be completed. Concatenation of the trees can easily be carried out, and a model of adsorbing trees (with only one colour interacting with the adsorbing plane) can be carried out using arguments similar to those in Section 5.4.6. In particular, there is a limiting free energy and an adsorption transition, and this model of branched copolymer adsorption will not be that much different from the adsorption of linear alternating copolymers (see Section 5.4.6).

Annealed models of adsorbing branched copolymers are obtained if every visit interacts with the adsorbing plane with probability p. Let $t_n^+(v)$ be the number of positive trees with v visits. Then the partition function is $t_n^+(z) = \sum_{v=0}^{n+1} t_n^+(v) \sum_{l=0}^{v} \binom{v}{l} (zp)^l (1-p)^{v-l}$. Evaluating the sum over l gives $t_n^+(z) = \sum_{v=0}^{n+1} t_n^+(v)(zp+1-p)^v$, and the limiting free energy is that of adsorbing branched homopolymers at activity $zp+1-p$: $\mathcal{T}_v^+(zp+1-p)$, where $T_v^+(z) = T_v^+(1,z)$ given in Theorem 6.23. The critical value of z in the annealed model can be read off from the above: it is given by $z_c^a = (z_c - 1)/p + 1$, where z_c is the critical activity for an adsorbing homopolymer. Thus, if $0 \leq p < 1$, then $1 < z_c < z_c^a$.

The situation is more interesting if random quenched models of adsorbing branched polymers are considered. These models are defined by labelling vertices in a tree. Let χ_i be a colour in the probability space Υ. Then a sequence $\chi = (\chi_1, \chi_2, \ldots, \chi_n)$ is in the probability space $\Omega = \Upsilon \times \Upsilon \times \Upsilon \times \cdots \times \Upsilon$. A quenched model of adsorbing trees is obtained by assigning the colours in χ to vertices in the trees. I shall use a lexicographic rule: order the vertices in a tree lexicographically, and assign χ_1 to the first vertex, χ_2 to the second, and so on. Such a model will be called a *lexicographically quenched* model of branched copolymers. The advantage of the above is that the labelling of trees by χ is preserved by concatenation, under certain circumstances.

Simplify the discussion by assuming that there are only two colours (two types of monomers) in χ, and let $t_n^+(r, b|\chi)$ be the number of trees lexicographically coloured by χ, with r red visits and b blue visits, and n edges. The partition function in this model is

$$t_n^+(z_r, z_b|\chi) = \sum_{r,b=0}^{n+1} t_n^+(r, b|\chi) z_r^r z_b^b. \tag{6.70}$$

The first issue is the existence of a limiting free energy in the averaged quenched ensemble. The proof that this exists relies on most popular arguments.

Theorem 6.40 *There exists a limiting free energy in the averaged quenched ensemble, defined by*

$$\mathcal{T}_{qu}^+(z_r, z_b) = \lim_{n\to\infty} \frac{1}{n} \langle \log t_n^+(z_r, z_b|\chi)\rangle_\chi,$$

where the average $\langle\cdot\rangle_\chi$ *is over all random sequences of two colours in* Ω.

Proof Let $t_n^+(r, b|\chi|[h_b h_t])$ be the number of positive trees coloured lexicographically by the sequence χ of two colours (say red and blue), with r red visits and b blue visits in the adsorbing plane $Z = 0$, and with its bottom vertex with Z-coordinate h_b, and top vertex with Z-coordinate h_t. A tree counted by $t_n^+(r - r_1, b - b_1|\chi_0|[hh])$ can be concatenated with a tree counted by $t_m^+(r_1, b_1|\chi_1|[hh])$ (by translating the second tree until its bottom vertex is one step in the X-direction from the top vertex of the first tree). The outcome of this construction is

$$\sum_{r_1, b_1 = 0}^{n+1} t_n^+(r - r_1, b - b_1|\chi_0|[hh]) t_m^+(r_1, b_1|\chi_1|[hh]) \leq t_{n+m+1}^+(r, b|\chi_0\chi_1|[hh]),$$

where $\chi_0\chi_1$ is a sequence of colours formed by concatenating χ_0 and χ_1, since every vertex in trees counted by $t_m^+(r_1, b_1|\chi_1|[hh])$ is lexicographically greater than every vertex in trees counted by $t_n^+(r - r_1, b - b_1|\chi_0|[hh])$. Multiply this by $z_r^r z_b^b$ and sum over r and b. The result is

$$t_n^+(z_r, z_b|\chi_0|[hh]) t_m^+(z_r, z_b|\chi_1|[hh]) \leq t_{n+m+1}^+(z_r, z_b|\chi_0\chi_1|[hh]).$$

Take logarithms, and average over all possible sequences of colours in Ω:

$$\langle \log t_n^+(z_r, z_b|\chi|[hh]) \rangle_\chi + \langle \log t_m^+(z_r, z_b|\chi|[hh]) \rangle_\chi \leq \langle \log t_{n+m}^+(z_r, z_b|\chi|[hh]) \rangle_\chi.$$

In other words, by Lemma A.1 in Appendix A there is a limiting free energy defined by

$$\mathcal{T}_{qu}^+(z; h) = \lim_{n\to\infty} \frac{1}{n} \langle \log t_n^+(z_r, z_b|\chi|[hh]) \rangle_\chi.$$

There is a most popular value of h, say h^*, so define

$$\mathcal{T}_{qu}^+(z_r, z_b) = \lim_{n\to\infty} \frac{1}{n} \langle \log t_n^+(z_r, z_b|\chi|[h^* h^*]) \rangle_\chi. \qquad (\dagger)$$

Next, consider the partition function $t_n^+(z_r, z_b|\chi|[h_b h_t])$, and let $[h_b^* h_t^*]$ be the most popular choices for $[h_b h_t]$. Then the following is true. In the first place,

$$t_n^+(z_r, z_b|\chi|[h^* h^*]) \leq t_n^+(z_r, z_b|\chi|[h_b^* h_t^*]),$$

and secondly,

$$\left[t_n^+(z_r, z_b|\chi|[h_b^* h_t^*])\right]^2 \leq t_{2n+1}^+(z_r, z_b|\chi|[h_b^* h_b^*]) \leq t_{2n+1}^+(z_r, z_b|\chi|[h^* h^*]),$$

and the first inequality in the last expression is found by taking two trees counted by $t_n^+(z_r, z_b|\chi|[h_b^* h_t^*])$, reflecting one through the hyperplane $X = 0$, and then concatenating them. Comparing this to eqn ($\dagger$), and noting that

$$t_n^+(z_r, z_b|\chi|[h_b^* h_t^*]) \leq t_n^+(z_r, z_b|\chi) \leq (n+1)^2 t_n^+(z_r, z_b|\chi|[h_b^* h_t^*])$$

shows that the limiting free energy in the averaged quenched ensemble exists as claimed.

□

The existence of the limiting free energy in the average quenched ensemble indicates that the self-averaging of adsorbing branched copolymers might be an interesting question. The key to a proof of self-averaging in this model is the following lemma.

Lemma 6.41 *Let χ_0 be a fixed sequence of colours in Ω. Then*

$$\liminf_{n\to\infty} \frac{1}{n} \log t_n^+(z_r, z_b|\chi_0) \geq \mathcal{T}_{qu}^+(z_r, z_b),$$

almost surely.

Proof Let $n = Nm + s$, and decompose the sequence χ_0 into m finite sequences χ_i of length N, so that $\chi_0 = \prod_{i=1}^{\infty} \chi_i$, where the product is taken as concatenation. Next, consider trees counted by $t_{N-1}^+(r_i, b_i|\chi_i|[hh])$, and concatenate them from $i = 1$ to $i = m$. Lastly, concatenate a tree counted by $t_s^+(r_s, b_s|\chi_s|[hh])$ on to this as well, where χ_s is the first s colours in the sequence χ_{m+1}. This shows that

$$t_n^+(r, b|\chi_0|[hh]) \geq \sum_{\{r_i\},\{b_i\}} \left[\prod_{i=1}^{m} t_{N-1}^+(r_i, b_i|\chi_i|[hh])\right] t_s^+(v_s|\chi_s|[hh])\delta_{(r-\sum r_i)}\delta_{(b-\sum b_i)}.$$

Multiply this by $z_r^r z_b^b$, and sum over r and b:

$$t_n^+(z_r, z_b|\chi_0|[hh]) \geq \left[\prod_{i=1}^{m} t_{N-1}^+(z_r, z_b|\chi_i|[hh])\right] t_s^+(z_r, z_b|\chi_s|[hh]).$$

Take logarithms, and divide by n, and take the lim inf as $n \to \infty$ of the left-hand side, with N fixed. Then $m \to \infty$, and

$$\liminf_{n\to\infty} \frac{1}{n} \log t_n(z_r, z_b|\chi_0|[hh]) \geq \liminf_{m\to\infty} \frac{1}{m} \sum_{i=1}^{m} \frac{1}{N} \log t_{N-1}^+(z_r, z_b|\chi_i|[hh]). \qquad (\dagger)$$

This is true for any choice of $[hh]$, and in particular, let $[h^*h^*]$ be the most popular choice of h. Next, let $[h_b^*h_t^*]$ be the most popular choices for $[h_bh_t]$ in $t_n(z_r, z_b|\chi_0|[h_b, h_t])$. Then

$$t_n(z_r, z_b|\chi_0|[h_b^*, h_t^*]) \leq t_n(z_r, z_b|\chi_0) \leq (n+1)^2 t_n(z_r, z_b|\chi_0|[h_b^*, h_t^*]), \qquad (\ddagger)$$

and

$$\begin{aligned} [t_n(z_r, z_b|\chi_0|[h^*h^*])]^2 &\leq [t_n(z_r, z_b|\chi_0|[h_b^*, h_t^*])]^2 \\ &\leq t_{2n+1}(z_r, z_b|\chi_0|[h_b^*, h_b^*]) \leq t_{2n+1}(z_r, z_b|\chi_0|[h^*h^*]). \end{aligned}$$

Thus, by eqn ($\ddagger$), choose the most popular values of $[hh]$ in eqn ($\dagger$). Then, by Theorem 6.40 and the law of large numbers,

$$\liminf_{n\to\infty} \frac{1}{n} \log t_n(z_r, z_b|\chi_0|[hh]) \geq \left\langle \frac{1}{N} t_{N-1}^+(z_r, z_b|\chi) \right\rangle_\chi \to \mathcal{T}_{qu}^+(z)$$

almost surely as $N \to \infty$, for almost all colourings χ_0. □

Lemma 6.41 gives a lower bound on $\liminf_{n\to\infty} \frac{1}{n} \log t_n(z_r, z_b|\chi_0)$. It turns out that Lemma 6.41 also implies that this model of lexicographically quenched adsorbing copolymers is self-averaging. This is proven in Theorem 6.42. Define μ_P to be the probability measure on the (probability) space Ω. Then $\mu_P(\Omega) = 1$ and by Theorem 6.40,

$$\mathcal{T}_{qu}^{+}(z_r, z_b) = \lim_{n\to\infty} \frac{1}{n} \int_{\Omega} d\chi \; \log t_n^{+}(z_r, z_b|\chi). \tag{6.71}$$

With this observation, self-averaging can be shown.

Theorem 6.42 *For almost all* $\chi_0 \in \Omega$,

$$\mathcal{T}_{qu}^{+}(z_r, z_b) = \lim_{n\to\infty} \frac{1}{n} \log t_n^{+}(z_r, z_b|\chi_0).$$

Proof Apply Fatou's lemma to eqn (6.71). This shows that

$$\mathcal{T}_{qu}^{+}(z_r, z_b) = \lim_{n\to\infty} \frac{1}{n} \int_{\Omega} d\chi \; \log t_n^{+}(z_r, z_b|\chi) \geq \int_{\Omega} d\chi \; \liminf_{n\to\infty} \frac{1}{n} \log t_n^{+}(z_r, z_b|\chi). \tag{$\dagger$}$$

Define the subsets $\{\Omega_-, \Omega_0, \Omega_+\}$ of Ω by

$$\liminf_{n\to\infty} \frac{1}{n} \log t_n^{+}(z_r, z_b|\chi) \begin{cases} < \mathcal{T}_{qu}^{+}(z_r, z_b), & \text{if } \chi \in \Omega_-; \\ = \mathcal{T}_{qu}^{+}(z_r, z_b), & \text{if } \chi \in \Omega_0; \\ > \mathcal{T}_{qu}^{+}(z_r, z_b), & \text{if } \chi \in \Omega_+. \end{cases}$$

By Lemma 6.41, $\mu_P(\Omega_-) = 0$. Suppose that $\mu_P(\Omega_+) = a > 0$. Then $\mu_P(\Omega_0) = 1 - a$, and therefore

$$\int_{\Omega} d\chi \; \liminf_{n\to\infty} \frac{1}{n} \log t_n^{+}(z_r, z_b|\chi) > a\, \mathcal{T}_{qu}^{+}(z_r, z_b) + (1-a)\, \mathcal{T}_{qu}^{+}(z_r, z_b) = \mathcal{T}_{qu}^{+}(z_r, z_b).$$

This is in contradiction with eqn ($\dagger$), unless $a = 0$. Thus, $\mu_P(\Omega_+) = 0$ and so

$$\mathcal{T}_{qu}^{+}(z_r, z_b) = \liminf_{n\to\infty} \frac{1}{n} \log t_n^{+}(z_r, z_b|\chi_0) \tag{$\ddagger$}$$

for all $\chi_0 \in \Omega_0$ where $\mu_P(\Omega_0) = 1$. It only remains to show that

$$\lim_{n\to\infty} \frac{1}{n} \log t_n^{+}(z_r, z_b|\chi_0)$$

exists for almost all χ_0. To see this, suppose that

$$\mathcal{T}_{qu}^{+}(z_r, z_b) < \limsup_{n\to\infty} \frac{1}{n} \log t_n^{+}(z_r, z_b|\chi)$$

for all $\chi \in U$, and where $\mu_P(U) > 0$. By definition of the lim sup, there is an $\epsilon_\chi > 0$, and an infinite set of integers n_i, such that

$$\mathcal{T}_{qu}^{+}(z_r, z_b) + [\epsilon_\chi]/2 < \frac{1}{n_i} \log t_{n_i}^{+}(z_r, z_b|\chi)$$

for every $\chi \in U$. Define $T_n(\chi) = \frac{1}{n_i} \log t_{n_i}^{+}(z_r, z_b|\chi)$ if $n_i \leq n < n_{i+1}$. Thus $\lim_{n\to\infty} T_n(\chi) = \limsup_{n\to\infty}[\log t_n^{+}(z_r, z_b|\chi)]/n$. In other words,

$$\mathcal{T}_{qu}^{+}(z_r, z_b) + [\epsilon_\chi]/2 < T_n(\chi)$$

for every $n \geq n_1$. Now take $n \to \infty$, and integrate χ over Ω. Furthermore, observe that $T_n(\chi)$ is measurable, and by the Lebesgue dominated convergence theorem

$$\int_\Omega d\chi \lim_{n\to\infty} T_n(\chi) = \lim_{n\to\infty} \int_\Omega d\chi\, T_n(\chi) = \mathcal{T}_{qu}^{+}(z_r, z_b)$$

by Theorem 6.40. Thus

$$\mathcal{T}_{qu}^{+}(z_r, z_b) > \mathcal{T}_{qu}^{+}(z_r, z_b) + \int_U d\chi\, [\epsilon_\chi]/2,$$

and this is a contradiction. Thus

$$\mathcal{T}_{qu}^{+}(z_r, z_b) \geq \limsup_{n\to\infty} \frac{1}{n} \log t_n^{+}(z_r, z_b|\chi)$$

for almost every $\chi \in \Omega$, and comparison to eqn ‡ shows that the limit exists as claimed. □

The existence of the limit $\lim_{n\to\infty} \frac{1}{n} \log t_n^{+}(z_r, z_b|\chi_0)$ is also a consequence of the local superadditive ergodic theorem of M.A. Akcoglu and U. Krengel [3]. Self-averaging in models of self-interacting trees and polygons was shown in reference [282].

Lastly, consider the limiting free energy in the case that only red vertices are interacting with the adsorbing plane: $\mathcal{T}_{qu}^{+}(\alpha, 0)$. The arithmetic–geometric inequality implies that $\log\langle t_n^{+}(z_r, 0|\chi)\rangle_\chi \geq \langle \log t_n^{+}(z_r, 0|\chi)\rangle_\chi$. This shows that the critical adsorption activity for an annealed model, z_c^a, is less than or equal to the critical adsorbing activity for the averaged lexicographically quenched model, z_c^q. Thus, the critical activities are related by $1 < z_c < z_c^a \leq z_c^q$.

6.5 Embeddings of graphs with specified topologies

An abtract graph consists of vertices (points) which are joined by line segments (called edges; but I shall refer to these as *branches* to avoid confusion with lattice edges in animals). A graph is connected if there is a path consisting of a sequence of branches between every pair of vertices. The *degree* of a vertex is the number of branches incident with it.[5] Let $\mathcal{G}$ be the union of the set of all abstract finite connected graphs with no

[5] Observe that there are no restrictions on the number of branches between a given pair of vertices, or with a branch joining a vertex to itself (and so forming a loop). Thus, the graphs considered here are not necessarily simple.

vertices of degree two, and the circle graph (this graph has one vertex and one branch). An embedding of a graph $g \in \mathcal{G}$ is a subcomplex of the lattice which represents g (that is, it is isotopic to g). Notice that the embedded graph is itself an abstract graph, but that it may contain vertices of degree two, namely the internal vertices in each branch. Examples of embedded graphs which have received attention in the literature are figure eight graphs [367], lattice animals with fixed cyclomatic index [156], trees with restricted branching [323], and brushes and combs [237].

The first important issue is the embeddability of a given graph. It appears that not every graph can be represented as a subcomplex in the hypercubic lattice. In particular, there are some graphs (non-planar graphs) which are embeddable in the cubic lattice, but not in the square lattice. Moreover, the degrees of vertices in a graph is an important restriction. In this respect the maximal degree of any vertex must be less than or equal to the degrees of lattice vertices. Below it will be proven that any graph in $\mathcal{G}$ with maximal vertex degree six can be embedded in the cubic lattice. A similar theorem shows that every planar graph of maximal degree four can be embedded in the square lattice.

Of particular interest in the study of embeddings of animals with a specified topology is the occurrence of knotted conformations [324], and a pattern theorem. A version of the Frisch–Wasserman–Delbruck conjecture (see Section 5.3.2) is known for embeddings of graphs in the cubic lattice. This issue will be considered below.

Embedded graphs may be restricted further by the requirement that they are *uniform*. In this context, uniform implies that the image of every branch in the graph is a self-avoiding walk of a given length n (that is, the edges in the embedded graph is uniformly distributed amongst the branches of the graph). Uniform embeddings of graphs have been studied as interacting models of branched polymers, especially with respect to adsorption and as models of branched polymers in confined geometries, such as slits, slabs, wedges, and so on [53, 55, 319, 320, 328, 365, 366]. Uniform embeddings of graphs are also said to be *monodisperse*, as opposed to *polydisperse* embeddings, when the lengths of branches are not fixed.

6.5.1 A pattern theorem for embedded graphs in the cubic lattice

Let $\sigma \in \mathcal{G}$ be an abstract graph with b branches. If σ has c cycles, then it has $v = 2+b-c$ vertices, and there are at most $\binom{v(v-1)/2}{b}$ abstract graphs on v vertices and with b branches. (Two abstract graphs are identical if they are isomorphic; in other words, there is a one-to-one correspondence between vertices which preserves the adjacencies in the graph.) Two graphs are *homeomorphic* if both can be obtained from the same graph by inserting vertices of degree two in the branches. Such graphs are also said to be of the same *homeomorphism-type*. Define $g_n(\sigma)$ to be the number of lattice animals (counted up to translation) which are (as graphs) homeomorphic to $\sigma \in \mathcal{G}$, and which contains a total of n edges, that is, they are (polydisperse) embeddings of σ. A graph σ is embeddable in the hypercubic lattice if there is an $n > 0$ such that $g_n(\sigma) > 0$. Not all graphs are embeddable in the square lattice (for example, non-planar graphs are not embeddable in the square lattice).

The number $g_n(\sigma)$ has been studied using conformal invariance techniques. In particular, if $g_n^m(\sigma)$ is the number of uniform embeddings of an abstract graph σ, and if the entropic exponents γ_σ are associated with polydisperse embeddings of σ, and γ_σ^m is associated with uniform embeddings of σ, then $g_n(\sigma) \simeq n^{\gamma_\sigma - 1}\mu_d^n$ and $g_n^m(\sigma) \simeq n^{\gamma_\sigma^m - 1}\mu_d^n$, where $\gamma_\sigma^m = \gamma_\sigma + 1 - N_a$, and where N_a is the number of branches in the embeddings [91]. For a uniform star with N arms, it is known that $\gamma_\sigma = [68 + 9N(3-N)]/64$ [95], and this has been tested by exact enumeration [370] and by Monte Carlo simulations [143]. These values are not rigorous, but are expected to be exact.

The cubic lattice cannot accommodate a representative of the same homeomorphism type as an abstract graph if it has a vertex of degree more than six. Let $\mathcal{G}_6$ be that subset of abstract graphs in $\mathcal{G}$ with maximal vertex degree less than or equal to six. Lemma 6.43 below shows that any graph in $\mathcal{G}_6$ is embeddable in the cubic lattice; this result is due to C.E. Soteros *et al.* [324]. Key ideas in the lemma are those of a top plane and a top line. The *top plane* of an embedded graph is the set of all vertices and edges in the graph with maximal Z-coordinate. The *top line* of an embedded graph is the set of all vertices and edges in the graph which are in the top plane, and has maximal Y-coordinates in the top plane. A vertex of degree one in a graph will be called an *end-vertex*; and a vertex of degree greater than two will be called a *branching point*.

Lemma 6.43 *Let $\sigma \in \mathcal{G}$. Then there is an embedding of σ in the cubic lattice if and only if $\sigma \in \mathcal{G}_6$. Moreover, there is also an embedding of σ in the cubic lattice such that at least one vertex, and at most one edge, in each branch is in the top plane, and every vertex of degree one in the embedded graph lies in the top line.*

Proof Suppose that T is an embedding of the graph σ, then the maximal vertex degree in T is at most six, and so $\sigma \in \mathcal{G}_6$. Now suppose that $\sigma \in \mathcal{G}_6$. The proof will do a double induction; first it is shown that trees are embeddable as stated above, and then it will do induction on the number of cycles. In turn, trees will be treated by doing induction on the number of branching points.

Let σ be a tree and suppose that it has b branching points. If $b = 1$, then σ is a star-graph with six or fewer arms. Such graphs can be embedded directly by construction, and moreover, by extending the arms to the top line in an obvious way, an embedding which satisfies Lemma 6.43 can be obtained. Suppose that Lemma 6.43 is true for all tree-graphs with B or fewer branching points. Select a tree τ from $\mathcal{G}_6$ with $B + 1$ branching points. Without loss of generality it may be supposed that every branching point in this tree has degree six (if not, then add branches until the degrees are six, embed it, and remove the branches to obtain an embedding of the original graph). The tree τ has $4(B + 1) + 1$ end-vertices, and so there is at least one branching point v which has five branches which ends in an end-vertex. Delete these five branches to obtain the graph τ' which has B branching points, and is embeddable in the cubic lattice. Let A be an embedding of τ' satisfying Lemma 6.43. Then the branching point v is now an end-vertex in the top line of A. Append a six-arm star to v as shown in Fig. 6.13, and extend the rest of the branches in the embedding so that all end-vertices are again in the top line. This finishes the lemma for embeddings of trees.

The theorem can be completed for any graph $\sigma \in \mathcal{G}_6$. Cut cycles in σ until it is a tree, and label the newly created end-vertices. Embed the resulting graph in the lattice with all end-vertices in the top line, and join the end-vertices with the same labels by adding edges. The edges and end-vertices in the top line and top plane can again be extended

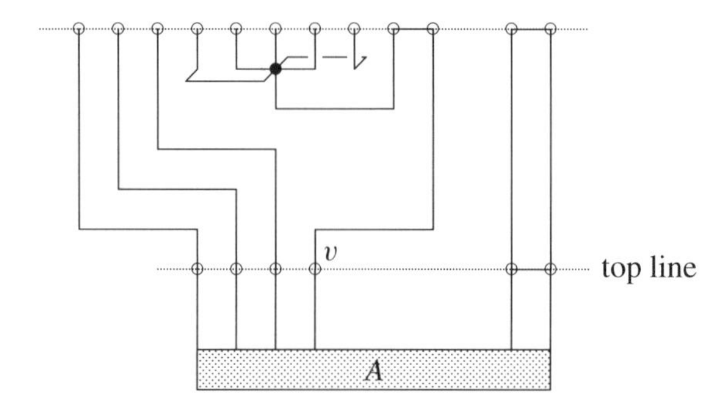

Fig. 6.13: A is a tree with B branching points which satisfies Lemma 6.43. A new branching point can be created by adding a six-armed star to any of the end-vertices of A in the top line. Moreover, by adding edges to the branches, it can be arranged that the new tree satifies Lemma 6.43.

in the Z-direction (as in Fig. 6.13) until an embedding compatible with the statement of Lemma 6.43 is found. □

The lemma above can be used to show that there is a limiting free energy in an interacting model of embedded graphs of fixed homeomorphism-type, provided that the limiting free energy exists (and is equal) for models of polygons and walks with the same interaction. To make this more precise, let $g_n(k|\sigma)$ be the number of embeddings of a graph $\sigma \in \mathcal{G}_6$ with n edges and energy k. The energy could be associated with features in the embedding, such as the number of visits to an adsorbing plane, or the number of right angles in the branches of the embedding. If σ is the circle graph (σ = pol), then $g_n(k|\text{pol}) = p_n(k)$, the number of polygons on n edges, and if σ is a line graph (σ = walk) then $g_n(k|\text{walk}) = c_n(k)$. In other words, these models are generalizations of the models of polygons and walks considered in Chapter 5. The partition function is defined by

$$g_n(z|\sigma) = \sum_{k\geq 0} g_n(k|\sigma) z^k, \tag{6.72}$$

and an immediate question is the existence of the limiting free energy. In at least the cases that σ = pol and σ = walk it is known that there is a limiting free energy in some cases (see Theorem 5.28).

Theorem 6.44 *Suppose that an energy is defined on models of embedded graphs, such that any embedded graph can be assigned an energy, and in particular such that embedded polygons and walks have defined energies. Suppose furthermore that the models of interacting walks and interacting polygons satisfy Assumptions 5.21, so that there are limiting free energies $\mathcal{F}_w(z)$ for models of interacting walks and $\mathcal{F}_p(z)$ for models of interacting polygons (and such that $\mathcal{F}_w(z) = \mathcal{F}_p(z)$). Then the limiting* *free energy*

$$\mathcal{F}_\sigma(z) = \lim_{n\to\infty} \frac{1}{n} \log g_n(z|\sigma)$$

exists, and moreover, $\mathcal{F}_\sigma(z) = \mathcal{F}_w(z) = \mathcal{F}_p(z)$.

Proof Let σ be any graph and let T be any embedding of σ containing $m > 0$ edges. Furthermore, let A be a polygon containing $n - m - 2$ edges. Let the top vertex of T be t_T and its top edge be e_T. Let the bottom vertex of A be b_A, and its bottom edge be e_A. The aim is to join T and A into a new animal with homeomorphism type σ, and which contains all the vertices in A. There are several cases to consider. Case (1): Suppose that the degree of t_T is one. Then translate A until $t_T + 3\hat{\imath} = b_A$. Add the edges $(t_T, t_T + \hat{\imath})$, $(t_T + \hat{\imath}, t_T + 2\hat{\imath})$, $(t_T + 2\hat{\imath}, t_T + 3\hat{\imath})$ to the resulting animal, and remove the bottom edge e_A of A. If the energy of T is l, and of A is $k - l$, then the resulting animal has size n and has energy $k + o(n)$ by Assumptions 5.21, since the polygon can be divided into two fl-walks. Case (2): Suppose that the degree of t_T is more than one, and that e_T and e_A are parallel edges. Translate σ and A so that $e_A = e_T + 2\hat{\imath}$, delete both e_A and e_T and add the edges $(t_T, t_T + \hat{\imath})$, $(t_T + \hat{\imath}, t_T + 2\hat{\imath})$, $(t_T - e_T, t_T - e_T + \hat{\imath})$ and $(t_T - e_T + \hat{\imath}, t_T - e_T + 2\hat{\imath})$. As before the resulting animal has size n and its total energy changed by at most $o(n)$. Case (3): Suppose that the degree of t_T is more than one, and that e_T and e_A are perpendicular. Deal with as in case (2), but rotate the polygon about the X-axis by 90° to make e_T and e_A parallel. Thus

$$\sum_{l=0}^{k} g_m(l|\sigma) p_{n-m-2}(k-l) \le 2 \sum_{i=-q}^{q} g_n(k+i|\sigma).$$

The factor of 2 appears because of case (3) in the above, and $q = o(n)$. Multiply by z^k and sum over k:

$$g_m(z|\sigma) p_{n-m-2}(z) \le 2[\phi(z)]^q g_n(z|\sigma),$$

where $\phi(z) = z + 1 + z^{-1}$. Take logarithms, divide by n and let $n \to \infty$. By Lemma 6.43 there is an $m > 0$ such that $g_m(z|\sigma) > 0$. Thus

$$\liminf_{n\to\infty} \frac{1}{n} \log g_n(z|\sigma) \ge \mathcal{F}_p(z),$$

by eqn (5.40).

On the other hand, the branches in the embedded graph contain a total of n edges. Let there be b branches labelled $\{1, 2, \ldots, b\}$, and suppose that branch i contains n_i edges and has energy k_i. Suppose that σ has v branching points and end-vertices. If it has c cycles, then $v = b + 2 - c$. If the branches are cut from T, then it is apparent that

$$g_n(k|\sigma) \le \binom{\binom{v}{2}}{b} \sum_{\{n_i\}} \sum_{\{k_i\}} \prod_{i=1}^{b} c_{n_i}(k_i) \delta_{(n-\sum_i n_i)} \delta_{(k-\sum_j k_j)}$$

where the combinatorial factor is an upper bound on the number of homeomorphism types of graphs with b branches. Multiply this by z^k and sum over k. This gives

$$g_n(z|\sigma) \leq \binom{\binom{v}{2}}{b} \sum_{n_1,n_2,\ldots,n_b} \prod_{i=1}^{b} c_{n_i}(z)\delta_{(n-\sum_i n_i)}. \qquad (\ddagger)$$

Now use the fact that the model of walks satisfies Assumptions 5.21. By Corollary 5.24, each $c_{n_i}(z)$ can be bounded from above by fl-walks such that $c_m(z) \leq e^{o(m)}[\phi(z)]^{o(m)} c_m^{fl}(z)$. Furthermore, by Assumption 5.21(1) and thus Assumptions 3.8(3), $\sum_{l=0}^{k} c_n^{fl}(l) c_m^{fl}(k-l) \leq \sum_{i=-q}^{q} c_{n+m}^{fl}(k+i)$ so that $c_n^{fl}(z)c_m^{fl}(z) \leq [\phi(z)]^q c_{n+m}^{fl}(z)$. Thus, by Lemma A.1, $c_m^{fl}(z) \leq [\phi(z)]^q e^{m\mathcal{F}_p(z)}$, where Theorem 5.28 was also used. Replace the c_{n_i} by these upper bounds, take the logarithm, divide by n and let $n \to \infty$. This shows that

$$\limsup_{n\to\infty} \frac{1}{n} \log g_n(z|\sigma) \leq \mathcal{F}_p(z),$$

and this completes the proof. □

The argument in the last paragraph of the proof of Theorem 6.44 is enough to prove a pattern theorem for interacting models of embedded graphs of fixed homeomorphism type. In particular, if $g_n(z; \leq\lfloor\epsilon n\rfloor P|\sigma)$ is the partition function of embedded animals of homeomorphism type σ and where the Kesten pattern P occurs at most $\lfloor\epsilon n\rfloor$ times, then eqn ($\ddagger$) shows that

$$\limsup_{n\to\infty} \frac{1}{n} \log g_n(z; \leq\lfloor\epsilon n\rfloor P|\sigma) \leq \mathcal{F}_p(z;\epsilon) < \mathcal{F}_p(z). \qquad (6.73)$$

Theorem 6.45 *If $g_n(z; \leq\lfloor\epsilon n\rfloor P|\sigma)$ is the partition function of embedded animals of homeomorphism type σ and where the Kesten pattern P occurs at most $\lfloor\epsilon n\rfloor$ times, then, if $\epsilon > 0$ is small enough,*

$$\limsup_{n\to\infty} \frac{1}{n} \log g_n(z; \leq\lfloor\epsilon n\rfloor P|\sigma) \leq \mathcal{F}_p(z;\epsilon) < \mathcal{F}_p(z). \qquad \square$$

It is possible to show that the limit in Theorem 6.45 exists, but this is not a critical issue. The growth constants of embedded animals of fixed homeomorphism types are obtained by putting $z = 1$ in the above arguments. In particular, they are equal to μ_3, the growth constant of walks and polygons, as can be seen directly from Theorem 6.44. The above arguments can be generalized to d dimensions, with a suitable class of graphs considered.

6.5.2 Knotted embeddings of graphs

It is not obvious what is meant by a knotted embedding of a graph. The embeddings of graphs examined in Section 6.6.1 are piecewise linear embeddings. This may be precisely defined as follows. Let $\sigma \in \mathcal{G}_6$ be an abstract graph, and suppose that $G \in \sigma$ is a representation of σ. A *knotted graph* is an embedding (injection) $k : G \to \mathcal{R}^3$ (the image of k is usually simply denoted by k). k is a 1-complex in $\mathcal{R}^3$, and it is piecewise linear if the map k is piecewise linear.

Two knotted embeddings k_1 and k_2 of G are equivalent if there exists an orientation-preserving homeomorphism $H : (\mathcal{R}^3, k_1) \to (\mathcal{R}^3, k_2)$. The equivalence class of embeddings is denoted by K, and k is a representative of K. The collection of all equivalence classes of embeddings k is denoted $\mathcal{K}(\sigma)$. Amongst the equivalence classes is the *unknot*. If σ is the circle graph, then there is an equivalence class of planar embeddings; and this is denoted the unknot. The notion of the unknot in the case of a general graph is not clear, and a definition will rely on the properties of regular projections.

A projection of an embedded graph is regular if the following two conditions are met:

(1) Each point in the projection is the image of at most two points in the embedding, and each double point in the image is the transverse intersection of two projected branches.

(2) The projected image of each branching point or end-vertex is the image of exactly one point (the branch point or end-vertex) in the embedding.

Transverse intersections of branches are denoted in Fig. 5.10 (see also Section 5.3.2). The crossing number of a regular projection is the number of double points. The crossing number of an equivalence class K is the minimal crossing number over all regular projections of all representatives of that equivalence class; this is denoted by $C(K)$. Define the minimal crossing number of homeomorphism-type σ by

$$N_{min}(\sigma) = \min_{K \in \mathcal{K}(\sigma)} C(K). \tag{6.74}$$

The *unknot* is an equivalence classes of embeddings K of σ which has $C(K) = N_{min}(\sigma)$. Generally, the unknot may not be unique.

If G is a representative of an abstract planar graph σ, then the unknot is the equivalence class of planar embeddings of G (these are in fact equivalent [256]). Some graphs are intrinsically complex with respect to embeddings in 3-space. For example, every embedding of the complete graph K_{4k+3} is chiral (and thus there is an even number of non-equivalent unknotted representatives of embeddings of K_{4k+3}), and every embedding of K_7 has an odd number of Hamiltonian cycles which are knotted circles, and so the unknotted embedding of K_7 will always contain knotted cycles!

It is now possible to prove a Frisch–Wasserman–Delbruck result (see Section 5.3.2) for animals of fixed homeomorphism type. Let Kesten patterns be defined as before, and suppose that $g_n(k; mP|\sigma)$ is the number of animals of fixed homeomorphism type σ, with n edges and energy k, and which contain the Kesten pattern P exactly m times in their branches. Assume as in Section 6.2.1 that the energy is such that the limiting free energies for polygons and walks exists, and that Assumptions 5.21 are met.

An animal has a cut-edge if there is a single edge which can be deleted to disconnect the animal. If an animal does not have a cut-edge, then it is said to be 2-edge-connected. The following result is true for 2-edge-connected animals of fixed homeomorphism type. Similar (but weaker) results are true for animals with cut-edges [324]. Let T be a tight trefoil Kesten pattern (see Fig. 5.9), and let T^q be a concatenated string of q suitably translated copies of T.

Theorem 6.46 *Let $\sigma \in \mathcal{G}_6$ be 2-edge-connected. Suppose that P is a Kesten pattern. Then for sufficiently large n, all but exponentially few embeddings of σ of size n in $\mathcal{Z}^3$ are knotted. Moreover, all but exponentially few embeddings contain at least $\lfloor \epsilon n \rfloor$ distinct knot-factors in their cycles. Moreover, if the activity z is conjugate to the energy of the embeddings, and if the limiting free energies for polygons and walks exist, and Assumptions 5.21 are satisfied, then at any finite value of z, the embeddings which contains at most $\lfloor \epsilon n \rfloor$ copies of T^q, make an exponentially small contribution to the partition function $g_n(z|\sigma)$, for some value of $\epsilon_0 > 0$, and all $0 < \epsilon < \epsilon_0$.*

Proof Take $q = \lceil 1 + N_{min}(\sigma)/3 \rceil$. Since T^q is a Kesten pattern, by Theorem 6.45,

$$\limsup_{n\to\infty} \frac{1}{n} \log g_n(z; \leq \lfloor \epsilon n \rfloor T^q | \sigma) < \mathcal{F}_p(z).$$

If an embedding contains a knotted cycle of knot type T^q, then every regular projection of that embedding has at least $3q$ crossings. Thus, such an embedding must have crossing number greater than or equal to $3q > N_{min}(\sigma)$, and so it must be knotted. This proves the second part of the theorem. The first part is obtained by putting $z = 1$. □

As a corollary of Theorem 6.45 the Frisch–Wasserman–Delbruck conjecture is obtained for embeddings of animals of fixed homeomorphism type in $\mathcal{Z}^3$. Issues related to the above are also discussed in [321, 324]. A review can be found in reference [322].

6.5.3 Walks and polygons in wedges and uniform animals

Uniform or monodisperse animals have branches which are all of the same length. A graph in $\mathcal{G}_6$ has a uniform embedding in $\mathcal{Z}^3$ with n edges in each branch (for n odd) if and only if it has no cycles of odd length. In other words, if n is odd, then a graph $g \in \mathcal{G}_6$ has a uniform embedding with n edges in each branch if and only if it is bipartite. If g is not bipartite, then even values of n are allowed. Notice that limits as $n \to \infty$ are taken through the sequence $nf \to \infty$, if there are f branches in the animal.

Uniform animals are best studied by using results and techniques for polygons and walks in confined geometries, and in the context here, confined to wedges. In particular, define an *f-wedge* by a sequence of non-negative functions $\{f_i\}_{i=2}^{d}$ as

$$W_f = \{(X_1, X_2, \ldots, X_d) \in \mathcal{Z}^d \,\big|\, 0 \leq X_i \leq f_i(X_1);\ \forall 2 \leq i \leq d\}. \tag{6.75}$$

In this wedge, count walks with an end-vertex at the origin (this orients the walk), and polygons which contain the origin (this roots the polygons). If there is a Y_0 such that $f_k(Y_0) = 0$ for all $k = 2, \ldots, d$, then a polygon of arbitrary length cannot be placed in the wedge. Thus, assume that $f_k(Y) > 0$ for at least one k, and all $Y \geq 0$. Let $c_n(f)$ and $p_n(f)$ be the number of walks and polygons confined to an f-wedge respectively, placed so that an end-vertex of the walk is at the origin, and so that the polygon contains the origin. The bottom and top vertices are the lexicographically least and most vertices in a polygon or a walk, and the bottom and top edges are those edges with lexicographically least and most mid-points in a polygon or a walk. It is also the case that

$$\liminf_{n\to\infty} \frac{1}{n} \log p_n(f) \leq \limsup_{n\to\infty} \frac{1}{n} \log c_n(f) \leq \log \mu_d, \tag{6.76}$$

since by removing the bottom edge of polygons counted by $p_n(f)$, a walk in the f-wedge of length $n+1$ which steps first in the X-direction is obtained. Thus $p_n(f) \leq c_{n-1}(f)$. Equation (6.76) shows that if it can be proven that $\liminf_{n\to\infty}[\log p_n(f)]/n \geq \log \mu_d$, then the corresponding limits exist. To see this, a most popular argument will be used.

Each positive polygon counted by p_N^+ can be fitted into a box of side-length at most N, and its bottom edge may have one of $(d-1)$ orientations, ϕ_b, and at most N^{d-1} positions for its mid-point, denoted by h_b, in the face of the box. Similarly, the orientation of the top edge, and the position of its mid-point, are denoted by ϕ_t and h_t. Let $p_N^+([h_b\phi_b, h_t\phi_t])$ be the number of positive polygons of length N and a bottom edge with orientation ϕ_b and with its mid-point at position h_b, and top edge with orientation ϕ_t and its mid-point at a position h_t. Each polygon counted by $p_N^+([h_b\phi_b, h_t\phi_t])$ can be fitted into a box of side-length N, and its bottom edge may have $(d-1)$ orientations and at most N^{d-1} positions in the face of the box. Let $[h_b^*\phi_b^*, h_t^*\phi_t^*]$ be the most popular values of heights and positions. Then

$$[(d-1)N^{d-1}]^2 p_N^+([h_b^*\phi_b^*, h_t^*\phi_t^*]) \geq p_N^+. \tag{6.77}$$

This inequality can be used to prove the following lemma.

Lemma 6.47 *It is the case that* $\liminf_{n\to\infty}[\log p_n(f)]/n \geq \log \mu_d$ *if and only if* $\lim_{X\to\infty} f_k(X) = \infty$ *for all* $2 \leq k \leq d$.

Proof Define the number $a(N) = \min\{X \mid f_k(X) \geq N, \forall k = 2, \ldots, d\}$. Choose polygons from $p_N^+([h_b^*\phi_b^*, h_t^*\phi_t^*])$ and concatenate them by first reflecting the second through the $X = 0$ hyperplane, and then translating it until the top edge of the first polygon is one step in the X-direction less than the reflected image of the top edge of the second. Delete the top edge of the first, and the reflected top edge of the second, and add two edges as in Fig. 1.4 to join the polygons. In the same way, M polygons chosen from $p_N^+([h_b^*\phi_b^*, h_t^*\phi_t^*])$ can be concatenated into a single polygon by reflecting every second polygon through the $X = 0$ plane. The resulting polygon has maximum height above the $Z = 0$ plane less than or equal to N, and contains NM edges. Translate the entire polygon until its least X-coordinate is equal to $a(N)+1$. Then it can be fitted into the W_f-wedge. It only remains to add edges to join it to the origin. Let the coordinates of the mid-point of the bottom edge be $\mathbf{b}$, and suppose that the bottom edge is in the ϕ_b direction. Then $X(\mathbf{b}) = a(N)+1$. Now construct the polygon $\{a(N)\hat{\imath}, Y(\mathbf{b})\hat{\jmath}, \ldots, Z(\mathbf{b})\hat{k}, \hat{\imath}, \phi_b, -\hat{\imath}, -Z(\mathbf{b})\hat{k}, \ldots, -Y(\mathbf{b})\hat{\jmath}, -a(N)\hat{\imath}, -\phi_b\}$ from the origin. This polygon contains the bottom edge of the polygon of length NM; delete that bottom edge. Then a polygon of length $NM - 2 + 2a(N) + 2P$ is obtained, where $P = |Y(\mathbf{b})| + \cdots + |Z(\mathbf{b})|$. Thus $p_n(f) \geq \left[p_N^+([h_b^*\phi_b^*, h_t^*\phi_t^*])\right]^M$, where $n = NM - 2 + 2a(N) + 2P$. Take logarithms, divide by n and take the lim inf of the left-hand side. On the right-hand side, fix N, and let $M \to \infty$ (with n). This shows that

$$\liminf_{n\to\infty} \frac{1}{n} \log p_n(f) \geq \frac{1}{N} \log p_N^+,$$

where eqn (6.75) was used. Since the left-hand side is independent of N, one may take $N \to \infty$ on the right-hand side. This shows that $\liminf_{n\to\infty}[\log p_n(f)]/n \geq \log \mu_d$, and so the proof is done. □

As a corollary to this lemma it follows that the growth constant of walks confined to an f-wedge is equal to μ_d if and only if the wedge can contain a d-dimensional box of arbitrary size.

Corollary 6.48 *Suppose that W_f is an f-wedge. Then*

$$\lim_{n\to\infty} \frac{1}{n} \log c_n(f) = \lim_{n\to\infty} \frac{1}{n} \log p_n(f) = \log \mu_d$$

if and only if $\lim_{X\to\infty} f_k(X) = \infty$ *for all* $k = 2, \ldots, d$. □

This result can be generalized to more arbitrary wedges. Let f_k and g_k for $k = 2, \ldots, d$ be non-negative functions. Define the fg-wedge as follows.

$$W_{fg} = \{(X_1, X_2, \ldots, X_d) \big| \, g_k(X_1) \leq X_k \leq f_k(X_1), \, \forall \, 2 \leq k \leq d\}. \tag{6.78}$$

The growth constant of polygons in an fg-wedge may be examined using the results for f-wedges above. The approach will be as follows. First examine polygons in the wedge defined by $0 \leq x_i \leq \alpha x_1$, for $i = 2, \ldots, d$. Choose that class with most popular X_1-span and height and orientations of the top edge. Concatenate two representatives in this most popular class to create a polygon in a generalized triangle (see Fig. 6.14). Copies of these polygons can be fitted into an fg-wedge, and then a similar argument to that in Lemma 6.47 can be used.

An α-wedge is defined by

$$W(\alpha) = \{(X_1, \ldots, X_d) \big| \, 0 \leq X_i \leq \lceil \alpha X_1 \rceil, \, \forall \, i = 2, \ldots, d\} \cup \{0, 1, 0, \ldots\}. \tag{6.79}$$

The point $\{0, 1, 0, \ldots\}$ is added to the wedge to make it possible for polygons containing the origin to fit into the wedge. By Corollary 6.48, the growth constant of polygons in an

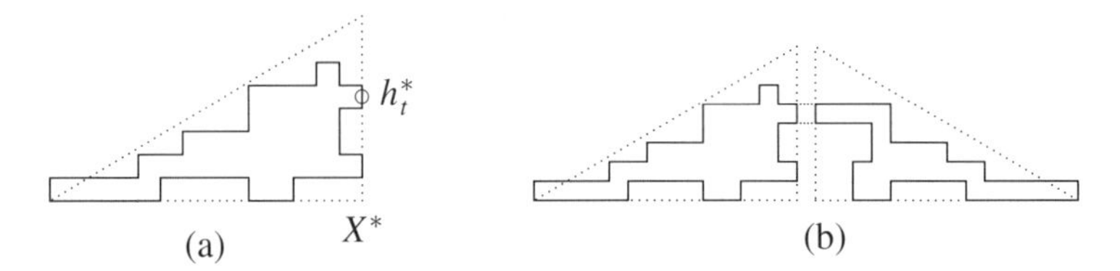

Fig. 6.14: (a) A polygon in an α-wedge has a most popular X-span, X^*, and most popular position and orientation $[h_t^* \phi_t^*]$ of its top edge. (b) Two polygons in the most popular class in an α-wedge can be concatenated to form a triangle polygon, which will grow at the same exponential rate as walks.

α-wedge is μ_d, for every $\alpha \in (0, \pi]$. Let $p_n^{\angle\alpha}$ be the number of polygons which contains the origin in an α-wedge. Define $p_n^{\angle\alpha}([h_t\phi_t X])$ to be the number of polygons counted by $p_n^{\angle\alpha}$ with X_1-span equal to X and position of the mid-point of the top edge at h_t, and its orientation ϕ_t. There are at most n^{d-1} choices for h_t, n choices for X, and $(d-1)$ for ϕ_t, so if $[h_t^*\phi_t^*X^*]$ are the most popular choices, then

$$(d-1)n^d p_n^{\angle\alpha}([h_t^*\phi_t^*X^*]) \geq p_n^{\angle\alpha} \geq p_n^{\angle\alpha}([h_t^*\phi_t^*X^*]). \tag{6.80}$$

Two polygons counted by $p_n^{\angle\alpha}([h_t^*\phi_t^*X^*])$ can be concatenated as shown in Fig. 6.14(b) to create a polygon which fits into a triangle with base of length $2X^*+1$ and angles at the base of size $\arctan\alpha$.[6] Note that X^* is a function of n, that the resulting polygon has length $2n$, and that if the number of such polygons is $p_{2n}^{\triangle}([2X^*+1])$, then

$$p_{2n}^{\triangle}([2X^*+1]) \geq [p_n^{\angle\alpha}([h_t^*\phi_t^*X^*])]^2. \tag{6.81}$$

In other words, by Corollary 6.48, and eqn (6.80), Lemma 6.49 is obtained. Polygons counted by $p_{2n}^{\triangle}([2X^*+1])$ are *triangle polygons* with base length $2X^*+1$.

Lemma 6.49 *Let $p_{2n}^{\triangle}([2X^*+1])$ be the number of triangle polygons of base length $2X^*+1$, and which passes through the edges $((0,\ldots,0),(0,1,0\ldots,0))$ and $((2X^*+1,0,\ldots,0),(2X^*+1,1,0,\ldots,0))$, and which is the result of concatenating two polygons of X_1-span X^* in the most popular class in an α-wedge as shown in Fig. 6.11(b). Then*

$$\lim_{n\to\infty} \frac{1}{2n} \log p_{2n}^{\triangle}([2X^*+1]) = \log \mu_d. \qquad \square$$

If $\alpha < \pi/4$, then two triangle polygons can be concatenated such that the bases of the triangles are parallel to one another, or at right angles with one another, as in Fig. 6.15(a). The triangle polygons can be fitted into an fg-wedge as in Fig. 6.15(b), and joined to the origin to create a polygon in an fg-wedge. This will show that the growth constant of polygons in fg-wedges is μ_d. An important property necessary in the statement and proof of the theorem is that both f and g are increasing functions of X.

Theorem 6.50 *Suppose that $\{f_i\}_{i=2}^d$ and $\{g_i\}_{i=2}^d$ are increasing and continuous functions defined on all non-negative real numbers, such that $0 \leq g_i(X) < f_i(X)$ for all $X \geq 0$ and $i = 2,3\ldots,d$. Suppose furthermore that $|f_i(X) - g_i(X)| \geq 1$ and $|f_i^{-1}(X) - g_i^{-1}(X)| \geq 1$ for at least one value of i.*

Let $c_n(fg)$ be the number of walks of length n with the origin as an end-point, and $p_n(fg)$ be the number of polygons of length n which passes through the origin, both confined to an fg-wedge. Then

$$\lim_{n\to\infty} \frac{1}{n} \log c_n(fg) = \lim_{n\to\infty} \frac{1}{n} \log p_n(fg) = \log \mu_d$$

if and only if $\lim_{X\to\infty} |f_i(X) - g_i(X)| = \infty$ and $\lim_{X\to\infty} |f_j^{-1}(X) - g_j^{-1}(X)| = \infty$ for every value of i and j in $\{2,3,\ldots,d\}$.

---

[6] The two points $\{0,1,0\ldots,0\}$ and $\{2X^*+1,1,0,\ldots,0\}$ are still considered part of this triangle, and the polygon passes through both these points and the origin, and the point $\{2X^*+1,0,\ldots,0\}$.

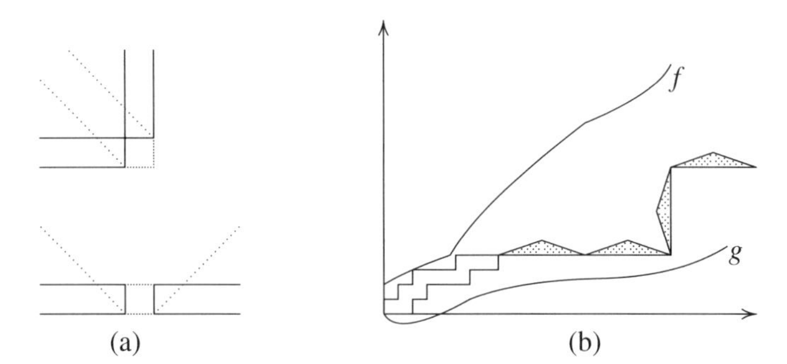

Fig. 6.15: (a) Two triangle polygons can be concatenated either with bases parallel, or with bases perpendicular. (b) By inserting triangle polygons in an fg-wedge, a lower bound on the number of polygons in the wedge is found.

Proof There are Kesten patterns which pass through every vertex in a d-dimensional cube. If either $|f_i(X) - g_i(X)|$ or $|f_j^{-1}(X) - g_j^{-1}(X)|$ does not increase to infinity with X, then Kesten patterns which contain d-dimensional cubes of arbitrary size cannot be fitted in the fg-wedge. Then Theorem 5.17 shows that $\lim_{n\to\infty} \frac{1}{n} \log p_n(fg) < \log \mu_d$.

To see the opposite, fix the number N, and consider the triangle polygons counted by $p_N^{\triangle}([2X^*+1])$. Then $X^* \leq N$. Choose a d-dimensional cube of size M with $M \geq 4X^*$. Since both $\lim_{X\to\infty} |f_i(X) - g_i(X)| = \infty$ and $\lim_{X\to\infty} |f_j^{-1}(X) - g_j^{-1}(X)| = \infty$, increase X until $|f_i(X) - g_i(X)| > M$ and $|f_j^{-1}(X) - g_j^{-1}(X)| > M$ for each i and j, and for all $X > X_0$. Translate a triangle polygon counted by $p_N^{\triangle}([2X^* + 1])$, with $\alpha < \pi/4$, and place it, with its base parallel to the X_1-axis, inside the wedge. Translate a second triangle polygon, and concatenate it to the first with its base parallel to X_1, provided that it fits in the wedge. If it intersects the wedge, then, since $X > X_0$, there is a direction (say Y) such that the second triangle polygon can be rotated with its base parallel to Y, and concatenated on the first. Continue this process, and join s such polygons into a polygon confined to the fg-wedge (see Fig. 6.15). By definition of the fg-wedge, there are two disjoint paths from the end-points of the bottom edge of the first triangle polygon to the origin. Delete the bottom edge of the first triangle polygon, add the two paths (say they contain R edges). Then a polygon of total length $R + sN$ edges is constructed, and moreover,

$$p_{R+sN}(fg) \geq [p_N^{\triangle}([2X^* + 1])]^s.$$

Choose $n = R + sN + q$, for fixed R and N, and with $q < N$ (and $n \geq R$). Then $p_n(fg) \geq p_{R+sN}(fg)$ (since $p_n(fg) \geq p_{n-2}(fg)$, and two edges can be added to the top edge of a polygon in the fg-wedge). Take logarithms of the equation above, divide by n, and take the lim inf as $n \to \infty$. This shows that

$$\liminf_{n\to\infty} \frac{1}{n} \log p_n(fg) \geq \frac{1}{N} \log p_N^{\triangle}([2X^* + 1]).$$

But now the left-hand side of the above is independent of N, and so take $N \to \infty$ and use Lemma 6.49 to complete the proof for polygons. Since $p_n(fg) \leq c_{n-1}(fg)$, the proof is then complete. □

The result on fg-wedges is useful in the study of uniform animals. This result has been applied to uniform animals adsorbing on the hyperplane $Z = 0$ in three dimensions. Let $h_n^+(v)$ be the number of uniform animals with f branches each of length n, v visits to the hyperplane $Z = 0$ and with the Z-coordinate of each vertex non-negative (that is, the number of positive embeddings). Moreover, every animal counted by $h_n^+(v)$ is assumed to have at least one vertex with Z-coordinate in the set $\{0, 1\}$ (that is, it is attached). If there are f branches, then the branches and vertices can be labeled in at most $2^f f!$ ways, thus, if the animal is bounded from above by independent positive walks, then,

$$h_n^+(v) \leq 2^f f! \sum_{\{v_i \geq 0\}} \prod_{j=1}^{f} c_n^+(v_j) \delta_{(\sum_k v_k - v)}. \tag{6.82}$$

Multiply this by z^v and sum over v to obtain

$$h_n^+(z) \leq 2^f f! [c_n^+(z)]^f. \tag{6.83}$$

Take logarithms, and divide by nf. Then let $n \to \infty$. Since the limiting free energy exists for adsorbing walks, this shows that

$$\limsup_{n \to \infty} \frac{1}{nf} \log h_n^+(z) \leq \mathcal{F}_v(z), \tag{6.84}$$

where $\mathcal{F}_v(z)$ is the limiting free energy of adsorbing polygons and walks. It only remains to find a lower bound to prove existence of the limiting free energy for adsorbing uniform graphs.

Theorem 6.51 *The limiting free energy of adsorbing uniform graphs with f branches exists in three dimensions, and is given by*

$$\lim_{n \to \infty} \frac{1}{nf} \log h_n^+(z) = \mathcal{F}_v(z),$$

where $\mathcal{F}_v(z)$ is the limiting free energy of a model of polygons or walks adsorbing in the hyperplane $Z = 0$.

Proof Consider the embedded animal in Fig. 6.16. Interchanging the Z-direction (North on the page) and the Y-direction (perpendicular to the page), by slightly extending the branches incident with the top line in the Y-direction, and adding ⊔-shaped extensions (one to each branch), a uniform embedding can be created with (say) m edges in each branch. Cut the quarter-space defined by the adsorbing plane and the top line into fg-wedges, and concatenate a polygon with v_i visits and n edges in each fg-wedge, where $\sum_{i=1}^{f} v_i = v$. Thus

$$h_{n+m}^+(v) \geq \prod_{i=1}^{f} p_n^+(v_i; f_i g_i),$$

where the i-th branch is attached to a polygon in the i-th wedge. Take logarithms, divide by $(n+m)f$ and let $n \to \infty$. This gives the desired result. □

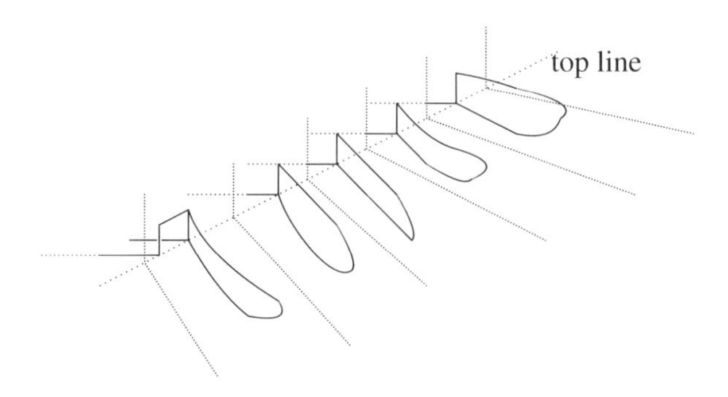

Fig. 6.16: An embedding of a uniform animal exists such that each end-vertex is in the top line, and every branch without an end-vertex has a vertex in the top line, and an edge in the top plane. By concatentaning polygons in narrow fg-wedges as shown, with edges in the top plane, and then by deleting the edges in the polygons which are in the top plane, a larger animal is obtained. Moreover, these polygons can expand in their own fg-wedges as the limit is taken. This shows that the limiting free energy in interacting models of uniform animals is determined by polygons which grow in fg-wedges, and so the limiting free energy is equal to that of adsorbing polygons.

Corollary 6.52 *There is an adsorption transition in a model of adsorbing uniform animals in three dimensions. Moreover, the free energy is given by the free energy of adsorbing walks, and the critical point is at the same location.* □

The situation is somewhat different in two dimensions. To understand this, consider the adsorption of a polygon in two dimensions, see Section 5.4. In three dimensions, the maximum number of visits in a polygon is n, in two dimensions, it is $n/2$. Thus, the limiting free energy of an adsorbing polygon is equal to the limiting free energy of an adsorbing walk in three dimensions, but not in two dimensions. A similar situation is true in the case of adsorbing uniform animals of fixed homeomorphism type in two dimensions. Since the maximum number of visits depends on the homeomorphism type, the limiting free energy is not equal to that of adsorbing walks, and it has not even been shown to exist [319].

The nature of walks and polygons in confined geometries was also discussed in references [360, 365, 366]. Uniform animals and branched polymers attached in wedge geometries have also been studied [55, 193, 328, 366], while the adsorption problem for uniform animals was discussed in reference [319]. Uniform animals in a slit geometry were considered in reference [327], where it is shown that the growth constant is dependent on the width of the slit. Similar results exists for slab geometries [53]. Lastly, the surface exponents γ_1 and γ_{11} for walks have been found in a wedge by conformal mappings [46], and for an excluded needle [345].

7

Lattice vesicles and surfaces

7.1 Introduction

In this chapter models of vesicles and surfaces will be examined. A general pattern theorem is not known for these models, although a very limited version of such a theorem has been proven [189]. This will be discussed in Section 7.5. In Section 7.2 two-dimensional models of vesicles are examined. These are square lattice polygons, counted by area, and lead naturally to the notion of a disk. Punctured disks are examined in Section 7.3, where the existence of growth constants, limiting free energies and density functions, and the asymptotic behaviour in these models are of particular interest. In three dimensions a disk is an orientable surface with genus zero and one boundary component; the adsorption problem for such disks is presented in Section 7.4. A crumpling transition in models of three-dimensional surfaces and vesicles is discussed in Section 7.5. Unrestricted models of surfaces are generally believed to be in the same universality class as branched polymers (and thus in the same class as lattice animals and trees). New phases are possible in interacting models; these include a "smooth" phase if there is a bending activity (which may also be sphere-like, disk-like or rod-like), an inflated phase if there is a volume activity, in addition to the branched polymer phase. The appearance of these phases depends on the particular ensemble under consideration, and the relation amongst them is only intuitively undestood. Generalizations to models of embedded manifolds in d-dimensions have also been considered [325]. This field is sometimes referred to as "statistical topology" or "statistical knot theory".

A review of the statistical mechanics of surfaces can be found in reference [267]. The Flory theory for surfaces was considered by J. Douglas [84], see also [254, 255], and branched polymer exponents are obtained (see Table 1.2). The scaling behaviour of surfaces was examined in reference [253], and the branched polymer character of both models of open and closed surfaces has been supported by numerical work and other arguments [6, 12, 138, 139, 266, 276, 308, 331, 334, 339].

7.2 Square lattice vesicles

A polygon in the square lattice can be used as a model of a two-dimensional vesicle. This model was examined by Fisher, Guttmann and Whittington [116], and numerically in references [114, 232]. The distinguishing feature in a polygonal model of vesicles (as opposed to the interacting models of lattice polygons in Chapter 5) is the fact that the enclosed area of the polygon plays an important role in the phase behaviour. In this context this model is similar to the models of composite polygons which were discussed in Section 5.6. The two-dimensional polygonal model of a lattice vesicle can also be generalized to hypersurface models in higher dimensions, which have (hyper)-area and (hyper)-volume; see for example reference [361].

Let $p_n(m)$ be the number of lattice polygons (in the square lattice) of length n, and enclosing an area of m unit squares, counted up to translation. The unit squares have vertices with integral coordinates, and are also called *plaquettes*. The number of polygons is $p_n = \sum_{m\geq 0} p_n(m)$, while the number of *disks* in the square lattice, of area m, is

$$D_m = \sum_{n=0}^{2m+2} p_n(m). \tag{7.1}$$

Models of disks in the square lattice have been studied [187, 199, 201], and I shall consider some of their properties in this section.

7.2.1 The Fisher–Guttmann–Whittington vesicle

The partition function of the vesicle in this model is defined by

$$p_n(z) = \sum_{m\geq 0} p_n(m) z^m, \tag{7.2}$$

where z is an area activity. Concatenation of two vesicles proceeds as in Fig. 1.4. This gives the supermultiplicative inequality

$$\sum_{m_1=0}^{m} p_{n_1}(m_1) p_{n_2}(m - m_1) \leq p_{n_1+n_2}(m+1). \tag{7.3}$$

In other words, this model satisfies the third condition of Assumptions 3.1; and there is a density function for the area, provided that it is finite (it may be infinite because the second condition of Assumptions 3.1 is not satisfied by this model). Multiplication of eqn (7.3) by z^m and summing over m gives

$$p_{n_1}(z) p_{n_2}(z) \leq [\phi(z)] p_{n_1+n_2}(z), \tag{7.4}$$

where $\phi(z) = z + 1 + z^{-1}$. Notice that if $z \leq 1$, then $p_n(z) \leq p_n = \mu_2^{n+o(n)}$, and so by Lemma A.1 there is a finite limiting free energy defined by

$$\mathcal{V}_a(z) = \lim_{n\to\infty} \frac{1}{n} \log p_n(z); \quad \text{if } z \in [0, 1]. \tag{7.5}$$

In addition, it will be important to note that Theorem A.1 also implies that

$$p_n(z) \leq [\phi(z)] e^{n\mathcal{V}_a(z)}. \tag{7.6}$$

If $z > 1$, then $p_{4k}(z) \geq z^{k^2}$, since there are square-shaped vesicles. This shows that $\limsup_{n\to\infty} [\log p_n(z)]/n = \infty$, and the free energy is infinite. Somewhat more is known about this regime, and the results are taken together in Theorem 7.1.

Theorem 7.1 *If $z \leq 1$ then there is a finite limiting free energy defined by*

$$\mathcal{V}_a(z) = \lim_{n\to\infty} \frac{1}{n} \log p_n(z).$$

Moreover, $p_n(z) \leq [\phi(z)]e^{n\mathcal{V}_a(z)}$, and if $z > 1$ then $\mathcal{V}_a(z) = \infty$ and

$$\lim_{n\to\infty} \frac{1}{n^2} \log p_n(z) = \frac{\log z}{16},$$

exists. In addition,

$$z\frac{d}{dz}\left[\lim_{n\to\infty} \frac{1}{n^2} \log p_n(z)\right] = \frac{1}{16}.$$

Proof It only remains to show that the last limit exists. If $n = 4k$, then the area of a square-shaped polygon is k^2; if n is not a multiple of 4, then the maximum area is $(n-2)(n+2)/16 = (n^2-4)/16$. Thus $\liminf_{n\to\infty}[\log p_n(z)]/n^2 \geq [\log z]/16$. On the other hand, since $z > 1$ it follows from eqn (7.1) that $p_n(z) \leq p_n z^{n^2/16}$, and so $\limsup_{n\to\infty}[\log p_n(z)]/n^2 \leq [\log z]/16$.

The expected area of a vesicle is

$$\langle m \rangle_n = z\frac{d}{dz}\big[\log p_n(z)\big].$$

Application of eqn (2.4) shows that $\log p_n(z)$ is a convex function of $\log z$, and the functions $zd([\log p_n(z)]/n^2)/dz$ are derivatives of a sequence of convex functions. The limit of this sequence is differentiable and by Theorem B.7,

$$\lim_{n\to\infty} z\frac{d}{dz}\left[\frac{\log p_n(z)}{n^2}\right] = z\frac{d}{dz}\left[\frac{\log z}{16}\right] = \frac{1}{16}.$$

This completes the proof. □

The free energy in this model is not a continuous function of z. By convexity, it is continuous in $[0, 1)$ (the right-continuity at 0 follows from the fact that $\log p_n(z)$ is both increasing and convex). Note that $\mathcal{V}_a(1) = \log \mu_2$, and continuity at $z = 1$ can also be shown.

Theorem 7.2 *The free energy $\mathcal{V}_a(z)$ is continuous if $z \in [0, 1]$, and infinite if $z > 1$.*

Proof It only remains to show left-continuity at $z = 1$. Let $\epsilon > 0$ and $\delta > 0$ be arbitrary and small. Since $[\log(p_n(z)/\phi(z))]/n \to \mathcal{V}_a(z)$, and by eqn (7.6), and since $[\log(p_n(z)/\phi(z))]/n$ is a continuous function for all $z \geq 0$, there exists an integer N_0 such that for all $n > N_0$,

$$\mathcal{V}_a(1) - [\log(p_n(1)/\phi(1))]/n < \epsilon/3;$$
$$[\log(p_n(1)/\phi(1))]/n - [\log(p_n(1-\delta)/\phi(1-\delta))]/n < \epsilon/3.$$

Moreover, by eqn (7.6),

$$[\log(p_n(1-\delta)/\phi(1-\delta))]/n \leq \mathcal{V}_a(1-\delta) + \frac{1}{n}\log\phi(z).$$

Take these together to find that

$$0 \leq \mathcal{V}_a(1) - \mathcal{V}_a(1-\delta) < \frac{2\epsilon}{3} + \frac{1}{n}\log\phi(z).$$

Increase n if necessary, until $[\log\phi(z)]/n < \epsilon/3$. Then $0 \leq \mathcal{V}_a(1) - \mathcal{V}_a(1-\delta) < \epsilon$. In other words, $\mathcal{V}_a(z)$ is left-continuous at $z = 1$. □

The singularity diagram of this model is obtained by studying its generating function, defined by

$$V_a(x, z) = \sum_{n=0}^{\infty} p_n(z)x^n. \tag{7.7}$$

The radius of convergence of $V_a(x, z)$ is $x_c(z) = e^{-\mathcal{V}_a(z)}$, and so it follows that $x_c(z) = 0$ if $z > 1$, and that $x_c(1) = \mu_2^{-1}$. Theorem 7.3 states more properties of $x_c(z)$.

Theorem 7.3 *The radius of convergence $x_c(z)$ of the generating function $V_a(x, z)$ has the following behaviour:*

$$x_c(z) \begin{cases} \simeq \sqrt{z^{-1}}, & \text{if } z < 1; \\ = \mu_2^{-1}, & \text{if } z = 1; \\ = 0, & \text{if } z > 1. \end{cases}$$

Moreover, $x_c(z)$ is continuous at all points $z \neq 1$.

Proof The continuity properties follow from Theorem 7.2. It only remains to show that $x_c(z) \simeq \sqrt{z^{-1}}$ if $z < 1$. Note that the minimal area of a polygon of length n is $(n-2)/2$. Thus, $p_n(z) \leq p_n z^{(n-2)/2}$, and thus $x_c(z) \geq \mu_2^{-1}\sqrt{z^{-1}}$. On the other hand, the polygons counted by $p_n((n-2)/2)$ can be concatenated to find

$$p_{n_1}((n_1-2)/2)p_{n_2}((n_2-2)/2) \leq p_{n_1+n_2}((n_1+n_2-2)/2),$$

as in Fig. 1.4; and thus $\lim_{n\to\infty}[\log p_n((n-2)/2)]/n = \log\nu$ exists and is finite and non-zero. Thus, since $p_n(z) \geq p_n((n-2)/2)z^{(n-2)/2}$, it follows that $x_c(z) \leq \nu^{-1}\sqrt{z^{-1}}$. This completes the theorem. □

The singularity diagram of $V_a(x, z)$ is illustrated in Fig. 7.1. Notice that $V_a(x, 1)$ is finite if $x < \mu_2^{-1}$, and that the derivatives of the generating function with respect to z are all finite along the phase boundary $z = 1$, $x < \mu_2^{-1}$:

$$\left[\frac{\partial^k}{\partial z^k} V_a(x, z)\right]\bigg|_{z=1} \leq \sum_{n=0}^{\infty} n^{2k} p_n(z)x^n, \tag{7.8}$$

since n^2 is larger than the area of any polygon of length n. This is finite if $x < \mu_2^{-1}$. Thus, the phase boundary along $z = 1$ and $x < \mu_2^{-1}$ is a line of essential singularities

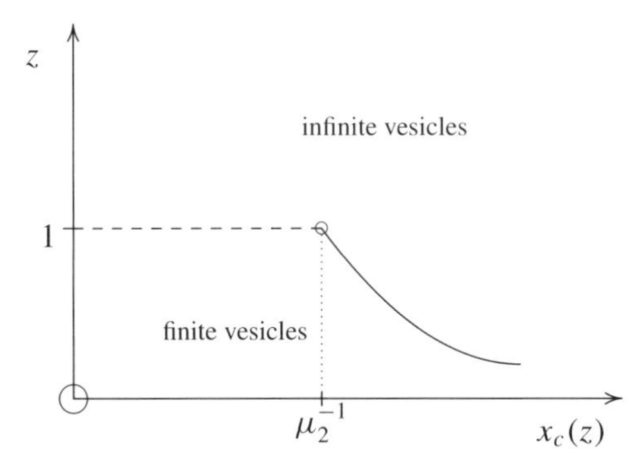

Fig. 7.1: The singularity diagram of the Fisher–Guttmann–Whittington vesicle. A line of essential singularities meets a curve of (presumably) simpler singularities at the point $(1, \mu_2^{-1})$, which is a tricritical point.

in $V_a(x, z)$. The points on this line correspond to droplet singularities at a first-order transition (point of condensation), see references [5, 184]. Along the critical curve for $z < 1$ it follows from eqn (7.6) that

$$V_a(x, z) \leq \frac{\phi(z)}{1 - e^{\mathcal{V}_a(z)} x}. \tag{7.9}$$

This suggests a curve of branch points in $V_a(x, z)$ which meets the line of essential singularities at the point $z = 1$, $x = \mu_2^{-1}$; thus by the notions of Chapter 2 this is a tricritical point.

The density function of this model exists by Theorem 5.93, and is defined by the existence of a sequence of integers $\sigma_n = o(n)$, and for any $\epsilon \in [1/2, \infty)$:

$$\begin{aligned} \mathcal{W}_a(\epsilon) &= \lim_{n \to \infty} \left[p_n(\lfloor \epsilon n \rfloor + \sigma_n) \right]^{1/n}; \\ \log \mathcal{W}_a(\epsilon) &= \inf_{z > 0} \{ \mathcal{V}_a(z) - \epsilon \log z \}, \end{aligned} \tag{7.10}$$

where the minimum value of ϵ is $\epsilon_m = 1/2$, since every vesicle of area n has perimeter at most $2n + 2$. Properties of the density function can be derived from the free-energy properties. For example, since the critical point in this asymmetric model is at $z_c = 1$, it follows from Lemma 3.20 that

$$\frac{d^+}{d\epsilon} \left[\log \mathcal{W}_v(\epsilon) \right] \Big|_{\epsilon = (1/2)^+} = 0. \tag{7.11}$$

In addition, by the proof of Theorem 7.3, and the pattern theorem for polygons, observe that $\mathcal{W}_a(1/2) = \nu < \mu_2$, and by the fact that $\log \mathcal{W}_a(\epsilon)$ is concave, $\mathcal{W}_a(\epsilon) \leq \nu$.

As in the case of composite polygons, more interesting variations on the density function can be defined. In particular, consider the number of polygons of perimeter n

enclosing area of size $\lfloor n^\epsilon \rfloor$, where $\epsilon \in [0, 2]$. Then if $\epsilon \in (0, 2)$, and if n is large enough (but finite), $1 \le p_n(\lfloor n^\epsilon \rfloor) \le p_n$, and so

$$1 \le \liminf_{n\to\infty} \left[p_n(\lfloor n^\epsilon \rfloor)\right]^{1/n} \le \limsup_{n\to\infty} \left[p_n(\lfloor n^\epsilon \rfloor)\right]^{1/n} \le \mu_2. \tag{7.12}$$

If $\epsilon = 2$, then $\lim_{n\to\infty} \left[p_n(\lfloor n^2 \rfloor)\right]^{1/n} = 0$ and if $\epsilon = 0$, then $\lim_{n\to\infty} \left[p_n(1)\right]^{1/n} = 0$. Define ϵ_{max} to be that least value of ϵ which maximizes $p_n(\lfloor n^\epsilon \rfloor)$. Then $\varepsilon = \limsup_{n\to\infty}[\epsilon_{max}/n] \le 2$ and the mean area enclosed in a polygon is proportional to n^ε. Numerical studies suggest that $\varepsilon \approx 1.5$ [42, 105].

Along the λ-line the generating function should behave as $V_a(x, z) \simeq (g_\lambda(t) - g)^{2-\alpha_+}$ (see eqn (2.12)). Since the value of z is less than one, this gives a deflated or branch polymer phase, and the vesicle should be in the same universality class as branched polymers. The exponent $2-\alpha_+$ can be guessed from the known branch polymer exponents. In particular, the generating function of lattice trees $\sum_{n\ge0} t_n x^n$ behaves as $|\log(\tau_2 x)|^{\theta-1}$, where θ is the entropic exponent; it is generally expected that $t_n \approx Cn^{-\theta}\tau_d^n$. Moreover, a dimensional reduction of lattice trees to the Ising model in two dimensions lower gave $\theta = 1$ in two dimensions [290], see also [114, 232]. Comparison to eqn (2.12) then shows that $2 - \alpha_+ = \theta - 1$, and in two dimensions, $2 - \alpha_+ = 0$. Numerical simulations suggested that $\nu = 0.65 \pm 0.04$ along the λ-line, consistent with the branched polymer value [232].

The behaviour of the generating function at the tricritical point is given by eqn (2.17) (plus lesser singular terms, plus analytic terms). Moreover, the cross-over exponent and the specific heat exponent are connected via the hyperscaling relation $2-\alpha = 1/\phi = d\nu$, where ν is the metric exponent of polygons at the tricritical point (compare this to eqns (1.30) and (2.28)). The important observation is that in this model it is uniformly weighted polygons which are found at the tricritical point. In other words, the metric exponent ν is that of (uniformly random) polygons, and this may be guessed from the Flory argument in Section 1.2.3. This gives $\nu = 3/4$, so that $\phi = 2/3$ and therefore $\alpha = 1/2$ (see also reference [6]). The entropic exponent of polygons (see eqn (1.6)), is equal to $1/2$, and by eqn (1.7), and eqn (2.16), the exponent $2 - \alpha_t = 3/2$ can be computed. Equation (2.18) then gives $2 - \alpha_u = 9/4$. In the case of inflated vesicles, $\nu = 1/2$. These numbers are expected to be exact, but no rigorous proofs are known, and they were computed using Coulomb gas [272] and conformal invariance techniques [47]. The exponents are listed in Table 7.1.

A connection was made between the tricritical zero-state Potts model, collapsing branched polymers and vesicles [346], and it was argued that compact animals (without "holes") are in the same universality class as vesicles, with $\phi = 2/3$. This shows that $\nu = 1/2$ for θ-animals with a cycle activity at the θ-point. These arguments are plausible

Table 7.1: Tricritical exponents for inflating two-dimensional vesicles

ϕ	α	$2-\alpha_t$	$2-\alpha_u$	μ_t	y_t	$2-\alpha_+$
$\frac{2}{3}$	$\frac{1}{2}$	$\frac{3}{2}$	$\frac{9}{4}$	$\frac{3}{4}$	$\frac{8}{9}$	0

if holes did not matter in the θ-animals, but this is not understood, and the connection between vesicles and θ-animals remains unresolved (see also [348]).

The derivative $z\partial \log V_a(x, z)/\partial z$ is the mean area of the vesicle, denoted by $\langle m \rangle$. This gives

$$\langle m \rangle = \frac{\sum_{n=0}^{\infty} \sum_{m=0}^{\infty} m\ p_n(m) x^n z^m}{\sum_{n=0}^{\infty} \sum_{m=0}^{\infty} p_n(m) x^n z^m}.$$

The denominator is dominated by its analytic term as x approaches the tricritical point, since $2 - \alpha_t = 3/2$. Thus, it should be expected from eqn (2.16) that $\langle m \rangle \simeq g^{2-\alpha_t-\Delta}$, where $\Delta = 1/\phi$ is the gap exponent. But as $g \to 0^+$ this is convergent, and so $2 - \alpha_t = 1/\phi$, exactly as obtained above. The dependence of the mean area as a function of the length of the polygon, $\langle m \rangle_n$, can also be determined. In particular, suppose that $\langle m \rangle_n \simeq n^{\varepsilon}$, where ε is an exponent to be determined. If it is assumed that almost all polygons in the ratio of sums above have area proportional to n^{ε}, then this assumption allows the calculation of $\langle m \rangle \simeq g^{2-\alpha_t-\varepsilon}$, as the tricritical point is approached. In other words, $\varepsilon = 1/\phi = 3/2$, as expected from numerical work [42, 105]. Thus, for a polygon $\langle m \rangle_n \simeq n^{3/2} \approx \langle R_n^2 \rangle$, so that its interior is "compact".

7.2.2 The perimeter of a vesicle

The previous paragraphs speculated that the area enclosed in a polygon could grow as a certain power of its perimeter. In this section, I shall consider the converse of this question; what is the relation between the area of a disk (two-dimensional vesicle) and its perimeter?

This question is best considered by defining the dual animal of a disk, which will be called a *surface-animal*. Consider a disk, and define the dual animal (in the dual square lattice) by joining vertices in the dual lattice if and only if they are the mid-points of adjacent plaquettes in the disk (see Fig. 7.2). Such a surface-animal is a site-animal, so that surface-animals are a subset in the set of site-animals. There are no nearest-neighbour contacts in site-animals, and each perimeter edge in the animal corresponds to an edge of the disk in its perimeter. Thus, the perimeter of the animal is the same size as the length of the perimeter of the disk, and by eqn (6.3), the length of the perimeter of the disk is given by

$$\pi = 2v + 2 - 2c = 2n + 4 - 4c, \tag{7.13}$$

if the dual animal has v vertices (and the disk has v plaquettes), n edges, and c cycles. Notice that every edge in a surface-animal is either a cut-edge, or is in a 4-cycle; a fundamental set of independent cycles can be chosen as a set with elements which are only 4-cycles.

By eqn (7.13) it appears that the perimeter of a disk is a maximum if the cyclomatic index of its dual surface-animal is a minimum. In other words, it seems important here to pay attention to the number of surface-animals with a density of 4-cycles. Thus, let $A_v(c)$ be the number of surface-animals with v vertices and c 4-cycles. Concatenation of

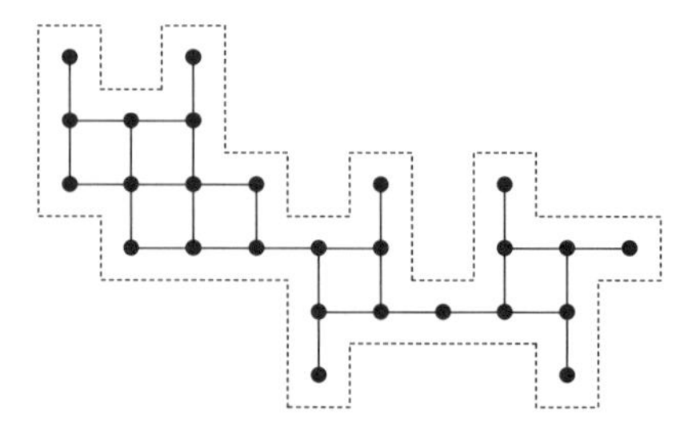

Fig. 7.2: A disk and its dual animal.

two surface-animals by inserting an extra vertex between them to prevent the formation of more cycles shows that they are supermultiplicative:

$$A_{v_1}(c_1)A_{v_2}(c_2) \leq A_{v_1+v_2+1}(c_1+c_2), \tag{7.14}$$

and thus $A_{v-1}(c)$ satisfies Assumptions 3.1 so that there is a density function

$$\mathcal{Q}_C(\epsilon) = \lim_{v\to\infty} [A_v(\lfloor \epsilon v \rfloor)]^{1/v}. \tag{7.15}$$

Moreover $\mathcal{Q}_C(\epsilon)$ is continuous in $(0, 1)$, since $\log \mathcal{Q}_C(\epsilon)$ is concave in $[0, 1)$. Since each site-animal is an animal with at least $v - 1$ and at most $2v$ edges, observe that

$$A_v(c) \leq \sum_{n=v-1}^{2v} a_n(c), \tag{7.16}$$

where $a_n(c)$ is the number of edge-animals with n edges and c cycles, see Section 6.3.1. Noting that $a_n(c) \leq a_{n+1}(c)$ (concatenate a single edge on the top vertex of the animals counted by $a_n(c)$), eqn (7.16) becomes

$$A_v(c) \leq (v+1)a_{2v}(c). \tag{7.17}$$

This relation has important consequences.

Theorem 7.4 *The density function $\mathcal{Q}_C(\epsilon)$ is bounded from below by $\mathcal{Q}_C(0) \geq \sqrt{\mu_2}$ and* $\lim_{\epsilon\to 1^-} \mathcal{Q}_C(\epsilon) = 1$.

Proof Notice that the density function $\mathcal{Q}_c(\epsilon)$ of cycles in an edge-animal was considered in Section 6.3.1. Theorem 6.15 states that $\mathcal{Q}_c(1/2) = 1$. By eqn (7.17), notice that $\mathcal{Q}_C(\epsilon) \leq [\mathcal{Q}_c(\epsilon/2)]^2$; this shows that $\lim_{\epsilon\to 1^-} \mathcal{Q}_C(\epsilon) = 1$. To see that $\mathcal{Q}_C(0) \geq \sqrt{\mu_2}$, notice that each self-avoiding walk of step-length two in the square lattice is a surface-animal. □

The important consequence of Theorem 7.4 is that there is a density of 4-cycles in surface-animals.

Corollary 7.5 *There exists an* $\epsilon_0 \in [0, 1)$ *such that*

$$\mathcal{Q}_C(\epsilon_0) = \sup_{\epsilon\in[0,1]} \{\mathcal{Q}_C(\epsilon)\} > \mathcal{W}_C(1^-). \qquad \square$$

By eqn (7.13) the mean perimeter length per plaquette of a disk with v plaquettes is given by

$$\langle \pi/v \rangle = \frac{\sum_{c=0}^{v}[(2v+2-2c)/v]A_v(c)}{\sum_{c=0}^{v} A_v(c)}. \tag{7.18}$$

But as v increases, both sums in the numerator and denominator above are dominated by the exponentially fastest growing terms, this is $A_v(\lfloor \epsilon_0 v \rfloor)$, which grows as $[\mathcal{Q}_C(\epsilon_0)]^{v+o(v)}$. Thus, the limiting behaviour of eqn (7.18) is determined by the terms with $c = \lfloor \epsilon_0 v \rfloor + o(v)$, and the result is the following theorem.

Theorem 7.6

$$\lim_{v\to\infty} \langle \pi/v \rangle = 2(1-\epsilon_0) > 0. \qquad \square$$

Thus, the perimeter of a disk is truly asymptotic to its area, $\pi \sim 2(1-\epsilon_0)v$ (see footnote 3, Chapter 1). This is in contrast to the speculations in Section 7.2.1, that the mean area of a polygon grows as the $3/2$ power of its perimeter.

7.3 Punctured disks in two dimensions

In the previous section a two-dimensional disk was (implicitly) defined as the connected interior of a polygon in the square lattice. It is composed of unit squares (plaquettes), glued at their edges. To define a punctured disk, more care is needed. I shall define a surface in the hypercubic lattice next; this will also define punctured disks in the square lattice.

Definition 7.7 (Surfaces) Two plaquettes are *adjacent* if they share an edge, but are otherwise disjoint. Two plaquettes are *connected* if they are in a sequence of plaquettes in which successive pairs are adjacent. If two plaquettes a and c share a vertex S, but are otherwise disjoint, then they are a *common pair* if there is a sequence $\{b_i\}_{i=1}^{M}$ of distinct plaquettes,[1] each plaquette b_i incident with the vertex S, such that $\{a, b_1\}$, $\{b_i, b_{i+1}\}$ for $i = 1, 2, \ldots, M-1$, and $\{b_M, c\}$ are all adjacent pairs. If there are no such b_i, then the pair $\{a, c\}$ is *uncommon*. With this terminology, it is possible to define a plaquette surface in the hypercubic lattice. A collection of plaquettes in the hypercubic lattice, or in the square lattice, is a *surface* if (1) every edge incident with any plaquette is incident with at most two plaquettes, (2) if every pair of plaquettes is connected, and (3) if there are no uncommon pairs of plaquettes. All the edges incident with exactly one plaquette form the *boundary* of the surface, and it is composed of a collection of (disjoint) polygons. Each such polygon is a *boundary component*. $\square$

In two dimensions a *punctured disk* is a plaquette surface with more than one boundary component. If it has only one boundary component, then it is a *disk*. Denote the number of punctured disks composed of n plaquettes and with h boundary components by $d_n(h)$; it is also said that the *area* of these disks is n. An interesting fact about disks is that they satisfy a submultiplicative relation, in addition to the expected supermultiplicative relation.

---

[1] In two dimensions, $M = 1$.

7.3.1 Submultiplicativity of disks

Disks in the square lattice can be concatenated as illustrated in Fig. 7.3. Define the top and bottom edges of a disk as those edges with the lexicographically most and least mid-points. Two disks can be concatenated by translating one to identify its top edge with the bottom edge of the second. This gives the supermultiplicative inequality

$$d_{n_1}(h_1)d_{n_2}(h_2) \leq d_{n_1+n_2}(h_1 + h_2 - 1), \tag{7.19}$$

so that $d_n(h-1)$ satisfies Assumptions 3.1. The consequence is that there is a density function for boundary components, and a limiting free energy with an activity z conjugate to the number of boundary components, defined by

$$\begin{aligned} \mathcal{W}_h(\epsilon) &= \lim_{n\to\infty} [d_n(\lfloor \epsilon n \rfloor)]^{1/n} \\ \mathcal{V}_h(z) &= \lim_{n\to\infty} \frac{1}{n} \log d_n(z), \end{aligned} \tag{7.20}$$

where $d_n(z) = \sum_{h\geq 0} d_n(h) z^h$ is the partition function. These results follow immediately from Theorems 3.4 and 3.17. Since each boundary component has length at least four, and is surrounded by eight plaquettes, it follows that $\epsilon < 1/3$. If $d_n = \sum_{h\geq 0} d_n(h)$ is the number of all punctured disks of area n, then the construction in Fig. 7.5 also shows that $d_{n_1} d_{n_2} \leq d_{n_1+n_2}$, with the result that

$$\lim_{n\to\infty} \frac{1}{n} \log d_n = \sup_{n>0} \frac{1}{n} \log d_n = \log \chi_2, \tag{7.21}$$

(from Lemma A.1), and where χ_2 is the growth constant of punctured disks in two dimensions. Similarly, there is a growth constant for disks defined by

$$\lim_{n\to\infty} \frac{1}{n} \log d_n(1) = \inf_{n>0} \frac{1}{n} \log d_n(1) = \log \beta_2. \tag{7.22}$$

It is an interesting fact that $d_n(1) = D_n$ (see eqn (7.1)) also satisfies a certain generalized submultiplicative inequality. This fact has a lengthy proof, and the first steps are constructions which will cut a disk into smaller disks. An important notion in the constructions is that of a *split-plaquette*, which are plaquettes which will cut a disk into two pieces if deleted. In Construction 7.9 I prove that if a disk has no split-plaquettes, then it can be cut into two disks of arbitrary sizes.

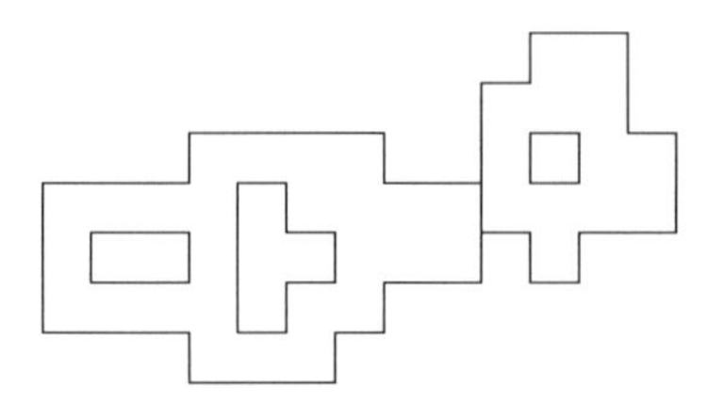

Fig. 7.3: Concatenating two punctured disks.

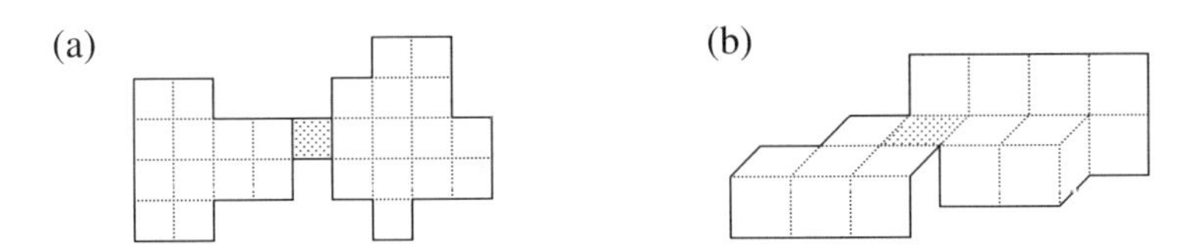

Fig. 7.4: Split-plaquettes in a disk in the square lattice, (a) and in a disk in the cubic lattice (b).

Definition 7.8 (Subdisks) Let A be a disk. If B is also a disk, and $B \subset A$, and if $A \setminus B$ is a disk; then B is a *subdisk* of A. □

In Constructions 7.9 and 7.10 I examine a process for cutting a disk into subdisks. These constructions are also valid in more than two dimensions; this will be important in Section 7.4.

Construction 7.9 If a disk of area n in the square lattice, or in the hypercubic lattice, has no split-plaquettes, then it can be cut into a disk of size m, and a disk of size $n - m$, for any $0 \leq m \leq n$.

Proof Consider a disk C of area n and suppose that $1 \leq m < n$. Any plaquette which has an edge in the boundary of C can be taken out of the disk, since there are no split plaquettes. The rest of the proof is by induction. Suppose that the disk has been cut along a piecewise linear curve ℓ, into a subdisk A of size $m - 1$, and a subdisk B of size $n - m + 1$. Select a plaquette p with exactly one edge in the curve ℓ in B (there is always such a plaquette; choose for example the lexicographically most plaquette in B with an edge in ℓ). If p is not a split-plaquette of the disk B, then it can be added to A, and the proof is finished. So, suppose that p is a split-plaquette of B. Then it must divide B in two (or more) smaller pieces if it is deleted, one of these (say B') with at least one edge in ℓ, and with $j < n - m + 1$ plaquettes (this follows from the fact that C has no split-plaquettes, and so p cannot split C if it is deleted). Select a plaquette p' in B', adjacent with ℓ. If p' is not a split-plaquette of B', then the proof is finished. Otherwise, it splits B' into pieces, one incident with ℓ and with $k < j$ plaquettes. Continue in this way, until an appropriate plaquette is found, or until eventually a plaquette is found incident with an edge of ℓ in the boundary of B. This plaquette cannot be a split-plaquette of B, since C has no split-plaquettes. Thus, it can be added to A. This completes the proof. □

If disks with split-plaquettes are considered, then the above construction will give a weaker result, and its proof is somewhat more complicated; it can then be shown that, given m, one can always cut a subdisk of area at least $\lceil m/3 \rceil$ and at most m from any given disk of size n.

Construction 7.10 Suppose that $n \geq 2$, and let m be any integer such that $m \geq 2$ be given such that $m \leq n$. Then it is always possible to cut a disk of size n plaquettes into two disks, one which has at least $\lceil m/3 \rceil$ and at most m plaquettes.

Proof The technique of proof is as follows: I shall first show that a subdisk of size $b_1 \geq \lceil m/3 \rceil$ can be cut from a disk of size n. Then, if $b_1 > m$, I shall show that a subdisk of area b_2 can be cut instead, where $b_1 > b_2 \geq \lceil m/3 \rceil$. By repeating this construction, a disk of desired area must finally be found. Let A be a disk of size n. If A has no split-plaquettes, then it is possible to cut it into two subdisks, one with m plaquettes, by Construction 7.9. Suppose that A has at least one split-plaquette (see Fig. 7.5). Incident with p are at most four subdisks $\{T_i\}_{i=1}^4$ of A (some possibly empty). Let T_i have t_i plaquettes each, with $1 \leq i \leq 4$. Note that $\sum_i t_i = n - 1$. Without loss of generality, suppose that $m \leq n/2$. If $t_i < m/2$ for each i, then $n - 1 = \sum_{i=1}^4 t_i < 2m \leq n$. This shows that in this case, $2m = n$. Thus $\sum_{i=1}^4 t_i = 2m - 1$. If $m = 2k$ is even, then the maximum of t_i is $k-1$, so that $\sum_{i=1}^4 t_i \leq 4k-4 < 2m-1$, a contradiction. If $m = 2k+1$ is odd, then the maximum of t_i is k, so that $\sum_{i=1}^4 t_i = 4k = 2m - 2 < 2m - 1$, another contradiction. Thus, the assumption that $t_i < m/2$ for each i is incorrect, and there is at least one i (say $i = 1$) so that $t_i \geq m/2$. Therefore, assume that $t_1 \geq \lceil m/2 \rceil$. Thus, there is always at least one of the T_i incident with a split-plaquette with area at least $\lceil m/3 \rceil$ plaquettes, provided that $m \leq \lfloor n/2 \rfloor$. If may still be the case that $t_1 > m$; and so further work is required to find a subdisk of the disk of size at most m, and at least $\lceil m/3 \rceil$.

Consider now the disk T_1 consisting of t_1 plaquettes, and joined to the rest of the surface at the split-plaquette p (see Fig. 7.5). If there are no split-plaquettes in T_1, then Construction 7.9 can be used to cut a subdisk of size m from it, which completes the proof. Thus, suppose that there is a split-plaquette p_1 in T_1 which divides the T_1 into four subdisks B_i of sizes u_i, and where one subdisk (say B_4) which contains p. There are now two subcases. Case (1): $\sum_{i=1}^3 u_i \geq m$. In this case there is an i (say $i = 1$) such that $u_1 \geq \lceil m/3 \rceil$; and note that $t_1 > u_1 \geq \lceil m/3 \rceil$. Case (2): $\sum_{i=1}^3 u_i < m$. There are two subcases. Case (2a): $\sum_{i=1}^3 u_i \geq \lceil m/3 \rceil - 1$. Add the split-plaquette p_1 to the subdisks of sizes u_1, u_2 and u_3 to obtain a subdisk of the desired size. Case (2b): $\sum_{i=1}^3 u_i < \lceil m/3 \rceil - 1$. In this case it is not enough to only add p_1 to B_1, B_2 and B_3 to obtain a subdisk of the desired size. Instead, plaquettes which are in B_4 must also be added; and for this Construction 7.9 is used. There is an important difference here though: at some stage a split-plaquette may be added, and this may increase the size from below $\lceil m/3 \rceil$ to above m. If this happens, let p_2 be that split-plaquette. Naturally, $p_2 \neq p$ (if $p_2 = p$, then $t_1 < \lceil m/3 \rceil$, a contradiction). Incident with p_2 are four subdisks C_i, one

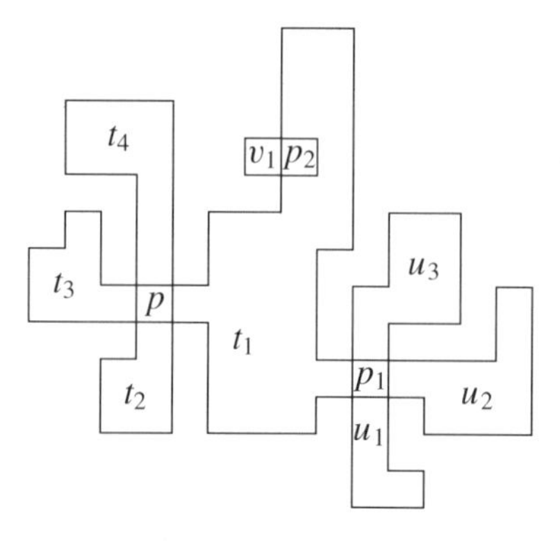

Fig. 7.5: p is a split-plaquette which splits the disk into four pieces containing t_i plaquettes each.

of which contains p (see Fig. 7.4). Let this be C_4, and let the remaining subdisks incident with p_2 contain v_i, $1 \leq i \leq 3$, plaquettes each. Now either p_1 is in the subdisk C_4, or it is not. If it is, then p and p_1 are both in C_4, and $t_1 > \sum_{i=1}^{3} v_i + 1 > m - \lceil m/3 \rceil = \lfloor 2m/3 \rfloor$, and so a new subdisk composed of p_2, C_1, C_2 and C_3 is found, and it has area less than t_1. If p_1 is not in the subdisk C_4, then p and p_1 are not in the same subdisk, say p_1 is in C_3. Thus $v_3 < \lceil m/3 \rceil$, and since $\sum_{i=1}^{3} v_i + 1 > m$, it follows that $\sum_{i=1}^{2} v_i + 1 + \lceil m/3 \rceil > \sum_{i=1}^{3} v_i + 1 > m$. Thus $\sum_{i=1}^{2} v_i \geq (m+2) - 1 - \lceil m/3 \rceil \geq \lfloor 2m/3 \rfloor + 1$. To satisfy this constraint, either v_1 or v_2 must be greater than or equal to $\lceil m/3 \rceil$, but smaller than t_1. □

There are the following consequences of Constructions 7.9 and 7.10.

Corollary 7.11 *Let A_{n+m} be a disk of area $n+m$, and let m be any positive integer, and suppose that $m \leq n$. Then k subdisks $\{B_i\}_{i=1}^{k}$ can be cut from A_{n+m}, where B_i has area m_i, such that*

(1) $\sum_{i=1}^{k} m_i = m$;

(2) $m - \sum_{i=1}^{\ell} m_i \geq m_{\ell+1} \geq \lceil (m - \sum_{i=1}^{\ell} m_i)/3 \rceil$, *if* $l \geq k$;

(3) $k \leq [\log m]/[\log(3/2)]$.

Proof By Construction 7.10, the maximum number of cuts of A_{n+m} (to remove subdisks) occurs if the smallest possible subdisk is cut from A_{n+m} at each cut. That is, when p plaquettes remain to be cut, then only $\lceil p/3 \rceil$ are actually cut. If p plaquettes must be removed at the j-th step, then the subdisk B_j has area $m_j = \lceil p/3 \rceil$, and $p - m_j = \lfloor 2p/3 \rfloor$ plaquettes remain to be removed in A_{n+m}. Iterating this process shows that at most $\lfloor 2 \lfloor 2 \lfloor \cdots \lfloor 2m/3 \rfloor \cdots \rfloor /3 \rfloor /3 \rfloor$ plaquettes remain to be removed at the j-th step (there are j factors of 2 and 3 in the expression). After k subdisks are removed, then m plaquettes are cut; this happens when the maximum number of plaquettes to be removed is zero at the $(k+1)$-th cut, and one at the k-th cut. But then $(2/3)^k m \geq \lfloor 2 \lfloor 2 \lfloor \cdots \lfloor 2m/3 \rfloor \cdots \rfloor /3 \rfloor /3 \rfloor = 1$, and so if $(2/3)^k m \leq 1$, then the value of k is an upper bound on the number of cuts. Thus $k \leq [\log m]/[\log(3/2)]$, and (3) is proven. To see (1), notice that if only one more plaquette is needed, then it can be cut, and so it is always possible to cut the exact number needed. (2) is a direct consequence of Construction 7.10. □

The submultiplicative nature of $D_n = d_n(1)$ can be found from Corollary 7.11. Cut $k \leq [\log m]/[\log(3/2)]$ subdisks of total area m from a disk A counted by D_{n+m}. These subdisks can be concatenated into a single disk of size m as in Fig. 7.3; let B be the resulting disk. B can be cut into k subdisks in at most $(4m)^k$ ways (choose $k-1$ edges from at most $4m$, and try to cut the disk in these edges). In addition, at most $[4(n+m)][4m]$ attempts may be made to put each of the subdisks back into A (select an edge in each, starting at that subdisk with least bottom edge, and identify them). Collecting these terms show that $D_{n+m} \leq [64(n+m)m^2]^k D_n D_m$. Put $\alpha = 1/\log(3/2)$, and from Corollary 7.11 notice that

$$D_{n+m} \leq [64(n+m)m^2]^{\alpha \log m} D_n D_m. \tag{7.23}$$

Thus, D_n satisfies a submultiplicative relation, and this can be tidied up as in Theorem 7.12.

Theorem 7.12 *If $n \geq 1$ and $m \geq 1$, then there exists a constant $\alpha > 0$ such that*

$$D_n D_m \geq (n+m)^{-\alpha \log(n+m)} D_{n+m}.$$

Proof This result is weaker than eqn (7.23); the factor of 64 was absorbed into α (by increasing α), and m was replaced by $n+m$ in the obvious places. □

The important consequence of Theorem 7.12 is found when Theorem A.3 is used, with eqn (7.22), to see that

$$\frac{\log D_n}{n} \geq \log \beta_2 + \frac{\alpha[\log n]^2}{n} - 4\alpha \sum_{m=2n}^{\infty} \frac{[\log m]^2}{m(m+1)}. \tag{7.24}$$

The last term can be bounded from above as follows:

$$\sum_{m=2n}^{\infty} \frac{[\log m]^2}{m(m+1)} \leq \sum_{m=2n}^{\infty} \left[\frac{\log m}{m}\right]^2 \leq \int_{m=2n-1}^{\infty} \left[\frac{\log x}{x}\right]^2 dx \leq \frac{[\log n]^2}{n}. \tag{7.25}$$

Substitution of this into eqn (7.24), followed by some simplification, shows that if $n \geq 2$, then

$$\frac{\log D_n}{n} \geq \log \beta_2 - \frac{3\alpha}{n}[\log n]^2. \tag{7.26}$$

Now replace 3α by α. Then this result gives the following bound on the number of disks.

Theorem 7.13 *If $n \geq 2$, then there exists a constant $\alpha > 0$ such that*

$$e^{-\alpha[\log n]^2} \beta_2^n \leq D_n \leq \beta_2^n.$$ □

Theorem 7.13 strongly suggests the asymptotic behaviour $D_n \sim Cn^{-\theta_1}\beta_2^n$. The entropic exponent θ_1 is believed to be the branched polymer entropic exponent; it is equal to 1 in two dimensions, and equal to $3/2$ in three dimensions [290]. The mean field value is $5/2$ (see Table 1.2). These values are supported by available numerical data [137, 138, 139, 140]. No proof of these facts is known, in fact it is not even known that $[\log(D_n \beta_2^{-n})]/[\log n] \sim -\theta_1$, for some constant θ_1.

7.3.2 The asymptotic behaviour of punctured disks

In this section the relationship between the entropic exponents of punctured disks in two dimensions will be considered. In particular, if the entropic exponent of a punctured disk with h boundary components is θ_h, then I shall show that $\theta_{h+1} = \theta_1 - h$. The proof that the claimed relationship exists relies on an assumption that $D_n = d_n(1)$ has the appropriate asymptotic behaviour. There are two ingredients in the proof. The first important ingredient is a density result for boundary components. Since disks are in one-to-one correspondence with a subclass of site-animals, the pattern theorem for animals can be amended suitably for the situation here. In particular, a new boundary component can be created in a disk by deleting all plaquettes in a 5×5 square which intersects the disk. Cover the centre 3×3 square with plaquettes, except for the single plaquette in the

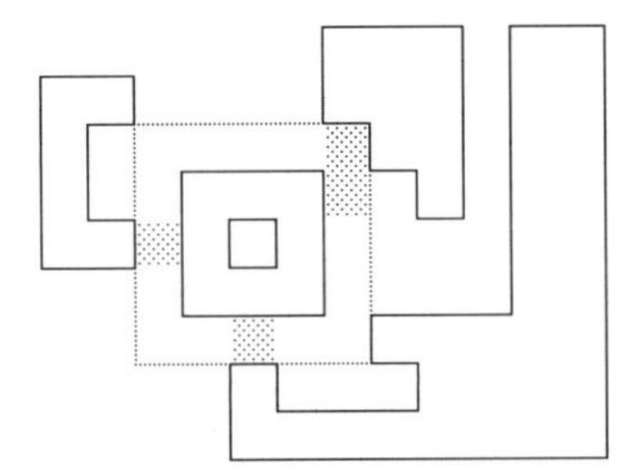

Fig. 7.6: A new boundary component can be created in a disk by deleting all plaquettes in a 5×5 square, inserting a 3×3 annulus in this square, and then adding plaquettes to reconnect the surface.

centre, which is left as a puncture. The disk is now reconnected by inserting plaquettes in the annulus between the 3×3 and 5×5 squares, as they are needed, see Fig. 7.6. Notice that all information about the disk inside the 5×5 square is lost; and that at most 2^{25} disks will give the same outcome. This bound is seen by noting that each plaquette in the 5×5 square is either present, or absent, before the onset of the construction. Since any covering of a disk of area n by 5×5 squares contains at least $\lfloor n/25 \rfloor$ squares, h new boundary components can be created in at least $\binom{\lfloor n/25 \rfloor}{h}$ ways. The area of the disk may change by as much as $25h$ in this construction. This gives Theorem 7.14.

Theorem 7.14 *Let $d_n(1) = D_n$ be the number of disks of area n in the square lattice. Then h boundary components can be created in any disk counted by D_n in at least $\binom{\lfloor n/25 \rfloor}{n}$ ways. The outcome is that*

$$\binom{\lfloor n/25 \rfloor}{h} d_n(1) \leq 2^{25h} \sum_{i=-25h}^{25h} d_{n+i}(h+1). \qquad \square$$

The second important ingredient is a construction which will reduce the number of boundary components in a disk, by cutting a linear strip between two boundaries from the disk.

Theorem 7.15 *There exists finite positive constants c and C such that*

$$d_n(h+1) \leq [Cn] d_{n+c}(h),$$

for all values of $h \geq 1$.

Proof Let A be a punctured disk with n plaquettes and $h+1$ boundary components. The boundary ∂A of A is composed of a collection of disjoint polygons. If one of these boundary components has length four, then it can be "plugged" by a single plaquette; so suppose that each boundary has length at least six, and so covers at least two unit squares. Select two mid-points v and w of edges in ∂A, and with v and w in different boundary components. Join v and w with a piecewise linear curve C_{vw} which passes throught the mid-point of every plaquette it traverses, and which passes from one plaquette to the next only through an interior point of an edge. Let the number of plaquettes C_{vw} crosses be $|C_{vw}|$. Vary v and w in ∂A until the least value of $|C_{vw}|$ is found. Let the plaquettes which C_{vw} traverses be $\{\pi_i\}_{i=1}^{M}$, with π_1 containing v. If π_1 is incident with only the boundary component containing v, then it may be deleted (in this case at most two adjacent edges of π_1 are in the boundary component). This may be continued until a plaquette with edges or vertices in two boundary components must be deleted. Without loss of generality, suppose that the plaquettes at $u - \hat{\imath}$ and u have been deleted, and suppose that the next plaquette to be deleted will be $u + \hat{\imath}$ (if $u \pm \hat{\jmath}$ is next instead, then a similar argument will complete the proof). Notice that the plaquettes at $u - \hat{\imath} \pm 2\hat{\jmath}$ are present; if not, then by deleting $u - \hat{\imath} \pm \hat{\jmath}$, a shorter path to a boundary can be found. Similarly, the plaquettes at $u \pm \hat{\jmath}$ are present, as are the plaquettes at $u + \hat{\imath} \pm \hat{\jmath}$. By our assumptions, there is either a vertex, or two vertices, or an edge, of $u + \hat{\imath}$ in the boundary. If the edge with mid-point $u + 3\hat{\imath}/2$ is in the boundary, then delete the plaquette at $u + \hat{\imath}$ to complete the theorem. Otherwise, suppose that the plaquette at $u + 2\hat{\imath}$ is present. In that case, at least one of the vertices $u + 3\hat{\imath}/2 \pm \hat{\jmath}/2$ is in the boundary; let the vertex $u + 3\hat{\imath}/2 + \hat{\jmath}/2$ be in the boundary. Then delete the two plaquettes at $u + \hat{\imath}$ and $u + \hat{\imath} + \hat{\jmath}$ instead to complete the theorem, *unless* the plaquette at $u + 2\hat{\jmath}$ is absent *and* the plaquette at $u + \hat{\imath} + 2\hat{\jmath}$ is present. But this case can be done by deleting $u + \hat{\jmath}$ instead. This covers all the cases, and a strip of plaquettes, C_{vw}, between two boundary components is removed, reducing the number of boundary components by one.

Lastly, concatenate the remaining part of D with a square disk of side-length 4 and area 16, and then with C_{vw}. This construction maps D into disks with h boundary components, and area $n + 16$. Since C_{vw} can be put back into D in at most $12n$ ways (fix an edge in the boundary of C_{vw} and incident with the first plaquette in C_{vw} in one of three ways, choose an edge in D in at most one of $4n$ ways, and identify them in attempts to recreate D). Thus,

$$d_n(h+1) \leq n\, d_{n+1}(h) + 12n\, d_{n+16}(h),$$

where the first term accounts for cases where a boundary of length four was plugged. Since $d_n(h) \leq d_{n+1}(h)$ by concatenation, the result follows. □

This result can be used to establish the next theorem [199, 201].

Theorem 7.16 *Suppose that the limit*

$$\lim_{n\to\infty} \frac{\log[d_n(1)\beta_2^{-n}]}{\log n} = -\theta_1$$

exists and is finite. Then the limit

$$\lim_{n\to\infty} \frac{\log[d_n(h)\beta_2^{-n}]}{\log n} = -\theta_h$$

exists for all $h > 1$, and moreover

$$\theta_{h+1} = \theta_1 - h.$$

Proof By Theorem 7.14,

$$\liminf_{n\to\infty} \frac{\log[d_n(h+1)\beta_2^{-n}]}{\log n} \geq -\theta_1 + \lim_{n\to\infty}\left[\frac{1}{\log n}\binom{\lfloor n/25\rfloor}{h}\right] = -\theta_1 + h. \qquad (\dagger)$$

On the other hand, Theorem 7.15 shows that

$$\limsup_{n\to\infty} \frac{\log[d_n(h+1)\beta_2^{-n}]}{\log n} \leq \limsup_{n\to\infty} \frac{\log[d_n(h)\beta_2^{-n}]}{\log n} + 1. \qquad (\ddagger)$$

First put $h = 1$ in the above. This shows that $-\theta_2 = -\theta_1 + 1$ and that θ_2 is well defined as the value of a limit. The rest of the θ_h follow inductively. □

Similar result are known for surfaces in three dimensions. Suppose for example that orientable surfaces in three dimensions are counted by genus and number of boundary components. Let $s_n(h,g)$ be the number of orientable surfaces with h boundary components and genus g. The hypothesis that $s_n(h,g) \simeq n^{-\theta_{h,g}}\beta_d^n$ then shows that $\theta_{h-\mu,g-\nu} \geq \theta_{h+1,g} \geq \theta_{1,0} - 2h - 4g$ [187]. For closed surfaces, let $s_n(0,g) \simeq n^{-\theta_g}\eta_d^n$; then the bound $\theta_{g-\nu} \geq \theta_g \geq \theta_0 - 4g$ is found instead [187]. In the case of θ_g, numerical work shows that $\theta_0 = 1.50 \pm 0.06$ (close to the branched polymer value), $\theta_1 = -0.5 \pm 0.5$ and $\theta_2 = -2.3 \pm 1.0$ [331].

7.3.3 Pattern theorems for punctured disks in two dimensions

The existence of a density function for the number of punctured disks with h boundary components was shown in eqn (7.20), see also reference [188]. In this section, some properties of this density function will be examined. Observe that Theorem 7.14 suggest a partial pattern theorem for boundaries; the construction leading to it creates boundary components of length four. Put $h = \lfloor \epsilon n \rfloor$. Then if $|\delta| \leq \epsilon$ is that choice of α which maximizes $\mathcal{W}_h(\epsilon/(1+25\alpha))$, and

$$\left[\frac{(1/25)^{1/25}}{\epsilon^{\epsilon}((1/25-\epsilon)^{1/25-\epsilon}}\right]\mathcal{W}_h(0) \leq [2^{25}\beta_2]^{\epsilon}\mathcal{W}_h(\epsilon/(1+25\delta)), \qquad (7.27)$$

then an $\epsilon_0 > 0$ can be chosen to maximize the left-hand side, and to prove that $\mathcal{W}_h(0) < \mathcal{W}_h(\epsilon_0)$. In other words, there seems to be a natural density for the number of boundary components (of length four) in a punctured disk. It is also possible to show that $\mathcal{W}_h(\epsilon)$ is strictly increasing if $\epsilon > 0$ is small enough, by applying the construction in Fig. 7.6 to punctured disks counted by $d_n(\lfloor \epsilon n \rfloor)$. Right-continuity of $\mathcal{W}_h(\epsilon)$ at $\epsilon = 0$ is not known.

It is possible to prove a pattern theorem for any particular shaped boundary component. Define a pattern to be any polygon in the square lattice. Let P be a polygon, and suppose that the X-span of P is X_P and the Y-span of P is Y_P. Then a density function for P can be defined by considering $d_n(mP)$, which is the number of punctured disks with $m+1$ boundary components in two dimensions, and with every component a translate of P, except for the external boundary. Thus, $d_n(h) = \sum_P d_n((h-1)P)$, where the sum is over all square lattice polygons which can be used as a boundary component.

Concatenation shows that $d_n(mP)$ satisfies Assumptions 3.1, and that there is a density function $\mathcal{W}_\chi(\epsilon)$ for P, defined by

$$\mathcal{W}_\chi(\epsilon) = \lim_{n\to\infty} [d_n(\lfloor \epsilon n \rfloor P)]^{1/n}. \tag{7.28}$$

Theorem 7.17 *Suppose that P is a polygon with X-span X_P and Y-span Y_P. Suppose that $s = (X_P + 4)(Y_P + 4)$. Then*

$$\binom{\lfloor n/s \rfloor}{m} d_n(0P) \leq \sum_{i=-ms}^{ms} d_{n+i}(mP).$$

Proof The proof follows the outlines of the proof of Theorem 6.10. Cover a disk counted by $d_n(0P)$ with rectangles of size $(X_P + 4) \times (Y_P + 4)$. At least $\lfloor n/s \rfloor$ are needed. Choose m of these, and fit the polygon P into them. Fill in the necessary plaquettes to create P, and additional plaquettes in the margin between P and the perimeter of the rectangle to prevent the creation of extra boundary components. This may change the number of plaquettes by at most ms. □

As an immediate consequence of Theorem 7.17, observe that

$$\binom{\lfloor n/s \rfloor}{m} d_n(0P) \leq (2ms + 1) \max_{-ms \leq q \leq ms} \{d_{n+q}(mP)\}. \tag{7.29}$$

Put $m = \lfloor \epsilon n \rfloor$, take the $(1/n)$-th power and take the lim sup as $n \to \infty$. If $|\delta| \leq \epsilon s$ is chosen to maximize $\mathcal{W}_\chi(\epsilon/(1+\delta))$, then

$$\left[\frac{(1/s)^{1/s} \beta_2^{-\epsilon s}}{\epsilon^\epsilon (1/s - \epsilon)^{1/s-\epsilon}} \right] \mathcal{W}_\chi(0) \leq \mathcal{W}_\chi(\epsilon/(1+\delta)). \tag{7.30}$$

In other words, there is an $\epsilon_0 > 0$ which maximizes the left-hand side above. Using the remarks in footnote 3 in Chapter 3, the result is the following corollary.

Corollary 7.18 *Suppose that P is a polygon with X-span X_P and Y-span Y_P. Suppose that $s = (X_P + 4)(Y_P + 4)$. Then there is a δ such that $|\delta| \leq 1/(1 + \beta_2)$, and*

$$(1 + \beta_2^{-1})^{1/s} \mathcal{W}_\chi(0) \leq \mathcal{W}_\chi\left(\frac{1/s}{(1+\beta_2)(1+\delta)} \right).$$

In other words, there is an $\epsilon_0 = 1/(s(1+\kappa_2)(1+\delta))$ such that $\mathcal{W}_\chi(0) < \mathcal{W}_\chi(\epsilon_0)$. □

More can be done by applying the arguments in Corollary 6.11 and Teorem 6.12. In the context here those theorems become the following pattern theorems.

Corollary 7.19 *There exists an $\epsilon_c > 0$ such that $\mathcal{W}_\chi(\epsilon_c) = \sup_{\epsilon>0} \mathcal{W}_\chi(\epsilon)$, and $\mathcal{W}_\chi(\epsilon) < \mathcal{W}_\chi(\epsilon_c)$ whenever $\epsilon < \epsilon_c$. In other words, since $\log \mathcal{W}_\chi(\epsilon)$ is concave, it is also strictly increasing in the interval $[0, \epsilon_c]$.* □

Theorem 7.20 *Let P be any polygon. Then there exists a $k > 0$, an $N_0 > 0$ and a number $\epsilon_0 > 0$ such that for all $\epsilon < \epsilon_0$,*

$$d_n(\lfloor \epsilon n \rfloor P) \leq d_n(\leq \lfloor \epsilon n \rfloor P) < e^{-kn} d_n,$$

for all $n > N_0$. □

Corollary 7.19 also implies a pattern theorem for punctured disks with a density of holes (of arbitrary shape). The techniques above can be used to prove the following for $\mathcal{W}_h(\epsilon)$.

Theorem 7.21 *There exists an $\epsilon_c > 0$ such that $\mathcal{W}_h(\epsilon_c) = \sup_{\epsilon > 0} \mathcal{W}_h(\epsilon) = \beta_2$, and $\mathcal{W}_h(\epsilon) < \mathcal{W}_h(\epsilon_c)$ whenever $\epsilon < \epsilon_c$. In other words, since $\log \mathcal{W}_h(\epsilon)$ is concave, it is also strictly increasing in the interval $[0, \epsilon]$.* □

7.4 Adsorbing disks in three dimensions

Surfaces were defined in Definition 7.7. As with all tame 2-manifolds, each surface is classified by orientability, genus, and number of boundary components. The embedding of each surface is further complicated in three dimensions by the possibility that boundary components may be knotted, and in four dimensions by the possibility that the embedded surface itself may be a knot. In this section I shall focus only on a model of orientable surfaces homotopic to a disk, with one (unknotted) boundary component if it is embedded in the cubic lattice (if the boundary component is a knot, then the surface is a torus, since its genus will be at least one).

The adsorption of a disk on a solid wall will be treated here in three dimensions. A little reflection will show that the existence of a limiting free energy is not immediate; concatenation of two disks (in a naive way) may give an outcome which is either a surface with two boundary components (a cylinder), or a surface with one boundary component, depending on the particular geometry of the disks involved. Thus, as a first step, it will be necessary to develop theorems which will give the existence of the limiting free energy (and thus of a density function).

A disk is *attached* if it has a vertex with Z-coordinate which is in the set $\{-1, 0, 1\}$. A *positive disk* is an attached disk with the Z-coordinate of every vertex non-negative. Denote the number of attached disks of area n and with v vertices in the $Z = 0$ plane (*visits*) by $D_n^>(v)$, and the number of positive disks of area n and with v visits by $D_n^+(v)$. The visit-activity will be denoted by z, and partition functions $D_n^>(z) = \sum_{v \geq 0} D_n^>(v) z^v$ and $D_n^+(z) = \sum_{v \geq 0} D_n^+(v) z^v$ are defined as before. The first important issue in this model is the existence of a limiting free energy. A key step in the proof that the limiting free energy exists is the unfolding of a disk.

7.4.1 Unfolded disks

The top edge and the bottom edge of a surface is defined by a lexicographic ordering of the edges in the surface by their mid-points. The top edge has at most $(d - 1)$ possible orientations. A disk is *unfolded* if its top edge is in its boundary, and it is *doubly unfolded* if both its top edge and its bottom edge are in its boundary. If D_n is the number of disks, let $D_n^\dagger$ be the number of unfolded disks, and let $D_n^\ddagger$ be the number of doubly unfolded disks. Notice that the number of disks with their bottom edges in the boundary is also given by $D_n^\dagger$. Two unfolded disks which have top edges of the same orientations can be concatenated by first reflecting the first one through the origin, and then by identifying the top edge of the second unfolded disk with the bottom edge of the first (reflected and unfolded) disk. This gives

$$D_n^\dagger D_m^\dagger \leq (d-1) D_{n+m}, \tag{7.31}$$

since the concatenation is only possible if the top edges have the same orientation. Thus, for every choice of a surface counted by $D_n^\dagger$, there are only $D_m^\dagger/(d-1)$ possible choices for the second surface. If doubly unfolded disks are concatenated, then for every disk counted by $D_n^\ddagger$ there are $D_m^\ddagger$ choices of disks counted by $D_m^\ddagger$. Thus

$$D_n^\ddagger D_m^\ddagger \leq (d-1) D_{n+m}^\ddagger, \tag{7.32}$$

so that the following theorem is found.

Theorem 7.22 *The growth constant of doubly unfolded disks in d dimensions is given by*

$$\lim_{n\to\infty} \frac{1}{n} \log D_n^\ddagger = \log \beta_d.$$

Moreover, $D_n^\ddagger \leq (d-1)\beta_d^n$.

Proof Since the supermultiplicative inequality in eqn (7.32) is already known, it only remains to bound $D_n^\ddagger$ from above by an exponential. Consider the lattice animal defined by the edges in the surface. These animals have less than $4n$ edges, and so the number of disks is at most $\sum_{m=1}^{4n} a_m$, and this number is bounded from above by C^n for some constant C, by Lemma 6.1. The existence of the limit then follows from Lemma A.1. □

The aim will be to prove that β_d is also the growth constant of disks. This requires a relationship involving disks and unfolded disks. Let A be a disk with area n plaquettes, and let the top edge of A be e_t. Suppose that e_t is not in the boundary of A. Then e_t is incident with at least one plaquette which is normal to the first direction. Let p_t be that plaquette incident with e_t and with the lexicographically most mid-point. This plaquette is the *top plaquette* of A.

Construction 7.23 (Unfolding of a disk) Let A be a disk of area n and with top plaquette p_t and top edge e_t. If e_t is in the boundary of A, then A is unfolded, and this completes the construction; thus, suppose that e_t is not in the boundary of A. Thus, there are two plaquettes incident with e_t, and at least one of them, the top plaquette p_t, is normal to the X-direction. If any edge of p_t is in the boundary of A, then a single plaquette can be appended on this edge. This will give a disk of area $n+1$, and with a new top edge in the boundary of the disk; this completes the unfolding. Thus, suppose that every edge of p_t is incident with a plaquette. Let the four edges of p_t be $\{e_i\}_{i=1}^4$, and let the four vertices of p_t be $\{v_i\}_{i=1}^4$, labelled cyclically from the top edge. Since the girth of the hypercubic lattice is four, there are also four distinct vertex disjoint paths, one from each v_i, to the boundary of A. Select those four paths p_i of the shortest length. These cut A into four pieces. Let the piece E_i contains the edge e_i. Collectively, the E_i contains $n-1$ plaquettes. One of them, say E_4, contains the bottom edge of A, and the remaining three, $\{E_1, E_2, E_3\}$, contains at most $n-1$ plaquettes. Thus, one of $\{E_1, E_2, E_3\}$ contain at most $\lfloor n/3 \rfloor$ plaquettes; let this be E_1. Cut E_1 along the paths P_i and the edge e_1 from A (by construction, its top edge is e_1), and since p_t is normal to the X-direction, E_1 can be reflected through the hyperplane normal to the X-direction, and containing p_t. Let the resulting surface be A_1. If either the top edge of A_1, or an edge in the top plaquette of A_1 is in the boundary of A_1, then it can be unfolded by adding a single plaquette. Otherwise, the construction above is repeated. However, the three pieces which do not contain the bottom plaquette of A_1 cannot (also) contain the top edge of A, and so they contain collectively at most $\lfloor n/3 \rfloor$ plaquettes. In other words, there is one which contains at most $\lfloor \lfloor n/3 \rfloor /3 \rfloor$ plaquettes. At the k-th repetition of the construction, a piece which has at most $\lfloor \cdots \lfloor \lfloor n/3 \rfloor /3 \rfloor \cdots /3 \rfloor$ plaquettes is reflected (with k factors of 3). This process must stop when this is zero; if $3^{-k} n < 1$, then an upper bound on the number of unfoldings is found. At the last step, a single plaquette is added to complete the unfolding. The maximum number of steps necessary is at most $[\log n]/[\log 3]$, and the total area of the unfolded surface is $n+1$. □

Corollary 7.24 *There exists a constant $\alpha > 0$ such that*

$$D_n \leq n^{\alpha \log n} D^{\dagger}_{n+1}.$$

Proof Construction 7.23 changes disks counted by D_n into disks counted by $D^{\dagger}_{n+1}$. This process may be reversed by first deleting the edge with lexicographically most mid-point in a disk counted by $D^{\dagger}_{n+1}$, and then choosing an edge (from at most $4n$) and, if it cuts the disk into two pieces, reflecting the piece containing the top vertex through the hyperplane normal to the X-direction and containing the chosen edge. This process may be repeated at most $[\log n]/[\log 3]$ times. Thus

$$D_n \leq \left[\sum_{k=0}^{[\log n]/[\log 3]} [4n]^k \right] D^{\dagger}_{n+1}.$$

If α is chosen equal to $1/[\log 3]$, then it is found that

$$D_n \leq (1 + [\log n]/[\log 3])[4n]^{\alpha \log n} D^{\dagger}_{n+1}.$$

Now increase α to absorb the factor of 4, and the factor $(1 + [\log n]/[\log 3])$. □

It is also possible to use Construction 7.23 to create a doubly unfolded disk from an unfolded disk. Reflect the unfolded disk so that its bottom edge is in the boundary, and apply Construction 7.23. This shows the following relationship between unfolded and doubly unfolded disks.

Corollary 7.25 *There exists a constant $\alpha > 0$ such that*

$$D^{\dagger}_n \leq n^{\alpha \log n} D^{\ddagger}_{n+1}.$$ □

Remark 7.26 In the context of positive disks, the important observation about Construction 7.23 is that the reflections is only in the X-direction; the Z-coordinates of all vertices remain unchanged. At most two extra visits may be created when the final plaquette is added. Thus, Corollaries 7.24 and 7.25 can be expressed as follows:

$$D_n^{>}(v) \le n^{\alpha \log n} \sum_{i=0}^{2} D_{n+1}^{>\dagger}(v+i); \quad D_n^{>\dagger}(v) \le n^{\alpha \log n} \sum_{i=0}^{2} D_{n+1}^{>\ddagger}(v+i), \tag{7.33}$$

for attached disks, and

$$D_n^{+}(v) \le n^{\alpha \log n} \sum_{i=0}^{2} D_{n+1}^{+\dagger}(v+i); \quad D_n^{+\dagger}(v) \le n^{\alpha \log n} \sum_{i=0}^{2} D_{n+1}^{+\ddagger}(v+i), \tag{7.34}$$

for positive disks. □

Corollaries 7.24 and 7.25 can be used to prove that there is a growth constant for disks. Notice that

$$D_{n-2} \le n^{\alpha \log n} D_{n-1}^{\dagger} \le n^{2\alpha \log n} D_n^{\ddagger} \le n^{2\alpha \log n} D_n. \tag{7.35}$$

Take logarithms of this, divide by n and let $n \to \infty$; by Theorem 7.22 the existence of a growth constant for D_n follows.

Theorem 7.27 *The growth constant for D_n is equal to β_d, as defined in Theorem 7.22:*

$$\lim_{n\to\infty} \frac{1}{n} \log D_n = \lim_{n\to\infty} \frac{1}{n} \log D_n^{\dagger} = \lim_{n\to\infty} \frac{1}{n} \log D_n^{\ddagger} = \log \beta_d. \quad \square$$

In the adsorption problem of disks, it is Remark 7.26 which will be the important result from this section. There is another consequence of Construction 7.23 which I shall consider in the next section: it is possible to find bounds on D_n by showing that it also satisfies a submultiplicative relation (in addition to the supermultiplicative relation in Theorem 7.22).

7.4.2 Bounds on D_n

Bounds on D_n can be computed by combining the dissection of a disk into subdisks in Constructions 7.9 and 7.10 with the unfolding of a disk in Construction 7.23. The key result is Corollary 7.11, where k subdisks of total area m have been removed from a disk counted by D_{n+m}, and where $k \le [\log m]/[\log(3/2)]$. Doubly unfold each of these subdisks and concatenate them in sequence into a doubly unfolded disk of area $m+2k$. This process can be reversed by taking a disk counted by $D_{m+2k}^{\ddagger}$, and by cutting it into k doubly unfolded subdisks in at most $[4(m+2k)]^k$ ways (select k edges from at most $4(m+2k)$ to do the cutting). Secondly, reverse the unfolding for each of the subdisks by cutting the one plaquette incident with the top edge and with the bottom edge from each, and then reverse the unfoldings in at most $[m^{2\alpha \log m}]^k$ ways. Lastly, try to fit these resulting disks back in the original disk in at most $[64(n+m)m^2]^k$ ways (see eqn (7.23)). Collecting all these factors give

$$D_{n+m} \le D_n \sum_{k=1}^{\lceil \log m/\log(3/2) \rceil} [64(n+m)m^2]^k [m^{2\alpha \log m}]^k (4m+2k)^k D^{\ddagger}_{n+2k}. \qquad (7.36)$$

Now use the fact that $D^{\ddagger}_n \le D^{\ddagger}_{n+1}$, and note that if $k = \lceil [\log m]/[\log(3/2)] \rceil$, then the factor $[m^{2\alpha \log m}]^{\lceil [\log m]/[\log(3/2)] \rceil}$ will dominate all the other terms if either m is large, or for all m if α is chosen sufficiently large. Thus, increase α to absorb all the constant factors. This gives

$$D_{n+m} \le [m^{\alpha [\log m]^2}] D_n D^{\ddagger}_{m+\lceil 2[\log m]/[\log(3/2)] \rceil} \qquad (7.37)$$

for some fixed constant $\alpha > 0$. This inequality is almost enough for our purposes, but to chance it into a more standard submultiplicative inequality discussed in Appendix A, one more lemma is needed.

Lemma 7.28 $D_{n+1} \le [4(2d-3)n] D_n$.

Proof Let A be a disk counted by D_{n+1}. If A has no split-plaquettes, then a plaquette on the boundary of A can be deleted to find a disk of area n. Suppose that A has at least one split-plaquette. Then A corresponds to an abstract tree T as follows. Let every split-plaquette in A correspond to a vertex in T. If p_1 and p_2 are vertices in T, then join them with an edge if the corresponding split-plaquettes are adjacent (share an edge), or if they can be joined by a path in A which does not pass through any split-plaquettes. Since T is a tree, it has a vertex of degree one. Let p be the split-plaquette corresponding to a vertex of degree one in T. If p is deleted, then A is cut into two or more subdisks, at least one of which contains no split-plaquettes; let this be B. If B consists of exactly one plaquette, then it may be deleted to give a disk of area n. If B consists of more than three plaquettes, then there is one plaquette in B not adjacent with p; it can be deleted to find a disk of area n. Lastly, the cases when B consists of two and three plaquettes can be checked explicitly; there is always a plaquette which can be deleted to give a disk of area n. Now notice that the number of ways this plaquette can be put back (by selecting an edge in a disk of area n in at most $4n$ ways, and appending a plaquette to this edge, in one of $2d-3$ ways) is at most $4(2d-3)n$. This completes the proof. □

Now use Lemma 7.28 with eqn (7.37). Since $D^{\ddagger}_m \le D_m$, the result is that

$$D_{n+m} \le [m^{\alpha [\log m]^2}] \left[4(2d-3)(m + 2[\log m]/[\log(3/2)])\right]^{\lceil 2[\log m]/[\log(3/2)] \rceil} D_n D_m. \qquad (7.38)$$

By increasing the constant α in the above, and by replacing m by $n+m$ whenever necessary, the outcome is the following theorem.

Theorem 7.29 *There exists a finite constant α such that*

$$D_{n+m} \le (m+n)^{\alpha [\log(n+m)]^2} D_n D_m. \qquad \square$$

By Theorem A.3 and Theorem 7.26, the following lower bound is found

$$\frac{\log D_n}{n} \geq \log \beta_d + \frac{\alpha[\log n]^3}{n} - 4\alpha \sum_{m=2n}^{\infty} \frac{[\log m]^3}{m(m+1)}. \tag{7.39}$$

The infinite series can be bounded from above by noting that

$$\sum_{m=2n}^{\infty} \frac{[\log m]^3}{m(m+1)} \leq \sum_{m=2n}^{\infty} \frac{[\log m]^3}{m^2} \leq \int_n^{\infty} \frac{[\log x]^3}{x^2}\, dx. \tag{7.40}$$

An upper bound can be found for the integral:

$$\int_n^{\infty} \frac{(\log x)^3}{x^2} dx \leq 16 \frac{[\log n]^3}{n}. \tag{7.41}$$

Substitute this into eqn (7.39), and absorb the constants into α. This gives the following theorem [187].

Theorem 7.30 *There exists constants α_1 and α_2 such that*

$$n^{\alpha_1 \log n} \beta_d^{n+2} \geq D_n \geq n^{-\alpha_2 (\log n)^2} \beta_d^n.$$

Proof The upper bound follows from Corollaries 7.24 and 7.25, and the lower bound is a result of Theorem 7.29, and eqns (7.37), (7.40) and (7.41). □

Of course, it is expected that $D_n \simeq n^{-\theta} \beta_d^n$, where θ is the branched polymer entropic exponent.

7.4.3 The free energy of positive disks

The most important results from Section 7.4.1 in the context of attached and positive disks are the inequalities in Remark 7.26. Consider first a model of doubly unfolded positive disks. The free energy of such disks with a visit activity z can be shown to exist by using a most popular argument. Thus, let $D_n^{+\ddagger}(v; [h_b\hat{\phi}_b], [h_t\hat{\phi}_t])$ be the number of positive and doubly unfolded disks, with a bottom edge with mid-point a height h_b above the $Z = 0$ hyperplane, and in one of $(d-1)$ orientations denoted by $\hat{\phi}_b$, and with a top edge with mid-point a height h_t above the $Z = 0$ hyperplane, and in the orientation denoted by $\hat{\phi}_t$. Disks counted by $D_n^{+\ddagger}(v; [h\hat{\phi}], [h\hat{\phi}])$ can be concatenated by translating one along the hyperplane $Z = 0$ until the top edge of the first is one step removed in the X-direction from the bottom edge of the second. Complete the concatenation by inserting a single plaquette between the top edge of the first and the bottom edge of the second. No new visits are created in this construction, even if the top and bottom edges are in the $Z = 0$ hyperplane. This shows that

$$D_n^{+\ddagger}(v_1; [h\hat{\phi}], [h\hat{\phi}]) D_m^{+\ddagger}(v_2; [h\hat{\phi}], [h\hat{\phi}]) \leq D_{n+m+1}^{+\ddagger}(v_1 + v_2; [h\hat{\phi}], [h\hat{\phi}]), \tag{7.42}$$

so that this model satisfies Assumptions 3.1, and there is a limiting free energy and a density function of visits defined by

$$\begin{aligned}\mathcal{V}_v^+(z) &= \lim_{n\to\infty} \frac{1}{n} \log D_n^{+\ddagger}(z; [h\hat{\phi}], [h\hat{\phi}]);\\ \mathcal{W}_v^+(\epsilon) &= \lim_{n\to\infty} \left[D_n^{+\ddagger}(\lfloor \epsilon n \rfloor; [h\hat{\phi}], [h\hat{\phi}])\right]^{1/n}.\end{aligned} \tag{7.43}$$

The partition function $D_n^{+\ddagger}(z; h\hat{\phi}, h\hat{\phi})$ is defined as usual. The existence of a limiting free energy in the full model is now proven using the techniques of Theorem 5.52. Indeed, since the boundaries of the disks are polygons, and the concatenation of doubly unfolded disks are done by joining the top and bottom edges which are in the boundaries, the proof of Theorem 5.52 can be used with minimal modifications here.

Theorem 7.31 *The limiting free energy and density function of a model of doubly unfolded positive disks exists, and is given by*

$$\begin{aligned}\mathcal{V}_v^+(z) &= \lim_{n\to\infty} \frac{1}{n} \log D_n^{+\ddagger}(z);\\ \mathcal{W}_v^+(\epsilon) &= \lim_{n\to\infty} \left[D_n^{+\ddagger}(\lfloor \epsilon n \rfloor)\right]^{1/n}.\end{aligned}$$

Moreover, $\mathcal{V}_v^+(z)$ is a convex function of $\log z$, and $\log \mathcal{W}_v^+(\epsilon)$ is a concave function of $\log z$. □

Theorem 7.32 *The limiting free energy and density function of a model of positive disks exists, and is given by*

$$\begin{aligned}\mathcal{V}_v^+(z) &= \lim_{n\to\infty} \frac{1}{n} \log D_n^{+}(z);\\ \mathcal{W}_v^+(\epsilon) &= \lim_{n\to\infty} \left[D_n^{+}(\lfloor \epsilon n \rfloor)\right]^{1/n}.\end{aligned}$$

Moreover, $\mathcal{V}_v^+(z)$ is a convex function of $\log z$, and $\log \mathcal{W}_v^+(\epsilon)$ is a concave function of $\log z$. □

There are adsorption transitions in both models of positive and attached disks; the existence of a critical point is proven in a way similar to the cases for walks or polygons. The important observations are the following. Since D_n counts disks up to translational equivalence, it follows that $D_n^+(0) = D_n$ and $\sum_{v\geq 1} D_n^+(v) = D_n$. In other words,

$$\sum_{v\geq 0} D_n^+(v) = 2\, D_n, \tag{7.44}$$

and by translating a disk counted by $D_n^+(v)$ one step in the positive Z-direction,

$$D_n^+(v) \leq D_n^+(0). \tag{7.45}$$

Thus, since any surface cannot have more than $2n + 2$ visits,

$$D_n^+(0) \leq \sum_{v=0}^{2n+2} D_n^+(v) z^v = D_n^+(z) \leq D_n^+(0)/(1+z), \quad \text{if } z < 1. \tag{7.46}$$

Now take the logarithm of this, divide by n, and let $n \to \infty$. Similar results are found for attached disks:

$$D_n^>(0) = 2D_n, \tag{7.47}$$

$$D_n^>(v) \leq D_n^>(0). \tag{7.48}$$

These equations shows that

$$D_n^>(0) \leq \sum_{v=0}^{2n+2} D_n^>(v) z^v = D_n^>(z) \leq D_n^>(0)/(1+z), \quad \text{if } z < 1. \tag{7.49}$$

Now take the logarithm of eqns (7.46) and (7.49), divide by n and let $n \to \infty$. The result is Theorem 7.33.

Theorem 7.33 *If $z \leq 1$ then the limiting free energies of attached and of positive disks is given by*

$$\mathcal{V}_v(z) = \mathcal{V}_v^+(z) = \log \beta_d. \qquad \square$$

The existence of a critical point corresponding to the adsorption of the attached disk and of the positive disk is shown by noting that both $\mathcal{V}_v(z)$ and $\mathcal{V}_v^+(z)$ are dependent on z for large enough values of z.

Theorem 7.34 *If $z > 1$, then*

$$\mathcal{V}_v(z) \geq \mathcal{V}_v^+(z) \geq \log \beta_{d-1} + 2 \log z.$$

Proof Notice that $D_n^>(z) \geq D_n^+(z)$. In addition, if all the vertices in a surface are adsorbed in the $Z = 0$ hyperplane, then eqn (6.3) implies that there are at most $2n + 2$ vertices (and thus visits).[2] In other words, keep only that term in $D_n^+(z)$ which corresponds to a disk completely adsorbed in the $Z = 0$ hyperplane, and with a maximum $2n + 2$ vertices (and thus visits). This shows that

$$D_n^+(z) \geq D_n^{(d-1)} z^{2n+2},$$

where $D_n^{(d-1)}$ is the number of disks of area n in $(d - 1)$ dimensions. Taking logarithms, dividing by n, and letting $n \to \infty$ gives the claimed inequalities. $\square$

Corollary 7.35 *There are non-analyticities at $z_c^>$ in $\mathcal{V}_v^>(z)$ and at z_c^+ in $\mathcal{V}_v^+(z)$ in the free energies $\mathcal{V}_v(z)$ and $\mathcal{V}_v^+(z)$ respectively. Moreoever, $1 \leq z_c^> \leq z_c^+ \leq (\log \beta_d - \log \beta_{d-1})/2$.*

---

[2] Consider the frame of the disk consisting only of edges and vertices. Each plaquette becomes a 4-cycle, and the number of cycles is equal to n. Each plaquette appended to a disk shares at least one edge with another plaquette, so the number of edges is at most $3n + 1$. Substition into eqn (6.2) shows that the number of vertices is at most $2n + 2$.

Proof That the non-analyticities are present is seen from Theorems 7.33 and 7.34. Moreover, $z_c^> \leq z_c^+$ since $\mathcal{V}_v(z) \geq \mathcal{V}_v^+(z)$. Next, $1 \leq z_c^>$ by Theorem 7.33. Lastly, since $\mathcal{V}_v^+(z) \geq \log \beta_d$ is non-decreasing, and since $\mathcal{V}_v^+(z) \geq \log \beta_{d-1} + 2 \log z$, it follows that $\mathcal{V}_v^+(z) > \log \beta_d$ whenever $z > (\log \beta_d - \log \beta_{d-1})/2$. □

These results establish the adsorption transition in disks. Moreover, since $1 \leq z_c^> \leq z_c^+ \leq (\log \beta_d - \log \beta_{d-1})/2$, Lemma 3.20 shows that

$$0 \geq \left[\frac{d^+}{d\epsilon} \log \mathcal{W}_v(\epsilon) \right] \Bigg|_{\epsilon=0^+} \geq \left[\frac{d^+}{d\epsilon} \log \mathcal{W}_v^+(\epsilon) \right] \Bigg|_{\epsilon=0^+} \geq -\big(\log[\beta_d/\beta_{d-1}] \big)/2. \tag{7.50}$$

As in the case of adsorbing polygons and animals, it is possible to show that $z_c^+ > 1$; this I shall do in the next section. The critical exponents of the adsorption transition in this model are those of adsorbing branched polymers, listed in Table 6.2.

7.4.4 The location of the adsorption transition

Let A be a disk counted by $D_n^+(v)$. Every visit of a vertex in A is incident with a plaquette entirely in the adsorbing plane, or with an edge entirely in the adsorbing plane. These edges and plaquettes are called "adsorbed edges" and "adsorbed plaquettes". An adsorbed edge has two visits as its end-point, and an adsorbed plaquette has four visits (its vertices). In Fig. 7.7 an adsorbed edge and an adsorbed plaquette are illustrated.

If A has v visits, then there are also at least $\lfloor v/4 \rfloor$ visits in A which is the bottom vertex of an adsorbed plaquette or an adsorbed edge. Choose a set of adsorbed plaquettes or edges such that the shortest distance between any two in the set is at least two steps. Since this requirement essentially surrounds every chosen adsorbed plaquette or edge by a 5×5 square, it implies that there is such a set of at least (say) $\lfloor v/100 \rfloor$ adsorbed plaquettes or edges. Choose a set of m of the adsorbed edges or adsorbed plaquettes, and perform the construction in Fig. 7.7 on each. The area of the disk increases by at most

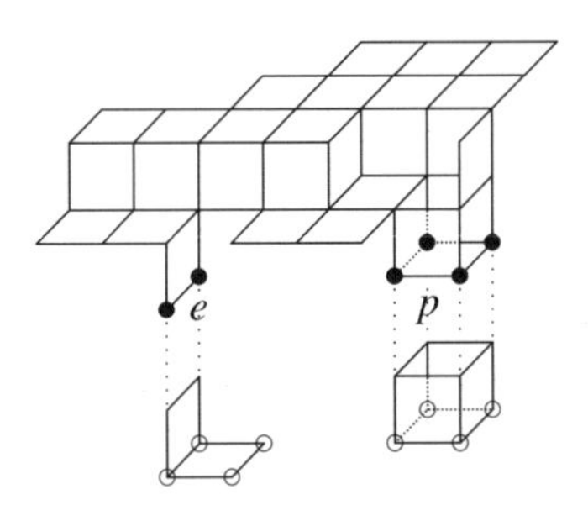

Fig. 7.7: An adsorbing disk with an adsorbed boundary edge e, and an adsorbed plaquette p. If the structures are appended one e and p as shown (and p is deleted), and the entire resulting surface is translated one step in the Z-direction, then the vertices marked by $\circ$ will become visits. The area increases by two or by four with every structure appended.

$4m$, and there are now $4m$ visits in the $Z = 0$ hyperplane. The resulting inequality is

$$\binom{\lfloor v/100 \rfloor}{m} D_n^+(v) \leq \sum_{j=0}^{4m} D_{n+j}^+(4m). \tag{7.51}$$

Choose $v = \epsilon n$ and $m = \lfloor \delta n \rfloor$, with $\delta < \epsilon/100$. Substitute in the above, take the $(1/n)$-th power and let $n \to \infty$. The outcome is Theorem 7.36.

Theorem 7.36 *If $\epsilon > 0$ is small enough, and $\delta < \epsilon/100$, then*

$$\left[\frac{(\epsilon/100)^{\epsilon/100}\beta_d^{4\delta}}{\delta^\delta(\epsilon/100-\delta)^{\epsilon/100-\delta}}\right]\mathcal{W}_v^+(\epsilon) \leq \mathcal{W}_v^+(4\delta/(1+4\delta)).$$

Moreover, if $\delta = \delta^* = \epsilon/(100(1+\beta_d^4))$, then

$$(1+\beta_d^{-4})^{\epsilon/100}\mathcal{W}_v^+(\epsilon) \leq \mathcal{W}_v^+(4\delta^*/(1+\delta^*)).$$

Proof Note that $D_n^+(v) \leq D_{n+4}^+(v)$. By taking the $(1/n)$-th power of eqn (7.51), and taking $n \to \infty$ (using Theorems 7.27 and 7.33), the first inequality is obtained. The second follows from footnote 3, Chapter 3. □

The inequalities in Theorem 7.36 are enough to prove that $z_c^+ > 1$. To see this, suppose that $z > z_c^+$. By Theorem 3.18,

$$\mathcal{V}_v^+(z) = \sup_{\epsilon>0}\{\log[\mathcal{W}_v^+(\epsilon)z^\epsilon]\} = \log[\mathcal{W}_v^+(\epsilon^*)z^{\epsilon^*}], \tag{7.52}$$

where $\epsilon_* > 0$ since $z > z_c^+$. (Note that ϵ^* is the fraction of visits, defined as the derivative of $\log \mathcal{V}_v^+(z)$, and that this derivative exists almost everywhere since the free energy is a convex function of $\log z$.) But by Theorem 7.36

$$\mathcal{V}_v^+(z) = \log[\mathcal{W}_v^+(\epsilon^*)z^{\epsilon^*}] \leq \log\left(\mathcal{W}_v^+(\delta^\dagger)z^{\delta^\dagger}\left[\frac{z^{\epsilon^*-\delta^\dagger}}{(1+\beta_d^{-4})^{\epsilon^*/100}}\right]\right), \tag{7.53}$$

where $\delta^\dagger = 4\delta^*/(1+\delta^*) > 0$. But this is a contradiction if

$$\left[\frac{z^{\epsilon^*-\delta^\dagger}}{(1+\beta_d^{-4})^{\epsilon^*/100}}\right] < 1, \tag{7.54}$$

since the supremum is attained as in eqn (7.52). Define α by $\delta^\dagger = \alpha\epsilon^*$; then $0 < \alpha < 1$. Solve for z from eqn (7.54):

$$\log z < \frac{\log(1+\beta_d^{-4})}{100(1-\alpha)}. \tag{7.55}$$

In other words, if z is so small that eqn (7.55) is true, then eqn (7.53) is a contradiction to eqn (7.52), *unless* $\epsilon^* = 0$. In other words, if eqn (7.55) is true, then the disk cannot be adsorbed, and so

$$\log z_c^+ \geq \frac{\log(1+\beta_d^{-4})}{100(1-\alpha)} > 0. \tag{7.56}$$

Thus, the adsorption transition for positive disks has a critical point $z_c^+ > 1$. The singularity diagram of the generating function of this model will contain a multicritical point which divides a critical curve into two. For values of $z < z_c^+$ the disk is desorbed, and the singularity in the generating function is determined by the branched polymer entropic exponents, since it is generally accepted that disks are in the same universality class as branched polymers. Along this critical curve (which may be called a λ-line), the singularity in the generating function is described by the exponent $2-\alpha_+$ (eqn (2.12)), and this gives $2-\alpha_+ = 1/2$, a fact derived from the entropic exponent of trees. (In three dimensions the entropic exponent has value $3/2$ [290], which means that $t_n \simeq Cn^{-5/2}\tau_3^n$. Thus $\sum_{n\geq 0} n^{-3/2}(\tau_3 x)^n \approx |\log(\tau_3 x)|^{1/2}$.) The cross-over exponent which described the cross-over in scaling from the desorbed to adsorbed phases is believed to have value $1/2$, which is also the case for trees. The adsorption of surfaces has also received attention in references [22] (real-space renormalization) and [285] (Monte Carlo). In the first case the cross-over exponent was estimated to be between 0.6 and $2/3$, while the second suggested that its value is 0.70 ± 0.06. These values are bigger than the branched polymer value. On the other hand the metric exponent was computed in reference [285] as well; in the desorbed phase it was found that $\nu = 0.506 \pm 0.008$, and in the adsorbed phase $\nu = 0.64 \pm 0.01$. These values agree well with the corresponding values for branched polymers (see for example eqn (1.35)).

7.5 Crumpling surfaces

In this section I shall consider a model of surfaces with no boundary components; such surfaces are called *closed surfaces*. Matters will be simplified if it is assumed that the surfaces are homeomorphic to a sphere, although a model of closed surfaces with unrestricted topology (for example, all orientable surfaces of arbitrary genus, or all non-orientable and orientable surfaces of arbitrary genus) may also be considered. The possibility of a crumpling transition in a model of surfaces was suggested by numerical simulations (see for example references [4, 189, 276, 279, 284, 285]). Moreover, it was conjectured that this is an asymmetric model, with a limiting free energy which is constant for small values of the activity, similar to the adsorption problem of the previous section. I shall rule out that possibility. Instead, a restricted model of surfaces will be found to be asymmetric, and to exhibit a crumpling transition.

The model will be restricted to the cubic lattice; higher-dimensional models may be treated similarly. A model of crumpling surfaces usually includes an activity conjugate to the total absolute curvature of the surface. In the context here, it is simplest to count the number of pairs of plaquettes which meet at right angles at a mutual edge. Any given edge in a closed surface is incident with two plaquettes. If these two plaquettes are perpendicular, then the edge is a *fold*. The total energy of a surface will be the number of folds it has; let $s_n(f)$ be the number of closed surfaces with n plaquettes and f folds. As before, the bottom and top plaquettes of a surface will be the first and last plaquettes

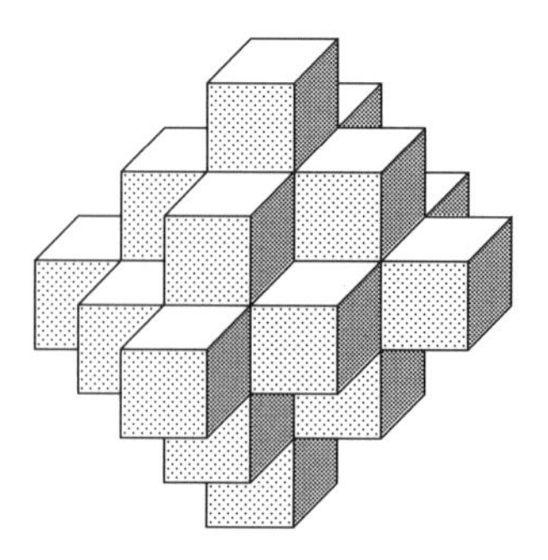

Fig. 7.8: A surface with area 78 and energy (number of folds) 156.

in a lexicographic ordering of the plaquettes by their mid-points. The *area* of a surface is the number of plaquettes it contains. The *volume* of a surface is the number of unit cubes in its interior. If a surface has n plaquettes, e edges and v vertices, then $e = 2n$ and by Euler's theorem, $v - n = 2$. In other words, a model of surfaces homeomorphic to a sphere counted by vertices is equivalent to a model of closed surfaces counted by area. The maximum energy (maximum number of folds) in a surface is $2n$, since it is possible that every edge in a given surface may be a fold, see Fig. 7.8.

The partition function of this model is $s_n(z) = \sum_{f\geq 0} s_n(f) z^f$. Moreover, any two surfaces can be concatenated by identifying the bottom plaquette of the first with the top plaquette of the second, and then deleting them. This construction may result in the loss of at most eight folds, or in the creation of at most eight extra folds. The result is that

$$s_{n_1}(f_1) s_{n_2}(f_2) \leq \sum_{i=-8}^{8} s_{n_1+n_2-2}(f_1 + f_2 + i). \tag{7.57}$$

Moreover, the number $s_n = \sum_{f\geq 0} s_n(f)$ is also bounded from above by an exponential. This is seen as follows. The *frame* of a surface is that lattice animal composed of all the edges in the surface. Thus, if a surface has area n, then its frame is composed of $2n$ edges. A surface may be recreated from its frame by adding n plaquettes on the edges in the frame in four possible directions, thus $s_n \leq a_{2n} 4^n$. But there exists a positive constant such that $a_{2n} < K^{2n}$ (Lemma 6.1). Thus, the number $s_n(f)$ satisfies Assumptions 3.8, and by Corollary 3.14 and Theorems 3.16 and 3.17 there exist integrated density functions, a density function, and a limiting free energy [361], defined by

$$\mathcal{W}_s(\leq\epsilon) = \lim_{n\to\infty} [s_n(\leq\lfloor\epsilon n\rfloor)]^{1/n}; \tag{7.58}$$

$$\mathcal{W}_s(\geq\epsilon) = \lim_{n\to\infty} [s_n(\geq\lfloor\epsilon n\rfloor)]^{1/n}; \tag{7.59}$$

$$\mathcal{W}_s(\epsilon) = \lim_{n\to\infty} [s_n(\lfloor\epsilon n\rfloor + \sigma_n)]^{1/n}; \tag{7.60}$$

$$\mathcal{V}_s(z) = \lim_{n\to\infty} \frac{1}{n} \log s_n(z) \tag{7.61}$$

where $\sigma_n = o(n)$ is a sequence of integers defined as in Theorem 3.16, and where $s_n(\leq g) = \sum_{f=0}^{g} s_n(f)$. Moreover,

$$\mathcal{W}_s(\epsilon) = \min\{\mathcal{W}_s(\leq\epsilon), \mathcal{W}_s(\geq\epsilon)\}, \tag{7.62}$$

and the growth constant of closed surfaces homeomorphic to a sphere in the cubic lattice is

$$\eta_3 = \lim_{n\to\infty} s_n^{1/n}. \tag{7.63}$$

Thus $\eta_3 = \sup_{\epsilon>0}\{\mathcal{W}_s(\leq\epsilon), \mathcal{W}_s(\geq\epsilon)\}$ (Corollary 3.15). In addition, these density functions and free energy satisfy the usual concavity and convexity conditions, and are continuous in the interiors of their domains. Observe also that Fig. 7.8 shows that $\epsilon \in (0, 2]$; and moreover, the density functions can be defined at $\epsilon = 0^+$ by taking a right limit and using continuity.

The first important issue to be considered is the number of bounds on $s_n(f)$; these will be useful in the calculation of some properties of the density function.

7.5.1 The density function of crumpling surfaces

An upper bound on $s_n(f)$ can be found by an application of the labelling scheme for animals in Lemma 6.1. The construction will first consider a surface and label its plaquettes, and then consider the constraints on constructing surfaces by building them from labelled plaquettes.

Lemma 7.37 *The number of surfaces with f folds and area n is bound from above by*

$$s_n(f) \leq \sum_{k=0}^{f} \binom{n}{k} 2^k.$$

Moreover, $s_n \leq 3^n$.

Proof Let S be a closed surface. Label the bottom plaquette of S with 1, and order the four edges incident with the bottom plaquette lexicographically by their mid-points. The plaquette incident with the least edge gets label 2, and the plaquette incident with the next least edge gets label 3, and so on. If j plaquettes have been labelled, and i is the least label so that plaquette i is incident with an unlabelled plaquette, then order the edges of i incident with unlabelled plaquettes lexicographically by their mid-points. Then label the plaquette incident with the least edge of i with $j + 1$, etc. This gives a canonical labelling for the plaquettes in S.

The surface S can also be constructed by recursively adding plaquettes to a boundary. Place a plaquette in the cubic lattice in one of three possible orientations, and label this plaquette with 1 (this will be the bottom plaquette of S). Order the edges of this plaquette lexicographically by their mid-points, and append plaquette 2 in one of three possible ways to the lexicographically least edge of plaquette 1, and then continue with the remaining edges of plaquette 1 by adding plaquettes 3, 4 and 5 in lexicographic order. Once j plaquettes have been added, let i be the smallest label such that plaquette i has an edge not paired with another plaquette. Order the unpaired edges of i lexicographically and append plaquettes $j+1$, $j+2$, ..., to them in order. Repeat this process until n plaquettes have been labelled. At each step in the construction there is a unique edge to append the next plaquette, which may be added in one of three possible orientations. Thus, the maximum number of surfaces that one may construct in this way is 3^n. In other words, $s_n \leq 3^n$. (Note that only a subclass of surfaces, which includes all closed surfaces, can be constructed in this way.) This bound is cited in reference [98], and was first shown in reference [135].

A bound on $s_n(f)$ can be obtained using the arguments in the previous two paragraphs as follows. Consider the construction of a surface S with f folds, and at the j-th step, let i be the smallest label with an unpaired edge and let $j+1$ be the label of the next plaquette to be added. There is a choice between (1) adding the plaquette with label $j+1$ on the unpaired edge of plaquette i with lexicographically least mid-point at right angles to the plaquette with label i (in one of two ways), this creates a fold; or, (2) adding the plaquette with label $j+1$ with the same orientation as the plaquette with label i. A plaquette is added n times, and the choice to create a fold is available a maximum of f times. If choice (1) is assumed to occur k times, then this can be done in $\binom{n}{k} 2^k$ ways, since a plaquette can be added in two ways to create a right angle. Note that a surface with f folds may be created by choosing $k \leq f$ from the n plaquettes for creating a fold. Thus

$$s_n(f) \leq \sum_{k=0}^{f} \binom{n}{k} 2^k.$$

Observe that this bound equals 3^n if $f = n$, as one would expect if there are no constraints on the number of folds. □

Since I shall be interested in the density functions and partial density functions, bounds on $s_n(\leq \lfloor \epsilon n \rfloor)$ will be of particular interest. In the next lemma lower bounds are derived.

Lemma 7.38 *For any $n \geq 1$ and every $\epsilon \in (0, 2]$, there exists a finite positive integer $m_0(\epsilon)$ such that for every fixed $m > m_0(\epsilon)$,*

$$s_{6nm^2-2(n-1)}(\leq \lceil \epsilon(6nm^2 - 2(n-1)) \rceil) \geq 4^{n-1}.$$

In addition, for every positive $\epsilon < 3/2$,

$$s_{6n^2+4\lfloor \epsilon n^2 \rfloor}(\leq 12(n + \epsilon n^2)) \geq \binom{6\lfloor (n-2)^2/4 \rfloor}{\lfloor \epsilon n^2 \rfloor}.$$

Proof I construct a family of surfaces of area $N = (6nm^2 - 2(n-1))$, and with fewer than ϵN folds. An m-cube is a surface of area $6m^2$, and with the geometry of a cube with side-length m in three dimensions. The number of folds in an m-cube is $12m$. A top plaquette t^{++} and a bottom plaquette b^{++} of an m-cube can be found by a lexicographic ordering of the plaquettes by their mid-points with respect to the directions $(\hat{\imath}, \hat{\jmath}, \hat{k})$. Similarly, if the lexicographic ordering is done with respect to the directions $(\hat{\imath}, -\hat{\jmath}, -\hat{k})$

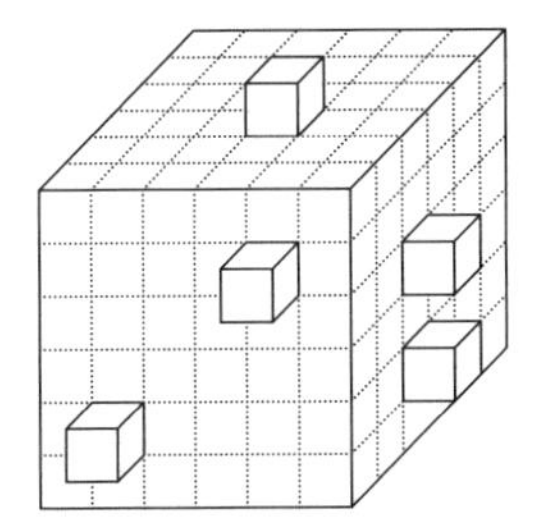

Fig. 7.9: An m-cube with 1-cubes fused to its outer surface.

instead, then the top and bottom plaquettes t^{--} and b^{--} are found. Lastly, if the ordering is done with respect to $(\hat{\imath}, \hat{\jmath}, -\hat{k})$, then the top and bottom plaquettes are t^{+-} and b^{+-}, and if the ordering is done with respect to $(\hat{\imath}, -\hat{\jmath}, \hat{k})$, then t^{-+} and b^{-+} are the top and bottom plaquettes. These top plaquettes (and bottom plaquettes) are distinct if $m > 1$. Two m-cubes can be "strung together" (if $m > 1$) by *identifying* either t^{++} on the first with b^{++} on the second, or t^{--} on the first with b^{--} on the second. Alternatively, one may identify t^{+-} with b^{+-}, or t^{-+} with b^{-+}. If n such m-cubes are strung together, then there are 4^{n-1} possible distinct outcomes. Each identification deletes two plaquettes from the m-cubes, but the number of folds remains unchanged. The total area is $6nm^2 - 2(n-1)$, and the total number of folds is $12nm$. Thus, $s_{6nm^2-2(n-1)}(12nm) \geq 4^{n-1}$. Now increase m (if necessary), until $\epsilon > 12m/(6m^2-2)$. Since $12m/(6m^2-2) \geq 12nm/(6nm^2-2(n-1))$ for any $n \geq 1$, this value of m is sufficient (put $m_0(\epsilon) = \lceil 1/\epsilon + \sqrt{1/\epsilon^2 + 1/3}\rceil$).

On the other hand, a 1-cube can be "fused" on the outside of an n-cube by identifying a plaquette on the 1-cube with a plaquette in the n-cube, and then deleting the plaquette (as illustrated in Fig. 7.9). Perform this construction by selecting l plaquettes disjoint with folds and with each other in the n-cube. This can be done in at least $\binom{6\lfloor (n-2)^2/4\rfloor}{l}$ ways. By counting the number of folds and plaquettes, the resulting surfaces have degree of folding $12n + 12l$, and area $6n^2 + 4l$. Thus, $s_{6n^2+4l}(\leq(12(n+l))) \geq \binom{6\lfloor (n-2)^2/4\rfloor}{l}$. Now put $l = \lfloor \epsilon n^2 \rfloor$, and the result follows. □

The following lemma will be useful if Lemma 7.37 is used to find an upper bound on the number of surfaces, see reference [250].

Lemma 7.39 *Let q be a finite positive real number, and suppose that $0 \leq \gamma \leq q/(q+1)$. Then*

$$\limsup_{n\to\infty} \left[\sum_{k=0}^{\lfloor \gamma n\rfloor} \binom{n}{k} q^k\right]^{1/n} \leq \frac{q^\gamma}{\gamma^\gamma (1-\gamma)^{1-\gamma}}.$$

Proof Put $p = q/(q+1)$; then $0 \leq \gamma < p < 1$, and so

$$\sum_{k=0}^{\lfloor \gamma n\rfloor} \binom{n}{k} q^k = \sum_{k=0}^{\lfloor \gamma n\rfloor} \binom{n}{k} p^k (1-p)^{n-k}.$$

Interpret p as the probability that a biased coin will land heads, and $(1-p)$ as the probability that it will land tails. Let η be the number of heads in n tosses of the coin. Then the probability that there are at most $\lfloor \gamma n \rfloor$ heads in n tosses is

$$P(\eta \leq \lfloor \gamma n \rfloor) = \sum_{k=0}^{\lfloor \gamma n \rfloor} \binom{n}{k} p^k (1-p)^{n-k}.$$

If ζ is a random variable which is equal to one if $\eta \leq \lfloor \gamma n \rfloor$, and zero otherwise, then $P(\eta \leq \lfloor \gamma n \rfloor) = E(\zeta)$, where $E(\zeta)$ is the expectation value of ζ. But

$$E(\zeta) \leq E(e^{t(\lfloor \gamma n \rfloor - \eta)}) = e^{t\lfloor \gamma n \rfloor} E(e^{-t\eta}) = e^{t\lfloor \gamma n \rfloor}[pe^{-t} + (1-p)].$$

This is a minimum if $\lfloor \gamma n \rfloor (1-p) = p(n - \lfloor \gamma n \rfloor)e^{-t}$. In other words, substitution of t gives

$$\sum_{k=0}^{\lfloor \gamma n \rfloor} \binom{n}{k} p^k (1-p)^{n-k} \leq \left(\frac{p}{\lfloor \gamma n \rfloor / n} \right)^{\lfloor \gamma n \rfloor} \left(\frac{1-p}{1 - \lfloor \gamma n \rfloor / n} \right)^{n - \lfloor \gamma n \rfloor}.$$

Replace p by the appropriate expressions in q. This gives

$$\sum_{k=0}^{\lfloor \gamma n \rfloor} \binom{n}{k} q^k \leq \left[\frac{q^{\lfloor \gamma n \rfloor / n}}{(\lfloor \gamma n \rfloor / n)^{\lfloor \gamma n \rfloor / n} (1 - \lfloor \gamma n \rfloor / n)^{1 - \lfloor \gamma n \rfloor / n}} \right]^n. \qquad (\ddagger)$$

Now take the $(1/n)$-th power and let $n \to \infty$. □

The properties of the density function $\mathcal{W}_s(\epsilon)$ are now summarized in Theorem 7.40.

Theorem 7.40 $\log \mathcal{W}_s(\epsilon)$ *is a concave function of* ϵ, *and is continuous in* $[0, 2)$. *Moreover,* $\lim_{\epsilon \to 0^+} \mathcal{W}_s(\epsilon) = \mathcal{W}_s(0) = 1$.

Proof Concavity of $\log \mathcal{W}_s(\epsilon)$ follows from the supermultiplicative nature of $s_n(f)$, and so it is continuous in $(0, 2)$. Continuity at 0 is obtained by defining $\mathcal{W}_s(0) = 1$. It then remains to be shown that $\lim_{\epsilon \to 0^+} \mathcal{W}_s(\epsilon) = 1$. This limit exists, since $\log \mathcal{W}_s(\epsilon)$ is a concave function in $(0, 2)$. By Lemma 7.38, for every $\epsilon > 0$ there exists a fixed number $m > 0$ such that $s_n(\leq \epsilon n) \geq 4^{(n-2)/(6m^2-2)} > 1$ for infinitely many values of n. Thus $\lim_{\epsilon \to 0^+} \mathcal{W}_s(\epsilon) \geq 1$. Otherwise, use Lemma 7.37: $s_n(\leq \epsilon n) \leq \sum_{f \leq \epsilon n} \sum_{k \leq f} \binom{n}{k} 2^k$. For small ϵ ($\epsilon \leq 1/2$ is sufficient), the sum over k can be bounded from above by $(f+1)\binom{n}{f} 2^f$, since the maximum is obtained by putting $k = f$. Thus $s_n(\leq \epsilon n) \leq \sum_{f \leq \epsilon n} (f+1) \binom{n}{f} 2^f \leq (\epsilon n + 1) \sum_{f \leq \epsilon n} \binom{n}{f} 2^f$. Take the power $1/n$ and the lim sup as $n \to \infty$. This shows that $\limsup_{n \to \infty} [s_n(\leq \lfloor \epsilon n \rfloor)]^{1/n} \leq 2^\epsilon / (\epsilon^\epsilon (1-\epsilon)^{1-\epsilon})$ by Lemma 7.39, provided that $0 \leq \epsilon \leq 2/3$. Thus

$$\mathcal{W}_s(\leq \epsilon) \leq \frac{2^\epsilon}{\epsilon^\epsilon (1-\epsilon)^{1-\epsilon}}. \qquad (\dagger)$$

As $\epsilon \to 0^+$, the result is $\lim_{\epsilon \to 0^+} \mathcal{W}_s(\leq \epsilon) \leq 1$. Lastly, $\mathcal{W}_s(\epsilon) \leq \mathcal{W}_s(\leq \epsilon)$. This completes the proof. □

From our perspective, the right-derivative of $\mathcal{W}_s(\epsilon)$ at $\epsilon = 0$ is important. This can be computed using Lemma 7.38 and Theorem 7.40. In particular, the second inequality in Lemma 7.38 implies that

$$\lim_{m\to\infty} \frac{1}{6m^2} \log s_{6m^2+4\lfloor \epsilon m^2 \rfloor}(\leq 12(m+\epsilon m^2)) \geq \lim_{m\to\infty} \frac{1}{6m^2} \log \binom{6\lfloor (m-2)^2/4 \rfloor}{\lfloor \epsilon m^2 \rfloor}, \tag{7.64}$$

and if $\epsilon < 3/2$, then the calculation of these limits gives

$$[\mathcal{W}_s(2\epsilon)]^2 \geq \left[\mathcal{W}_s(2\epsilon/(1+2\epsilon/3))\right]^{1+2\epsilon/3} \geq \frac{(1/4)^{1/4}}{(\epsilon/6)^{\epsilon/6}(1/4-\epsilon/6)^{1/4-\epsilon/6}}. \tag{7.65}$$

Together with eqn (†) in the proof of Theorem 7.40, this becomes

$$\left[\frac{27}{\epsilon^\epsilon (3-\epsilon)^{3-\epsilon}}\right]^{1/24} \leq \mathcal{W}_s(\epsilon) \leq \frac{2^\epsilon}{\epsilon^\epsilon (1-\epsilon)^{1-\epsilon}}, \tag{7.66}$$

where the lower bound is valid if $\epsilon < 2$, and the upper bound is valid if $\epsilon < 1/2$. By the squeeze theorem for limits,

$$\left[\frac{d^+}{d\epsilon}\mathcal{W}_s(\epsilon)\right]\bigg|_{\epsilon=0^+} = \infty. \tag{7.67}$$

Thus, the density function $\mathcal{W}_s(\epsilon)$ approaches 1 with infinite slope as $\epsilon \to 0^+$. This observation rules out an asymmetric model; in fact, it implies that the limiting free energy $\mathcal{V}_s(z)$ is a strictly increasing function of z. These results are in fact a limited pattern theorem for closed surfaces (see Theorem 5.19 and Corollary 6.11); there is a natural density for folds.

7.5.2 Bounds on $\mathcal{V}_s(z)$

The limiting free energy of a model of crumpling surfaces can be bounded from above and below by using the combinatorial properties of $s_n(f)$. If $z \geq 1$, then an upper bound can be found with some ease. Note that $s_n(z) = \sum_{f\geq 0} s_n(f) z^f \leq s_n \sum_{l=0}^{2n} z^f \leq 2ns_n z^{2n}$. By taking the logarithm of this inequality, dividing by n and letting $n \to \infty$, the following theorem is obtained.

Theorem 7.41 *The limiting free energy of crumpling surfaces is bound from above by*

$$\mathcal{V}_s(z) \leq \log \eta_3 + 2\log z, \qquad \textit{provided that } z \geq 1,$$
$$\mathcal{V}_s(1) = \log \eta_3$$

where η_3 is the growth constant of surfaces with the topology of a sphere. □

It is more difficult to find a lower bound. I shall use the surface in Fig. 7.8 to construct one. This surface has width equal to 5, area $n = 78$ and $f = 2n = 156$ folds. In general, if the width of this surface is p (where p is odd), then it has area $3(p^2+1)$,

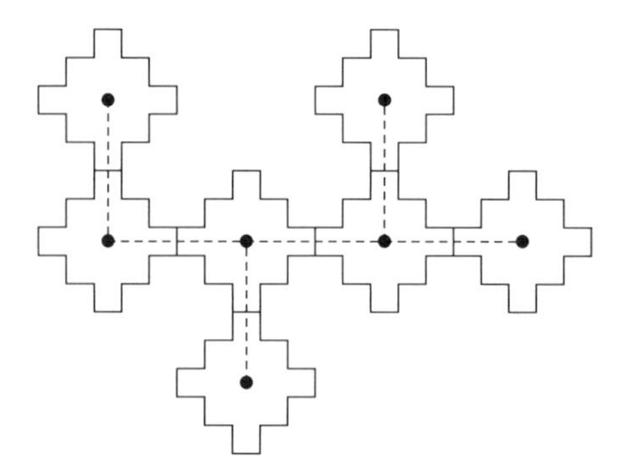

Fig. 7.10: A cross-section of a site tree generated by glueing together the surfaces in Fig. 7.7. The vertices are separated by 5 units.

and $6(p^2+1)$ folds. One can join these blocks together, as illustrated in Fig. 7.10, by identifying rightmost and leftmost plaquettes, and so on.

The number of ways in which the surfaces in Fig. 7.8 can be put together as in Fig. 7.10 is related to the number of site-trees in a sublattice of $\mathcal{Z}^3$ as follows. Let the mid-point (barycentre) of each surface be a vertex in a site tree on the dual lattice of $\mathcal{Z}^3$, and let two vertices be adjacent if the corresponding surfaces are joined. The length of the edge between these vertices is p, and the number of different ways in which q surfaces can be joined into a surface is T_q, the number of site trees in $\mathcal{Z}^3$ with q vertices. Each time that two surfaces are joined, two plaquettes are lost, as well as eight folds. Thus, in a surface built from q of the surfaces in Fig. 7.8, the area is $n = 3(p^2+1)q - 2(q-1) = (3p^2+1)q+2$ and the number of folds is $f = 6(p^2+1)q - 8(q-1) = (6p^2-2)q+8$. Hence, $s_{(3p^2+1)q+2}(z) \geq T_q z^{(6p^2-2)q+8}$. Fix p and let $q \to \infty$. Then $n \to \infty$, and

$$\mathcal{V}_s(z) \geq \lim_{q\to\infty} \left(\frac{1}{(3p^2+1)q+2} \log T_q + \frac{(6p^2-2)q+8}{(3p^2+1)q+2} \log z \right). \tag{7.68}$$

The number of site-trees in three dimensions is known to be $T_q = \Lambda_s^{q+o(q)}$ (this result follows directly from the concatenation of site-trees [216]), where $\Lambda_s \geq 3$ is the growth constant of site-trees.[3] Consequently, from (7.68):

$$\mathcal{V}_s(z) \geq \left[\frac{1}{3p^2+1}\right] \log \Lambda_s + \left[\frac{6p^2-2}{3p^2+1}\right] \log z, \tag{7.69}$$

for odd $p \geq 1$ provided that $z \geq 1$. If $p = 1$ in (7.69), then $\mathcal{V}_s(z) \geq [\log 3]/4 + \log z$, and if $p \to \infty$ in (7.69), then $\mathcal{V}_s(z) \geq 2 \log z$.[4] No improvement is gained from other values

[3] This bound is seen by noting that the number of self-avoiding walks, with steps only in the positive lattice directions, grows as 3^n.

[4] Note that $p = 1$ corresponds to a "deflation" of the surfaces, and that the surfaces "inflate" as $p \to \infty$. The bound $[\log 3]/4 + \log z$ can be improved with better lower bounds on Λ_s.

of p; one can show that all the lines defined by $y = [\log 3 + (6p^2 - 2)\log z]/(3p^2 + 1)$ intersect in the point $([\log 3]/4, [2\log 3]/4)$. This gives the following lower bounds.

Theorem 7.42 *If $z \geq 1$, then*

$$\mathcal{V}_s(z) \geq \begin{cases} \dfrac{\log 3}{4} + \log z, & \text{if } 0 \leq \log z \leq \dfrac{\log 3}{4}, \\ 2\log z, & \text{if } \log z \geq \dfrac{\log 3}{4}. \end{cases}$$

□

If $z \leq 1$ then bounds on the limiting free energy may be obtained by using the generating function $V_s(x, z)$. Its radius of convergence is $x_c(z)$ and $\mathcal{V}_s(z) = -\log x_c(z)$. Now notice the following.

Theorem 7.43 *For every $z \in [0, 1]$:*

$$x_c(z) \leq \min\{1, \eta_3^{-1} z^{-2}\},$$

and thus $x_c(z) < 1$ if $\sqrt{\eta_3^{-1}} \leq z \leq 1$.

Proof Note that

$$V_s(x, z) \geq \sum_{n\geq 0}\sum_{f=0}^{\epsilon n} s_n(f) z^f x^n \geq \sum_{n\geq 0}\sum_{f=0}^{\epsilon n} s_n(f) z^{\epsilon n} x^n \geq \sum_{n\geq 0} s_n(\leq \epsilon n) z^{\epsilon n} x^n.$$

But by eqns (7.58) and (7.62), if ϵ is small, $s_n(\leq \epsilon n) = [\mathcal{W}_s(\leq\epsilon)]^{n+o(n)}$, and so $V_s(x, z) \geq \sum_n [\mathcal{W}(\leq\epsilon)]^{n+o(n)} z^{\epsilon n} x^n$. The radius of convergence of the last series is an upper bound on $x_c(z)$, thus $x_c(z) \leq [\mathcal{W}(\leq\epsilon)]^{-1} z^{-\epsilon}$. Now take $\epsilon \to 0$ to obtain $x_c(z) \leq 1$. Alternatively, bound $V_s(x, z)$ from below as follows: $V_s(x, z) \geq \sum_{n\geq 0}\sum_{l=0}^{2n} s_n(l) z^{2n} x^n$, since $z \leq 1$. Thus $V_s(x, z) \geq \sum_{n\geq 0} s_n z^{2n} x^n$, and the radius of convergence of this last series is likewise an upper bound on $x_c(z)$. Thus $x_c(z) \leq \eta_3^{-1} z^{-2}$. This bound is better than the first if $\sqrt{\eta_3^{-1}} \leq z \leq 1$, otherwise the first is better. □

Since the right-derivative of the density function is infinite at $\epsilon = 0^+$, it follows that $x_c(z) < 1$ for all $z > 0$. An explicit upper bound on $x_c(z)$ can be derived from eqn (7.66), which was obtained by considering the surfaces in Fig. 7.9. The isolated 1-cubes exploring large smooth areas in the surfaces are responsible for driving the value of $x_c(z)$ away from one (and thus $\mathcal{V}_s(z)$ away from zero). In Section 7.5.3 a model which suppresses these "excitations" will be considered. That model will be asymmetric, with a crumpling transition.

Proposition 7.44 *For every positive $z \leq 1$, $x_c(z) \leq e^{-z^{24}/8e} < 1$.*

Proof Let $z \leq 1$. Observe that

$$V_s(x, z) \geq \sum_{n=0}^{\infty} \sum_{l \leq \epsilon n} s_n(l) z^l x^n \geq \sum_{n=0}^{\infty} s_n(\leq \epsilon n) z^{\epsilon n} x^n = \sum_{n=0}^{\infty} [\mathcal{W}(\leq \epsilon)]^{n+o(n)} z^{\epsilon n} x^n.$$

Thus, by using the bound in eqn (7.66)

$$V_s(x, z) \geq \sum_{n=0}^{\infty} \left[\frac{27}{\epsilon^\epsilon (3-\epsilon)^{3-\epsilon}} \right]^{n/24+o(n)} z^{\epsilon n} x^n,$$

if $\epsilon \leq 2$. The factor in the square brackets above is bounded from below by $3/\epsilon$. Hence

$$x_c(z) \leq \left(\frac{\epsilon/3}{z^{24}} \right)^{\epsilon/24} < 1 \quad \text{if } \epsilon < 3z^{24}.$$

Thus, for every positive $z \leq 1$ there exists such an $\epsilon > 0$. The minimum upper bound is derived by taking

$$\frac{d}{d\epsilon} \left[\frac{\epsilon}{24} \log(\epsilon/3z^{24}) \right] = 1/24 + [\log(\epsilon/3z^{24})]/24 = 0.$$

This gives the upper bound $e^{-z^{24}/8e}$ on $x_c(z)$, for $z \in [0, 1]$. □

The bounds in Theorem 7.43 and Proposition 7.44 translate into lower bounds on $\mathcal{V}_s(z)$.

$$\mathcal{V}_s(z) \geq \begin{cases} \log \eta_3 + 2 \log z & \text{if } -0.275\ldots \leq \log z \leq 0, \\ z^{24}/8e & \text{if } \log z \leq -0.275\ldots. \end{cases} \tag{7.70}$$

The two lower bounds on $\mathcal{V}_s(z)$ are equal if $\log z \approx -0.275$; if $\log z$ is larger than this number, then the first bound is better, otherwise the second bound is better.

A lower bound on $x_c(z)$ is found by writing $V_s(x, z)$ as the sum of two terms which will be bounded from above:

$$\begin{aligned} V_s(x, z) &= \sum_{n=0}^{\infty} \sum_{f=0}^{2n} s_n(f) z^f x^n \\ &= \sum_{n=0}^{\infty} \sum_{f \leq \epsilon n} s_n(f) z^f x^n + \sum_{n=0}^{\infty} \sum_{f > \epsilon n} s_n(f) z^f x^n \\ &\leq \sum_{n=0}^{\infty} \sum_{f \leq \epsilon n} \left(\sum_{k=0}^{f} \binom{n}{k} 2^k \right) z^f x^n + \sum_{n=0}^{\infty} \sum_{f > \epsilon n} s_n(f) z^f x^n, \end{aligned} \tag{7.71}$$

where Lemma 7.37 was used, and where ϵ will be put equal to one-half. The following bound will prove useful.

Lemma 7.45 *Let $n > 1$ and l be integers such that $0 < l < n$. Then*

$$\frac{n^n}{l^l(n-l)^{n-l}} \leq \frac{72\sqrt{2\pi}}{110}\sqrt{n}\binom{n}{l}.$$

Proof Use the Stirling approximation to $k!$. For every $k \geq 1$ (see for example reference [37]):

$$\left|\frac{k!}{k^k e^{-k}\sqrt{2\pi k}} - 1\right| \leq \frac{1}{11k}.$$

Hence

$$\frac{11}{12}\frac{e^k}{\sqrt{2\pi k}} \leq \frac{k^k}{k!} \leq \frac{11}{10}\frac{e^k}{\sqrt{2\pi k}}.$$

If these bounds on k^k are substituted in $n^n(l^l(n-l)^{(n-l)})$ for $k = n, l$ and $(n-l)$, then the upper bound is proven. □

The second term in (7.71) is bounded as follows:

$$\sum_{n=0}^{\infty}\sum_{f=\lfloor n/2\rfloor+1}^{2n} s_n(f)z^f x^n \leq \sum_{n=0}^{\infty}(2n - \lfloor n/2\rfloor)3^n z^{\frac{n}{2}} x^n, \tag{7.72}$$

since $s_n \leq 3^n$ by Lemma 7.37, and this is finite if $x < 1/3\sqrt{z}$. The first term requires somewhat more work and the use of Lemma 7.45:

$$\begin{aligned}
\sum_{n=0}^{\infty}\sum_{f=0}^{\lfloor n/2\rfloor}\left(\sum_{k=0}^{f}\binom{n}{k}2^k\right)z^f x^n &\leq \sum_{n=0}^{\infty}\sum_{f=0}^{\lfloor n/2\rfloor}\left[\frac{2^{f/n}}{(f/n)^{f/n}(1-f/n)^{1-f/n}}\right]^n z^f x^n,\\
&= \sum_{n=0}^{\infty}\sum_{f=0}^{\lfloor n/2\rfloor}\left[\frac{n^n}{f^f(n-f)^{n-f}}\right](2z)^f x^n,\\
&\leq \sum_{n=0}^{\infty}\frac{72\sqrt{2\pi}}{110}\sqrt{n}\left[\sum_{f=0}^{\lfloor n/2\rfloor}\binom{n}{f}(2z)^l\right]x^n,\\
&\leq \frac{72\sqrt{2\pi}}{110}\sum_{n=0}^{\infty}\sqrt{n}(1+2z)^n x^n,
\end{aligned} \tag{7.73}$$

where the fact that $f \leq n/2$ and Lemma 7.39 was used with $\gamma = 1/2$ in the first inequality, and where Lemma 7.45 was used in the second inequality. This is finite if $x < 1/(1+2z)$. Thus, by comparing eqns (7.72) and (7.73), the following lower bound on $x_c(z)$ is obtained.

Theorem 7.46 *$x_c(z)$ is bounded from below in $[0, 1]$ as*

$$x_c(z) \geq \begin{cases} 1/3\sqrt{z}, & \text{if } z \in [\frac{1}{4}, 1];\\ 1/(1+2z), & \text{if } z \in [0, \frac{1}{4}]. \end{cases}$$

□

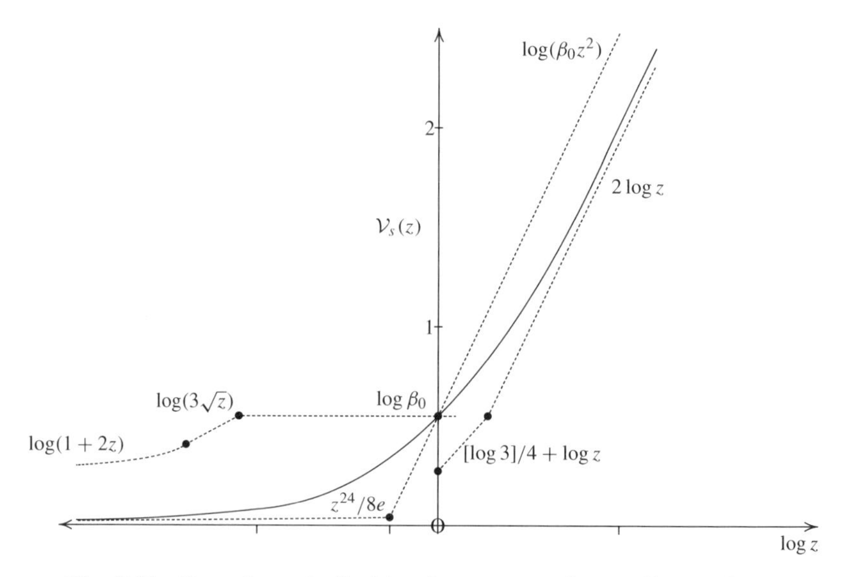

Fig. 7.11: Bounds on the limiting free energy of crumpling surfaces.

The lower bounds in Theorem 7.46 are upper bounds on $\mathcal{V}_s(z)$:

$$\mathcal{V}_s(z) \leq \begin{cases} \log 3 + [\log z]/2, & \text{if } 1/4 \leq z \leq 1; \\ \log(1+2z), & \text{if } 0 \leq z \leq 1/4. \end{cases} \tag{7.74}$$

The bounds derived in this section are plotted in Fig. 7.11 against $\log z$. Note that $\mathcal{V}_s(z)$ is asymptotic to the line $C_0 + 2\log z$, where C_0 is the limiting entropy. It is indeed possible to show that $C_0 > 0$; this follows from a construction similar to that of Fig. 7.10, but where the surfaces are attached with "necks" of length one. The result is a surface with every edge a fold, and as before, the entropy is at least that of site-trees in a sublattice of the cubic lattice. This argument shows that $\mathcal{W}_2(2) > 1$; and since $\mathcal{V}_s(z) = \sup_{\epsilon>0}\{\log \mathcal{W}_s(\epsilon) + \epsilon \log z\}$, if follows that $C_0 = \log \mathcal{W}_2(2) > 0$. Numerical estimates shows that $\log \eta_3 \approx 0.55$ [138, 139, 140], and since $\eta_3 \geq \mathcal{W}_2(\epsilon)$, it follows that $C_0 \leq 0.55$. For $\log z < 0$, $\mathcal{V}_s(z)$ is asymptotic to 0. These results have implications for the average degree of folding in the $n \to \infty$ limit. This is defined as

$$\lim_{n\to\infty} [\langle l \rangle / n] = \lim_{n\to\infty} z \frac{d}{dz} \frac{[\log s_n(z)]}{n}$$

(if this limit exists). Since $[\log s_n(z)]/n$ is a sequence of convex functions, and the limit of this sequence is a convex function $\mathcal{V}_s(z)$, it follows that

$$z\frac{d^-}{dz}\mathcal{V}_s(z) \leq \liminf_{n\to\infty} z\frac{d^-}{dz}\frac{[\log s_n(z)]}{n} \leq \limsup_{n\to\infty} z\frac{d^+}{dz}\frac{[\log s_n(z)]}{n} \leq z\frac{d^+}{dz}\mathcal{V}_s(z)$$

and moreover, the derivative

$$z\frac{d}{dz}\mathcal{V}_s(z) = \lim_{n\to\infty} z\frac{d}{dz}\frac{[\log s_n(z)]}{n}$$

Table 7.2: Tricritical exponents for crumpling surfaces

ϕ	α	$2-\alpha_t$	$2-\alpha_u$	ν_t	y_t	$2-\alpha_+$
$\frac{2}{3}$	$\frac{1}{2}$	≈ 0.78	≈ 1.17	≈ 0.26	≈ 2.56	$\frac{1}{2}$

exists almost everywhere (see Theorem B.7). With increasing z, the derivatives of $\mathcal{V}_s(z)$ approach 2, and the average density of folds is 2 in the limit. Similarly, if z approaches 0, then the average density of folds is 0 in the limit. Numerical simulations [284] suggest that $x_c(z)$ (which is equal to $e^{-\mathcal{V}_s(z)}$)) approaches very close to 1 for modest values of z. In particular, $x_c(0.635) = 0.987 \pm 0.010$. This is slightly smaller than the upper bound proved in Theorem 7.47, which predicts that $x_c(0.635) \leq 0.99999915\ldots$. This upper bound is due to inflated surfaces (those in Fig. 7.9) with isolated folds exploring large smooth surface areas. In this regime, it seems that the limiting theory is one of inflated surfaces, even though a limiting theory of "disk-like" surfaces have also been proposed [284], with small "excitations" which explores the large smooth areas as in Fig. 7.9. It may be speculated that the critical curve $x_c(z)$ in this case corresponds to a curve of essential singularties in the generating function $V_s(x, z)$ [284]. These essential singularities are interpreted as first-order transitions.

For larger values of z the upper bound on $x_c(z)$ is given by Theorem 7.43, and this bound is due to the "tree-like" surfaces such as in Fig. 7.10. It seems likely that branched polymers are found as the limiting theory. However, there is a qualification which must be taken into account. The bound for large values of z is derived by taking $p \to \infty$ in eqn (7.69), thus "inflating" the surfaces which are schematically illustrated in Fig. 7.10 (this is only done after the $n \to \infty$ limit is already taken). This inflation should not affect the branched polymer character of the limiting theory. In analogy with the discussion for disks, it should be expected that the generating function has branch points along the critical curve $x_c(z)$ for these values of z. Since the limiting theory seems to be one of branched polymers, the exponent $2-\alpha_+$ along the λ-line (see eqn (2.12)) should be given by branched polymer exponents. The entropic exponent in three dimensions of branched polymers is $\theta = 3/2$, and so $2 - \alpha_+ = \theta - 1 = 1/2$. There seems to be no reason to suspect that the branched polymer character of the surfaces will change at the tricritical point; if this is so, then $2 - \alpha_t = 1/2$ as well. This is not supported by the numerical simulations in reference [284]. In particular, the entropic exponent of tricritical crumpling surfaces is found to be 1.78 ± 0.03, while the metric exponent is $\nu = 0.4825 \pm 0.0025$. These estimates show that $2-\alpha_t = \theta - 1 = 0.78 \pm 0.03$, while the cross-over exponent can be computed from the hyperscaling relation $3\nu = 1/\phi$, which then gives $\phi = 0.6908 \pm 0.0023$. Since $\nu = 1/2$ for trees, it seems reasonable to suspect that $\phi = 2/3$. Furthermore, it follows that $2 - \alpha_u = 1.17 \pm 0.05$. These values are uncertain; but they do point to the possibility that tricritical crumpling surfaces are in a universality class of their own, as speculated in reference [284]. The tricritical nature of the crumpling transition is of course a hypothesis, although it has been argued for in the case of a crumpling transition in tethered membranes [368]. The exponents are listed in Table 7.2.

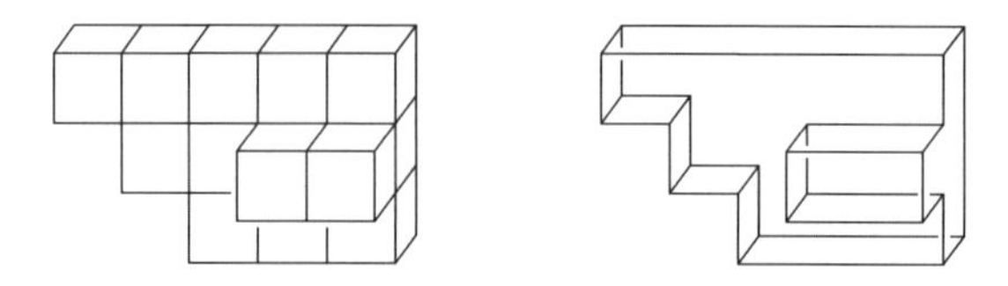

Fig. 7.12: A c-surface of its skeleton. The skeleton is a lattice animal.

7.5.3 A crumpling transition in surfaces with connected skeletons

Let S be a surface with area n and degree of folding f. The *skeleton* of S is the set of all edges which are folds in S. In general, the skeleton of a surface is a set of lattice animals. Some surfaces have a connected skeleton, such as a q-cube, or the surface in Fig. 7.8, but many surfaces have skeletons which are not connected, such as the example in Fig. 7.9. Let $s_n^c(f)$ be the number of surfaces with a *connected* skeleton, with area n and f folds. These surfaces are called *c-surfaces*, and one is illustrated in Fig. 7.12.

By using the same constructions as in Section 7.5.2, it can be shown that

$$\lim_{n\to\infty} [s_n^c(\leq \epsilon n)]^{1/n} = \mathcal{W}_S(\leq \epsilon), \quad \text{for every } \epsilon \in [0, 2]. \tag{7.75}$$

The density function $\mathcal{W}_S(\epsilon)$ can be defined similarly to eqn (7.60). $\log \mathcal{W}_S(\epsilon)$ is a concave function of ϵ, is continuous in $[0, 2)$, monotonic non-decreasing with ϵ and $\mathcal{W}_S(0) = 1$ (this is seen from eqn (7.66)). Also note that $\mathcal{W}_S(\epsilon) \leq \eta_3$. The upper bound in eqn (7.66) can be improved.

Theorem 7.47 *If λ_3 is the growth constant of edge-animals in the cubic lattice, then*

$$\mathcal{W}_S(\epsilon) \leq (4\lambda_3)^\epsilon.$$

Proof Observe that any surface with a connected skeleton and f folds can be mapped into lattice animals weakly embedded in the cubic lattice, with f edges (to see this, remove all plaquettes from the surface, and just leave behind its skeleton). Suppose that A is a lattice animal with the property that every edge in A is in a planar (two-dimensional) polygon. Then A may be converted into a closed surface by filling in the planar polygons with planar sheets of plaquettes. About every edge in A there are potentially four possible directions for adding the sheet of plaquettes. If there are a_l animals with l edges, then this implies that $s_n^c(f) \leq 4^f a_f$. Consequently, $\mathcal{W}_S(\epsilon) \leq (4\lambda_3)^\epsilon$, as can be seen from Corollary 6.2. □

The density function $\mathcal{W}_S(\epsilon)$ of folds in c-surfaces has many of the same properties as $\mathcal{W}_s(\epsilon)$, as noted above, but it differs in one important respect: it does not approach one as $\epsilon \to 0^+$ with infinite slope. Since $\mathcal{W}_S(\epsilon)$ is monotonic non-decreasing it follows that

$$\begin{aligned} 0 &\leq \liminf_{\epsilon\to 0^+} [\log \mathcal{W}_S(\epsilon) - \log \mathcal{W}_S(0)]/\epsilon \\ &\leq \limsup_{\epsilon\to 0^+} [\log \mathcal{W}_S(\epsilon) - \log \mathcal{W}_S(0)]/\epsilon \\ &\leq \log(4\lambda). \end{aligned} \tag{7.76}$$

In fact, these two limits are equal, since the right-derivative of a concave function exists (see Theorem B.7). Thus

$$\frac{d^+}{d\epsilon}\left[\log \mathcal{W}_c(\epsilon)\right]\bigg|_{\epsilon=0^+} \leq \log 4\lambda_3, \tag{7.77}$$

as opposed to the result for all surfaces in eqn (7.66).

The partition function for this model is $s_n^c(z)$, and the generating function is

$$V_S(x,z) = \sum_{n=0}^{\infty} s_n^c(z) x^n. \tag{7.78}$$

The existence of a limiting free energy per plaquette, $\mathcal{V}_S(z)$, and its convexity and continuity follows by the same methods as in Section 7.4.3. Note that $\mathcal{V}_S(z) \leq \mathcal{V}_s(z)$, so that every upper bound on $\mathcal{V}_s(z)$ is an upper bound on $\mathcal{V}_S(z)$. Moreover, the lower bounds in eqn (7.40) are also lower bounds on $\mathcal{V}_S(z)$ if $z \geq 1$, since the surfaces in Figs 7.8 and 7.10 have connected skeletons.

By Theorem 7.42 it is apparent that $\mathcal{V}_S(z) > 0$ if $z > 1$, since the surfaces which gives those lower bounds have connected skeletons. On the other hand, the following theorem is an immediate consequence of eqn (7.77), Theorem 3.13 and Lemma 3.20.

Theorem 7.48 *There exists a critical value* $z_c \geq 1/4\lambda_3$, *and* $z_c \leq 1$, *such that* $\mathcal{V}_S(z) = 0$ *if* $z \leq z_c$. □

Since $\mathcal{V}_S(z)$ is strictly positive if $z > 1$, the derivative of $\mathcal{V}_S(z)$ (which exists almost everywhere) is positive if $z > z_c$ and zero if $z < z_c$. The singularity in $\mathcal{V}_S(z)$ at $z = z_c$ corresponds to a phase transition in this model from a phase of smooth vesicles with a zero density of folds, to a phase of vesicles with a positive density of folds. It is not known whether this transition is first order or continuous. The generating function $V_S(x,z)$ has a multicritical point at $(z_c, x_c(z_c))$, and this separates the critical curve $x_c(z)$ into two. There is a straight line $x_c(z) = 1$ if $z < z_c$, and this corresponds to surfaces with a small number of folds. It seems that this may correspond to a phase of smooth surfaces, possibly inflated. In analogy with the FGW vesicle, this may be a curve of essential singularities in $V_S(x,z)$. On the other hand, the transition along the curve $x_c(z)$ for $z > z_c$ is believed to be a phase of branched polymers (which could be inflated, as observed above). It seems likely that this is a curve of branch cuts in $V_S(x,z)$; and so it may be conjectured that the point $(z_c, x_c(z_c))$ is a tricritical point. Moreover, the singularity diagram of this model has the (presumed) appearance of the singularity diagram of the FGW vesicle (see Fig. 7.1), and this is an asymmetric model. As far as the phases and critical exponent of this model are concerned, the same comments can be made here as were made in Section 7.5.2.

7.5.4 Inflating and crumpling c-surfaces

In Section 7.5.3 I showed that c-surfaces undergo a crumpling transition at a critical value of the folding activity. In this section I shall extend this model by the introduction of a second activity, conjugate to the volume of the c-surface.

Let $s_n^c(m, l)$ be the number of c-surfaces with area n, volume m and l folds. Then the partition function is $s_n^c(y, z) = \sum_{m=0}^{\infty} \sum_{l=0}^{\infty} s_n^c(m, l) y^m z^l$. The generating function of this model is defined by

$$V_S(x, y, z) = \sum_{n=0}^{\infty} s_n^c(y, z) x^n. \tag{7.79}$$

Concatenation of two vesicles gives

$$\sum_{i=0}^{m} \sum_{j=0}^{l} s_{n_1}^c(m - i, l - j) s_{n_2}^c(i, j) \leq \sum_{k=-2}^{2} s_{n_1+n_2+2}^c(m + 1, l + 2k). \tag{7.80}$$

Multiplying by $y^m z^l$ and summing over m and l gives the following supermultiplicative relation for $s_n^c(y, z)$:

$$s_{n_1}^c(y, z) s_{n_2}^c(y, z) \leq y^{-1} \left[\sum_{k=-2}^{2} z^{2k} \right] s_{n_1+n_2+2}^c(y, z). \tag{7.81}$$

By Lemma A.3 there is a limiting free energy.

Theorem 7.49 *There exists a limiting free energy in a model of c-surfaces with partition function $s_n^c(y, z)$. That is*

$$\lim_{n \to \infty} \frac{1}{n} \log s_n^c(y, z) = \mathcal{V}_S(y, z)$$

exists for every y and z (but may be infinite). □

$\log s_n^c(y, z)$ is a convex function of both $\log y$ and $\log z$. In these circumstances $\log s_n^c(y, z)$ is a *doubly convex* function of $\log y$ and $\log z$, and it is continuous for $y \in (0, \infty)$ and $z \in (0, \infty)$ (since it is finite). The free energy $\mathcal{V}_S(y, z)$ is the limit of a sequence of doubly convex functions, and is also continuous for $y \in (0, \infty)$ and $z \in (0, \infty)$ *provided that it is finite*. It will also turn out that $\mathcal{V}_S(y, z)$ is a non-analytic function, and its non-analyticities will correspond to phase transitions in the c-surface. The radius of convergence of the generating function in eqn (7.79) is $x_c(y, z) = e^{-\mathcal{V}_S(y,z)}$.

If $y = 1$, then $s_n^c(1, z)$ is the partition function of a c-surfaces with a folding activity z. This model was considered in Section 7.5.3 and there exists a critical value z_c such that $x_c(1, z) = 1$ if $z \leq z_c$, and $x_c(1, z) < 1$ otherwise (Theorem 7.48). In other words, $\mathcal{V}_S(1, z) = 0$ if $z \leq z_c$ and $0 < \mathcal{V}_S(1, z) < \infty$ otherwise. Thus, there is a non-analyticity in $\mathcal{V}_S(y, z)$ if $y = 1$ at $z = z_c$. The density of folds is the partial derivative

$$z \frac{\partial}{\partial z} \mathcal{V}_S(1, z) \quad \begin{cases} = 0, & \text{if} z < z_c; \\ > 0, & \text{otherwise,} \end{cases} \tag{7.82}$$

which exists almost everywhere. A model of c-surfaces is "smooth" (with a zero density of folds) if $z < z_c$ and $y = 1$. For larger values of z the c-surface has a positive density of folds. It is expected, as stated before, that c-surfaces collapse to branched polymers if $z > z_c$, and to a phase of smooth disk-like or inflated c-surfaces of $z < z_c$ [284]. A tricritical point separates these phases at $z = z_c$.

Theorem 7.50 *If $y > 1$ then $\mathcal{V}_S(y, z) = \infty$.*

Proof Suppose that $y > 1$. A q-cube is a c-surface which is a geometric cube of side-length q. It has area $6q^2 = n$, volume $q^3 = (n/6)^{3/2}$ and number of folds $12q = 12(n/6)^{1/2}$. If the limit in Theorem 7.49 is taken through q-cubes, then

$$\mathcal{V}_S(y, z) \geq \lim_{q\to\infty} \frac{1}{6q^2}(q^3 \log y + 12q \log z) = \infty,$$

since $y > 1$. □

Thus, the limiting free energy is divergent if $y > 1$, regardless of the value of z. Since $\mathcal{V}_S(1, z) < \infty$ for any z, there is a line of phase transitions along the line $y = 1$. If $z < z_c$, then the transition is from a phase of a smooth c-surface to an inflated c-surface, and if $z > z_c$, then the transition is from a phase of branched polymers to an inflated c-surface.

To see that there is indeed a phase of an inflated c-surface if $y > 1$, argue as follows. First note that (the limit is taken through a sequence of q-cubes)

$$\liminf_{q\to\infty} \frac{1}{q^3} \log s^c_{6q^2}(y, z) \geq \lim_{q\to\infty} \frac{1}{q^3}(q^3 \log y + 12q \log z) = \log y, \tag{7.83}$$

On the other hand, note that $s^c_n(y, z) \leq s^c_n y^{(n/6)^{3/2}}[\max_{f\geq 0} z^f]$, where we have replaced m and f by values which maximize the powers, and where the sums over m and f have been executed. Since s^c_n grows at most exponentially with n, and $0 \leq f \leq 2n$,

$$\begin{aligned}\limsup_{n\to\infty} \frac{1}{(n/6)^{3/2}} \log s^c_n(y, z) &\leq \lim_{n\to\infty} \frac{1}{(n/6)^{3/2}}\left(\log\left(s^c_n y^{(n/6)^{3/2}}\left[\max_{f\geq 0} z^f\right]\right)\right) \\ &= \log y,\end{aligned} \tag{7.84}$$

and so it follows that

$$\lim_{n\to\infty} \frac{[\log s^c_n(y, z)]}{n^{3/2}} = \left(\frac{1}{6}\right)^{3/2} \log y \quad \text{if } y > 1.$$

By (7.83) and (7.84), the partition function is dominated by surfaces of volume $(n/6)^{3/2}$; that is, inflated c-surfaces. Note that the derivatives

$$\left[\frac{\partial^N}{\partial y^N} \mathcal{V}_S(y, z)\right]\Bigg|_{y=1} < \infty$$

are all finite. In other words, there is an essential singularity in the generating function $V_S(x, y, z)$ along the line $y = 1$. The transition to an inflated c-surface is therefore a first-order transition, and one can argue along the lines of references [5,116] (see also Section 7.2) that these are "droplet singularities" in analogy with condensation of a fluid along the line $y = 1$. These droplets coalesce as y increases from $y = 1$ into the inflated phase.

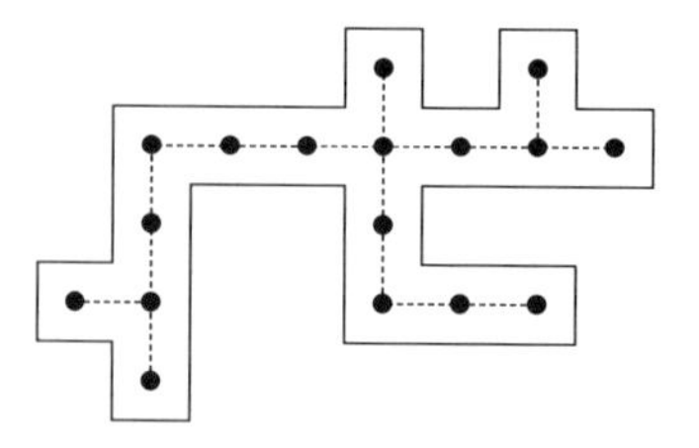

Fig. 7.13: A c-surface with minimal volume is dual to a site tree.

The situation is more complicated when $y < 1$. Note that $\mathcal{V}_S(y, z) \leq \mathcal{V}_S(1, z) < \infty$ in this case (by Theorem 7.49). In order to derive bounds on $\mathcal{V}_S(y, z)$, define *deflated c-surfaces* to be that set of c-surfaces with volume $m = (n-2)/4$. Let the number of such c-surfaces be $s_n^d(f) = s_n^c((n-2)/4, f)$. A two-dimensional cut through such a surface with volume $(n-2)/4$ is illustrated in Fig. 7.13. These c-surfaces are dual to a subset of site-trees, and by concatenation, and the techniques of Theorem 7.49, it is seen that the limiting free energy $\mathcal{V}_d(z)$ of deflated c-surfaces exists and is defined by

$$\mathcal{V}_d(z) = \lim_{n\to\infty} \frac{1}{n} \log \sum_{f=0}^{\infty} s_n^d(f) z^f. \tag{7.85}$$

Note that the minimum value of f in eqn (7.85) is $n+6$, in which case the c-surface is a long thin (capped) cylinder or a rod. Moreover, using Theorem 7.46 and noting that $\mathcal{V}_S(z) \leq \mathcal{V}_d(z)$, it follows that there exists a critical z_d such that

$$\mathcal{V}_d(z) \begin{cases} > 0, & \text{if } z > z_d, \\ = 0, & \text{otherwise.} \end{cases} \tag{7.86}$$

Since $s^d(f) \leq s^c(f)$, it follows that $z_d \geq z_c$. Observe that

$$\mathcal{V}_d(z) \geq \lim_{n\to\infty} \frac{1}{n} \log(z^{n+6} s_n^d(n+6)) \geq \log z \tag{7.87}$$

if $z > 1$, thus z_d is finite. A lower bound on the free energy $\mathcal{V}_S(y, z)$ can be derived as follows if $y < 1$:

$$\begin{aligned} s_n^c(y, z) &\geq \sum_{f=0}^{\infty} s_n^d(f) y^{(n-2)/4} z^f, \quad \text{taking } m = (n-2)/4, \\ &= y^{(n-2)/4} \sum_{f=0}^{\infty} s_n^d(f) z^f. \end{aligned} \tag{7.88}$$

Using (7.85) it is then found that

$$\mathcal{V}_S(y, z) \geq [\log y]/4 + \mathcal{V}_d(z), \quad \text{if } y < 1. \tag{7.89}$$

On the other hand, an upper is found more easily:

$$s_n^c(y, z) \leq \sum_{m=0}^{\infty} \sum_{f=0}^{\infty} s_n^c(m, f) y^{(n-2)/4} z^f, \qquad \text{since } \min(m) = (n-2)/4,$$

$$= y^{(n-2)/4} \sum_{f=0}^{\infty} s_n^c(f) z^f. \tag{7.90}$$

Taking logarithms, dividing by n and letting $n \to \infty$ gives

$$\mathcal{V}_S(y, z) \leq [\log y]/4 + \mathcal{V}_S(1, z). \tag{7.91}$$

These arguments are enough to show the following.

Theorem 7.51 *Suppose that* $y \leq 1$. *Then the limiting free energy of crumpling and inflating c-surfaces is constrained to satisfy*

$$\mathcal{V}_S(y, z) \begin{cases} = [\log y]/4, & \text{if } z < \min\{z_c, z_d\}, \\ \geq [\log y]/4 + \log z, & \text{if } z > 1. \end{cases}$$

Proof If $z < \min\{z_c, z_d\}$, then if follows from eqs (7.86) and (7.89) that $\mathcal{V}_S(y, z) \geq [\log y]/4$. From eqn (7.91) and since $\mathcal{V}_S(1, z) = \mathcal{V}_S(z)$, it follows from Theorem 7.48 that if $z < z_c$ then $\mathcal{V}_S(y, z) \leq [\log y]/4$. The lower bound if $z > 1$ follows directly from eqns (7.89) and (7.87). □

Thus, for every $y < 1$ there is a non-analyticity in $\mathcal{V}_S(y, z)$ corresponding to a critical value in z (and thus to a "crumpling transition", where the c-surface becomes smooth). These results can be taken together by drawing the phase diagram of a c-surface, which is Fig. 7.14.

If the osmotic pressure is positive, then the c-surface inflates to a phase with large volume; the presence or absence of a folding energy does not affect this. This was also shown in reference [361]. The transition is a first-order transition. If the volume activity is put equal to 1 ($y = 1$), then there are two phases. The first is the smooth phase of c-surfaces (believed to be cubical or disk-like [284]) at low values of the folding activity, and this was discussed in Sections 7.5.2 and 7.5.3. Secondly, there is a "flaccid phase" or branched polymer phase at higher values of the folding activity. This phase was similarly discussed in Section 7.5.2, and it was noted that it may be a phase of surfaces which have branched polymer characteristics, but which is also inflated. If the volume activity is smaller than 1 ($y < 1$, or at negative osmotic pressure), then there are at least two phases. The first is a smooth phase, which is suggested to be a phase of rod-like vesicles by eqns (7.89) and (7.91). It is not clear that this phase is different from the cubical or disk-like vesicles encountered in the smooth phase when $y = 1$, but the deflation of cubical or disk-like vesicles to rod-like vesicles seems to be likely to be a transition. The second phase is a phase of deflated crumpled vesicles, with branch polymer characteristics. This is in contrast to the deflated branched vesicles obtained here for $y < 1$.

In summary, there seems to be three broadly defined cases. The first is at $y > 1$, when a phase of inflated vesicles is encountered. The second is at $y = 1$, where two phases of

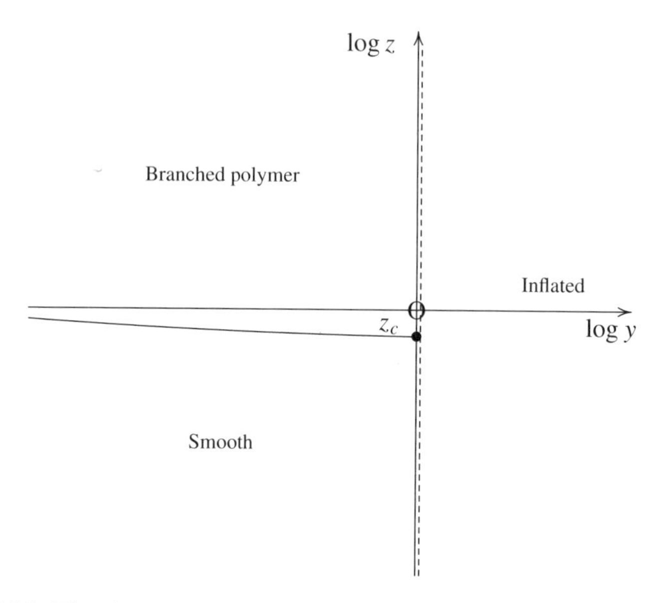

Fig. 7.14: The phase diagram of c-surfaces. The log z axis is a line of first-order transitions to inflated c-surfaces. A second line of crumpling transition separate the branched polymer (crumpled) and smooth phases. The line of absorption transitions ends in a critical end-point in the first-order transitions of the inflating vesicle.

inflated smooth and *inflated* crumpled vesicles are encountered. The third is at $y < 1$, where two phases of *deflated* smooth and *deflated* crumpled vesicles are encountered. The line of adsorption transitions ends in a critical end-point in the first-order transitions of the inflating vesicle. The cross-over exponent for the inflation is $\phi = 1$. In the deflated regime, $\nu = 1/2$ (this is the branched polymer value) and this is generally supported by numerical simulations [331].

Appendix A: Subadditive functions

Let a_n be a function from the natural numbers $\mathcal{N}$ to integers $\mathcal{Z}$. Then a_n is *subadditive* if

$$a_n + a_m \geq a_{n+m}. \tag{A.1}$$

There are two generalizations of subadditivity, namely

$$a_n + a_m \geq a_{n+f(m)}, \tag{A.2}$$

where $f(m)$ is a function of m such that $\lim_{m\to\infty} f(m)/m = 1$, and

$$a_n + a_m \geq a_{n+m} - g(n+m), \tag{A.3}$$

where $g(n)$ is non-decreasing and satisfies some further conditions. The results in this appendix are adapted from references [162, 182, 369] and [249].

Lemma A.1 *Let Z be an infinite subset of $\mathcal{N}$ closed under addition. If a_n is a subadditive function defined from Z to $\mathcal{Z}$, then*

$$\lim_{n\to\infty} \frac{a_n}{n} = \inf_{n\geq 1} \frac{a_n}{n} = \upsilon,$$

where $\upsilon \in [-\infty, \infty)$, and where the limit is taken in Z. Moreover, $a_n \geq n\upsilon$.

Proof Fix $k \in Z$, and choose $A_k = \max\{a_l \big| l \leq k, l \in Z\}$. Let $n \in Z$ and define the natural number $j = \lfloor n/k \rfloor$. Then, for some $0 \leq r \leq k$, $n = jk + r$. By subadditivity, $a_n = a_{jk+r} \leq ja_k + A_k$. Hence, $a_n/n \leq ja_k/(jk+r) + A_k/(jk+r)$. Now take $n \to \infty$ in Z by letting $j \to \infty$ in $\mathcal{N}$. This gives $\limsup_{n\to\infty}[a_n/n] \leq a_k/k$. Now take the infimum over k in Z to obtain the theorem. Since a_k/k is finite for every k, the limit is not $+\infty$, but it could be $-\infty$. □

Next consider the generalized subadditivity relation in eqn (A.2). Lemma A.2 is due to J.B. Wilker and S.G. Whittington [369].

Lemma A.2 *Let a_n be a function from $\mathcal{N}$ to $\mathcal{Z}$. If there exists a positive function $f : \mathcal{N} \to \mathcal{Z}$ such that* $\lim_{n\to\infty}[f(m)/m] = 1$, *and such that $a_n + a_m \geq a_{n+f(m)}$. Then* $\lim_{n\to\infty}[a_n/n] = \upsilon$ *exists, and $a_n \geq \upsilon \cdot f(n)$.*

Proof Fix $n \geq m > 0$ and let p be the largest integer such that $n - m \geq (p-1)f(m)$. Then there exists a function $r(m)$, $0 \leq r(m) \leq f(m)$, such that $n = (p-1)f(m) + m + r(m)$. Repeated applications of the generalized subadditive inequality gives $a_n = a_{(p-1)f(m)+m+r(m)} \leq (p-1)a_m + a_{m+r(m)}$. That is,

$$\frac{a_n}{n} \leq \frac{(p-1)a_m}{(p-1)f(m)+m+r(m)} + \frac{a_{m+r(m)}}{(p-1)f(m)+m+r(m)}.$$

Now let $p \to \infty$. Define

$$\upsilon = \limsup_{n\to\infty} \frac{a_n}{n} \leq \frac{a_m}{f(m)} = \frac{a_m}{m}\frac{m}{f(m)}.$$

Take the lim inf of this last equation to obtain the existence of the limit, and the inequality follows from the definition of υ. □

Lastly, consider eqn (A.3). The following theorem is due to J.M. Hammersley [162].

Theorem A.3 *Let a_n be a function from $\mathcal{N}$ to $\mathcal{Z}$. Suppose that there exists a function g_n such that $a_n + a_m + g_{n+m} \geq a_{n+m}$, for all $n \geq M$ and $m \geq M$. If g_n is non-decreasing for all $n \geq N_0 > 0$, then $\sum_{n=N_0}^{\infty}[g_n/(n(n+1))] < \infty$ if and only if $\lim_{n\to\infty}[a_n/n]$ exists, and*

$$\frac{a_n}{n} \geq \lim_{m\to\infty}\frac{a_m}{m} + \frac{g_n}{n} - 4\sum_{m=2n}^{\infty}\left[\frac{g_m}{m(m+1)}\right].$$

Proof Suppose that

$$\sum_{n=N_0}^{\infty}\left[\frac{g_n}{n(n+1)}\right] = \infty.$$

Define the function

$$h_n = \begin{cases} g_n - g_{N_0} & \text{if} n \geq N_0, \\ 0 & \text{otherwise.} \end{cases}$$

Then define

$$\phi_n = \left\lceil n \sin \sum_{m=1}^{n-1}\left[\frac{h_m}{m(m+1)}\right]\right\rceil.$$

Since the sum above oscillates to infinity with n, $\lim_{n\to\infty}[\phi_n/n]$ does not exist. To see that ϕ_n satisfies the generalized subadditivity relation above, argue as follows. Note that $|\sin x - \sin(x-\delta)| = |\delta\cos(x-\theta\delta)| \leq |\delta|$ (by the mean value theorem). Let n and

m be natural numbers, and put $N = n + m$. Define the rational numbers $p = n/N$ and $q = m/N$, so that $p+q = 1$. Then (note that $p\lceil a\rceil \le \lceil p(a+1)\rceil$ and $\lceil a-b\rceil \ge \lceil a\rceil - \lceil b\rceil$)

$$
\begin{aligned}
& p\phi_N - \phi_{pN} \\
&\quad = p\left\lceil N \sin\left(\sum_{j=1}^{N-1}\left[\frac{h_j}{j(j+1)}\right]\right)\right\rceil \\
&\qquad - \left\lceil pN \sin\left(\sum_{j=1}^{N-1}\left[\frac{h_j}{j(j+1)}\right] - \sum_{j=pN}^{N-1}\left[\frac{h_j}{j(j+1)}\right]\right)\right\rceil \\
&\quad \le \left\lceil pN \sin\left(\sum_{j=1}^{N-1}\left[\frac{h_j}{j(j+1)}\right] + p\right)\right\rceil \\
&\qquad - \left\lceil pN \sin\left(\sum_{j=1}^{N-1}\left[\frac{h_j}{j(j+1)}\right] - \sum_{j=pN}^{N-1}\left[\frac{h_j}{j(j+1)}\right]\right)\right\rceil \\
&\quad \le \left\lceil pN \sin\left(\sum_{j=pN}^{N-1}\left[\frac{h_j}{j(j+1)}\right] + p\right)\right\rceil \\
&\quad \le \left\lceil pNh_N\left(\sum_{j=pN}^{N-1}\left[\frac{1}{j(j+1)}\right] + p\right)\right\rceil \qquad \text{since } h_n \text{ is non-decreasing,} \\
&\quad \le \left\lceil pNh_N\left[\frac{1}{pN} - \frac{1}{N}\right] + p\right\rceil \qquad \text{since} \sum_{j=pN}^{N-1}\left[\frac{1}{j(j+1)}\right] = \int_{pN}^{N}\frac{dz}{z^2} = \frac{1}{pN} - \frac{1}{N} \\
&\quad = \lceil (1-p)h_N + p\rceil \\
&\quad \le \lceil (1-p)h_N\rceil + 1 \\
&\quad \le qh_N + 2.
\end{aligned}
$$

If p and q are interchanged, then $q\phi_N - \phi_{qN} \le ph_N + 2$. Add these last equations to obtain $\phi_{n+m} - \phi_n - \phi_m \le h_{n+m} + 4$. In other words, since $h_n = g_n - g_{N_0}$, $\phi_n + 4 - g_{N_0}$ satisfies the generalized subadditive inequality for large enough n, but the limit does not exist.

To see that if $\sum_{n=N_0}^{\infty}\left[g_n/(n(n+1))\right] < \infty$, and if the generalized subadditive inequality is satisfied, then the limit $\lim_{n\to\infty}[a_n/n]$ exists and the bound claimed holds, argue as follows. Define $M_0 = \max\{N_0, M\}$, and let $\phi_n = a_n + g_{N_0}$. Then $\phi_{n+m} \le \phi_n + \phi_m + h_{n+m}$ for all $n \ge M_0$, $m \ge M_0$ and $h_n \ge 0$ and non-decreasing if $n \ge M_0$. Let $p > q > 2M_0 > 0$ be integers, and define

$$M(p,q) = \max\{\phi(q), \phi(q+1), \ldots, \phi(p)\}.$$

Hence $\phi(n) \le M(p,q)$ if $q \le n \le p$. Let r and s be natural numbers with $r > M_0$ and $s > 5r$, and define $s = nr + z$, $3r \le z < 5r$, and $n = 2^{n_1} + 2^{n_2} + \cdots + 2^{n_k}$ with natural numbers $n_1 > n_2 > \cdots > n_k \ge 0$. By the generalized subadditive inequality and the bound on ϕ_n,

$$\begin{aligned}\phi_s &\le \phi_{nr} + \phi_z + h_{nr+z} \\ &\le \phi_{nr} + M(3r, 4r) + h_{4nr},\end{aligned}$$

since $5r \le s \le nr + z < (n+4)r$ which implies $n \ge 2$, and $(n+4)r \le 4nr$. Since h_n is non-decreasing, and $1/x = \sum_{l=x}^{\infty} \left[1/l(l+1)\right]$, it follows that

$$h_{4nr} = 4nr h_{4nr} \sum_{l=4nr}^{\infty} \left[\frac{1}{l(l+1)}\right] \le 4nr \sum_{l=4nr}^{\infty} \left[\frac{h_l}{l(l+1)}\right].$$

Repeated application of the generalized subadditivity relation gives (redefining $\phi_n = \phi(n)$ gives a more convenient notation),

$$\begin{aligned}\phi_{nr} &= \phi((2^{n_1} + 2^{n_2} + \cdots + 2^{n_k})r) \\ &\le \phi(2^{n_1}r) + \phi((2^{n_2} + \cdots + 2^{n_k})r) + h((2^{n_1} + 2^{n_2} + \cdots + 2^{n_k})r) \\ &\le \phi(2^{n_1}r) + \phi((2^{n_2} + \cdots + 2^{n_k})r) + h(2^{n_1+1}r) \\ &\le \cdots \\ &\le \sum_{j=1}^{k} \left[\phi(2^{n_j}r) + h(2^{n_j+1}r)\right]. \qquad (\dagger)\end{aligned}$$

In the above, the term $h(2^{n_k+1}r)$ was inserted, and since this is positive or zero, this is allowed, and the fact that h is non-decreasing was used. The generalized subadditive relation implies that $\phi(2^\nu r) \le 2\phi(2^{\nu-1}r) + h(2^\nu r)$. Multiply this by $2^{n_j-\nu}$ to obtain:

$$\phi(2^\nu r)2^{n_j-\nu} \le 2 \cdot 2^{n_j-\nu}\phi(2^{\nu-1}r) + 2^{n_j-\nu}h(2^\nu r).$$

Sum this last equation over $\nu = 1, 2, \ldots, n_j$. Then

$$\begin{aligned}\sum_{\nu=1}^{n_j} 2^{n_j-\nu}\phi(2^\nu r) &\le \sum_{\nu=1}^{n_j} 2^{n_j-\nu}[2\phi(2^{\nu-1}r) + h(2^\nu r)] \\ &= \sum_{\nu=0}^{n_j-1} 2^{n_j-\nu}\phi(2^{\nu-1}r) + \sum_{\nu=1}^{n_j} 2^{n_j-\nu}h(2^\nu r).\end{aligned}$$

The result is that

$$\phi(2^{n_j}r) \le 2^{n_j}\phi(r) + \sum_{\nu=1}^{n_j} 2^{n_j-\nu}h(2^\nu r).$$

The above is trivially valid if $n_j = 0$, and so is valid for all $n_j \ge 0$. Now substitute this into eqn (†) above. Then

$$\begin{aligned}
\phi(nr) &\le \sum_{j=1}^{k}\left[(2^{n_j}\phi(r) + h(2^{n_j+1}) + \sum_{\nu=1}^{n_j} 2^{n_j-\nu}h(2^\nu r)\right] \\
&\le \sum_{j=1}^{k} 2^{n_j}\phi(r) + \sum_{m=0}^{n_1}\left[h(2^{m+1}r) + \sum_{\nu=1}^{m} 2^{m-\nu}h(2^\nu r)\right] \\
&\le n\phi(r) + \sum_{m=0}^{n_1+1} 2^{n_1+1-m}h(2^m r) \\
&\le n\phi(r) + 2n\sum_{m=1}^{n_1+1} 2^{-m}h(2^m r).
\end{aligned}$$

Lastly, note that

$$\begin{aligned}
\sum_{m=1}^{n_1+1} 2^{-m}h(2^m r) &= \sum_{m=1}^{n_1+1} h(2^m r)2r\left(\sum_{l=2^m r}^{(2^{m+1}r-1)}\left[\frac{1}{l(l+1)}\right]\right) \\
&\le 2r\sum_{m=1}^{n_1+1}\left(\sum_{l=2^m r}^{(2^{m+1}r-1)}\left[\frac{h(l)}{l(l+1)}\right]\right) \\
&= 2r\sum_{l=2r}^{(2^{n_1+2}r-1)}\left[\frac{h(l)}{l(l+1)}\right] \le 2r\sum_{l=2r}^{4nr-1}\left[\frac{h(l)}{l(l+1)}\right].
\end{aligned}$$

Taken together,

$$\phi_s \le n\phi_r + 4nr\sum_{l=2r}^{\infty}\left[\frac{h_l}{l(l+1)}\right] + M(3r, 4r).$$

Divide this by $s = nr + z$ and let $n \to \infty$ with r and z fixed. Then $s \to \infty$ and

$$\limsup_{s\to\infty}\frac{\phi_s}{s} \le \frac{\phi_r}{r} + 4\sum_{l=2r}^{\infty}\left[\frac{h_l}{l(l+1)}\right].$$

If $r \to \infty$, then the existence of the limit is established, and the limit is finite, since the infinite sum above is finite. But $a_n = \phi_r - g(N_0)$, thus

$$\lim_{n\to\infty}\frac{a_n}{n} \le \frac{a_r}{r} + 4\sum_{l=2r}^{\infty}\left[\frac{h_l}{l(l+1)}\right] - \frac{g(N_0)}{r}.$$

Notice that $g(n)$ is non-decreasing, so replace $g(N_0)$ by $g(r)$ in the above to obtain the claimed upper bound on the limit. □

There is a generalization of Lemma A.2, which is useful in some models.

Theorem A.4 *Suppose that a_n is a function from $\mathcal{N}$ into $\mathcal{Z}$, such that $a_n - a_{n-1} = u_n$ where $|u_n| = o(n)$, and suppose that a_n satisfies the generalized subadditivity relation*

$$a_n + a_m \geq a_{n+f_{n,m}}.$$

Suppose furthermore that there is a function ϕ_m, and a constant ϕ such that

$$\sup_{n>0}\{f_{n,m}\} = \phi_m, \qquad \sup_{m>0}\{\phi_m/m\} = \phi$$

and a function ψ_m defined by

$$\inf_{n>0}\{f_{n,m}\} = \psi_m, \qquad \textit{where } \lim_{m\to\infty}[\psi_m/m] = 1.$$

If in addition $\inf_{n>0}\{a_n/n\} \geq c_0$, where c_0 is a constant, then the limit

$$\lim_{n\to\infty}[a_n/n] = \nu$$

exists, and is finite. Moreover, $a_n \geq \psi_n \nu$.

Proof Fix $m > 0$ and recursively define $m_0 = m$, $m_p = m_{p-1} + f_{m_{p-1},m_0}$ for $p = 1, 2, \ldots$. Then

$$a_{m_p} = a_{m_{p-1}} + f_{m_{p-1},m_0} \leq a_{m_{p-1}} + a_{m_0} \leq \cdots \leq p\, a_{m_0}.$$

Choose an n, and let p be the largest integer such that $n \geq m_p$, say $n = m_p + r$. Then $r \leq f_{m_p,m_0} \leq \psi_{m_0}$. Since $a_n - a_{n-1} = u(n) = o(n)$, it also follows that $a_n = a_{n-r} + v_n$, where $v_n = \sum_{k=n-r+1}^{n} u_k = o(n)\, m_0\phi$ (notice that $r \leq m_0\phi$). Thus

$$a_n = a_{n-r} + v_n = a_{m_p} + v_n \leq p\, a_{m_0} + v_n. \qquad (\dagger)$$

Observe that $m_p = \sum_{i=0}^{p-1} f_{m_i,m_0} \geq p\,\psi_{m_0}$. Divide eqn ($\dagger$) by n, and use this lower bound on m_p. This gives

$$\frac{a_n}{n} \leq \frac{a_{m_0}}{\psi_{m_0}} + \frac{v_n}{p\psi_{m_0}}.$$

Take the lim sup as $n \to \infty$ with m_0 fixed. Then $p \to \infty$, and since the p-dependence of v_n is $o(n) = o(m_p) = o(p)$, it follows that

$$\limsup_{n\to\infty} \frac{a_n}{n} \leq \frac{a_{m_0}}{\psi_{m_0}}.$$

Now take the lim inf as $m_0 \to \infty$ on the right-hand side to show the existence of the limit as claimed. Next, the last equation then shows that $a_m \geq \psi_m \nu$, and the finiteness of ν follows from the bound $\inf_{n>0}\{a_n/n\} \geq c_0$. □

Appendix B: Convex functions

Convex (and concave) functions play a special role in the study of the free energies of various models in statistical mechanics. In this appendix, I review the properties of convex functions. If $f(x)$ is a concave function, then $-f(x)$ is convex, and so, with an appropriate reinterpretation, all the results also apply to concave functions. The interested reader can consult the book by G.H. Hardy, J.E. Littlewood and G. Pólya [178], or references [21, 112, 141] and [309] for more details.

A function is said to be *convex* on a closed interval $[a, b]$ in the real line, if $\lambda f(x) + (1-\lambda)f(y) \geq f(\lambda x + (1-\lambda)y)$, for every $a \leq x < y \leq b$, whenever $0 \leq \lambda \leq 1$. Convex functions are continuous if they are finite and defined on a open set; this is shown in the next lemma.

Lemma B.1 *If $f(x)$ is a convex function in $[a, b]$, and $f(x)$ is bounded above in some open interval $I \subseteq (a, b)$, then there exists an interval $(c, d) \supseteq I$ such that $f(x)$ is bounded above and continuous in (c, d), and $f(x) = +\infty$ in $(a, b) \setminus (c, d)$.*

Proof Suppose that $I = (c, d)$ and suppose there exists an e such that $f(e)$ is bounded above and $a < e < c$. Let $z \in (e, c]$ and choose a real number $\mu > 1$ such that $\chi = e+\mu(z-e)$ is in (c, d). Then by the convexity of $f(x)$, $f(z) = f(1/\mu\chi+(1-1/\mu)e) \leq f(\chi)/\mu + (1-1/\mu)f(e)$, and so $f(z)$ is bounded above. So put $c = e$, and continue this process until an $e \in (a, c)$ such that $f(e)$ is bounded above cannot be found. The same process between d and b will show the existence of $(c, d) \subseteq (a, b)$, such that $f(x)$ is bounded above in (c, d) and is $+\infty$ in $(a, b) \setminus (c, d)$.

Let $(p, q) \subseteq (c, d)$ and suppose that $f(x)$ is bounded above by C in (p, q). Let $n > m$ be two natural numbers. Let $x \in (p, q)$ and choose a $\delta > 0$ such that $x \pm n\delta \in (p, q)$. Note that $f(x+\delta) = f(x-\delta) \geq 2f(x)$ by convexity. Then $f(x + m\delta) = f([m(x+n\delta)+(n-m)x]/n) \leq (m/n)f(x+n\delta)+(1-m/n)f(x)$, and thus $[f(x+n\delta) - f(x)]/n \geq [f(x+m\delta) - f(x)]/m$. Replace δ by $-\delta$ and put $m = 1$. Using the fact that C bounds $f(x)$ from above in (p, q) gives

$$\frac{C - f(x)}{n} \geq f(x+\delta) - f(x) \geq f(x) - f(x-\delta) \geq \frac{f(x) - C}{n}.$$

Now let $n \to \infty$ and $\delta \to 0$ such that $x \pm n\delta$ is in (p, q). Then $f(x)$ is continuous in (c, d). □

In applications, it often only proven that a function satisfies the relation $f(x) + f(y) \geq 2f((x+y)/2)$. These functions are convex as well, as the next lemma shows.

Lemma B.2 *Suppose that $f(x) + f(y) \geq 2f((x+y)/2)$ for all $a \leq x \leq y \leq b$. If $f(x)$ is bounded above in some open interval $I \subset (a, b)$, then there exists an interval*

$(c, d) \supseteq I$ *such that* $f(x)$ *is convex and bounded above in* (c, d), *and* $f(x) = +\infty$ *in* $(a, b)\backslash(c, d)$.

Proof Let $m = 2^n$, and consider a set $\{x_1, x_2, \ldots, x_m\}$. Then the hypothesis implies that

$$f(x_1) + f(x_2) + \cdots + f(x_m) \geq mf((x_1 + x_2 + \cdots + x_m)/m). \qquad (*)$$

If this relation is true for $m - 1$, then it will also be true for m equal to any natural number. So suppose that $\{x_1, x_2, \ldots, x_{m-1}\}$ are given, and suppose the inequality is true for m. Define $x_m = (x_1 + x_2 + \cdots + x_{m-1})/(m - 1)$ and observe that

$$\begin{aligned} f(x_m) &= f([(m-1)x_m + x_m]/m) \\ &= f((x_1 + x_2 + \cdots + x_m)/m) \\ &\leq \frac{1}{m}(f(x_1) + f(x_2) + \cdots + f(x_m)). \end{aligned}$$

Thus, eqn $(*)$ is true for any natural number m. If $p+q = m$, and $x = x_1 = x_2 = \cdots = x_p$ and $y = x_{p+1} = x_{p+2} = \cdots = x_m$, then an immediate consequence of eqn $(*)$ is that

$$\lambda_p f(x) + (1 - \lambda_p) f(y) \geq f(\lambda_p x + (1 - \lambda_p) y), \qquad (**)$$

for any rational number $\lambda_p = p/m$. If $f(x)$ is continuous, then the proof is finished, since the limit can be taken through the rationals to any real number. Since $f(x)$ is not necessarily continuous, more work is necessary.

The interval (c, d) can be enlarged as follows. Suppose that there exists a number e, such that $f(e) < \infty$ and $a < e < c$. Let $x \in (e, c)$, and choose integers $p > q$ such that $\chi = e + p(x - e)/q$ is in (c, d). Then, by $(**)$,

$$f(x) = f((q\chi + (p - q)e)/p) \leq (q/p)f(\chi) + ((p - q)/p)f(e) < \infty.$$

Thus $f(x) < \infty$, and put $c = e$ to find $f(x)$ bounded in an enlarged interval (c, d). Similarly, "move" d towards b until $f(x)$ becomes infinite. Hence, one can assume that $f(x)$ is bounded inside (c, d) and infinite in $(a, b) \setminus (c, d)$.

To see that $f(x)$ is continuous in (c, d), argue as follows. Suppose that $f(x) < C$ in (c, d), and let $n > m$ be two natural numbers. Let $\delta > 0$ be a small real number, such that $x + n\delta \in (c, d)$ if $x \in (c, d)$. Then $f(x + m\delta) = f([m(x + n\delta) + (n - m)x]/n) \leq (m/n)f(x + n\delta) + ((n - m)/n)f(x)$. Thus, $[f(x + n\delta) - f(x)]/n \geq [f(x + m\delta) - f(x)]/m$. If $\delta = -\delta$, then $[f(x) - f(x - m\delta)]/m \geq [f(x) - f(x - n\delta)]/n$ instead. Using the bound on $f(x)$ the result is that (while noting that $f(x + \delta) + f(x - \delta) \geq 2f(x)$)

$$\frac{C - f(x)}{n} \geq f(x + \delta) - f(x) \geq f(x) - f(x - \delta) \geq \frac{f(x) - C}{n},$$

and where $m = 1$. Now take $\delta \to 0$ and $n \to \infty$ such that $x \pm n\delta \in (c, d)$. Then $f(x + \delta) \to f(x)$ and $f(x - \delta) \to f(x)$, so $f(x)$ is continuous in (c, d). Finally, let $\{m/n\}$ be a sequence of rational numbers in $[0, 1]$ converging to $\lambda \in [0, 1]$. Since $f(x)$ is continuous, the limit of $(m/n)f(x) + (1 - m/n)f(y) \geq f((m/n)x + (1 - m/n)y)$

can be taken to obtain $\lambda f(x) + (1-\lambda) f(y) \geq f(\lambda x + (1-\lambda) y)$. Hence $f(x)$ is convex in (c, d). □

Finite convex functions have left- and right-derivatives everywhere.

Lemma B.3 *If $f[a, b] \to \mathcal{R}$ is a convex function and is finite in (a, b), then $d^- f(x)/dx$ and $d^+ f(x)/dx$ exist everywhere in (a, b). Moreover,*

$$-\infty < \frac{d^-}{dx} f(x) \leq \frac{d^+}{dx} f(x) < \infty$$

and both the left- and right-derivatives are non-decreasing with increasing x.

Proof Suppose that $h > 0$ and define $q(x, h) = f(x+h) - f(x)$. Then

$$\begin{aligned} q(x + h/n, h) - q(x, h) &= \sum_{i=0}^{n-1} (f(x + (i+2)h/n) - 2f(x + (i+1)h/n) \\ &\qquad + f(x + ih/n)) \\ &\geq 0, \end{aligned}$$

since $f(x)$ is convex. Repeating this gives a sequence of inequalities $q(x, h) \leq q(x + h/n, h) \leq \cdots \leq q(x + mh/n, h)$. Let m/n be a sequence of rational numbers which converge to δ/h, for some $\delta > 0$. Then by the continuity of $f(x)$, one finds $q(x, h) \leq q(x + \delta, h)$, and $q(x, h)$ is an non-decreasing function of x. Observe now that $q(x, h/n) \leq q(x + h/n, h/n) \leq \cdots \leq q(x + (n-1)h/n, h/n)$, so that for $0 < m < n$:

$$\frac{1}{m} \sum_{i=0}^{m-1} q(x + ih/n, h/n) \leq \frac{1}{n} \sum_{i=0}^{n-1} q(x + ih/n, h/n).$$

Evaluating the above gives

$$\frac{f(x + mh/n) - f(x)}{mh/n} \leq \frac{f(x+h) - f(x)}{h}.$$

Let $0 < g < h$ and suppose that m/n is a sequence of rational numbers which converges to g/h. Then from the above and for $x < y$, and since $q(x, h)$ is increasing with x,

$$\frac{f(x+g) - f(x)}{g} \leq \frac{f(x+h) - f(x)}{h} \leq \frac{f(y+h) - f(y)}{h}.$$

For fixed x, $[f(x+h) - f(x)]/h$ is a non-decreasing function of h and $\lim_{h \to 0^+} [f(x+h) - f(x)]/h$ exists and may be infinite, for every $x \in (a, b)$. By taking $g \to 0^+$, $d^+ f(x)/dx < \infty$.

If $h < 0$, then an analysis similar to the above gives, for $0 > g > h$ and $x > y$

$$\frac{f(x+g) - f(x)}{g} \geq \frac{f(x+h) - f(x)}{h} \geq \frac{f(y+h) - f(y)}{h}.$$

Take $h \to 0^-$ to see that $\lim_{h \to 0^-} [f(x+h) - f(x)]/h$ exists. On the other hand, if $g \to 0^-$, then $d^- f(x)/dx > -\infty$.

Finally, if $g > 0$, then

$$\frac{f(x-g)-f(x)}{g} \leq \frac{f(x)-f(x+g)}{g} = \frac{f(x+g)-f(x)}{g}.$$

Take $g \to 0^+$ to get

$$-\infty < \frac{d^-}{dx} f(x) \leq \frac{d^+}{dx} f(x) < \infty. \qquad \square$$

Convergent sequences of convex functions have a convex limit.

Lemma B.4 *Suppose that $f_n : [a,b] \to \mathcal{R}$, $n = 1, 2, \ldots$, is a sequence of convex functions converging pointwise to a limit $f : [a,b] \to \mathcal{R}$. Then $f(x)$ is convex in $[a,b]$.*

Proof Suppose that $f(x)$ is not convex, in other words, there exist points c and d in $[a,b]$ such that $f(c) + f(d) < 2f((c+d)/2)$. Define $A = 2f((c+d)/2) - f(c) - f(d) > 0$. Since $f_n(x)$ is convex, $f_n(c) + f_n(d) \geq 2f_n((c+d)/2)$. For every $\epsilon > 0$ there exist natural numbers N_c, N_d and N such that, (1) for all $n > N_c$, $|f_n(c) - f(c)| < \epsilon$, (2) for all $n > N_d$, $|f_n(d) - f(d)| < \epsilon$, and (3) for all $n > N$, $|f_n((c+d)/2) - f((c+d)/2)| < \epsilon$. But $|f_n(c) - f(c)| + |f_n(d) - f(d)| \geq |f_n(c) + f_n(d) - f(c) - f(d)| \geq 2[f_n((c+d)/2) - f((c+d)/2)] + A$. Take $n > \max\{N_c, N_d, n\}$, and maximize the first term and minimize the last term in the previous expression. This gives $2\epsilon \geq A - 2\epsilon$. Now take $n \to \infty$. Then $\epsilon \to 0^+$ which implies $A \to 0^+$, hence $f(x)$ is convex on $[a,b]$. $\square$

Convex functions have another property which is important in applications to statistical mechanics. They are differentiable almost everywhere. The proof of this fact relies on the notion of a Vitali covering, and on Vitali's theorem. Vitali's theorem is a basic result in measure theory, and I shall not prove it here.

Let $I = [a,b]$ be an interval. The length of I is $l(I) = b - a$. A set E is covered *in the sense of Vitali*, if there is an (infinite) collection of intervals $\mathcal{I} = \{I_i\}$ such that $E \subset \bigcup_i I_i$ and for every $x \in E$ and $\epsilon > 0$ there is an $I \in \mathcal{I}$ such that $x \in I$ and $l(I) < \epsilon$. Vitali's theorem is an important result on measurable sets in the real line (see for example reference [309]).

Theorem B.5 (Vitali's thoerem) *Let E be a finite measure subset of the real line, and let $\mathcal{I}$ be a collection of intervals which cover E in the sense of Vitali. Given an $\epsilon > 0$, there is a* finite, disjoint *collection of intervals in $\mathcal{I}$, say $\{I_1, I_2, \ldots, I_N\}$, such that*

$$\mu^*\big[E \setminus \cup_{i=1}^N I_n\big] < \epsilon. \qquad \square$$

In other words, in a Vitali covering of a finite measure set, there is a finite disjoint collection of intervals which almost covers. In the next theorem a proof that all non-decreasing real-valued functions are differentiable almost everywhere is given. As a corollary it follows that convex functions are differentiable almost everywhere. To see this, observe that if $x \leq y \leq z$, and f is a convex function such that $f(x) \leq f(y)$, then $f(y) \leq f(z)$. To demonstrate this, put $\lambda = (y-z)/(x-z)$ and consider $\lambda f(x) + (1-\lambda) f(z) \geq f(y)$. Now replace $f(x)$ by $f(y)$ in this to find that $f(y) \leq f(z)$.

A similar argument can be made for the case that $x \geq y \geq z$ and $f(x) \leq f(y)$. Thus, a convex function is either non-increasing, or non-decreasing or first non-increasing, and then non-decreasing. Thus, it is enough to consider non-decreasing functions in what follows. The proof below follows the general outline in the book by H.L. Royden [309].

Theorem B.6 *Let f be an non-decreasing real-valued function on $[a, b]$. Then f is differentiable almost everywhere. Moreover, f' is measurable.*

Proof Define the following (Dini)-derivatives of f:

$$D^+ f(x) = \limsup_{h \to 0^+} \frac{f(x+h) - f(x)}{h}, \qquad D^- f(x) = \limsup_{h \to 0^+} \frac{f(x) - f(x-h)}{h};$$

$$D_+ f(x) = \liminf_{h \to 0^+} \frac{f(x+h) - f(x)}{h}, \qquad D_- f(x) = \liminf_{h \to 0^+} \frac{f(x) - f(x-h)}{h}.$$

Since f is non-decreasing, $D^+ f(x) \geq D^- f(x)$, etc. For convex functions, Lemma B.3 states that $D^+ f(x) = D_+ f(x)$ and $D^- f(x) = D_- f(x)$ almost everywhere. To proceed, define the sets

$$E_{u,v} = \{x \big| D^+ f(x) > u > v > D_- f(x)\},$$

where u, v are rational numbers. The union of all these sets is the set where $D^+ f(x) > D_- f(x)$. I shall show that the Lebesgue outer measure of each $E_{u,v}$ is zero, and hence, $D^+ f(x) > D_- f(x)$ only on a set of zero measure. The other combinations of the Dini-derivatives are similarly dealt with.

Let $\mu^*(E_{u,v}) = s$ be the Lebesgue outer measure of $E_{u,v}$, and let $\epsilon > 0$ be given. Then there is an open set U containing $E_{u,v}$ such that $\mu^*(U) < s + \epsilon$. For each $x \in E_{u,v}$ there is an interval $[x - h, x] \subset U$ such that (by definition of $D_- f(x)$),

$$f(x) - f(x-h) < vh.$$

The collection of these intervals is a Vitali covering, and so there is a finite disjoint collection $\{I_1, I_2, \ldots, I_N\}$ which almost covers $E_{u,v}$. The interiors of these I_i cover a subset A of U such that $\mu^*(A) > s - \epsilon$. Let $l(I_i) = h_i$ and sum over the intervals:

$$\sum_{i=1}^{N} [f(x_i) - f(x_i - h)] < v \sum_{i=1}^{N} h_n < v\mu^*(A) < v(s + \epsilon).$$

Each point $y \in A$ is the left-hand point of an arbitrarily small open interval $(y, y + k)$ which is contained in some I_n, and for which

$$f(y+k) - f(y) > uk,$$

by definition of $D^+ f(y)$. These intervals are a Vitali covering of A, and so there is a finite disjoint collection $\{J_1, J_2, \ldots, J_M\}$ which almost covers A; their union covers a subset of A of Lebesgue outer measure $s - 2\epsilon$. If $l(J_i) = k_i$, then

$$\sum_{i=1}^{M}[f(y_i + k_i) - f(y_i)] > u \sum_{i=1}^{M} k_i > u(s - 2\epsilon).$$

Each of the J_i is contained in some interval I_n; now sum over those $J_i \subset I_n$. This gives

$$\sum_{i=1}^{N}[f(y_i + k_i) - f(y_i)] \le f(x_n) - f(x_n - h_n),$$

since f is increasing. Thus

$$\sum_{i=1}^{N}[f(x_i) - f(x_i - h)] \ge \sum_{i=1}^{M}[f(y_i + k_i) - f(y_i)],$$

and so $v(s + \epsilon) \ge u(s - 2\epsilon)$. Take $\epsilon \to 0^+$. This shows that $vs \ge us$. Since $u > v$, this implies that $s = 0$, so that $\mu^*(E_{u,v}) = 0$.

Thus, the function

$$g(x) = \lim_{h \to 0} \frac{f(x + h) - f(x)}{h},$$

is defined almost everywhere, and f is differentiable almost everywhere, whenever g is finite. Define $g_n(x) = n[f(x + 1/n) - f(x)]$, and set $f(x) = f(b)$ if $x \ge b$. Then $g_n(x) \to g(x)$ for almost all x, thus g is measurable. Since $f(x)$ is increasing, it also follows that $g_n(x) \ge 0$. By Fatou's lemma

$$\begin{aligned}
\int_a^b g \le \liminf_{n \to \infty} \int_a^b g_n &= \liminf_{n \to \infty} n \int_a^b [f(x + 1/n) - f(x)] \\
&= \liminf_{n \to \infty} \left[n \int_b^{b+1/n} f - n \int_a^{a+1/n} f \right] \\
&= \liminf_{n \to \infty} \left[f(b) - n \int_a^{a+1/n} f \right] \\
&\le f(b) - f(a)
\end{aligned}$$

Thus, g is integrable, and therefore measurable. Since $f' = g$ almost everywhere in $[a, b]$, this shows that f' is measurable. □

The fact that a convex function is differentiable almost everywhere has a useful consequence. If a sequence of convex functions $\{f_i(x)\}$ has a convex limit $f(x)$, then the derivative of $f(x)$ exists almost everywhere, and it is equal to the sequence of derivatives $\{df_i(x)/dx\}$, which exists almost everywhere, see reference [112].

Theorem B.7 *Suppose that the sequence of convex functions $\{f_n\}$ approaches a limit f. Then the right- and left-derivatives satisfy*

$$\frac{d^-}{dx}f(x) \leq \liminf_{n\to\infty} \frac{d^-}{dx}f_n(x) \leq \limsup_{n\to\infty} \frac{d^+}{dx}f_n(x) \leq \frac{d^+}{dx}f(x).$$

Moreover, this implies that

$$\lim_{n\to\infty} \frac{d}{dx}f_n(x) = \frac{d}{dx}f(x)$$

almost everywhere.

Proof By Lemma B.4, f is convex and is differentiable almost everywhere; thus f_n and f have right- and left-derivatives everywhere. Choose a $\delta > 0$ and an $\epsilon > 0$, and notice that there exists an N_0 such that for all $n > N_0$, the Cauchy condition on the convergent sequence of functions implies

$$\begin{aligned} |f_n(x+\delta) - f(x+\delta)| &\leq \epsilon; \\ |f_n(x) - f(x)| &\leq \epsilon; \\ |f_n(x-\delta) - f(x-\delta)| &\leq \epsilon. \end{aligned}$$

But since f_n is convex, for all $n > N_0$,

$$\begin{aligned} \frac{d^-}{dx}f_n(x) &\geq \frac{f_n(x) - f_n(x-h)}{h} \geq \frac{f(x) - f(x-h) + 2\epsilon}{h}; \\ \frac{d^+}{dx}f_n(x) &\leq \frac{f_n(x+h) - f_n(x)}{h} \leq \frac{f(x+h) - f(x) - 2\epsilon}{h}. \end{aligned}$$

Let $n \to \infty$, and then take $\epsilon \to 0^+$. If $h \to 0^+$ as well, then

$$\begin{aligned} \liminf_{n\to\infty} \frac{d^-}{dx}f_n(x) &\geq \frac{d^-}{dx}f(x); \\ \limsup_{n\to\infty} \frac{d^+}{dx}f_n(x) &\leq \frac{d^+}{dx}f(x). \end{aligned}$$

This completes the proof. □

Appendix C: Asymptotics for q-factorials

In this appendix the asymptotic behaviour of a q-product $(t; q)_\infty$ as q approaches 1^- is examined. Various sources exists for the results here, see for example [265], and [294]. The strategy is as follows. Note that $\log(t; q)_\infty = \sum_{n=1}^{\infty} \log(1 - tq^n)$. Estimate this infinite series by using the Euler–MacLaurin formula (see for example [177]) and examine the remainder term.

Lemma C.1 (Euler–MacLaurin) *If $f \in C^{2m}[0, N]$ for an integer $N > 0$, then*

$$\sum_{n=0}^{N} f(N) = \int_0^N f(x)dx + [f(0) - f(N)]/2$$
$$+ \sum_{n=1}^{m-1} \frac{B_{2n}}{(2n)!}\left(f^{(2n-1)}(N) - f^{(2n-1)}(0)\right) + R_m,$$

where

$$R_m = \int_0^N \frac{B_{2m} - B_{2m}(x - [x])}{(2m)!} f^{(2m)}(x)dx.$$

$B_n(x)$ is the n-th Bernoulli polynomial, and $B_n = B_n(0)$. m can be taken to be any positive integer. □

The functions $B_m(x)$ are Bernoulli polynomials [141, 177], and in this context they are defined as

$$B_m(x) = \sum_{k=0}^{m} \binom{m}{k} B_k x^{m-k},$$

where the numbers B_k are Bernoulli numbers as defined in the expansion $s/(e^s - 1) = \sum_{k=0}^{\infty}[B_k s^n/n!]$. Thus, $B_0(x) = 1$, $B_1(x) = x + 1/6$ and $B_2(x) = x^2 + x/3 + 1/30$.

There already exist asymptotic formulas for q-products of the form $(q; q)_\infty$, due to G.H. Hardy[1] [265].

[1] See for example reference [176]; the following conjugate modulus transformation is found there:

$$(q; q)_\infty (r; r)_\infty = \left(\frac{r}{q}\right)^{1/24} \sqrt{\frac{2\pi}{-\log q}},$$

where $r = \exp(4\pi^2/\log q)$ and $0 < q < 1$.

Lemma C.2 *Let $r = \exp(4\pi^2/\log q)$, where $0 < q < 1$. Then*

$$(q;q)_\infty = (r/q)^{1/24} \sum_{n=-\infty}^{\infty} \left[r^{n(6n+1)} - r^{(3n+1)(2n+1)}\right] \sqrt{\frac{2\pi}{-\log q}}.$$

Taking logarithms gives the following asymptotic approximation:

$$\log(q;q)_\infty = \frac{\pi^2}{6\log q} + \frac{1}{2}\log\left[\frac{2\pi}{-\log q}\right] + O(\log q). \qquad \square$$

Apply Lemma C.1 to $\log(t;q)_N$ to find an asymptotic expansion: I use the approach due to T. Prellberg [294], but only keep the first term in the Euler–MacLaurin series. In other words, put $m = 1$ in Lemma C.1, and examine the remainder term R_1.

Lemma C.3 *If $0 < t < 1$, and $0 < q < 1$, then*

$$\log(t;q)_{N+1} = \frac{1}{\log q}\big(\mathcal{L}i_2(t) - \mathcal{L}i_2(tq^N)\big) + \frac{1}{2}\big(\log(1-t) - \log(1-tq^N)\big) + R_1,$$

where

$$|R_1| \le \frac{4}{3}|t|^2 |\log q| \left|\frac{1-q^N}{(1-t)(1-tq^N)}\right|.$$

Proof Direct application of the Euler–MacLaurin formula gives

$$\log(t;q)_{N+1} = \int_0^N \log(1-tq^x)dx + \frac{1}{2}\big[\log(1-t) + \log(1-tq^N)\big] + R_1,$$

where

$$R_1 = \int_0^N \frac{1}{2}\big[B_2 - B_2(x - \lfloor x \rfloor)\big]\left[\frac{d^2}{dx^2}\log(1-tq^x)\right]dx.$$

The integral over $\log(1-tq^x)$ can be done by first expanding the logarithm and then integrating term by term; the result is a sequence of terms which gives the dilogarithms and logarithms in the result, and all that is left is to examine the remainder term. R_1 is given by

$$R_1 = \int_0^N \big([B_2 - B_2(x - \lfloor x \rfloor)]/2\big)\left[\frac{d^2}{dx^2}\log(1-tq^x)\right]dx.$$

First note that $\big|B_2 - B_2(x - \lfloor x \rfloor)\big| = \big|-(x-\lfloor x \rfloor)/3 - (x - \lfloor x \rfloor)^2\big| \le 4/3$ if $x \in [0, N]$.

Evaluating the second derivative,

$$\begin{aligned}|R_1| &\le \int_0^N \left| \left([B_2 - B_2(x - \lfloor x \rfloor)]/2 \right) \left[\frac{tq^x (\log q)^2}{(1 - tq^x)^2} \right] \right| dx \\ &\le \frac{2}{3} \int_0^N \left| \frac{tq^x (\log q)^2}{(1 - tq^x)^2} \right| dx \\ &= \frac{2}{3} t \log q \int_{tq^N}^t \frac{dz}{(1 - z)^2}, \\ &= \frac{2}{3} t^2 \log q \left[\frac{1 - q^N}{(1 - t)(1 - tq^N)} \right].\end{aligned}$$

For fixed $t < 1$, this term vanishes as $q \nearrow 1^-$. □

In the limit that $N \to \infty$ in the above lemma, the asymptotic behaviour of $\log(t; q)_\infty$ is given by Corollary C.4.

Corollary C.4 *For every t, such that $0 \le t \le 1 - \epsilon$, where $0 < \epsilon < 1$, and for $q \in (0, 1)$,*

$$\log(t; q)_\infty = (\mathcal{L}i_2(t) - \pi^2/6)/\log q + \log(1 - t)/2 + R_1,$$

or if the relation $\mathcal{L}i_2(t) + \mathcal{L}i_2(1 - t) = \pi^2/6 - \log t \log(1 - t)$ is used, then

$$\log(t; q)_\infty = -[\mathcal{L}i_2(1 - t) - \log(t)\log(1 - t)]/\log q + \log(1 - t)/2 + R_1,$$

where R_1 approaches zero uniformly as $q \nearrow 1^-$ for any $t < 1-\epsilon$ (and where $1 > \epsilon > 0$). □

The asymptotic properties and behaviour of the q-product $(t; q)_N$, as $N \to \infty$, is also of interest. In the first instance, if $q < 1$ and if $0 < tq^i < 1$ for all $y \ge 1$, then the limit $(t; q)_\infty$ is itself non-zero.

Lemma C.5 *If $q < 1$ and $0 < tq^i < 1$ for all $i \ge 1$, then*

$$0 < (t; q)_\infty \le 1.$$

Proof Surely, $0 \le (t; q)_N \le 1$ for all $N \ge 0$, and $(t; q)_N$ is monotone decreasing, so the limit exists and is less than or equal to 1. To see that it is not zero, use the following. For any positive $p < 1$,

$$\log(1 - p) \ge \frac{-p}{1 - p}.$$

Moreover, for any small $\epsilon > 0$, there exists a natural number N_0 such that $0 < 1 - \epsilon < 1 - tq^i$ if $i > N_0$, since $q < 1$ and since $t \ne 0$. Thus,

$$\log(1 - tq^i) \ge \frac{-tq^i}{1 - \epsilon}, \qquad \text{if } i > N_0.$$

Thus,

$$
\begin{aligned}
(t;q)_N &= \exp\left[\sum_{i=0}^{N-1} \log(1-tq^i)\right] \\
&\geq \exp\left[\sum_{i=0}^{N_0} \log(1-tq^i) + \frac{-t}{1-\epsilon}\sum_{i=N_0+1}^{N-1} q^i\right], \qquad \text{if } N > N_0+1 \\
&= \exp\left[\sum_{i=0}^{N_0} \log(1-tq^i) + \frac{-tq}{(1-\epsilon)}\frac{(1-q^{N-N_0-1})}{(1-q)}\right]
\end{aligned}
$$

By taking $N \to \infty$, one obtains

$$
\lim_{N\to\infty} (t;q)_N \geq \exp\left[\sum_{i=0}^{N_0} \log(1-tq^i) + \frac{-tq}{(1-\epsilon)(1-q)}\right]
$$

which is bigger than zero. □

Appendix D: Bond or edge percolation

In this appendix I review the essential theory of percolation, as it relates to application elsewhere in this monograph. Extensive reviews of percolation can be found in references [106, 330] and in the book by G. Grimmett [147].

Let $p \in [0, 1]$ and let $q = 1 - p$. Let the set of all unit-length edges in $\mathcal{Z}^d$ be $\mathcal{E}^d$. An edge in $\mathcal{E}^d$ is *open* with probability p and *closed* otherwise. The parameter p is also called the *density* of the process. In other words, let μ_b be the Bernoulli measure on the set $\{0, 1\}$, and define the space $\Pi = \prod_{e \in \mathcal{E}^d} \{0, 1\}$. If $\tau \in \Pi$ then $\mu_b(\tau(e) = 1) = p$ and $\mu_b(\tau(e) = 0) = q$, where it is understood that $P_p = \prod_{e \in \mathcal{E}^d} \mu_b$ is the product measure with density p on the sample space Π. If the σ-algebra $\mathcal{F}$ of subsets of Π is defined, then the probability space $(\Pi, \mathcal{F}, P_p)$ is obtained. The random graph consisting of all the vertices in $\mathcal{Z}^d$ and all the open edges in $\mathcal{E}^d$ consists of components which are called *percolation clusters*. The cluster containing a vertex x will be indicated by $C(x)$, and since the probability measure P_p is translationally invarariant, the distribution of $C(x)$ is independent of x. Thus, one should only be concerned with the open cluster $C(0) = C$ at the origin. The number of edges in the open cluster at the origin is its size, denoted by $|C|$, and the order of the cluster is the number of vertices its contains, denoted by $\|C\|$.

The *percolation probability* $\theta(p)$ is the probability that the origin (or any other vertex) belongs to a cluster of infinite size. If the probability $P_n(p) = P_p(|C| = n)$ that the origin belongs to a cluster of size n is defined, then

$$\theta(p) = P_\infty(p) = 1 - \sum_{n=0}^{\infty} P_n(p). \tag{D.1}$$

The fundamental theorem of percolation states that in two or higher dimensions there exists a $p_c \in (0, 1)$ such that $\theta(p) = 0$ if $p < p_c$ and $\theta(p) > 0$ if $p > p_c$ [36]. The probability p_c is called the *critical percolation probability*, and is defined by

$$p_c = \sup\{p \,|\, \theta(p) = 0\}. \tag{D.2}$$

In d dimensions the percolation probability is denoted by $\theta_d(p)$, and the critical percolation probability is denoted by $p_c(d)$. Since the d-dimensional lattice is a sublattice of the $(d+1)$-dimensional lattice, the existence of an infinite cluster in d dimensions implies an infinite cluster in $(d+1)$ dimensions, so that

$$p_c(d+1) \leq p_c(d), \quad \forall d \geq 1. \tag{D.3}$$

Notice also that $p_c(1) = 1$.

The mean size of an open cluster is defined by $\chi(p) = E_p|C|$, where E_p is the expectation at density p. In terms of $|C|$, this becomes

$$\chi(p) = \infty \cdot P_\infty(p) + \sum_{n=0}^{\infty} n P_n(p). \tag{D.4}$$

Thus, $\chi(p) = \infty$ if $p > p_c$.

Theorem D.1 (Fundamental theorem of percolation) *If $d \geq 2$ then there exists a critical percolation probability $p_c(d)$ in $(0, 1)$.*

Proof Since $p_c(d+1) \leq p_c(d)$, it is only necessary to show that $p_c(2) < 1$ and $p_c(d) > 0$. To see that $p_c(2) < 1$ it is only necessary to show that $\theta(p) > 0$ if p is close enough to 1. The *planar dual* of the lattice $\mathcal{E}^2$ is found by creating vertices in the centre of each face (open unit square) in $\mathcal{E}^2$, and by joining these vertices if they are in adjacent faces. These new edges are in one-to-one correspondence with the edges in $\mathcal{E}^2$, and they are called *dual edges*. Let C be again the open cluster at the origin, and suppose that C is finite. The *perimeter edges* of C are all those closed edges which have at least one end-point in C, while *solvent* edges are all those closed edges with exactly one end-point in C. The union of the dual edges of the solvent edges of C is a lattice polygon in the dual lattice which contains C in its interior. Let $p_n(C)$ be the number of polygons of length n in the dual lattice and which contains the origin in its interior. Since each polygon counted by $p_n(C)$ passes through the x-axis, it must contain a point of the form $(k + 1/2, 1/2)$, where $|k| \leq n$. Choose this point as the root in the polygon. Since k can take at most $2n$ values, $p_n(C)$ is less than the number of rooted polygons of length n (denoted by $p_n(r)$) multiplied by $2n$: $p_n(C) \leq 2np_n(r)$. If p_n is the number of polygons of length n counted modulo translation (and with no root), then $p_n(r) \leq np_n$. Thus $p_n(C) \leq 2n^2 p_n$. Let A be a polygon in the dual lattice which contains the origin. The probability that A has finite length (and so is closed) is at most

$$\sum_{n=1}^{\infty} (1-p)^n p_n(C) \leq \sum_{n=1}^{\infty} 2n^2 p_n (1-p)^n.$$

Since $p_n = \mu_2^{n+o(n)}$ (by Theorem 1.2), this is finite if $(1-p) < \mu_2^{-1}$, and the sum approaches zero as $(1-p) \to 0^+$. In other words, there is a number $\epsilon < 1$ such that $\sum_{n=1}^{\infty} (1-p)^n p_n(C) < 1/2$ if $p > \epsilon$. Thus $P_\infty(p) \geq 1 - \sum_A P(A \text{ is closed}) \geq 1/2$ if $p > \epsilon$. Thus, $p_c(2) \leq \epsilon < 1$. This argument involving polygons around finite clusters at the origin is called a "Peierls argument".

Next, it must be shown that $p_c(d) > 0$. This is done by showing that $\theta(p) = 0$ whenever $p > 0$ is close enough to 0. Let $M(n)$ be the number of paths from the origin composed of open edges. Since each path is open with probability p^n, the expected number of open paths in $M(n)$ is $E_p(M(n)) = p^n c_n$, where c_n is the number of walks from the origin. If the origin is in an infinite open cluster, then there exists open paths of all lengths from the origin, and $\theta(p) \leq E_p(M(n)) = p^n c_n$, for all n. Since $c_n = \mu_d^{n+o(n)}$, $\theta(p) \leq p^n c_n \to 0$ if $p < \mu_d^{-1}$. Thus $p_c(d) \geq \mu_d^{-1} > 0$ for all d. □

In the applications of percolation to models of animals, more than the above is needed. In particular, the limit $\lim_{n\to\infty}[P_n(p)]^{1/n}$ must be studied, and thus the asymptotic behaviour of $P_n(p)$ is really the concern. Since $1 \geq \theta(p) > 0$ if $p > p_c$, it is already apparent that

$$\limsup_{n\to\infty}[P_n(p)]^{1/n} = 1, \quad \forall p > p_c, \tag{D.5}$$

but not much is known yet if $p < p_c$. The existence of the limit in eqn (D.5) for all values of p follows from concatenation, which is done in Theorem D.2. The proof is due to H. Kunz and B. Souillard [221], and it is adapted here to the ensemble of interest here.

Theorem D.2 *There exists a finite function* $\zeta(p) \geq 0$ *such that*

$$\zeta(p) = -\lim_{n\to\infty}\frac{1}{n}\log P_n(p).$$

In addition,

$$P_n(p) \leq (n+1)e^{-n\zeta(p)}, \quad \textit{for all } n \geq 1.$$

Proof Let $P(C)$ be the probability for a cluster C of open edges. Define the top and bottom vertices in C as the lexicographically most and least vertices respectively. Let σ be a cluster with n open edges and with its top vertex t equal to the origin, and let τ be a cluster with m open edges and with its bottom vertex b equal to the origin. Concatenation of σ and τ gives $P(\sigma * \tau) = P(\sigma)P(\tau)$, where $\sigma * \tau$ is the cluster of $n+m$ open edges, composed of the open edges of σ and τ. If the sum over all σ with n open edges and with the top vertex equal to the origin is taken, then the probability that the open cluster at the origin has n edges *and* has its top vertex at the origin is found. Let this probability be $P'_n(p) = \sum_\sigma P(\sigma)$. Note that $\sum_{\sigma,\tau} P(\sigma * \tau) \leq P'_{n+m}(p)$, since not all open clusters with $n+m$ edges and its top vertex at the origin, can be decomposed into a σ and a τ as above. Thus $P'_n(p)P'_m(p) \leq P'_{n+m}(p)$, which shows that $P'_n(p)$ is a supermultiplicative inequality. Lemma A.1 shows that $\zeta(p) = -\lim_{n\to\infty}[\log P'_n(p)]/n$ exists, and since $P'_n(p) \leq 1$, $\zeta(p) \geq 0$. Note also that $P'_n(p) \geq p^n(1-p)^{2n+4}$, so that $\zeta(p) < \infty$. A further consequence of Lemma A.1 is that $P'_n(p) \leq e^{-n\zeta(p)}$, and since $(n+1)P'_n(p) \geq P_n(p)$, the claimed bound is obtained. Lastly, notice also that $P'_n(p) \leq P_n(p)$. This shows that $\zeta(p) = -\lim_{n\to\infty}[\log P_n(p)]/n$. □

Thus, from eqn (D.5) it is also the case that $\zeta(p) = 0$ if $p > p_c$. To see that $\zeta(p) > 0$ if $p < p_c$ requires much more work. The standard proof is due to M. Aizenman and C.M. Newman [2, 147]. First, *events* must be defined, and in particular, the notion of an *increasing event* will be important. Secondly, the BK inequality will play a key role. Thirdly, I shall shift the focus from $P_n(p)$ (the probability that the cluster at the origin has n edges) to $P_v(p)$ (the probability that the cluster at the origin has v vertices, this is defined by $P_v(p) = P_p(\|C\| = v)$). Notice that

$$P_n(p) = \sum_{k,c} v a_n(k,c) p^n q^\rho, \tag{D.6}$$

where $a_n(k,c)$ is the number of lattice animals with n edges, k nearest-neighbour contacts and c cycles (and consequently with $v = n+1-c$ vertices and $\rho = 2dv - 2n - k$

perimeter edges). Now notice that any animal with n edges has at least $\lfloor n/d \rfloor$ vertices, and at most $n+1$ vertices. Thus, if the probability that the cluster at the origin has v vertices is $P_v(p)$, then $P_n(p) \leq \sum_{v=\lfloor n/d \rfloor}^{n+1} P_v(p)$. Let v_M be that value of v which maximizes $P_v(p)$ in $\{\lfloor n/d \rfloor, \lfloor n/d \rfloor + 1, \ldots, n+1\}$. Define $\limsup_{n\to\infty} v_M/n = \alpha_m$; then

$$\lim_{n\to\infty} \frac{1}{n} \log P_n(p) \leq \alpha_m \lim_{v\to\infty} \frac{1}{v} \log P_v(p), \tag{D.7}$$

since the limit exists by the same construction as in Lemma D.2. On the other hand, if an animal has v vertices, c cycles and k contacts, then it has at least $v-1$ edges, and at most $(d-1)v$ edges, so that $P_v(p) \leq \sum_{n=v-1}^{(d-1)v} P_n(p)$. Again, define n_M as that value of n in $\{v-1, v, \ldots, (d-1)v\}$ which maximizes $P_n(p)$, and define $\limsup_{v\to\infty} n_M/v = \alpha_M$. Then

$$\lim_{v\to\infty} \frac{1}{v} \log P_v(p) \leq \alpha_M \lim_{n\to\infty} \frac{1}{n} \log P_n(p). \tag{D.8}$$

Thus, from eqns (D.7) and (D.8) it follows that $\lim_{n\to\infty}[\log P_n(p)]/n = 0$ if and only if $\lim_{v\to\infty}[\log P_v(p)]/v = 0$. By Theorem D.2 the following result is obtained.

Lemma D.3 $\zeta(p) = 0$ *if and only if*

$$\lim_{v\to\infty} \frac{1}{v} \log P_v(p) = 0,$$

where $P_v(p)$ is the probability that the cluster at the origin has v vertices at density p. □

By Lemma D.3, it is only necessary to study the function $P_v(p)$. Consider the probability space $(\Pi, \mathcal{F}, P_p)$. The space Π is partially ordered in the sense that if two states α and β are related by $\alpha(e) \leq \beta(e)$ for all subsets in $\mathcal{F}$, then it is said that $\alpha \leq \beta$ (notice that $\alpha(e) = 1$ if e is an open edge, and is zero otherwise). An *event* A is a collection of edges, and is said to occur if every edge in A is open. This is best defined by using an indicator function I_A, such that $I_A(\alpha) = 1$ whenever $\alpha(e) = 1$ for all $e \in A$. An event A is *increasing* if $I_A(\alpha) \leq I_A(\beta)$ whenever $\alpha \leq \beta$.

The probability of an event is defined as follows. Let X be a collection of random variables $(X(e), e \in \mathcal{E}^d)$ which are independent and uniform in $[0, 1]$. Let $\eta_p(e) = 1$ if $X(e) < p$, and $\eta_p(e) = 0$ otherwise. Let $I_A(\eta_p)$ be the indicator that A occurs. Then the probability that A occurs is $P_p(A) = E_p(I_A(\eta_p))$. Notice that $P_{p_1}(A) \leq P_{p_2}(A)$ if $p_1 \leq p_2$ and if A is an increasing event. Increasing events satisfy the FKG inequality. Let A and B be increasing events, and let $A \cap B$ be the event that both A and B occurs. Then

$$P_p(A \cap B) \geq P_p(A) P_p(B). \tag{D.9}$$

In other words, increasing events are positively correlated. The BK inequality concerns the probability that A and B occur *disjointly*. In particular, denote by $A \circ B$ the event that A and B occur on disjoint edge sets (for example, A and B could be edge-disjoint open paths). Then the BK inequality states that

$$P_p(A \circ B) \leq P_p(A) P_p(B). \tag{D.10}$$

More generally,

$$P_p(A_1 \circ A_2 \circ \cdots \circ A_n) \leq \prod_{i=1}^{n} P_p(A_i), \tag{D.11}$$

if all the A_i are increasing events. Proofs for the FKG and BK inequalities can be found in the book by G. Grimmett [147].

The fourth ingredient in the proof that $\zeta(p) > 0$ if $p < p_c$ is some results from graph theory. A *skeleton* is a tree with all its vertices of degree 3 or of degree 1, and where the vertices of degree 1 are all labelled by $\{0, 1, \ldots, m\}$. The labelled vertices of degree 1 are called *end-vertices*, while those of degree 3 will be called *internal vertices*. If S is a skeleton, then its set of end-vertices will be indicated by $D(S)$ and its internal vertices will be indicated by $I(S)$. Note that if $|D(S)| = m$, then $|I(S)| = m - 2$, and there are $2m - 3$ edges in S. Let the number of skeletons with m end-vertices be w_m. By selecting any one of the $2m - 3$ edges, and adding a new edge wu so that w subdivides the chosen edge, and assigning to u the label $m + 1$, a skeleton with $m + 1$ end-vertices can be created in $2m - 3$ ways. Thus $w_{m+1} = (2m - 3)w_m$ and since $w_3 = 1$ it follows that

$$w_{m+1} = \frac{(2m-3)!}{2^{m-1}(m-1)!}. \tag{D.12}$$

The following lemma can be applied to skeletons, and will later prove to be very useful.

Lemma D.4 *If G is a connected graph and $W = \{v_0, v_1, \ldots, v_m\}$ are vertices in G, then there exists a vertex $v_i \in W$ such that $G - v_i$ (this is the graph obtained by deleting v_i and all edges incident with it in G) has one component which contains all the vertices in the set $W - \{v_i\}$.*

Proof Let T be a spanning tree of G, and let u be a root of T. Define $d(u, v_i)$ to be the number of edges in the path between u and v_i. Choose i such that $d(u, v_i)$ is a maximum. Then $T - v_i$ has two components, and one component contains all the vertices in $W - \{v_i\}$. Thus, since the connected component of $T - v_i$ which contains all the vertices in $W - \{v_i\}$ is a subgraph of $G - v_i$, the lemma follows. □

The next important constituent in the proof is the notion of *connectivity functions* in the percolation process. Let $I_{v_0 \leftrightarrow v_1}$ be the indicator function of the event that the vertices v_0 and v_1 belong to the same open cluster. Then the order of the open cluster which contains the vertex v_0 is $\|C(v_0)\| = \sum_{v_1} I_{v_0 \leftrightarrow v_1}$, and the pair-connectivity function is defined by

$$\tau_p(v_0, v_1) = E_p(I_{v_0 \leftrightarrow v_1}). \tag{D.13}$$

Evidently, the mean order of the cluster containing v_0 is

$$\chi'(p) = \sum_{v_1} \tau_p(v_0, v_1) = E_p(\|C(v_0)\|). \tag{D.14}$$

The connectivity function of the vertices $\{v_0, v_1, \ldots, v_n\}$ is defined by

$$\tau_p(v_0, v_1, \ldots, v_n) = P_p(v_0, v_1, \ldots, v_n \text{ belongs to the same open cluster}). \quad \text{(D.15)}$$

Thus, by using transitivity,

$$\begin{aligned} E_p(\|C(v_0)\|^m) &= \sum_{v_1, v_2, \ldots, v_m} \tau_p(v_0, v_1, \ldots, v_m) \\ &= \sum_{v_1, v_2, \ldots, v_m} I_{v_0 \leftrightarrow v_1} I_{v_0 \leftrightarrow v_2} \ldots I_{v_0 \leftrightarrow v_m} \end{aligned} \quad \text{(D.16)}$$

There is a basic inequality relating the three-point connectivity function $\tau_p(v_0, v_1, v_2)$ to a sum over products of pair connectivity functions. This is shown as follows. Let $C(v_0)$ be the cluster of open edges containing the vertex v_0, and let $W = \{v_0, v_1, v_2\}$. By Lemma D.4 there is a path π of open edges between two of the vertices (say v_0 and v_1) in W which is edge-disjoint with the third (say v_2). Moreover, since $C(v_0)$ is connected, there is a path from v_2 to a first vertex in the path π, say v. This path is edge-disjoint with the rest of the open edges in the cluster. Notice that if $v_0 = v_1 = v_2$ or $v_0 = v_1$, then one may pick $v = v_0$ to find the same situation. Thus

$$\tau_p(v_0, v_1, v_2) \leq \sum_v P_p(v \leftrightarrow v_0 \circ v \leftrightarrow v_1 \circ v \leftrightarrow v_2), \quad \text{(D.17)}$$

where $v \leftrightarrow v_i$ is the event that the vertices v_0 and v_i are in the same open cluster. Apply the BK inequality to this to obtain

$$\tau_p(v_0, v_1, v_2) \leq \sum_v \tau_p(v, v_0)\tau_p(v, v_1)\tau_p(v, v_2). \quad \text{(D.18)}$$

If the sums over v_1 and v_2 are executed, while eqns (D.14) and (D.16) are used, then

$$E_p(\|C\|^2) \leq [\chi'(p)]^3. \quad \text{(D.19)}$$

This is sometimes called a *skeleton-inequality*, and it will be generalized to find upper bounds on $E_p(\|C\|^m)$.

Consider the connectivity function $\tau_p(v_0, v_1, \ldots, v_m)$. A cluster contributing to this function will be mapped to a skeleton S by a map $\Phi_{\mathbf{v}}$ (where $\mathbf{v} = (v_0, v_1, \ldots, v_m)$ are the vertices incident with open edges, and will be mapped to the labelled end-vertices of the skeleton). In particular, define $\Phi_{\mathbf{v}}$ such that (1) the end-vertex of S with label i is mapped to the edge v_i: $\Phi(i) = v_i$, and (2) each of the $2m - 1$ edges of S corresponds to edge-disjoint open paths joining the edges v_i in the cluster. To see that this mapping can be found, do induction on the number of end-vertices (of which there are $m + 1$ in the above). If $m = 2$, then S has three end-vertices and one internal vertex; this case has been considered above. Suppose that there is such a mapping if there are m end-vertices in S. By Lemma D.4 there is a label j such that the edges in the set $\{v_0, \ldots, v_{j-1}, v_{j+1}, \ldots, v_m\}$ are in the same connected component of the cluster if v_j is removed. Without loss of generality, suppose that $j = m$. If all the v_i are distinct from v_m, then first use the induction hypothesis to construct edge-disjoint open paths and a skeleton S for the edges $\{v_0, v_1, \ldots, v_{m-1}\}$. Then add the edge v_m and a path π which meets the skeleton in a first edge as was done for the case that $m = 2$. If v_m is equal to another edge, say v_0, then the path π will consist of only a single vertex, and the outcome is still the same. The internal vertices of S can be labelled $\{m+1, m+2, \ldots, 2m-1\}$, and edges of S can then be indicated by (i, j). Under the mapping such an edge becomes a path $(\Phi_{\mathbf{v}}(i), \Phi_{\mathbf{v}}(j))$ in the cluster. Consider now all the possible skeletons, and possible mappings that can be defined in this way; together they give an upper bound on the connectivity function: $\tau_p(v_0, v_1, \ldots, v_m)$ is less than or equal to $\sum_S \sum_{\Phi_{\mathbf{v}}} P_p$(there are edge-disjoint paths joining $(\Phi_{\mathbf{v}}(i)$ and $\Phi_{\mathbf{v}}(j))$ if $(i, j) \in S$). By the BK inequality this becomes

$$\tau_p(v_0, v_1, \ldots, v_m) \leq \sum_S \sum_{\Phi_{\mathbf{v}}} \prod_{(i,j)\in S} \tau_p(\Phi_{\mathbf{v}}(i), \Phi_{\mathbf{v}}(j)). \tag{D.20}$$

In other words, if eqn (D.16) is used, and the sums are executed over the external edges labelled $\{\Phi_{\mathbf{v}}(1), \Phi_{\mathbf{v}}(2), \ldots, \Phi_{\mathbf{v}}(m)\}$ in the above, then

$$E_p(\|C\|^m) \leq \sum_S \sum_{\Phi_{\mathbf{v}}} \prod_{\substack{(i,j)\in S \\ i,j>m}} \tau_p(\Phi_{\mathbf{v}}(i), \Phi_{\mathbf{v}}(j))[\chi'(p)]^m. \tag{D.21}$$

Next execute the sum over all possible mappings: this gives

$$E_p(\|C\|^m) \leq \sum_S [\chi'(p)]^{2m-1} = \left[\frac{(2m-3)!}{2^{m-1}(m-1)!}\right][\chi'(p)]^{2m-1}. \tag{D.22}$$

But note now that

$$\begin{aligned} E_p(\|C\|e^{t\|C\|}) &= \sum_{m=0}^{\infty} \frac{t^m}{m!} E_p(\|C\|^{m+1}) \\ &\leq \chi'(p)\left(1 + \sum_{m=1}^{\infty}\left[\frac{t^m}{m!}\frac{(2m-3)!}{2^{m-1}(m-1)!}\right][\chi'(p)]^{2m}\right) \\ &= \frac{\chi'(p)}{\sqrt{1-2t[\chi'(p)]^2}}. \end{aligned} \tag{D.23}$$

But Markov's inequality gives that

$$P_p(\|C\| \geq v) = P_p(\|C\|e^{t\|C\|} \geq ve^{tv}) \leq \frac{1}{ve^{tv}} E_p(\|C\|e^{t\|C\|}) \tag{D.24}$$

if $t \geq 0$. Thus

$$P_p(\|C\| \geq v) \leq \frac{\chi'(p)}{ve^{tv}} \frac{1}{\sqrt{1-2t[\chi'(p)]^2}}. \tag{D.25}$$

Choose

$$t = \frac{1}{2[\chi'(p)]^2} - \frac{1}{2v} \qquad \text{if } v > [\chi'(p)]^2.$$

Then $t > 0$, and the above simplifies to

$$P_p(\|C\| \geq v) \leq \sqrt{\frac{e}{v}}\, e^{-v/2[\chi'(p)]^2}. \tag{D.26}$$

In other words, $P_v(p) = P_p(\|C\| = v) \leq P_p(\|C\| \geq v) \leq \sqrt{e/v}\, e^{-v/2[\chi'(p)]^2}$, and thus it follows that $\zeta(p) > 0$ if $p < p_c$, since $\chi(p) < \infty$ (and thus $\chi'(p) < \infty$) if $p < p_c$.

Theorem D.5 *There exists a finite function $\zeta(p) \geq 0$ such that*

$$\zeta(p) = -\lim_{n\to\infty} \frac{1}{n} \log P_n(p),$$

and moreover, $\zeta(p) = 0$ if $p > p_c$ and $\zeta(p) > 0$ if $p < p_c$. □

Bibliography

[1] Adler J., Meir Y., Harris A.B., Aharony A. and Duarte J.A.M.S. (1988). Series study of random animals in general dimensions. *Physical Review B*, **38**, 4941–4954.

[2] Aizenmann M. and Newman C.M. (1984). Tree graph inequalities and critical behaviour in percolation models. *Journal of Statistical Physics*, **36**, 107–143.

[3] Akcoglu M.A. and Krengel U. (1981). Ergodic theorems for superadditive processes. *Journal für die Reine und Angewandte Mathematik*, **323**, 53–67.

[4] Ambjørn J., Durhuus B. and Jonsson T. (1989). Kinematical and numerical study of the crumpling transition in crystalline surfaces. *Nuclear Physics B*, **316**, 526–558.

[5] Andreev A.F. (1964). Singularity of thermodynamic quantities at a first order phase transition point. *Soviet Physics-JETP*, **18**, 1415–1416.

[6] Banavar J.R., Maritan A. and Stella A.L. (1991). Geometry, topology, and universality of random surfaces. *Science*, **252**, 825–827.

[7] Barber M.N. (1970). Asymptotic results for self-avoiding walks on a Manhattan lattice. *Physica*, **48**, 237–241.

[8] Barber M.N. (1973). Scaling relations for critical exponents of surface properties of magnets. *Physical Review B*, **8**, 407–409.

[9] Barber M.N. (1973). Finite size scaling. In *Phase Transitions and Critical Phenomena*, **8**, 145–266. Eds. C. Domb and J.L. Lebowitz. Academic Press (London).

[10] Bastolla U. and Grassberger P. (1997). Phase transitions of single semi-stiff polymer chains. *Journal of Statistical Physics*, **89**, 1061–1078.

[11] Batchelor M.T. and Yung C.M. (1995). Exact results for the adsorption of a flexible self-avoiding polymer chain in 2 dimensions. *Physical Review Letters*, **74**, 2026–2029.

[12] Baumgärtner A. and Romero A. (1992). Microcanonical simulations of self-avoiding surfaces. *Physica A*, **187**, 243–248.

[13] Belavin A.A., Polyakov A.M. and Zamolodchikov A.B. (1984). Infinite conformal symmetries of critical fluctuations in 2 dimensions. *Journal of Statistical Physics*, **34**, 763–774.

[14] Bender E. (1974). Convex n-ominoes. *Discrete Mathematics*, **8**, 31–40.

[15] Bennett-Wood D., Brak R., Guttmann A.J., Owczarek A.L. and Prellberg T. (1994). Low-temperature two dimensional polymer partition function scaling: Series analysis results. *Journal of Physics A: Mathematical and General*, **27**, L1–L8.

[16] Bennett-Wood D., Enting I.G., Gaunt D.S., Guttmann A.J., Leask J.L., Owczarek A.L. and Whittington S.G. (1997). Exact enumeration study of free energies of interacting polygons and walks in two dimensions. *Journal of Physics A: Mathematical and General*, **31**, 4725–4741.

[17] Bennett-Wood D. and Owczarek A.L. (1996). Exact enumeration results for self-avoiding walks on the honeycomb lattice attached to a surface. *Journal of Physics A: Mathematical and General*, **29**, 4755–4768.

[18] Berretti A. and Sokal A.D. (1985). New Monte Carlo method for the self-avoiding walk. *Journal of Statistical Physics*, **40**, 483–531.

[19] Binder K. (1983). Critical behaviour at surfaces. In *Phase Transitions and Critical Phenomena*, **8**, 1–144. Eds. C. Dombs and J.L Lebowitz. Academic Press (London).

[20] Blöte H.W. and Hilhorst H.J. (1984). Spiralling self-avoiding walks: An exact solution. *Journal of Physics A: Mathematical and General General*, **17**, L111–L115.

[21] Boas Jr. R.P. (1960). *A Primer of Real Functions*. Carus Mathematical Monographs (Mathematical Association of America). Wiley (New York).

[22] Bouchard E. and Bouchard J.-P. (1989). Self-avoiding surfaces at interfaces. *Journal de Physique*, **50**, 829–841.

[23] Bousquet-Mélou M. (1992). Convex polyominoes and heaps of segments. *Journal of Physics A: Mathematical and General General*, **25**, 1925–1934.

[24] Bousquet-Mélou M. (1994). Codage des polyominos convexes et èquations pour l'ènumèration suivant l'aire. *Discrete Applied Mathematics*, **48**, 21–43.

[25] Bousquet-Mélou M. and Guttmann A.J. (1997). Three dimensional self-avoiding convex polygons. *Physical Review E*, **55**, R6323–R6326.

[26] Bousquet-Mélou M. and Viennot X.G. (1992). Empilements de segments et q-énumération de polyominos convexes dirigés, *Journal of Combinatorial Theory, Serial A*, **60**, 196–224.

[27] Bovier A., Fröhlich J. and Glaus U. (1984). Mathematical aspects of the physics of disordered systems. *In Critical Phenomena, Random Systems, Gauge Theories (Les Houghes, Session XLIII)*. Eds. K. Osterwalder and R. Stora. Elsevier (Amsterdam).

[28] Brak R. and Essam J.M. (1999). Directed compact percolation near a wall III: Exact results for the mean length and number of contacts. *Journal of Physics A: Mathematical and General General*, **32**, 355–367.

[29] Brak R., Essam J.M. and Owczarek A.L. (1998). New results for directed vesicles and chains near an attractive wall. *Journal of Statistical Physics*, **93**, 155–192.

[30] Brak R. and Guttmann A.J. (1990). Exact solution of the staircase and row-convex polygon perimeter and area generating function. *Journal of Physics A: Mathematical and General General*, **23**, 4581–4588.

[31] Brak R., Guttmann A.J. and Whittington S.G. (1991). On the behaviour of collapsing linear and branched polymers. *Journal of Mathematical Chemistry*, **8**, 255–267.

[32] Brak R., Guttmann A.J. and Whittington S.G. (1992). A collapse transition in a directed walk model. *Journal of Physics A: Mathematical and General General*, **25**, 2437–2446.

[33] Brak R. and Owczarek A.L. (1995). On the analyticity properties of scaling functions in models of polymer collapse. *Journal of Physics A: Mathematical and General General*, **28**, 4709–4725.

[34] Brak R., Owczarek A.L. and Prellberg T. (1993). A scaling theory of the collapse transition in geometric cluster models of polymers and vesicles. *Journal of Physics A: Mathematical and General General*, **26**, 4565–4579.

[35] Brak R., Owczarek A.L. and Prellberg T. (1994). Exact scaling behaviour of partially convex vesicles. *Journal of Statistical Physics*, **76**, 1101–1128.

[36] Broadbent S.R. and Hammersley J.M. (1957). Percolation processes I. Crystals and mazes. *Proceedings of the Cambridge Philosophical Society*, **53**, 629–641.

[37] Buck R. (1965). *Advanced Calculus*, second edition. McGraw-Hill (New York).

[38] Burde G. and Zieschang H. (1985). *Knots*. de Gruyter Studies in Mathematics **5**. de Gruyther (Berlin).

[39] Burkhardt T.W. (1982). Bond-moving and variational methods in real-space renormalisation. In *Real Space Renormalisation*. Eds. T.W. Burkhardt and J.M.J. van Leeuwen, Springer (Berlin).

[40] Burkhardt T.W., Eisenriegler E. and Guim I. (1989). Conformal theory of energy correlations in the semi-infinite two dimensional $O(n)$ model. *Nuclear Physics B*, **316**, 559–572.

[41] Burkhardt T.W. and Guim I. (1991). Self-avoiding walks that cross a square. *Journal of Physics A: Mathematical and General*, **24**, L1221–L1228.

[42] Camacho C.J. and Fisher M.E. (1990). Tunable fractal shapes in self-avoiding polygons and planar vesicles. *Physical Review Letters*, **65**, 9–12.

[43] Caracciolo S., Causo M.S. and Pelissetto A. (1998). High precision determination of the critical exponent γ for self-avoiding walks. *Physical Review E*, **57**, R1215–R1218.

[44] Caracciolo S. and Glaus U. (1985). A new Monte Carlo simulation for two models of self-avoiding lattice trees in two dimensions. *Journal of Statistical Physics*, **41**, 95–114.

[45] Caracciolo S., Pelissetto A. and Sokal A.D. (1990). Universal distance ratios for two dimensional self-avoiding walks: Corrected conformal invariance predictions. *Journal of Physics A: Mathematical and General*, **23**, L969–L974.

[46] Cardy J.L. (1984). Conformal invariance and surface critical behaviour. *Nuclear Physics B*, **240**, 514–532.

[47] Cardy J.L. (1984). Conformal invariance. In *Phase Transitions and Critical Phenomena*, **11**. Eds. C. Domb and J.L. Lebowitz. Academic Press (London).

[48] Cardy J.L. and Guttmann A.J. (1993). Universal amplitude combinations for self-avoiding walks, polygons and trails. *Journal of Physics A: Mathematical and General*, **26**, 2485–2494.

[49] Cardy J.L. and Saleur H. (1989). Universal distance ratios for two dimensional polymers. *Journal of Physics A: Mathematical and General*, **22**, L601–L604.

[50] Carroll L. (1939). The Hunting of the Snark. In *Complete Works of Lewis Carroll*, 680–681, Nonesuch (London).

[51] Chang I.S. and Meirovitch H. (1993). Collapse transition of self-avoiding walks on a square lattice in the bulk and near a linear wall: The universality classes of the θ and θ' points. *Physical Review E*, **48**, 3656–3660.

[52] Chayes J.T. and Chayes L. (1986). Ornstein–Zernike behaviour for self-avoiding walks at all non-critical temperatures. *Communications in Mathematical Physics*, **105**, 221–238.

[53] Chee M.-N. and Whittington S.G. (1987). The growth of uniform star polymers in a slab geometry. *Journal of Physics A: Mathematical and General*, **20**, 4915–4921.

[54] Cherayil B.J., Douglas J.F. and Freed K.F. (1985). Effect of residual interactions on polymer properties near the θ-point. *Journal of Chemical Physics*, **83**, 5293–5310.

[55] Colby S.A., Gaunt D.S., Torrie G.M. and Whittington S.G. (1987). Branched polymers attached in a wedge geometry. *Journal of Physics A: Mathematical and General*, **20**, L515–L520.

[56] Coniglio A. (1983). Potts model formulation of branched polymers in a solvent. *Journal of Physics A: Mathematical and General*, **16**, L187–L191.

[57] Conway A.R. and Guttmann A.J. (1996). Square lattice self-avoiding walks and corrections to scaling. *Physical Review Letters*, **77**, 5284–5287.

[58] Cozzarelli N.R. (1992). Evolution of DNA topology: Implications for its biological roles. *Proceedings of the Symposium in Applied Mathematics*, **45**, 1–16.

[59] De'Bell K. and Essam J.W. (1981). Mean field theory of percolation with application to surface effects. *Journal of Physics A: Mathematical and General*, **14**, 1993–2008

[60] De'Bell K. and Lookman T. (1993). Surface phase transitions in polymer systems. *Reviews in Modern Physics*, **65**, 87–114.

[61] De'Bell K., Lookman T. and Zhao D. (1991). Exact results for trees attached to a surface and estimates of the critical Boltzman factor for surface adsorption. *Physical Review A*, **44**, 1390–1392.

[62] de Gennes P.-G. (1975). Collapse of a polymer chain in poor solvents. *Journal de Physique Letters*, **36**, L55–L57.

[63] de Gennes P.-G. (1976). Scaling theory of polymer adsorption. *Journal de Physique*, **37**, 1445–1452.

[64] de Gennes P.-G. (1979). *Scaling Concepts in Polymer Physics*. Cornell University Press (Ithaca).

[65] de Gennes P.-G. (1984). Tight knots. *Macromolecules*, **17**, 703–704.

[66] Deguchi T. and Tsurusaki K. (1993). Topology of closed random polygons. *Journal of the Physical Society (Japan)*, **62**, 1411–1414.

[67] Delbrück M. (1962). Knotting problems in biology. *Proceedings of the Symposium in Applied Mathematics*, **14**, 55–63.

[68] Delest M.-P. (1988). Generating functions for column-convex polyominoes. *Journal of Combinatorial Theory, Serial A*, **48**, 12–31.

[69] Delest M.-P. and Viennot X.G. (1984). Algebraic languages and polyominoe enumeration. *Theoretical Computer Science*, **34**, 169–206.

[70] de Queirez S.L.A. (1995). Surface crossover exponent for branched polymers in two dimensions. *Journal of Physics A: Mathematical and General*, **28**, 6315–6321.

[71] Derbez E. and Slade G. (1998). The scaling limit of lattice trees in high dimensions. *Communications in Mathematical Physics*, **193**, 69–104.

[72] Derrida B. (1981). Phenomenological renormalisation of the self-avoiding walk in three dimensions. *Journal of Physics A: Mathematical and General*, **14**, L5–L9.

[73] Derrida B. and de Seze L. (1982). Application of the phenomenological renormalisation to percolation and lattice animals in dimensions two. *Journal de Physique*, **43**, 475–483.

[74] Derrida B. and Hermann H.J. (1983). Collapse of branched polymers. *Journal de Physique*, **44**, 1365–1376.

[75] Derrida B. and Stauffer D. (1985). Corrections to scaling and phenomenological renormalisation of two dimensional percolation and lattice animal problems. *Journal de Physique*, **46**, 1623–1630.

[76] des Cloisseaux J. and Jannink G. (1990). *Polymers in Solution, Their Modelling and Structure*. Oxford University Press.

[77] Dhar D. (1987). The collapse of directed animals. *Journal of Physics A: Mathematical and General*, **20**, L847–L850.

[78] Di Francesco P., Saleur H. and Zuber J.-B. (1987). Modular invariance in non-minimal two-dimensional conformal theories. *Nuclear Physics B*, **285**, 454–480.

[79] Di Francesco P., Saleur H. and Zuber J.-B. (1987). Relations between the coulomb gas picture and conformal invariance in two-dimensional critical models. *Journal of Statistical Physics*, **49**, 57–79.

[80] Dickman R. and Shieve W.C. (1986) Collapse transition and asymptotic behaviour of lattice animals: Low temperature expansion. *Journal of Statistical Physics*, **44**, 465–489.

[81] Domany E. and Kinzel W. (1984). Equivalence of cellular automata to Ising models and directed percolation. *Physical Review Letters*, **53**, 311–314.

[82] Domb C. (1976). Lattice animals and percolation. *Journal of Physics A: Mathematical and General*, **9**, L141–L148.

[83] Domocos V. (1996). A combinatorial method for the enumeration of column-convex polyominoes. *Discrete Mathematics*, **152**, 115–123.

[84] Douglas J.F. (1986). Swelling and growth of polymers, membranes and sponges. *Physical Review E*, **54**, 2677–2689.

[85] Douglas J.F., Cherayil B.J. and Freed K.F. (1985). Polymers in two dimensions: Renormalisation group study using the three parameter model. *Macromolecules*, **18**, 2455–2463.

[86] Douglas J.F. and Freed K.F. (1985). Polymer contraction below the θ-point: A renormalisation group description. *Macromolecules*, **18**, 2445–2454.

[87] Doye J.P.K., Sear R.P. and Frenkel D. (1998). The effect of chain stiffness on the phase behaviour of isolated homopolymers. *Journal of Chemical Physics*, **108**, 2134–2142.

[88] Duplantier B. (1982). Lagrangian tricritical theory of polymer chains solution near the θ-point. *Journal de Physique*, **43**, 991–1019.

[89] Duplantier B. (1986). Tricritical polymer chains in or below three dimensions. *Europhysics Letters*, **1**, 491–498.

[90] Duplantier B. (1986). Exact critical exponents for two dimensional dense polymers. *Journal of Physics A: Mathematical and General*, **19**, L1009–L1014.

[91] Duplantier B. (1986). Polymer network of fixed topology: renormalisation, exact critical exponent γ in two dimensions. *Physical Review Letters*, **57**, 941–944.

[92] Duplantier B. (1987). Geometry of polymer chains near the θ-point and dimensional regularisation. *Journal of Chemical Physics*, **86**, 4233–4244.

[93] Duplantier B. (1989). Fractals in two dimensions and conformal invariance. *Physica D*, **38**, 71–87.

[94] Duplantier B. (1990). Renormalisation and conformal invariance for polymers. In *Fundamental Problems in Statistical Mechanics VII*, 171–223. Ed. H. van Beijeren. Elsevier Science Publishers B.V. (Amsterdam).

[95] Duplantier B. and Saleur H. (1986). Exact surface and wedge exponents for polymers in two dimensions. *Physical Review Letters*, **57**, 3179–3182.

[96] Duplantier B. and Saleur H. (1987). Exact tricritical exponents for polymers at the θ-point in two dimensions. *Physical Review Letters*, **59**, 539–542.

[97] Duplantier B. and Saleur H. (1987). Exact critical properties of two-dimensional dense self-avoiding walks. *Nuclear Physics B*, **290** [FS20], 291–326.

[98] Durhuus B., Frölich J. and Jònsson T. (1983). Self-avoiding and planar random surfaces on the lattice. *Nuclear Physics B*, **225** [FS9], 185–203.

[99] Durrett R. (1984). Oriented percolation in two dimensions. *Annals of Probability*, **12**, 999–1040.

[100] Edwards S.F. (1965). The statistical mechanics of polymers with excluded volume. *Proceedings of the Physical Society*, **85**, 613–624.

[101] Edwards S.F. (1967). Statistical mechanics with topological constraints I. *Proceedings of the Physical Society*, **91**, 513–519.

[102] Edwards S.F. (1968) Statistical mechanics with topological constraints II. *Journal of Physics A (Proceedings of the Physical Society), Serial 2*, **1**, 15–28.

[103] Ellis R.S. (1985). *Entropy, Large Deviations and Statistical Mechanics*. Springer (New York).

[104] Enting I.G. and Guttmann A.J. (1989). Polygons on the honeycomb lattice. *Journal of Physics A: Mathematical and General*, **22**, 1371–1384.

[105] Enting I.G. and Guttmann A.J. (1990). On the area of square lattice polygons. *Journal of Statistical Physics*, **58**, 475–484.

[106] Essam J.W. (1980). Percolation theory. *Reports on Progress in Physics*, **43**, 833–912.

[107] Essam J.W., De'Bell K. and Adler J. (1986). Analysis of extended series for bond percolation on the directed square lattice. *Physical Review B*, **33**, 1982–1986.

[108] Essam J.W., Guttmann A.J. and De'Bell K. (1988). On two dimensional directed percolation. *Journal of Physics A: Mathematical and General*, **21**, 3815–3832.

[109] Essam J.W., Lin J.-C. and Taylor P. (1995). Potts model on the Bethe lattice with mixed interactions. *Physical Review E*, **52**, 44–52.

[110] Feller W. (1968). *An Introduction to Probability Theory and its Applications*. Wiley (New York).

[111] Finsy R., Janssens M. and Bellemans A. (1975). Internal transitions in an infinitely long polymer chain. *Journal of Physics A: Mathematical and General*, **8**, L106–L109.

[112] Fisher M.E. (1965). Correlation functions and the coexistence of phases. *Journal of Mathematical Physics*, **6**, 1643–1653.

[113] Fisher M.E. (1978). Yang–Lee edge singularity and ϕ^3 field theories. *Physical Review Letters*, **40**, 1610–1613.

[114] Fisher M.E. (1989). Fractal and non-fractal shapes in two dimensional vesicles. *Physica D*, **38**, 112–118.

[115] Fisher M.E. and Gaunt D.S. (1964). Ising model and self-avoiding walks on hypercubical lattices and "high-density" expansions. *Physical Review A*, **133**, 224-239.

[116] Fisher M.E., Guttmann A.J. and Whittington S.G. (1991). Two dimensional lattice vesicles and polygons. *Journal of Physics A: Mathematical and General*, **24**, 3095–3106.

[117] Fisher M.E. and Sykes M.F. (1959). Excluded volume problem and the Ising model of ferromagnetism. *Physical Review*, **114**, 45–58.

[118] Flesia S. and Gaunt D.S. (1992). Lattice animals contact models of a collapsing branched polymer. *Journal of Physics A: Mathematical and General*, **25**, 2127–2137.

[119] Flesia S., Gaunt D.S., Soteros C.E. and Whittington S.G. (1992). Models for collapse in trees and c-animals. *Journal of Physics A: Mathematical and General*, **25**, 3515–3521.

[120] Flesia S., Gaunt D.S., Soteros C.E. and Whittington S.G. (1993). Collapse transition in animals and vesicles. *Journal of Physics A: Mathematical and General*, **26**, L993–L997.

[121] Flesia S., Gaunt D.S., Soteros C.E. and Whittington S.G. (1994). Statistics of collapsing lattice animals. *Journal of Physics A: Mathematical and General*, **27**, 5831–5846.

[122] Flory P.J. (1955). Statistical thermodynamics of semi-flexible chain molecules. *Proceedings of the Royal Society (London) A*, **234**, 60–73.

[123] Flory P.J. (1969). *Statistical Mechanics of Chain Molecules*. Wiley Interscience (New York).

[124] Forgacs G., Privman V. and Frisch H.L. (1989). Adsorption–desorption of polymer chains interacting with a surface. *Journal of Chemical Physics*, **90**, 3339–3345.

[125] Fortuin C. and Kasteleyen P. (1972). On the random cluster model. I: Introduction and relation to other models. *Physica*, **57**, 536–564.

[126] Foster D. (1990). Exact evaluation of the collapse phase boundary for two dimensional directed polymers. *Journal of Physics A: Mathematical and General*, **23**, L1135–L1138.

[127] Foster D., Orlandini E. and Tesi M.C. (1992). Surface critical exponents for models of polymer collapse and adsorption: The universality of the θ and θ' points. *Journal of Physics A: Mathematical and General*, **25**, L1211–L1217.

[128] Foster D. and Yeomans J. (1991). Competition between self-attraction and adsorption in directed self-avoiding polymers. *Physica A*, **177**, 443–452.

[129] Frisch H.L. and Wasserman E. (1968). Chemical topology. *Journal of the American Chemical Society*, **83**, 3789–3795.

[130] Fuller F.B. (1971). The writhing number of a space curve. *Proceedings of the National Academy of Science (USA)*, **68**, 815–819.

[131] Gaunt D.S. and Flesia S. (1990). Modelling the collapse transition in branched polymers. *Physica A*, **168**, 602–608.

[132] Gaunt D.S. and Flesia S. (1991). The collapse transition for lattice animals. *Journal of Physics A: Mathematical and General*, **24**, 3655–3670.

[133] Gaunt D.S., Peard P.J., Soteros C.E. and Whittington S.G. (1994). Relationships between the growth constants for animals and trees. *Journal of Physics A: Mathematical and General*, **27**, 7343–7351.

[134] Gessel I.M. (1986). A probabilistic method for lattice path enumeration. *Journal of Statistical Planning and Inference*, **14**, 49–58.

[135] Ginibre J., Grossman A. and Ruelle D. (1966). Condensation of lattice gases. *Communications in Mathematical Physics*, **3**, 187–193.

[136] Giri M., Stephan M.J. and Grest G.S. (1977). Spin models and cluster distribution for bond and site percolation models. *Physical Review B*, **16**, 4971–4977.

[137] Glaus U. (1985). Monte Carlo test of dimensional reduction in branched polymers in three dimensions. *Journal of Physics A: Mathematical and General*, **18**, L609–L615.

[138] Glaus U. (1986). Monte Carlo simulation of self-avoiding surfaces in three dimensions. *Physical Review Letters*, **56**, 1996–1999.

[139] Glaus U. (1988). Monte Carlo study of self-avoiding surfaces. *Journal of Statistical Physics*, **50**, 1141–1166.

[140] Glaus U. and Einstein T.L. (1987). On the universality class of planar self-avoiding surfaces with fixed boundary. *Journal of Physics A: Mathematical and General*, **20**, L105–L111.

[141] Goffman C. (1966). *Introduction to Real Analysis.* Harper and Row (New York).

[142] Gradshteyn I.S. and Ryzhik I.M. (1994). *Table of Integrals, Series and Products*, fifth edition. Academic Press (New York).

[143] Grassberger P. (1994). Non-uniform star polymers in two dimensions. *Journal of Physics A: Mathematical and General*, **27**, L721–L725.

[144] Grassberger P. (1995). The Bak–Snippen model for punctuated evolution. *Physics Letters A*, **200**, 277–282.

[145] Grassberger P. and Hegger R. (1995). Simulations of θ polymers in two dimensions. *Journal de Physique I: General Physics, Condensed Matter, Cross-disciplinary Physics*, **5**, 597–606.

[146] Grassberger P. and Hegger R. (1995). Simulation of three dimensional θ polymers. *Journal of Chemical Physics*, **102**, 6881–6899.

[147] Grimmett G. (1989). *Percolation.* Springer-Verlag (New York).

[148] Grossberg A., Izrailev S. and Neachev S. (1994). Phase transition in a heteropolymer chain at a selective interface. *Physical Review E*, **50**, 1912–1921.

[149] Guida R. and Zinn-Justin J. (1998). Critical exponents of the N-vector model. *Journal of Physics A: Mathematical and General*, **31**, 8103–8121.

[150] Guttmann A.J. (1983). Bounds on connective constants for self-avoiding walks. *Journal of Physics A: Mathematical and General*, **16**, 2233–2238.

[151] Guttmann A.J. (1989). On the critical behaviour of self-avoiding walks II. *Journal of Physics A: Mathematical and General*, **22**, 2807–2813.

[152] Guttmann A.J. and Enting I.G. (1988). The size and number of rings on the square lattice. *Journal of Physics A: Mathematical and General*, **21**, L165–L172.

[153] Guttmann A.J. and Enting I.G. (1988). The number of convex polygons on the square and honeycomb lattices. *Journal of Physics A: Mathematical and General*, **21**, L467–L474.

[154] Guttmann A.J. and Hirschhorn M. (1984). Comment on the number of spiral self-avoiding walks. *Journal of Physics A: Mathematical and General*, **17**, 3613–3614.

[155] Guttmann A.J. and Torrie G.M. (1984). Critical behaviour at an edge for the SAW and Ising model. *Journal of Physics A: Mathematical and General*, **17**, 3539–3552.

[156] Guttmann A.J. and Whittington S.G. (1978). Two-dimensional lattice embeddings of connected graphs of cyclomatic index two. *Journal of Physics A: Mathematical and General*, **11**, 721–729.

[157] Guttmann A.J. and Wormald N.C. (1984). On the number of spiral self-avoiding walks. *Journal of Physics A: Mathematical and General*, **17**, L271–L274.

[158] Hammersley J.M. (1957). Percolation processes II: The connective constant. *Mathematical Proceedings of the Cambridge Philosophical Society*, **53**, 642–645.

[159] Hammersley J.M. (1960). Limiting properties of numbers of self-avoiding walks. *Physical Review*, **118**, 656–656.

[160] Hammersley J.M. (1961). On the rate of convergence to the connective constant of the hypercubical lattice. *Quarterly Journal of Mathematics, Oxford* (2), **12**, 250–256.

[161] Hammersley J.M. (1961). The number of polygons on a lattice. *Mathematical Proceedings of the Cambridge Philosophical Society*, **57**, 516–523.

[162] Hammersley J.M. (1962). Generalization of the fundamental theorem on sub-additive functions. *Mathematical Proceedings of the Cambridge Philosophical Society*, **58**, 235–238.

[163] Hammersley J.M. and Broadbent S.R. (1957). Percolation processes I: Crystals and mazes. *Mathematical Proceedings of the Cambridge Philosophical Society*, **53**, 629–641.

[164] Hammersley J.M. and Morton K.W. (1954). Poor man's Monte Carlo. *Journal of the Royal Statistical Society B*, **16**, 23–38.

[165] Hammersley J.M., Torrie G.M. and Whittington S.G. (1982). Self-avoiding walks interacting with a surface. *Journal of Physics A: Mathematical and General*, **15**, 539–571.

[166] Hammersley J.M. and Welsh D.J.A. (1962). Further results on the rate of convergence to the connective constant of the hypercubic lattice. *Quarterly Journal of Mathematics, Oxford* (2), **13**, 108–110.

[167] Hammersley J.M. and Whittington S.G. (1985). Self-avoiding walks in wedges. *Journal of Physics A: Mathematical and General*, **18**, 101–111.

[168] Hara T. and Slade G. (1990). The lace expansion for self-avoiding walk in five or more dimensions. *Reviews in Mathematical Physics*, **4**, 235–327.

[169] Hara T. and Slade G. (1990). On the upper critical dimension of lattice trees and lattice animals. *Journal of Statistical Physics*, **59**, 1469–1510.

[170] Hara T. and Slade G. (1991). Critical behaviour of self-avoiding walks in five or more dimensions. *Bulletin of the American Mathematical Society*, **25**, 417–423.

[171] Hara T. and Slade G. (1992). Self-avoiding walk in five or more dimenions. I: The critical behaviour. *Communications in Mathematical Physics*, **147**, 101–136.

[172] Hara T. and Slade G. (1992). The number and size of branched polymers in high dimensions. *Journal of Statistical Physics*, **67**, 1009–1038.

[173] Hara T. and Slade G. (1993). New lower bounds on the self-avoiding walk connective constant. *Journal of Statistical Physics*, **72**, 479–517.

[174] Hara T. and Slade G. (1995). The self-avoiding walk and percolation critical points in high dimensions. *Combinatorial Probability and Computation*, **4**, 197–215.

[175] Hara T., Slade G. and Sokal A.D. (1993). New lower bounds on the self-avoiding walk connective constant. *Journal of Statistical Physics*, **72**, 479–517.

[176] Hardy G.H. (1940). *Ramanujan*. Cambridge University Press.

[177] Hardy G.H. (1991). *Divergent Series*, second (textually unaltered) edition. Chelsea Publishing Company (New York).

[178] Hardy G.H., Littlewood J.E. and Pòlya G. (1952). *Inequalities*, second edition. Cambridge University Press.

[179] Hardy G.H. and Ramanujan S. (1918). Asymptotic formulae in combinatory analysis. *Proceedings of the London Mathematical Society, Serial 2*, **17**, 75–115.

[180] Hegger R. and Grassberger P. (1994). Chain polymers near an adsorbing surface. *Journal of Physics A: Mathematical and General*, **27**, 4069–4081.

[181] Henkel M. and Seno F. (1996). Phase diagrams of branched polymer collapse. *Physical Review E*, **53**, 3662–3672.

[182] Hille E. (1948). *Functional Analysis and Semi-Groups. AMS Colloquim Publications*, **31**, American Mathematical Society (Providence, Rhode Island).

[183] Isaacson J. and Lubensky T.C. (1980). Flory exponents for generalised branched polymer problems. *Journal de Physique Letters*, **41**, L469–L471.

[184] Isakov S.N. (1984). Non-analytic features of the first order transitions in the Ising model. *Communications in Mathematical Physics*, **95**, 427–443.

[185] Ishinabe T. (1987). Examination of the θ point from exact enumeration of self-avoiding walks II. *Journal of Physics A: Mathematical and General*, **20**, 6435–6453.

[186] Janse van Rensburg E.J. (1992). On the number of trees in $\mathcal{Z}^d$. *Journal of Physics A: Mathematical and General*, **25**, 3523–3528.

[187] Janse van Rensburg E.J. (1992). Surfaces in the hypercubic lattice. *Journal of Physics A: Mathematical and General*, **25**, 3529–3547.

[188] Janse van Rensburg E.J. (1994). The statistical mechanics and topology of surfaces in $\mathcal{Z}^d$. *Journal of Knot Theory and its Ramifications*, **3**, 365–378.

[189] Janse van Rensburg E.J. (1997). Crumpling self-avoiding surfaces. *Journal of Statistical Physics*, **88**, 177–200.

[190] Janse van Rensburg E.J. (1998). Collapsing and adsorbing polygons. *Journal of Physics A: Mathematical and General*, **31**, 8295–8306.

[191] Janse van Rensburg E.J. (1999). Models of composite polygons. *Journal of Physics A: Mathematical and General*, **32**, 4351–4372.

[192] Janse van Rensburg E.J. (1999). Adsorbing staircase walks and staircase polygons. *Annals of Combinatorics*, **3**, 451–473.

[193] Janse van Rensburg E.J., Guillet J.E. and Whittington S.G. (1989). Exciton migration on polymers. *Macromolecules*, **22**, 4212–4220.

[194] Janse van Rensburg E.J. and Madras N. (1992). A non-local Monte Carlo algorithm for lattice trees. *Journal of Physics A: Mathematical and General*, **25**, 303–333.

[195] Janse van Rensburg E.J. and Madras N. (1997). Metropolis Monte Carlo simulation of lattice animals. *Journal of Physics A: Mathematical and General*, **30**, 8035–8066.

[196] Janse van Rensburg E.J., Orlandini E., Sumners D.W., Tesi M.C. and Whittington S.G. (1993). The writhe of a self-avoiding polygon. *Journal of Physics A: Mathematical and General*, **26**, L981–L986.

[197] Janse van Rensburg E.J., Orlandini E. and Tesi M.C. (1999). Collapsing animals. *Journal of Physics A: Mathematical and General*, **32**, 1567–1584.

[198] Janse van Rensburg E.J., Sumners D.W., Wasserman E. and Whittington S.G. (1992). Entanglement complexity of self-avoiding walks. *Journal of Physics A: Mathematical and General*, **25**, 6557–6566.

[199] Janse van Rensburg E.J. and Whittington S.G. (1989). Self-avoiding surfaces. *Journal of Physics A: Mathematical and General*, **22**, 4939–4958.

[200] Janse van Rensburg E.J. and Whittington S.G. (1990). Punctured discs on the square lattice. *Journal of Physics A: Mathematical and General*, **23**, 1287–1294.

[201] Janse van Rensburg E.J. and Whittington S.G. (1990). Self-avoiding surfaces with knotted boundaries. *Journal of Physics A: Mathematical and General*, **23**, 2495–2505.

[202] Janse van Rensburg E.J. and Whittington S.G. (1990). The knot probability in lattice polygons. *Journal of Physics A: Mathematical and General*, **23**, 3573–3690.

[203] Janse van Rensburg E.J. and You S. (1998). Adsorbing and collapsing trees. *Journal of Physics A: Mathematical and General*, **31**, 8635–8651.

[204] Janssens H.K. and Lyssy A. (1994). Surface adsorption of branched polymers: Mapping onto the Yang–Lee edge singularity and exact results for three dimensions. *Physical Review E*, **50**, 3784–3796.

[205] Johnston D.A. (1998). Thin animals. *Journal of Physics A: Mathematical and General*, **31**, 9405–9417.

[206] Joyce G.S. (1984). An exact formula for the number of spiral self-avoiding walks. *Journal of Physics A: Mathematical and General*, **17**, L463–L467.

[207] Joyce G.S. and Brak R. (1985). An exact solution for a spiral self-avoiding walk model on the triangular lattice. *Journal of Physics A: Mathematical and General*, **18**, L293–L298.

[208] Joyce G.S. and Guttmann A.J. (1994). Exact results for the generating function of directed column-convex animals on the square lattice. *Journal of Physics A: Mathematical and General*, **27**, 4359–4367.

[209] Kantor Y., Kardar M. and Nelson D. (1986). Statistical mechanics of tethered surfaces. *Physical Review Letters*, **57**, 791–794.

[210] Kesten H. (1963). On the number of self-avoiding walks. *Journal of Mathematical Physics*, **4**, 960–969.

[211] Kesten H. (1964) On the number of self-avoiding walks II. *Journal of Mathematical Physics*, **5**, 1128–1137.

[212] Kholodenko A.L. and Vilgis T.A. (1996). Path integral calculation of the writhe for circular semi-flexible polymers. *Journal of Physics A: Mathematical and General*, **29**, 939–948.

[213] Kholodenko A.L. and Vilgis T.A. (1998). Some geometrical and topological problems in polymer physics. *Physics Reports*, **298**, 251–370.

[214] Kim D. (1988). The number of convex polyominoes with given perimeter. *Discrete Mathematics*, **70**, 47–51.

[215] Klarner D. (1965). Some results concerning polyominoes. *Fibonacci Quarterly*, **3**, 9–20.

[216] Klarner D. (1967). Cell growth problems. *Canadian Journal of Mathematics*, **19**, 851–863.

[217] Klarner D. and Rivest R. (1973). A procedure for improving the upper bound for the number of n-ominoes. *Canadian Journal of Mathematics*, **25**, 585–602.

[218] Klarner D. and Rivest R. (1974). Asymptotic bounds for the number of convex n-ominoes. *Discrete Mathematics*, **8**, 31–40.

[219] Klein D.J. (1981). Rigorous results for branched polymer models with excluded volume. *Journal of Chemical Physics*, **75**, 5186–5189.

[220] Koniaris K. and Muthukumar M. (1991). Self-entanglement in ring polymers. *Journal of Chemical Physics*, **95**, 2873–2881.

[221] Kunz H. and Souillard B. (1978). Essential singularity in percolation problems in asymptotic behaviour of cluster size distribution. *Journal of Statistical Physics*, **19**, 77–106.

[222] Kurze D.A. and Fisher M.E. (1979). Yang–Lee edge singularities at high temperatures. *Physical Review B*, **20**, 2785–2796.

[223] Lacher R.C. and Sumners D.W. (1991). Data structures and algorithms for the computation of topological invariants of entanglements: links, twist and writhe. In *Computer Simulations of Polymers*. Ed. R.J. Roe. Prentice-Hall (Englewood-Cliffs, New Jersey).

[224] Lam P.M. (1987). Specific heat and collapse transition of branched polymers. *Physical Review B*, **36**, 6988–6992.

[225] Lawrie I.D. and Sarlbach S. (1984). Tricriticality. In *Phase Transitions and Critical Phenomena*, **9**, 65–161. Eds. C. Domb and J.L. Lebowitz. Academic Press (London).

[226] Le Guillou J.C. and Zinn-Justin J. (1977). Critical exponents for the n-vector model in three dimensions from field theory. *Physical Review Letters*, **39**, 95–98.

[227] Le Guillou J.C. and Zinn-Justin J. (1980). Critical exponents from field theory. *Physical Review B*, **21**, 3976–3998.

[228] Le Guillou J.C. and Zinn-Justin J. (1985). Accurate critical exponents from the ϵ-expansion. *Journal de Physique Letters*, **46**, L137–L141.

[229] Le Guillou J.C. and Zinn-Justin J. (1989). Accurate critical exponents from field theory. *Journal de Physique*, **50**, 1365–1370.

[230] Leibler S. (1986). Curvature instability in membranes. *Journal de Physique*, **47**, 507–516.

[231] Leibler S. (1988). Equilibrium statistical mechanics of fluctuating films and membranes. In *Statistical Mechanics of Membranes and Surfaces*. Eds. D. Nelson, T. Piran, S. Weinberg. *Jerusalem Winter School for Theoretical Physics*, **5**, 45–103, World Scientific (Singapore).

[232] Leibler S., Singh R.R.P. and Fisher M.E. (1987). Thermodynamic behaviour of two dimensional vesicles. *Physical Review Letters*, **59**, 1989–1992.

[233] Li B., Madras N. and Sokal A.D. (1995). Critical exponents, hyperscaling, and universal amplitude ratios for two and three dimensional self-avoiding walks. *Journal of Statistical Physics*, **80**, 661–754.

[234] Lin K.Y. (1985). Spiral self-avoiding walks on a triangular lattice. *Journal of Physics A: Mathematical and General*, **18**, L145–L148.

[235] Lin K.Y. and Chang S.J. (1988). Rigorous results for the number of convex polygons on the square and honeycomb lattices. *Journal of Physics A: Mathematical and General*, **21**, 2635–2642.

[236] Lin K.Y. and Liu K.C. (1986). On the number of spiral self-avoiding walks on a triangular lattice. *Journal of Physics A: Mathematical and General*, **19**, 585–589.

[237] Lipson J.E.G., Gaunt D.S., Wilkinson M.K. and Whittington S.G. (1987). Lattice models of branched polymers. *Macromolecules*, **20**, 186–190.

[238] Lipson J.E.G. and Whittington S.G. (1983). Lattice trees with a restricted number of branched points. *Journal of Physics A: Mathematical and General*, **16**, 3119–3125.

[239] Liu K.C. and Lin K.Y. (1985). Spiral self-avoiding walks on a triangular lattice: End-to-end distance. *Journal of Physics A: Mathematical and General*, **18**, L647–L650.

[240] Livne S. and Meirovitch H. (1988). Computer simulation of long polymers adsorbed on a surface I: Corrections to scaling in an ideal chain. *Journal of Chemical Physics*, **88**, 4498–4506.

[241] Livne S. and Meirovitch H. (1988). Computer simulation of long polymers adsorbed on a surface II: Critical behaviour of a single self-avoiding walk. *Journal of Chemical Physics*, **88**, 4507–4515.

[242] Lubensky T.C. and Isaacson J. (1979). Statistics of lattice animals and dilute branched polymers. *Physical Review*, **20**, 2130–2146.

[243] Ma S.K. and Lin K.Y. (1988). Anisotropic spiral self-avoiding loops. *International Journal of Modern Physics B*, **2**, 287–299.

[244] Madras N. (1988). End patterns of self-avoiding walks. *Journal of Statistical Physics*, **53**, 689–701.

[245] Madras N. (1991). Bounds on the critical exponents of self-avoiding polygons. In *Random Walks, Brownian Motion and Interacting Particle Systems*. Eds. R. Durrett and H. Kesten. Birkhauser (Boston).

[246] Madras N. (1995). Critical behaviour of self-avoiding walks that cross a square. *Journal of Physics A: Mathematical and General*, **28**, 1533–1547.

[247] Madras N. (1995). A rigorous bound on the critical exponents for the numbers of lattice trees, animals and polygons. *Journal of Statistical Physics*, **78**, 681–699.

[248] Madras N. (1999). A pattern theorem for lattice clusters. *Annals of Combinatorics*, **3**, 357–384.

[249] Madras N. and Slade G. (1993). *The Self-Avoiding Walk*. Birkhäuser (Boston).

[250] Madras N., Soteros C.E. and Whittington S.G. (1988). Statistics of lattice animals. *Journal of Physics A: Mathematical and General*, **21**, 4617–4635.

[251] Madras N., Soteros C.E., Whittington S.G., Martin J.L., Sykes M.F., Flesia S. and Gaunt D.S. (1990). The free energy of a collapsing branched polymer. *Journal of Physics A: Mathematical and General*, **23**, 5327–5350.

[252] Maes D. and Vanderzande C. (1990). Self-avoiding rings at the θ-point. *Physical Review A*, **41**, 3074–3080.

[253] Maritan A. and Stella A.L. (1984). Scaling behaviour of self-avoiding random surfaces. *Physical Review Letters*, **53**, 123–126.

[254] Maritan A. and Stella A.L. (1987). Some exact results for self-avoiding random surfaces. *Nuclear Physics B*, **280**, 561–575.

[255] Maritan A. and Stella A.L. (1987). Hierarchical random surfaces. *Physical Review Letters*, **59**, 300–303.

[256] Mason W.K. (1969). Homeomorphic continuous curves in 2-space are isotopic in 3-space. *Transaction of the American Mathematical Society*, **142**, 269–290.

[257] McCrea W.H. and Whipple F.J.W. (1940). Random paths in two and three dimensions. *Proceedings of the Royal Society of Edinburgh*, **60**, 281–298.

[258] McDonald, Hunter D.L., Kelly K. and Jan N. (1993). Self-avoiding walks in two to five dimensions: exact enumeration and series study. *Journal of Physics A: Mathematical and General*, **25**, 1429–1440.

[259] Meirovitch H. and Lim H.A. (1989). The collapse transition of self-avoiding walks in the square lattice: A computer simulation study. *Journal of Chemical Physics*, **91**, 2544–2554.

[260] Meirovitch H. and Lim H.A. (1989). θ point exponents of polymers in $d = 2$. *Physical Review Letters*, **62**, 2640–2640.

[261] Meirovitch H. and Lim H.A. (1990). Computer simulation study of the θ point in three dimensions. I: Self-avoiding walks on a simple cubic lattice. *Journal of Chemical Physics*, **92**, 5144–5154.

[262] Melzak Z.A. (1962). Partition functions and spiralling in the plane random walk. *Canadian Mathematical Bulletin*, **6**, 231–237.

[263] Michelletti C. and Yeomans J.M. (1993). Adsorption transition of directed vesicles in two dimensions. *Journal of Physics A: Mathematical and General*, **26**, 5705-5712.

[264] Michels J.P.J. and Wiegel F.W. (1986). On the topology of a polymer ring. *Proceedings of the Royal Society (London) A*, **403**, 269–284.

[265] Moak D.S. (1984). The q-analogue of Stirling's formula. *Rocky Mountain Journal of Mathematics*, **14**, 403–413.

[266] Mutz M. and Bensimon D. (1991). Observation of toroidal vesicles. *Physical Review A*, **43**, 4525–4527.

[267] Nelson D., Piran T. and Weinberg S. (1989). *Statistical Mechanics of Membranes and Surfaces*. World Scientific, Singapore.

[268] Nemirovsky A.M., Freed K.F., Ishinabe T. and Douglas J.F. (1992). Marriage of exact enumeration and $1/d$ expansion methods: Lattice model of dilute polymers. *Journal of Statistical Physics*, **67**, 1083–1108.

[269] Nidras P.P. (1996). Grand canonical simulation of the interacting self-avoiding walk model. *Journal of Physics A: Mathematical and General*, **29**, 7929–7942.

[270] Niederhausen H. (1986). The enumeration of restricted random walks by Sheffer polynomials with applications to statistics. *Journal of Statistical Planning and Inference*, **14**, 95–114.
[271] Nienhuis B. (1982). Exact critical point and critical exponents of $O(n)$ models in two dimensions. *Physical Review Letters*, **49**, 1062–1065.
[272] Nienhuis B. (1984). Critical behaviour of two-dimensional spin models and charge asymmetry in the Coulomb gas. *Journal of Statistical Physics*, **34**, 731–761.
[273] Nienhuis B. (1984). Coulomb gas formulation of two-dimensional phase transitions. In *Phase Transitions and Critical Phenomena*, **11**, 1–53. Eds. C. Domb and J.L. Lebowitz. Academic Press (London).
[274] Nienhuis B. (1990). Critical spin-1 vertex models and $O(N)$ models. *International Journal of Modern Physics B*, **4**, 929–942.
[275] O'Brien G.L. (1990). Monotonicity of the number of self-avoiding walks. *Journal of Statistical Physics*, **59**, 969–979.
[276] O'Connell J., Sullivan F., Libes D., Orlandini E., Tesi M.C., Stella A.L. and Einstein T.L. (1991). Self-avoiding random surfaces: Monte Carlo study using oct-tree data-structure. *Journal of Physics A: Mathematical and General*, **24**, 4619–4635.
[277] Olami Z., Procaccia I. and Zeittav R. (1994). Theory of self-organised interface depinning. *Physical Review E*, **49**, 1232–1237.
[278] Olver F.W.J. (1974). *Asymptotics and Special Functions*. Academic Press (New York).
[279] Orlandini E. and Tesi M.C. (1992). Monte Carlo study of three dimensional vesicles. *Physica A*, **185**, 160–165.
[280] Orlandini E. and Tesi M.C. (1998). Knotted polygons with curvature in $\mathcal{Z}^3$. *Journal of Physics A: Mathematical and General*, **31**, 9441–9454.
[281] Orlandini E., Tesi M.C. and Whittington S.G. (1999). A self-avoiding model of random copolymer adsorption. *Journal of Physics A: Mathematical and General*, **32**, 469–477.
[282] Orlandini E., Tesi M.C. and Whittington S.G. (1999). Self-averaging in models of polymer collapse. To appear in *Journal of Physics A: Mathematical and General*.
[283] Orlandini E., Tesi M.C., Whittington S.G., Sumners D.W. and Janse van Rensburg E.J. (1994). The writhe of the self-avoiding walk. *Journal of Physics A: Mathematical and General*, **27**, L333–L338.
[284] Orlandini E., Stella A.L., Einstein T.L., Tesi M.C., Beichl I. and Sullivan F. (1996). Bending rigidity driven transition and crumpling point scaling of lattice vesicles. *Physical Review E*, **53**, 5800–5807.
[285] Orlandini E., Stella A.L., Tesi M.C. and Sullivan F. (1993). Vesicle adsorption on a plane: Scaling regimes and crossover phenomena. *Physical Review E*, **48**, R4203–R4206.
[286] Owczarek A.L. and Prellberg T. (1993). Exact solution of the discrete $(1+1)$-dimensional SOS model with field and surface interactions. *Journal of Statistical Physics*, **70**, 1175–1194.
[287] Owczarek A.L., Prellberg T., Bennett-Wood D., Guttmann A.J. (1994). Universal distance ratios for interacting two dimensional polymers. *Journal of Physics A: Mathematical and General*, **27**, L919–L925.
[288] Owczarek A.L., Prellberg T. and Brak R. (1993). New scaling form for the collapsed polymer phase. *Physical Review Letters*, **70**, 951–953.
[289] Owczarek A.L., Prellberg T. and Brak R. (1993). The tricritical behaviour of self-interacting partially directed walks. *Journal of Statistical Physics*, **72**, 737–772.
[290] Parisi G. and Sourlas N. (1981). Critical behaviour of branched polymers and the Yang–Lee edge singularity. *Physical Review Letters*, **46**, 871–874.

[291] Paulhus M. (1994). *A study of computer simulations of combinatorial structures with applications to lattice animals models of branched polymers*. Ph.D. Thesis, Department of Mathematics and Statistics, University of Saskatchewan.

[292] Pippenger N. (1989). Knots in random walks. *Discrete Applied Mathematics*, **25**, 273–278.

[293] Pólya G. (1969). On the number of certain lattice polygons. *Journal of Combinatorial Theory*, **6**, 102–105.

[294] Prellberg T. (1994). Uniform q-series asymptotics for staircase polygons. *Journal of Physics A: Mathematical and General*, **28**, 1289–1304.

[295] Prellberg T. and Brak R. (1995). Critical exponents from nonlinear functional equations for partially directed cluster models. *Journal of Statistical Physics*, **78**, 701–730.

[296] Prellberg T. and Owczarek A.L. (1995). Models of polymer collapse in three dimensions: Evidence from kinetic growth simulations. *Physical Review E*, **51**, 2142–2149.

[297] Prellberg T. and Owczarek A.L. (1995). Stacking models of vesicles and compact clusters. *Journal of Statistical Physics*, **80**, 755–779.

[298] Prentice J.J. (1991). Renormalisation theory of self-avoiding walks which cross a square. *Journal of Physics A: Mathematical and General*, **24**, 5097–5103.

[299] Privman V. (1983). Spiral self-avoiding walks. *Journal of Physics A: Mathematical and General*, **16**, L571–L573.

[300] Privman V. (1984). Convergence and extrapolation in finite size scaling renormalisation. *Physica A*, **123**, 428–442.

[301] Privman V. (1986). Study of the θ point by enumeration of self-avoiding walks on the triangular lattice. *Journal of Physics A: Mathematical and General*, **19**, 3287–3297.

[302] Privman V., Forgacs G. and Frisch H.L. (1988). New solvable model of polymer chains adsorption at a surface. *Physical Review B*, **37**, 9897–9900.

[303] Privman V. and Kurtze D.A. (1986). Partition function zeros in two dimensional lattice models of the polymer θ-point. *Macromolecules*, **19**, 2377–2379.

[304] Privman V. and Švrakić N.M. (1988). Difference equations in statistical mechanics I. Cluster statistics models. *Journal of Statistical Physics*, **51**, 1091–1110.

[305] Rajesh R. and Dhar D. (1998). An exactly solvable anisotropic directed percolation model in three dimensions. *Physical Review Letters*, **81**, 1646–1649.

[306] Read R.C. (1962). Contributions to the cell growth problem. *Canadian Journal of Mathematics*, **14**, 1–20.

[307] Redner S. (1979). Mean end-to-end distance of branched polymers. *Journal of Physics A: Mathematical and General*, **12**, L239–L244.

[308] Redner S. (1985). Enumeration study of self-avoiding random surfaces. *Journal of Physics A: Mathematical and General*, **18**, L723–L726.

[309] Royden H.L. (1987). *Real Analysis*, third edition. Prentice Hall (Englewood-Cliffs, New Jersey).

[310] Sackmann E., Duwe H.-P. and Engelhardt H. (1986). Membrane bending elasticity and its role for shape fluctuations and shape transformations of cells and vesicles. *Faraday Discussions of the Chemical Society*, **81**, 281–290.

[311] Seitz W.A. and Klein D.J. (1981). Excluded volume effects for branched polymers: Monte Carlo results. *Journal of Chemical Physics*, **75**, 5190–5193.

[312] Seno F. and Stella A.L. (1988). θ point of a linear polymer in two dimensions: A renormalisation group analysis of Monte Carlo enumerations. *Journal de Physique*, **49**, 739–748.

[313] Seno F. and Vanderzande C. (1994). Non-universality in the collapse of two dimensional polymers. *Journal of Physics A: Mathematical and General*, **27**, 5813-5830. Correction in *Journal of Physics A: Mathematical and General*, **27**, 7937–7938.

[314] Shaw S.Y. and Wang J.C. (1993). Knotting of a DNA chain during ring closure. *Science*, **260**, 533–536.

[315] Slade G. (1988). Convergence of self-avoiding random walk to brownian motion in high dimensions. *Journal of Physics A: Mathematical and General*, **21**, L417–L420.

[316] Slade G. (1995). Bounds on the self-avoiding walk connective constant. *Journal of Fourier Analysis and Applications*. Kahane special edition. 526–533.

[317] Slomson A. (1991). *An Introduction to Combinatorics*. Chapman and Hall, London.

[318] Sokal A.D. (1995). Monte Carlo methods for the self-avoiding walk. In *Monte Carlo and Molecular Dynamics simulations in Polymer Science*. Ed. K. Binder. Oxford University Press.

[319] Soteros C.E. (1992). Adsorption of uniform lattice animals with specified topology. *Journal of Physics A: Mathematical and General*, **25**, 3153–3173.

[320] Soteros C.E. (1993). Lattice models of branched polymers with specified topologies. *Journal of Mathematical Chemistry*, **14**, 91–102.

[321] Soteros C.E. (1993). Random knots in uniform branched polymers. *Mathematical Modelling and Scientific Computing*, **2**, 747–752.

[322] Soteros C.E. (1998). Knots in graphs in subsets of $\mathcal{Z}^3$. *Topology and Geometry in Polymer Science*. Eds. S.G. Whittington, D.W. Sumners and T. Lodge. IMA Volumes in Mathematics and its Applications, **103**, 101–134.

[323] Soteros C.E., Narayanan K.S.S, De'Bell K. and Whittington S.G. (1996). Polymers with restricted branching. *Physical Review E*, **53**, 4545–4553.

[324] Soteros C.E., Sumners D.W. and Whittington S.G. (1992). Entanglement complexity of graphs in $\mathcal{Z}^3$. *Mathematical Proceedings of the Cambridge Philosophical Society*, **111**, 75–91.

[325] Soteros C.E., Sumners D.W. and Whittington S.G. (1999). Linking of random p-spheres in $\mathcal{Z}^d$. *Journal of Knot Theory and its Ramifications*, **8**, 49–70.

[326] Soteros C.E. and Whittington S.G. (1988). Critical exponents for lattice animals with fixed cyclomatic index. *Journal of Physics A: Mathematical and General*, **21**, 2187–2193.

[327] Soteros C.E. and Whittington S.G. (1988). Polygons and stars in a slit geometry. *Journal of Physics A: Mathematical and General*, **21**, L857–L861.

[328] Soteros C.E. and Whittington S.G. (1989). Lattice models of branched polymers: Effect of geometrical constraints. *Journal of Physics A: Mathematical and General*, **22**, 5259–5270.

[329] Stanley R.P. (1986). *Enumerative Combinatorics*, Volume 1. Wadsworth & Brooks/Cole (Monterey, California).

[330] Stauffer D. (1979). Scaling theory of percolation clusters. *Physics Reports*, **54**, 1–74.

[331] Stella A.L., Orlandini E., Beichl I., Sullivan F., Tesi M.C. and Einstein T.L. (1992). Self-avoiding surfaces, topology and lattice animals. *Physical Review Letters*, **69**, 3650–3653.

[332] Stella A.L. and Vanderzande C. (1989). Scaling and fractal dimensions of Ising clusters and the $d = 2$ critical point. *Physical Review Letters*, **62**, 1067–1070.

[333] Stephen J.M. (1975). Collapse of a polymer chain. *Physics Letters A*, **53**, 363–364.

[334] Sterling T. and Greensite J. (1983). Entropy of self-avoiding surfaces on the lattice. *Physics Letters. B*, **121**, 345–348.

[335] Stratychuk L.M. and Soteros C.E. (1996). Statistics of collapsed lattice animals: Rigorous results and Monte Carlo simulations. *Journal of Physics A: Mathematical and General*, **29**, 7067–7087.

[336] Sumners D.W. and Whittington S.G. (1988). Knots in self-avoiding walks. *Journal of Physics A: Mathematical and General*, **21**, 1689–1694.

[337] Sun S.-T., Nishio I., Swislow G. and Tanaka T. (1980). The coil–globule transition: Radius of gyration of polystyrene in cyclohexane. *Journal of Chemical Physics*, **73**, 5971–5975.

[338] Takács L. (1986). Some asymptotic formulas for lattice paths. *Journal of Statistical Planning and Inference*, **14**, 123–142.

[339] Tasaki H. and Hara T. (1985). Collapse of random surfaces in the connected plaquettes model. *Physics Letters A*, **112**, 115–118.

[340] Temperley H.N.V. (1956). Combinatorial problems suggested by the Statistical Mechanics of domains and of rubber-like molecules. *Physical Review*, **103**, 1–16.

[341] Tesi M.C., Janse van Rensburg E.J., Orlandini E. and Whittington S.G. (1995). Interacting self-avoiding walks and polygons in three dimensions. *Journal of Physics A: Mathematical and General*, **29**, 2451–2463.

[342] Tesi M.C., Janse van Rensburg E.J., Orlandini E. and Whittington S.G. (1996). Monte Carlo study of the interacting self-avoiding walk model in three dimensions. *Journal of Statistical Physics*, **82**, 155–181.

[343] Tesi M.C., Janse van Rensburg E.J., Orlandini E. and Whittington S.G. (1997). Torsion of polygons in $\mathcal{Z}^3$. *Journal of Physics A: Mathematical and General*, **30**, 5179–5194.

[344] van der Hofstad R., den Hollander F. and Slade G. (1998). A new inductive approach to the lace expansion for self-avoiding walks. *Probabality Theory and Related Fields*, **111**, 253–286.

[345] Vanderzande C. (1990). Self-avoiding walks near an excluded line. *Journal of Physics A: Mathematical and General*, **23**, 563–566.

[346] Vanderzande C. (1993). Vesicles, the tricritical 0-state Potts model, and the collapse of branched polymers. *Physical Review Letters*, **70**, 3595–3598.

[347] Vanderzande C. (1995). On knots in a model for the adsorption of ring polymers. *Journal of Physics A: Mathematical and General*, **28**, 3681–3700.

[348] Vanderzande C. (1998). *Lattice Models of Polymers*. Cambridge University Press.

[349] Veal A.R., Yeomans J.M. and Jug G. (1991). The effect of attractive monomer–monomer interactions on the adsorption of a polymer chain. *Journal of Physics A: Mathematical and General*, **24**, 827–849.

[350] Vologodski A.V., Levene S.D., Klenin K.V., Frank-Kamenetskii M.D. and Cozzarelli N.R. (1992). Conformational thermodynamic properties of supercoiled DNA. *Journal of Molecular Biology*, **227**, 1224–1243.

[351] Vologodski A.V., Lukashin A.L., Frank-Kamenetskii M.D. and Anshelevich V.V. (1974). The knot problem in statistical mechanics of polymer chains. *Soviet Physics-JETP*, **39**, 1059–1063.

[352] Vrbová T. and Whittington S.G. (1996). Adsorption and collapse of self-avoiding walks and polygons in three dimensions. *Journal of Physics A: Mathematical and General*, **29**, 6253–6264.

[353] Vrbová T. and Whittington S.G. (1998). Adsorption and collapse of self-avoiding walks in three dimensions: A Monte Carlo study. *Journal of Physics A: Mathematical and General*, **31**, 3989–3998.

[354] Vrbová T. and Whittington S.G. (1998). Adsorption and collapse of self-avoiding walks at a defect plane. *Journal of Physics A: Mathematical and General*, **31**, 7031–7041.

[355] Wakefield A.J. (1951). Some problems in statistical equilibria. D. Phil. Thesis. Oxford University.

[356] Wall H.S. (1967). *Analytic Theory of Continued Fractions*. Chelsea Publishing Company (New York).

[357] Welsh D.J.A. (1993). Percolation in the random cluster process and the q-state Potts model. *Journal of Physics A: Mathematical and General*, **26**, 2471–2483.
[358] Whittaker E.T. and Watson G.N. (1990). *A Course in Modern Analysis*, fourth edition. Cambridge University Press.
[359] Whittington S.G. (1987). Self-avoiding walks and related systems. *Proceedings of the International Course and Conference in Interfaces between Mathematics, Chemistry and Computer Science*. Ed. R.C. Lacher. *Studies in Physical and Theoretical Chemistry*, **54**, 285–296.
[360] Whittington S.G. (1987). Self-avoiding walks in restricted geometries. *Proceedings of the International Course and Conference in Interfaces between Mathematics, Chemistry and Computer Science*. Ed. R.C. Lacher. *Studies in Physical and Theoretical Chemistry*, **54**, 297–306.
[361] Whittington S.G. (1993). Statistical mechanics of three dimensional vesicles. *Journal of Mathematical Chemistry*, **14**, 103–110.
[362] Whittington S.G. (1998). A directed walk model of copolymer adsorption. *Journal of Physics A: Mathematical and General*, **31**, 8797–8803.
[363] Whittington S.G. (1998). A self-avoiding walk model of copolymer adsorption. *Journal of Physics A: Mathematical and General*, **31**, 3769–3775.
[364] Whittington S.G. and Guttmann A.J. (1990). Self-avoiding walks which cross a square. *Journal of Physics A: Mathematical and General*, **23**, 5601–5609.
[365] Whittington S.G. and Soteros C.E. (1991). Polymers in slabs, slits and pores. *Israel Journal of Chemistry*, **31**, 127–133.
[366] Whittington S.G. and Soteros C.E. (1992). Uniform branched polymers in confined geometries. *Macromolecular Reports A*, **29** (Suppl. 2), 195–199.
[367] Whittington S.G., Valleau J.P. and Torrie G.M. (1977). How many figure eights are there? Some bounds. *Journal of Physics A: Mathematical and General*, **10**, L111–L112.
[368] Wiese K.J. and David F. (1995). Self-avoiding tethered membranes at the tricritical point. *Nuclear Physics B*, **450**, 495–557.
[369] Wilker J.B. and Whittington S.G. (1979). Extension of a theorem on super-multiplicative functions. *Journal of Physics A: Mathematical and General*, **12**, L245–L247.
[370] Wilkinson M.K., Gaunt D.S., Lipson J.E.G. and Whittington S.G. (1986). Lattice models of branched polymers: Uniform combs in two dimensions. *Journal of Physics A: Mathematical and General*, **19**, L811–L816.
[371] Wright M.E. (1968). Stacks. *Quarterly Journal of Mathematics (Oxford)*, **19**, 313–320.
[372] Wu F.Y. (1978). Percolation and the Potts Model. *Journal of Statistical Physics*, **18**, 115–123.
[373] Wu F.Y. (1982). The Potts model. *Reviews in Modern Physics*, **54**, 235–268.
[374] Yeomans J.M. (1992). *Statistical Mechanics of Phase Transitions*. Oxford University Press.

Index